W9-CMN-005

NONLINEAR AND RANDOM VIBRATIONS

FLORA DINCĂ CRISTIAN TEODOSIU
Center of the Mechanics of Solids, Bucharest

NONLINEAR AND RANDOM VIBRATIONS

Editura Academiei Republicii Socialiste România

București

Academic Press, Inc.
A Subsidiary of
Harcourt Brace Jovanovich,
Publishers

New York and London

1973

TRANSLATED BY CRISTIAN TEODOSIU

The English version is the revised and updated translation of the Romanian work "Vibrații neliniare și aleatoare" published by Editura Academiei, str. Gutenberg 3 bis, București, 1970

For Academic Press, Inc.
I.S.B.N. : 0-12-216750-3
Library of Congress Catalog Card Number : 73-16596

All rights reserved
PRINTED IN ROMANIA

ENGR. LIBRARY
QA
935
.D4613
1397326

Contents

PART III

FORCED VIBRATIONS OF LINEAR AND NONLINEAR SYSTEMS ACTED ON BY RANDOM EXCITATIONS

Preface

Both Nonlinear and Random Vibration necessarily belong among the more advanced topics of Vibration Analysis. While elementary results are to be found in both fields a proper presentation is scarcely possible within the scope of an elementary general text; they cannot easily figure in the basic training of engineers. Yet, because both have important applications in practical engineering problems, many engineers find it necessary to become familiar with them. Books which combine a considerable mathematical sophistication with a consciousness of professional needs in these fields are not numerous, and one can only welcome a new book which brings, as this does, a fresh approach to both subjects.

In writing the preface to the Romanian version of this book four years ago I complained facetiously of one major fault — that it was written in a language which was difficult for me to understand. This complaint has been attended to with gratifying speed, and the fact that the English language version has so quickly found a publisher is some index of the quality of the book. The quality of the translation is a matter in which the authors have been fortunate.

This edition is not simply a translation of the earlier Romanian text. Certainly, readers familiar with the Romanian version will find few changes in the chapters which deal with Nonlinear Vibration, but the authors have taken the opportunity to rethink the whole presentation of Random Vibration. This appears in an entirely new form, though the interesting results on optimisation of a single degree of freedom system still form its culmination.

A particular feature of the new presentation is that the basic results are developed with a much greater formality. Like others working in this field, the authors have evidently become aware of the contribution which careful formulation can make to a deeper understanding of the subject. There is, of course, a penalty to be paid for this policy in that unsophisticated readers may well be deterred by such a development: nevertheless its long-term advantages are real and the reader must be advised in his own interest not to be deterred.

The authors have done me honour in inviting me to write this preface: I hope that the book will have the success it deserves.

The University, Glasgow, March, 1973

J. D. ROBSON

Introduction

The mathematical theory of nonlinear vibrations was grounded by the end of the last century through the works of POINCARÉ [146], who laid the foundations of the topological and perturbation methods for the study of weakly nonlinear systems. The same period witnessed important steps in the consideration of vibration stability, owing to the work done by LYAPUNOV [111]. The profound ideas of these two scientists exerted a determining influence on further research in this field.

After 1920, studies undertaken by DUFFING [55], MANDELSTAM and PAPALEXI [114] and VAN DER POL [147] brought the first definite solutions into the theory of nonlinear vibrations and drew attention to the importance of this theory for physics and engineering.

The late thirties and the postwar years saw a great development of the Russian school of nonlinear vibrations. Among the important works published in this period in the Soviet Union, we mention those of KRYLOV and BOGOLYUBOV [92], ANDRONOV and CHAIKIN [3], BOGOLYUBOV and MITROPOLSKY [16], MITROPOLSKY [123—124], and MALKIN [113].

In the last twenty years efforts were also made for collecting the main results concerning nonlinear vibrations in monographs. In this respect we mention the books by MINORSKY [121, 122], STOKER [184], MC LACHLAN [116, 117], KAUDERER [82], HAYASHI [71, 72], CESARI [32], HALANAY [65], SANSONE and CONTI [177], and ROSEAU [172].

The theory of random vibrations had a comparatively late development. From the publication in 1905 of EINSTEIN's study on Brownian motion [56], considered by him as a particular type of random vibration, more than two decades elapsed until the appearance of the first works applying the theory of random processes to the study of bar and string vibrations [96]. The introduction of the correlation function by TAYLOR [189] in 1920 and of the spectral density by WIENER [192] and KHINCHIN [86] in the early thirties opened new prospects for progress in the theory of random vibrations.

The papers by LIN [103] and RICE [167], published between 1943 and 1945, preceded a series of articles dedicated to various technical applications.

Finally, the monographs of CRANDALL and MARK [36] and ROBSON [170] achieved the work of systematization of the existing knowledge in the theory of linear random vibrations at the beginning of the last decade.

The study of nonlinear vibrations raised great mathematical difficulties; to overcome these, three different principal tools have been proposed: the use of the FOKKER-PLANCK-KOLMOGOROV equations, the method of statistical linearization and the perturbation method. The possible application of the FOKKER-PLANCK-KOLMOGOROV equations for the study of random vibrations was indicated for the first time by ANDRONOV et al. [4], but actually little use has been made of this method before 1954. It was about this year that there appeared the papers by BOOTON [18, 19] and KAZAKOV [84, 85] introducing the method of statistical linearization, which has been later independently developed by CAUGHEY [26]. Finally, CRANDALL [39] proposed and applied a perturbation technique, which is an extension to random vibrations of the perturbations method used for weakly nonlinear deterministic systems.

Presently there is a great number of papers concerning the theory of linear and nonlinear random vibrations. For an extensive bibliography on this subject we refer the reader to the review articles by CRANDALL [37] and CAUGHEY [31].

The present book is devoted to nonlinear and random vibrations of *mechanical single-degree-of-freedom systems.* A large class of these vibrations may be mathematically described by the differential equation:

$$\ddot{x} + f(\dot{x}) + g(x) = p(t), \tag{I.1}$$

where x is the displacement of the oscillator from its equilibrium position, $\dot{x}$ is the velocity, and $-f(\dot{x})$, $-g(x)$, and $p(t)$ denote, respectively, the damping force, the elastic restoring force, and the external force per unit mass of the oscillator. For the study of deterministic vibrations, we shall associate Eq. (I.1) with initial conditions of Cauchy type, i.e.,

$$x(t_0) = x_0, \qquad \dot{x}(t_0) = v_0. \tag{I.2}$$

If in (I.1), $f(\dot{x}) \equiv 0$, the oscillator is said to be *conservative*, and if $\dot{x} f(\dot{x}) > 0$ for any $\dot{x} \neq 0$, it is called *dissipative*. In this book we deal only with conservative and dissipative systems, since most of the mechanical oscillators belong to one or the other of these two types. We leave aside relaxation systems and systems with parametric excitation, for they are usually used to model electrical systems and only seldom mechanical systems. This restriction of the subject matter is compensated for by detailed study of conservative and dissipative systems, as well as by a grouping of the results concerning nonlinear and random vibrations for the first time within the same monograph.

We make certain mathematical assumptions on the functions $f(\dot{x})$, $g(x)$, and $p(t)$, which are justified by the physical properties of the systems considered. Since we shall often refer to these hypotheses, we will state them from the very beginning:

H_1: $f(\dot{x})$ and $g(x)$ are continuous in $(-\infty, \infty)$; $p(t)$ is continuous and bounded in $[t_0, \infty)$.

$H_2 : f(\dot{x})$ and $g(x)$ are continuously differentiable in $(-\infty, \infty)$, with the possible exception of a finite number of points, at which they possess, however, one-sided derivatives. [1]

$H_3 : f(0) = 0$; $\dot{x} f(\dot{x}) > 0$ for $\dot{x} \neq 0$; $g(0) = 0$; $xg(x) > 0$ for $x \neq 0$.

$H_4 : f'(\dot{x}) > 0$, with the possible exception of a finite number of values of $\dot{x}$, for which it may vanish; $g'(x) > 0$.

$H_5 : \lim_{\dot{x} \to \pm\infty} f(\dot{x}) = \pm\infty, \qquad \lim_{x \to \pm\infty} g(x) = \pm\infty.$

The graphs of the functions $f(\dot{x})$ and $g(x)$ are called, respectively, the damping and elastic characteristics of the system. The above conditions are fulfilled if the characteristics are piecewise smooth curves that pass through the origin and have, possibly, a finite number of discontinuities, and if the functions $f(\dot{x})$ and $g(x)$ are strictly increasing and tend to $-\infty$, or to $+\infty$, as their arguments tend to $-\infty$, and $+\infty$, respectively. Such characteristics are illustrated in Fig. 2.

As we shall later see, hypotheses H_1 and H_2 assure the existence and the uniqueness of a solution to Eq. (I.1) satisfying the initial conditions (I.2). The other three hypotheses play a definite role in considering the boundedness of solutions, as well as the existence, uniqueness, and stability of periodic solutions. Obviously, they are not completely independent. For instance, from $f(0) = 0$ and $f'(\dot{x}) > 0$ it follows that $\dot{x} f(\dot{x}) > 0$ for $\dot{x} \neq 0$. Nevertheless, we will keep the above form in order to facilitate the statement of various theorems whose proof does not need the simultaneous fulfillment of all these hypotheses.

It should be also mentioned that, although the hypotheses $H_1 - H_5$ describe in the main the properties of the oscillators considered in this book, they do not absolutely circumscribe the class of these oscillators. To assure the existence of some stronger properties, e.g., asymptotic stability in the large of periodic solutions, we shall sometimes adopt supplementary regularity hypotheses. We shall also consider oscillators for which the hypotheses above are not completely satisfied. As a first case of this situation, we have *the conservative systems*, whose equation of motion is

$$\ddot{x} + g(x) = p(t). \tag{I.3}$$

For such systems we assume, as before, that $g(x)$ and $p(t)$ satisfy the hypotheses $H_1 - H_5$. However, since now $f(\dot{x}) \equiv 0$, the hypotheses $H_3 - H_5$ concerning $f(\dot{x})$ are no longer fulfilled. Another important exception will be *the dissipative systems with dry friction*, whose equation of motion is

$$\ddot{x} + F(\dot{x}) + R \operatorname{sgn} \dot{x} + g(x) = p(t), \tag{I.4}$$

where

$$\operatorname{sgn} \dot{x} = \begin{cases} +1 & \text{for } \dot{x} > 0, \\ -1 & \text{for } \dot{x} < 0. \end{cases}$$

[1] From this hypothesis it follows that $f(\dot{x})$ and $g(x)$ satisfy a LIPSCHITZ condition in any finite interval.

In this case the function $f(\dot{x}) = F(\dot{x}) + R \operatorname{sgn} \dot{x}$ is discontinuous at $\dot{x} = 0$, irrespective of the value ascribed to this function for $\dot{x} = 0$, thus contradicting hypothesis H_1.

Before expounding the contents of the book, we will show how the equations of motion of a particular oscillator, namely a road vehicle, are set up. We choose this physical model, since it provides a good example of a system with linear or nonlinear characteristics, which may perform free or forced, deterministic or random vibrations.

The frame and the body of the vehicle are usually considered as a *rigid* structure referred to as the *sprung mass*; the wheels, axles, and the suspension mechanism are referred to as the *unsprung mass*. If motions parallel to a horizontal plane of the sprung and unsprung masses are neglected, the vehicle may be reduced to a seven-degree-of-freedom system. Three degrees of freedom correspond to the motion of the sprung mass, namely the vertical translational motion (bounce), the angular motion around the lengthwise horizontal axis, and the angular motion around the transversal horizontal axis (pitch). The other four degrees of freedom correspond to the motion of the unsprung masses, namely the vertical translational motion (wheel hop) and the angular motion of the unsprung masses around the lengthwise horizontal axis. The oscillator scheme may be further reduced to that of a single-degree-of-freedom oscillator if we adopt the following simplifying hypotheses:

a) It is assumed that the irregularities of the road strips covered by the right and left wheels are identical. Taking into account the geometric and mechanical symmetry of the vehicle, this hypothesis allows one to neglect the angular motion of the sprung and unsprung masses around the lengthwise horizontal axis.

b) The pitching motion of the sprung mass is neglected. Since one seeks to diminish the pitch by choosing a suitable vehicle geometry and mass repartition, its neglection is generally admissible.

c) Since the unsprung mass is small relative to the sprung mass, the effect of the unsprung mass may be neglected in determining the bouncing motion of the sprung mass without introducing a large error. Furthermore, the elasticity of the tires, and hence the wheel hop, may be neglected also, thus assuming that the axles move parallel to the road surface. Experiments on vehicle suspensions with linear characteristics show that the order of magnitude of the error introduced by this hypothesis is about 10–15%.

By first using only the simplifying hypotheses a) and b), we may reduce the vibration of the vehicle to that of the two-degree-of-freedom oscillator schematically represented in Fig. 1*a*. Let m_1 be the sprung mass carried by the front or by the rear wheels and m_2 the corresponding unsprung mass. The springs R_1, R_2, and the dashpot P schematically represent the suspension springs, the tires, and, respectively, the shock absorbers corresponding to the axle considered. Next, we denote by x_1 and x_2 the displacements of the masses m_1 and m_2 at time t, measured from their positions of static equilibrium, by

$$x(t) \equiv x_1(t) - x_2(t) \tag{I.5}$$

the displacement of the sprung mass relative to the unsprung mass, and by $x_0(t)$ the height above some fixed level of the road irregularity surmounted by the wheel at time t. The positive sense of measuring the quantities $x_0(t)$, $x_1(t)$, and $x_2(t)$ is chosen upward, as shown in Fig. 1*b*.

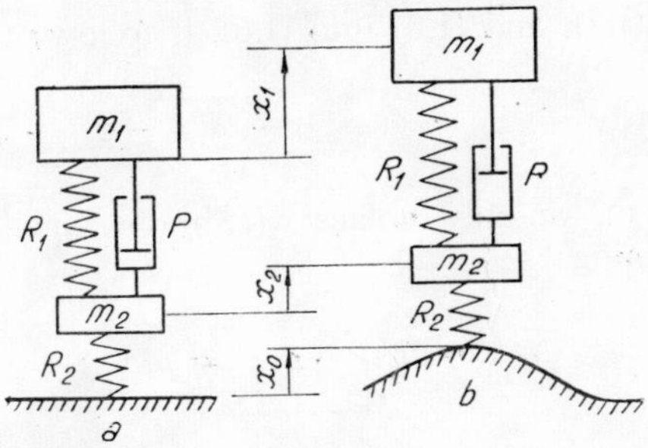

Fig. 1

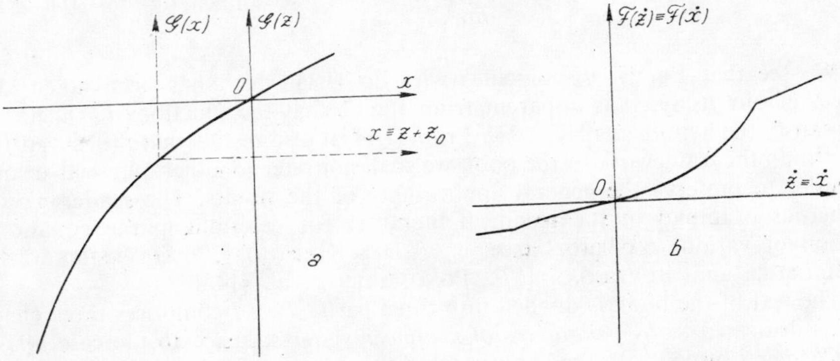

Fig. 2

Assume that the spring R_1 and the shock absorber have nonlinear characteristics of the type shown in Fig. 2, where z denotes the increment of the distance between the sprung and unsprung masses and $\dot{z}$ the rate of this change. Let $-\mathscr{F}(\dot{z})$ and $-\mathscr{G}(z)$ be the damping and the elastic restoring force, respectively. The equation of motion of the mass m_1 is

$$m_1\ddot{x}_1 = m_1g - \mathscr{F}(\dot{z}) - \mathscr{G}(z), \tag{I.6}$$

where g is the gravity acceleration. Let now $-z_0$ be the shortening of the spring R_1 under the action of the weight m_1g, that is the static deflection of the suspension springs. Hence

$$m_1g = \mathscr{G}(-z_0). \tag{I.7}$$

On the other hand, by considering Fig. 1*b* and Eq. (I.5), the extension of the spring R_1 at time t may be written as

$$x(t) = x_1(t) - x_2(t) - z_0 = x(t) - z_0. \tag{I.8}$$

By substituting (I.5), (I.7), and (I.8) into (I.6), it follows that

$$m_1\ddot{x}_1 + m_1\ddot{x}_2 = \mathscr{G}(-z_0) - \mathscr{G}(x - z_0) - \mathscr{F}(\dot{x}). \tag{I.9}$$

Finally, by hypothesis c), we may replace $x_2(t)$ by the function $x_0(t)$, which is taken as known, thus obtaining

$$\ddot{x} + f(\dot{x}) + g(x) = p(t), \tag{I.10}$$

where

$$f(\dot{x}) \equiv \frac{1}{m_1}\mathscr{F}(\dot{x}), \quad g(x) \equiv \frac{1}{m_1}[\mathscr{G}(x - z_0) - \mathscr{G}(z_0)], \quad p(t) \equiv -\ddot{x}_0(t). \tag{I.11}$$

We see that Eq. (I.10) coincides with Eq. (I.1) which has been taken as the object of our study. It is apparent from Fig. 2 that the functions $f(\dot{x})$, $g(x)$, and $p(t)$ satisfy the hypotheses $H_1 - H_5$. From (I.11) it also results that $f(0) = g(0) = 0$.

Within the framework of the book we shall not refer to other physical examples in order to preserve the general applicability of the results. The reader may find numerous examples in the technical literature for determining the equations of motion of various oscillators (see, e.g., Den Hartog [70], Timoshenko [191], Hamburger and Buzdugan [67], Ponomarev et al. [148].

The text of the book is divided into three parts. *Part I* comprises three chapters and is dedicated to *free vibrations of nonlinear systems*, hence to the case $p(t) \equiv 0$.

Chapter I begins with the exposition of the main theorems concerning the existence and uniqueness of the solution of the equation

$$\ddot{x} + f(\dot{x}) + g(x) = 0, \tag{I.12}$$

which satisfies the initial conditions (I.2).

It is proved that in the case of dissipative systems any solution of Eq. (I.1) is bounded in the future and exists in the whole interval $[t_0, \infty)$. Moreover, it is shown that the displacement and the velocity tend to 0 as $t \to \infty$. The notions of stability, asymptotic stability, and asymptotic stability in the large, which essentially intervene in the study of the qualitative properties of the solutions, are then briefly dealt with.

Next, the topological method, based on the analysis of the phase trajectories, is set forth. For dissipative systems whose characteristics satisfy the hypotheses $H_1 - H_4$, the only singular point is the origin of the phase plane ($x = \dot{x} = 0$). Denoting $2h = f'(0)$ and $\omega^2 = g'(0)$, it is proved that the singular point is always an attractor, namely a focus if $0 \leqslant h < \omega$, a node with one tangent if $0 < h = \omega$, and a node with two tangents if $0 < \omega < h$.

Sometimes it is much easier to find the equation of the phase trajectory than the solution $x(t)$. It is then necessary to approximately determine the interval of time corresponding to a known arc of the phase trajectory. For solving this problem SIMPSON's method, the method of LEGENDRE's polynomials, and a graphical method are proposed.

There are only a few cases in which an exact solution of Eq. (I.12) may be found. This is why great attention has been paid to approximate methods. The graphical and graphic-analytical methods have the advantage of being still applicable when the nonlinearity is strong. In exchange, they do not permit the derivation of analytic formulas, the drawings must be remade from the very beginning for other values of the parameters, and thus the study of the qualitative properties of the solutions is practically impossible. Of the analytical methods existing in the literature for the study of free vibrations, the perturbation method and the asymptotic method of KRYLOV and BOGOLYUBOV have been chosen. Both these methods are applicable to weakly nonlinear systems.

The reader primarily interested in the applications of the theory may retain from the first chapter only the definitions and the results, omitting at a first reading the proofs and insisting on the approximate analytical methods. This suggestion holds for the first chapters of the other two parts of the book, too.

Chapter II is devoted to free vibrations of conservative systems. For such systems the equation of the phase trajectories may be obtained explicitly. If $g(x)$ satisfies hypothesis H_3, the singular point (the origin of the phase plane) is a center, and the phase trajectories are closed curves, on which the representative point performs periodic motions. The possible type of elastic characteristics and their influence on the period of vibration are examined.

Great attention is paid to the conservative oscillator with cubic elastic restoring force, whose exact vibration expressed by elliptic integrals is compared with approximate solutions obtained by the perturbation method and by the method of KRYLOV and BOGOLYUBOV.

Chapter III deals with free vibrations of dissipative systems. After indicating the particular use of the method of KRYLOV and BOGOLYUBOV for the study of these systems, several dissipative systems, beginning with the systems with dry friction, are considered.

Due to their wide practical use, the systems with quadratic damping and those whose damping characteristic may be approximated by linear segments and/or by parabolic arcs are minutely examined. The results obtained are then applied to the analysis of free vibrations of road vehicles equipped with hydraulic shock absorbers.

Part II is dedicated to *forced vibrations of nonlinear systems acted on by deterministic excitations.* In all applications examined in this part, the exciting force in Eq. (I.1) is assumed periodic with minimum period T.

Chapter IV comprises a survey of the main available qualitative results concerning Eq. (I.1), namely the existence, uniqueness, and stability of periodic solutions. Also in this chapter, harmonic, subharmonic, superharmonic, and supersubharmonic solutions are defined.

Next, the use of the perturbation method and of the method of KRYLOV and BOGOLYUBOV for obtaining approximate solutions of Eq. (I.1) and for studying the stability of periodic solutions according to the first approximation is minutely explained. Then, the technique of three other analytical methods for the determination of approximate periodic solutions, the method of finite sums of trigonometric functions, GALERKIN's method, and RAUSCHER's method, is briefly expounded.

At the end of Chapter IV, a general study of subharmonic vibrations is undertaken. It is shown how nonlinear equations that admit given subharmonic solutions may be built by means of CHEBYSHEV's polynomials. Then, the method of successive approximations, which was often successfully applied for the analysis of subharmonic vibrations, is mathematically justified. Finally, the particular use of the method of KRYLOV and BOGOLYBOV for the study of subharmonic vibrations is indicated.

Chapter V deals with forced vibrations of conservative systems. Since a cubic characteristic may approximate any elastic characteristic given by an odd analytic function, provided $|x|$ is sufficiently small, much attention is paid to DUFFING's equation:

$$\ddot{x} + \omega^2 x(1 + \beta x^2) = P \cos \nu t. \tag{I.13}$$

Various methods are used to study harmonic solutions of this equation near the principal "resonance", as well as the stability of the periodic solutions and the jump phenomena occurring in the variation of the response amplitude, when P or ν varies continuously. The subharmonic solutions of order 1/3 are analyzed and the supplementary effects with respect to the linear superposition that occur when the oscillator is acted on by two harmonic forces with different frequencies are pointed out.

In the last section of Chapter V, forced vibrations of an oscillator with piecewise linear elastic characteristic are investigated.

Chapter VI is devoted to forced vibrations of dissipative systems. Besides DUFFING's equation with damping

$$\ddot{x} + 2h\dot{x} + \omega^2 x(1 + \beta x^2) = P \cos \nu t, \tag{I.14}$$

which is minutely examined, the forced vibrations of systems with quadratic damping and linear or nonlinear elastic restoring forces are also analyzed. In each case the resonance curves are graphically represented for various typical values of the parameters, and the stability of the solutions is explored.

Part III deals with *forced vibrations of linear and nonlinear systems acted on by random excitations.* The main difference against the second part of the book is that the function $p(t)$ in the right-hand side of Eq. (I.1) is now a random function.

Chapter VII comprises a survey of the basic notions used for the study of random functions. Highest generality is deliberately avoided and only mathematical elements strictly necessary for the analysis of random vibrations, such as the notions of random variable and random function, the convergence of sequences of random functions, as well as the continuity, differentiability, and integrability of random functions, are discussed. All these notions are defined on a probability field. In spite of some difficulties connected with the axiomatic introduction of the probability field, this approach proves to be very useful for defining precisely various notions.

Special attention is paid to the stationary random functions, for which the WIENER-KHINCHIN relations are proved. Chapter VII ends with the exposition of some elements of the theory of selection that allow to obtain the correlation function and the spectral density from the collection of sample functions.

Chapter VIII begins with a discussion of the meaning to be given to the notion of solution, depending on the type of convergence adopted. Then, the main theorems concerning the existence and uniqueness of the solution are indicated. Various types of stability are introduced, also depending on the type of convergence adopted. Special stress is laid upon asymptotic stability in the large, which play an essential role in the study of randomly excited oscillators by the correlation theory.

Next, the WIENER-KHINCHIN relations are used to examine the behavior of the linear oscillator by the correlation theory. Finally, two approximate methods that allow the reduction of the study of the nonlinear to linear systems are expounded: the method of equivalent linearization and the perturbation method.

Chapter IX contains the analysis of random vibrations of the linear oscillator whose equation of motion is

$$\ddot{x} + 2\zeta\nu_n\dot{x} + \nu_n^2 x = p(t), \tag{I.15}$$

where $p(t)$ is a stationary random function. The cases when $p(t)$ may be approximated by a pure white noise, by a band-limited white noise, or by some more general representations are explored. Then, the behavior of the square means of the steady-state solution and of its derivatives is minutely examined.

Chapter X deals with the optimization of the dissipation in linear and nonlinear systems under the assumption that the spectral density of $p(t)$ may be obtained by limiting the spectral band corresponding to correlation functions of the form $\sum_i A_i \, e^{-\alpha_i|\tau|} \cos \beta_i\tau$. Three main particular cases are considered, namely

a) $A_1 = 1, \quad \beta_1 = 0; \quad A_i = 0 \quad \text{for} \quad i > 1.$

b) $A_2 = 1, \quad A_i = 0 \quad \text{for} \quad i \neq 2.$

c) $A_1 + A_2 = 1, \quad \beta_1 = 0, \quad \beta_2 \neq 0; \quad A_i = 0 \quad \text{for} \quad i > 2.$

One of the important practical problems is the determination of the optimum damping corresponding to a given random excitation. This optimization must be achieved with respect to certain criteria that depend on the physical problem considered. For instance, in the case of road vehicles, optimization means minimum dynamic loading and maximum confort. A necessary condition for the fulfilment of these two requirements is that the mean square acceleration be minimum. It is then proved that there exists an optimum value of the damping ratio ζ which minimizes the mean square acceleration. It is also shown that a certain phenomenon of "resonance" may appear when one of the dominant frequencies of the excitation is close to the natural frequency ν_n.

Finally, the optimization problem is also solved for an oscillator with nonlinear damping, whose governing equation is

$$\ddot{x} + k\,|\dot{x}|^\alpha \operatorname{sgn} \dot{x} + \omega^2 x = p(t), \tag{I.16}$$

where $p(t) = -\ddot{x}_0(t)$ is a stationary random function, and k and α are constants.

Studies that make use of the FOKKER-PLANCK-KOLMOGOROV equations have been deliberately omitted, since the description of this method needs a mathematical apparatus that goes beyond the framework of the present book.

The authors endeavored to use a unified system of notations. However, since they tried at the same time to preserve as much as possible agreement with standard notations, some inconsistencies between the notations used in the study of random vibrations and those employed in the first two parts of the book could not be avoided. Nevertheless, misunderstandings can be easily eluded, by taking into account the explicit mentioning of the changes introduced.

ACKNOWLEDGMENTS

The authors express their deep gratitude to Professor J. D. ROBSON, from the University of Glasgow, for his valuable comments on topics related to random vibrations and for accepting to write the preface to the book. They are particularly indebted to their colleague T. SIRETEANU for his criticism and suggestions on the manuscript, as well as for his co-operation in performing some of the calculations for the third part of the book.

PART I

Free vibrations of nonlinear systems

The first part of the book is devoted to the free vibrations of nonlinear systems, described by the equation of motion

$$m\ddot{x} + \mathscr{F}(\dot{x}) + \mathscr{G}(x) = 0,$$

with the initial conditions

$$x(t_0) = x_0, \qquad \dot{x}(t_0) = v_0$$

Chapter I

General methods of investigation

§ 1. GENERAL THEOREMS ON THE PROPERTIES OF THE SOLUTIONS

a) Existence and uniqueness of solutions

Introducing the new functions

$$f(\dot{x}) \equiv \frac{1}{m}\mathscr{F}(\dot{x}), \qquad g(x) \equiv \frac{1}{m}\mathscr{G}(x), \tag{1.1}$$

the equation of motion becomes

$$\ddot{x} + f(\dot{x}) + g(x) = 0. \tag{1.2}$$

The second-order differential equation (1.2) is equivalent to the system of first-order differential equations

$$\dot{x} = v, \qquad \dot{v} = -f(v) - g(x), \tag{1.3}$$

for which the initial conditions

$$x(t_0) = x_0, \qquad v(t_0) = v_0 \tag{1.4}$$

define a Cauchy problem.

We indicate now without proof two existence theorems for the solution of system (1.3) satisfying the initial conditions (1.4). Demonstrations of these and other more general theorems may be found in standard books on differential equations (see, e.g., NEMYTSKY and STEPANOV [131]).

Theorem 1. *If $f(v)$ and $g(x)$ are continuous in a closed and bounded region $\overline{D}$ of the plane (x, v), then at least one solution of the Cauchy problem (1.3), (1.4) exists,* *which is defined on the segment* $[t_0 - d/M\sqrt{2},\ t_0 + d/M\sqrt{2}]$, *where d is the distance of the point* (x_0, v_0) *to the boundary of* $\overline{D}$, *and*

$$M = \sup_{(x,\,v)\in\overline{D}} \{|f(v)|,\ |g(x)|\}.$$

The preceding theorem assures the existence of a solution in a finite interval of time. Sometimes, however, we are interested in the existence of solutions for all $t \geqslant t_0$. An important case when the solution displays this property is given by the following theorem:

Theorem 2. *If the representative point* $(x(t), v(t))$ *of the motion remains within a closed region* $\overline{D}$ *as* $t \to \infty$, $\overline{D}$ *contained in an open region of the plane* (x, v) *in which* $f(v)$ *and* $g(x)$ *are continuous, then the solution of the Cauchy problem* (1.3), (1.4) *may be continued in the whole interval* $[t_0, \infty)$.

In particular, if f and g satisfy the condition H_1, i.e., they are continuous for any value of their arguments, it is sufficient to demonstrate that $x(t)$ and $v(t)$ are uniformly bounded for all values of t for which they exist. Ziemba [195] proved this under the hypothesis that f and g are odd analytic functions of their arguments. We shall slightly generalize his result by showing that it is enough to assume that f and g satisfy the weaker conditions H_1 and H_3.

Theorem 3. *If* $f(v)$ *is continuous in* $(-\infty, \infty)$, $f(0) = 0$, $vf(v) > 0$ *for* $v \neq 0$, *and if* $g(x)$ *is continuous in* $(-\infty, \infty)$, $g(0) = 0$, $xg(x) > 0$ *for* $x \neq 0$, *then any solution of* (1.3), (1.4) *remains bounded as* $t \to \infty$.

Proof. Let

$$G(x) \equiv \int_0^x g(s)\mathrm{d}s.$$

The assumptions about g in the hypothesis of the theorem imply

$$G'(x) \equiv g(x) \begin{cases} > 0 & \text{for} \quad x > 0, \\ = 0 & \text{for} \quad x = 0, \\ < 0 & \text{for} \quad x < 0. \end{cases} \tag{1.5}$$

Hence $G(x)$ has an absolute minimum for $x = 0$, i.e., $G(x) \geqslant G(0) = 0$ for all $x \in (-\infty, \infty)$.

By multiplying now Eq. $(1.3)_2$ by $v\mathrm{d}t$ and integrating between the limits t_0 and t, we obtain

$$E(t) \equiv \frac{v^2}{2} + G(x) = \frac{v_0^2}{2} + G(x_0) - \int_{t_0}^{t} v(t)f(v(t))\mathrm{d}t. \tag{1.6}$$

Since $G(x) \geqslant 0$ and $vf(v) \geqslant 0$, it follows that

$$v^2 \leqslant 2E_0 \qquad G(x) \leqslant E_0, \tag{1.7}$$

where $E_0 = v_0^2/2 + G(x_0)$ represents the total energy of the system per unit mass at time t_0. By $(1.7)_1$ we have

$$|v(t)| \leqslant \sqrt{2E_0}. \tag{1.8}$$

Let η_1 and η_2 be the inverses of the function G on the intervals $(-\infty, 0]$ and $[0, \infty)$, respectively [1]). From (1.5) we deduce

$$\eta_1'(G) < 0, \qquad \eta_2'(G) > 0,$$

wherefrom, by $(1.7)_2$, it follows that

$$x \equiv \eta_1(G(x)) \leqslant \eta_1(E_0) \qquad \text{for} \quad x \leqslant 0,$$

$$x \equiv \eta_2(G(x)) \leqslant \eta_2(E_0) \qquad \text{for} \quad x \geqslant 0,$$

hence

$$|x(t)| \leqslant \max \{\eta_1(E_0), \quad \eta_2(E_0)\}. \tag{1.9}$$

Relations (1.8) and (1.9) demonstrate the stated theorem.

As regards the uniqueness of the solution, this is assured by the hypothesis H_2 and the following theorem (see, e.g., NEMYTSKY and STEPANOV [131]):

Theorem 4. *If $f(v)$ and $g(x)$ satisfy the Lipschitz condition in $\bar{D}$, then there exists a unique solution of system* (1.3) *that satisfies the initial conditions* (1.4).

Our assumption H_2 is more restrictive than the hypothesis of the theorem above because a function whose first derivative is bounded and piecewise continuous always satisfies a Lipschitz condition.

By summarizing the results comprised in Theorems 1—4 we can state the following *theorem of existence and uniqueness*:

Theorem 5. *If f and g satisfy hypotheses H_1, H_2, and H_3, then system* (1.3) *has a unique solution $x(t)$, $v(t)$ that satisfies the initial conditions* (1.4). *This solution exists for all $t \geqslant t_0$.*

b) The behavior of the solution as t → ∞

From the uniqueness theorem we infer that the only solution of Eq. (1.2) satisfying homogeneous initial conditions, i.e., $x_0 = v_0 = 0$, is the trivial solution $x(t) \equiv 0$.

1) The existence of η_1 and η_2 is assured by (1.5)

If f and g satisfy hypotheses $H_1 - H_4$, then all solutions of Eq. (1.2) approach zero as $t \to \infty$, that is, the oscillator asymptotically approaches its equilibrium position, irrespective of the initial values x_0 and v_0. To prove this we shall first demonstrate five lemmas, which give a complete description of the possible asymptotic behavior of the solution. Throughout this section we assume that at least one of the values x_0, v_0 is not zero and that f and g satisfy hypotheses $H_1 - H_4$. We shall always denote by $x(t)$ the unique solution of Eq. (1.2) that satisfies the initial conditions (1.4).

Lemma 1. *The zeros of $x(t)$ and $\dot{x}(t)$, if any, are simple and do not admit of accumulation points at finite distance.*

The first part of the lemma results from the fact that if $x(t_1) = 0$, then $\dot{x}(t_1) \neq 0$, and if $\dot{x}(t_1) = 0$, then $\ddot{x}(t_1) \neq 0$, because, otherwise, from Eq. (1.2) and the uniqueness theorem it would follow that $x(t) \equiv 0$, and this would contradict the initial conditions (1.4).

The zeros of $x(t)$ cannot have an accumulation point t_1 at finite distance because in the first case we would have $x(t_1) = \dot{x}(t_1) = 0$, and in the second one $\dot{x}(t_1) = \ddot{x}(t_1) = 0$, and this would contradict the first part of the lemma.

Lemma 2. *If t_1 and t_2 are two consecutive zeros of $\dot{x}(t)$, i.e.,*

$$\dot{x}(t_1) = \dot{x}(t_2) = 0, \qquad \dot{x}(t) \neq 0 \quad \textit{for} \quad t \in (t_1, t_2),$$

then $x(t)$ vanishes at exactly one interior point of (t_1, t_2).

Indeed, we have in this case by (1.2)

$$\ddot{x}(t_1)\, g(x(t_1)) < 0, \qquad \ddot{x}(t_2)\, g(x(t_2)) < 0,$$

wherefrom, by making use of H_3, it follows that

$$\ddot{x}(t_1)\, x(t_1) < 0, \qquad \ddot{x}(t_2)\, x(t_2) < 0.$$

Moreover, since $\dot{x}(t) \neq 0$ for $t \in (t_1, t_2)$ we see that $x(t_1)$ and $x(t_2)$ cannot be both maxima or both minima, so that $\ddot{x}(t_1)\ \ddot{x}(t_2) < 0$. Hence $x(t_1)x(t_2) < 0$ and $x(t)$ should vanish at at least one interior point of (t_1, t_2). On the other hand, $x(t)$ cannot vanish at two interior points of (t_1, t_2) since, otherwise, by Rolle's theorem, $\dot{x}(t)$ ought to vanish once more between these points, in contradiction with the hypothesis.

From the above inequalities we infer also that the points where $\dot{x}(t) = 0$ are points of maximum if $x(t) > 0$, are points of minimum if $x(t) < 0$, and thus the maxima and the minima of $x(t)$ are alternate if the solution is oscillatory.

Lemma 3. *If t_1 and t_2 are two consecutive zeros of $\ddot{x}(t)$, i.e.,*

$$\ddot{x}(t_1) = \ddot{x}(t_2) = 0, \qquad \ddot{x}(t) \neq 0 \quad \textit{for} \quad t \in (t_1, t_2),$$

then $x(t)$ vanishes at exactly one interior point of (t_1, t_2).

The proof follows along the same lines as that of the preceding lemma. By differentiating (1.2) we obtain

$$\dddot{x}(t) + f'(x)\,\ddot{x}(t) + g'(x)\,\dot{x}(t) = 0$$

and, since $g'(x) > 0$, we have under the conditions above

$$\dddot{x}(t_1)\,\dot{x}(t_1) < 0, \qquad \dddot{x}(t_2)\,\dot{x}(t_2) < 0.$$

But $\dot{x}(t_1)$ and $\dot{x}(t_2)$ cannot be both maxima or both minima because $\ddot{x}(t) \neq 0$ for $t \in (t_1, t_2)$. Therefore $\dddot{x}(t_1)\,\dddot{x}(t_2) < 0$, hence $\dot{x}(t_1)\;\dot{x}(t_2) < 0$, and $x(t)$ vanishes at at least one interior point of (t_1, t_2). On the other hand, $\dot{x}(t)$ cannot vanish at two interior points of (t_1, t_2) since, otherwise, according to Rolle's theorem, $\ddot{x}(t)$ ought to vanish once more between these two points, in contradiction with the hypothesis.

Lemma 4. *If there is a $t_1 \geqslant t_0$ such that $x(t) \neq 0$ for $t > t_1$, then $x(t)$ and $\dot{x}(t)$ become monotonic after a finite time and approach zero as $t \to \infty$.*

Proof. Since $x(t) \neq 0$ for $t > t_1$, there exists by Lemma 2 at most one value $t_2 > t_1$ such that $\dot{x}(t_2) = 0$. Since now $\dot{x}(t) \neq 0$ for $t > t_2$, according to Lemma 3 there exists at most one time $t_3 > t_2$, at which $\ddot{x}(t)$ vanishes.

From the considerations above it follows that $x(t)$ and $\dot{x}(t)$ are monotonic for $t > t_3$, and, since they are also bounded (Theorem 3), we deduce that there exist

$$\lim_{t\to\infty} x(t) = x_\infty, \qquad \lim_{t\to\infty} \dot{x}(t) = v_\infty,$$

with x_∞ and v_∞ finite. On the other hand, we can have neither $v_\infty > 0$ nor $v_\infty < 0$ because this would imply $x_\infty = \infty$ or $x_\infty = -\infty$, respectively. Hence

$$\lim_{t\to\infty} \dot{x}(t) = 0.$$

From (1.2) it follows now that

$$\lim_{t\to\infty} \ddot{x}(t) = a_\infty \quad \text{with} \quad a_\infty \quad \text{finite,}$$

and we deduce in a similar manner that

$$\lim_{t\to\infty} \ddot{x}(t) = 0.$$

Finally, we infer from (1.2) and (1.6) that

$$\lim_{t\to\infty} x(t) = 0, \qquad \lim_{t\to\infty} E(t) = 0.$$

The possible forms of the graph of $x(t)$ under the conditions of Lemma 4 are shown schematically in Fig. 3, where the motion is at first oscillatory then aperiodic, and in Fig. 4, where the motion is aperiodic from the very beginning. As we see, it may well happen that $t_0 < t_1 < t_2 < t_3$ (Fig. 3), or $t_0 = t_1 < t_2 < t_3$ (Fig. 4*a*), or $t_0 = t_1 = t_2 < t_3$ (Fig. 4*b*), or $t_0 = t_1 = t_2 = t_3$ (Fig. 4*c*).

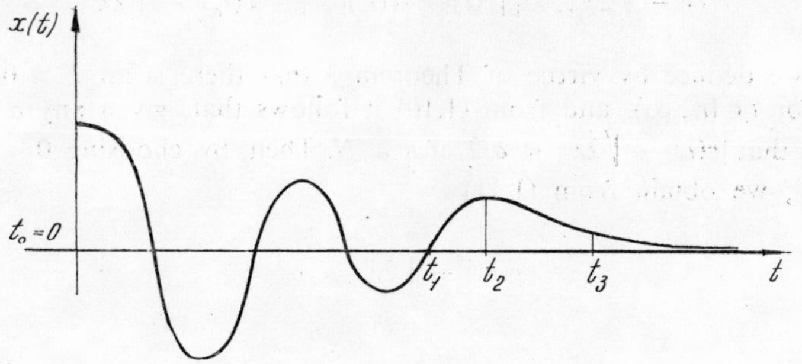

Fig. 3

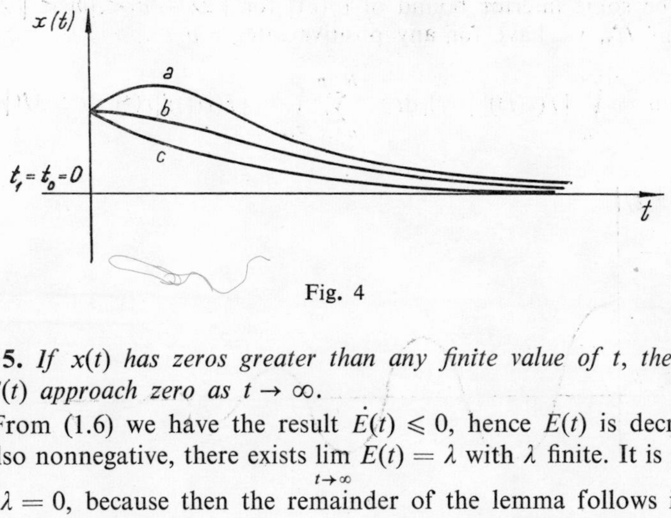

Fig. 4

Lemma 5. *If* $x(t)$ *has zeros greater than any finite value of* t, *then* $x(t)$, $\dot{x}(t)$, $\ddot{x}(t)$, *and* $E(t)$ *approach zero as* $t \to \infty$.

Proof. From (1.6) we have the result $\dot{E}(t) \leqslant 0$, hence $E(t)$ is decreasing and, since it is also nonnegative, there exists $\lim_{t\to\infty} E(t) = \lambda$ with λ finite. It is sufficient to prove that $\lambda = 0$, because then the remainder of the lemma follows immediately from (1.8), (1.9), and (1.2). We will prove this by contradiction. In order to make a choice let us assume that $\lambda > 0$.

From the infinite set of zeros of $x(t)$ we extract a sequence $\{t_n\}$ such that $\lim_{n\to\infty} t_n = \infty$ and $t_n > t_{n-1} + \rho$ for any positive integer n, where $\rho > 0$ is a fixed number to be made precise below. Since $x(t_n) = 0$, we obtain from (1.6)

$$|v(t_n)| = \sqrt{2E(t_n)},$$

and hence

$$\lim_{n\to\infty} |v(t_n)| = \sqrt{2\lambda}. \tag{1.10}$$

We can write now for every $t \in (t_n - \rho, t_n)$:

$$\big| |v(t)| - \sqrt{2\lambda} \big| \leqslant \big| |v(t)| - |v(t_n)| \big| + \big| |v(t_n)| - \sqrt{2\lambda} \big|. \tag{1.11}$$

From (1.2) we deduce by virtue of Theorem 3 that there is an $L > 0$ such that $|\dot{v}(t)| < L$ for $t \in [t_0, \infty)$, and from (1.10) it follows that, given any $\sigma > 0$, there is an N such that $|v(t_n) - \sqrt{2\lambda}| < \sigma/2$, if $n \geqslant N$. Then, by choosing $0 < \rho < \sigma/2L$, $0 < \sigma < \sqrt{2\lambda}$, we obtain from (1.11)

$$\big| |v(t)| - \sqrt{2\lambda} \big| < \sigma,$$

and hence

$$\sqrt{2\lambda} - \sigma < |v(t)| < \sqrt{2\lambda} + \sigma.$$

Let now M be some inferior bound of $|f(v)|$ for $\sqrt{2\lambda} - \sigma < |v| < \sqrt{2\lambda} + \sigma$. By making use of H_3, we have for any positive integer p

$$\int_{t_0}^{\infty} f(v(t))v(t)\,\mathrm{d}t = \int_{t_0}^{\infty} |f(v(t))|\,|v(t)|\,\mathrm{d}t \geqslant \sum_{n=N}^{N+p} \int_{t_n-\rho}^{t_n} |f(v(t))||v(t)|\,\mathrm{d}t > M(\sqrt{2\lambda}-\sigma)p\rho.$$

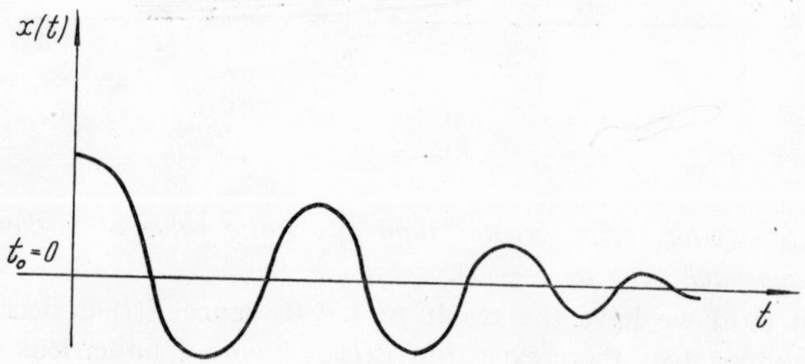

Fig. 5

Since p is arbitrary, and ρ and σ are fixed, it follows that the integral on the left-hand side of this inequality ought to be divergent. On the other hand, we deduce from (1.6) that the same integral must be less than $E_0 - \lambda$, a contradiction. We conclude that $\lambda = 0$.

The possible forms of the graph of $x(t)$, under the conditions of Lemma 5, are sketched in Fig. 5, where the motion is oscillatory for all $t \geqslant t_0$, and in Fig. 6, where the motion becomes oscillatory for $t \geqslant t_1 > t_0$.

From the last two lemmas we deduce:

Theorem 6. *If f and g satisfy hypotheses $H_1 - H_4$, then any solution of* (1.2), *as well as its first two derivatives and $E(t)$, approaches zero as $t \to \infty$.*

c) Stability [1)]

Before studying the stability of the solutions of system (1.3) we shall give some definitions concerning the solutions of the more general system:

$$\dot{x} = X(x, v, t), \qquad \dot{v} = V(x, v, t). \tag{1.12}$$

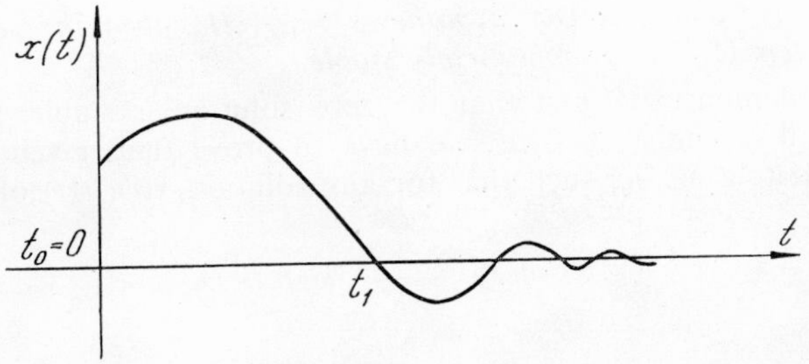

Fig. 6

We assume that the functions X and V satisfy regularity conditions assuring the existence and uniqueness of a solution for the initial conditions (1.4). We may suppose without restricting the generality that x and v are nondimensional quantities. This can always be achieved by taking the values of x and v divided by some constant values of them as new dependent variables; we consider in the following that this substitution has already been performed, and we still maintain the same notations.

Let $\bar{x}(t)$, $\bar{v}(t)$ be a solution of (1.12) that can be continued in the maximal interval (t^-, ∞) as shown at the beginning of this section. We adopt the following definitions for various types of stability *in the sense of* LYAPUNOV:

Definition 1. *A solution $\bar{x}(t)$, $\bar{v}(t)$ of* (1.12) *is said to be* stable *for $t \to \infty$ if, given any t_0 and ε, $t_0 > t^-$, $\varepsilon > 0$, there exists a $\delta(\varepsilon, t_0) > 0$ such that for any solution $x(t)$, $v(t)$ of* (1.12) *satisfying*

$$|x(t_0) - \bar{x}(t_0)| < \delta, \qquad |v(t_0) - \bar{v}(t_0)| < \delta, \tag{1.13}$$

the inequalities

$$|x(t) - \bar{x}(t)| < \varepsilon, \qquad |v(t) - \bar{v}(t)| < \varepsilon \tag{1.14}$$

hold for all $t_0 \leqslant t < \infty$. A solution that is not stable is called unstable. *If $\delta(\varepsilon, t_0)$ depends only on ε, then the solution $\bar{x}(t)$, $\bar{v}(t)$ is called* uniformly stable.

1) For a more comprehensive treatment of the notions expounded here see, e.g., SANSONE and CONTI [177], chapters I and IX, and HALANAY [65], chapter I.

Definition 2. *A solution* $\bar{x}(t)$, $\bar{v}(t)$ *of* (1.12) *is said to be* asymptotically stable *for* $t \to \infty$ *if it is stable and if, given any* $t_0 > t^-$, *there is a* $\delta(t_0) > 0$ *such that*

$$\lim_{t\to\infty} |\bar{x}(t) - x(t)| = 0, \qquad \lim_{t\to\infty} |\bar{v}(t) - v(t)| = 0 \tag{1.15}$$

for any solution $x(t)$, $v(t)$ *satisfying* $|x(t_0) - \bar{x}(t_0)| < \delta(t_0)$, $|v(t_0) - \bar{v}(t_0)| < \delta(t_0)$ *If* $\delta(t_0)$ *does not depend on* t_0, *i.e., it may be fixed for all* $t_0 > t^-$, *then the solution* $\bar{x}(t)$, $\bar{v}(t)$ *is said to be* uniformly asymptotically stable.

We can now demonstrate the following theorem.

Theorem 7. *If f and g satisfy hypotheses* $H_1 - H_4$, *then the solution* $\bar{x}(t) \equiv 0$, $\bar{v}(t) \equiv 0$ *of system* (1.3) *is asymptotically stable.*

Proof. We demonstrate first that the zero solution is stable. Since $\bar{x}(t_0) = 0$, $\bar{v}(t_0) = 0$ for all t_0 and $t^- = -\infty$, we have to prove that, given any t_0 and any $\varepsilon > 0$, there exists a $\delta(\varepsilon, t_0)$ such that for any solution $x(t)$, $v(t)$ of (1.3) satisfying

$$|x(t_0)| < \delta, \qquad |v(t_0)| < \delta, \tag{1.16}$$

the inequalities

$$|x(t)| < \varepsilon, \qquad |v(t)| < \varepsilon \tag{1.17}$$

hold for all $t_0 \leqslant t < \infty$. Let $x(t_0) = x_0$, $v(t_0) = v_0$. By taking into account (1.8) and (1.9), it follows that the inequalities (1.17) are satisfied if

$$E_0 < \frac{\varepsilon}{2}, \qquad |\eta_1(E_0)| < \varepsilon, \qquad |\eta_2(E_0)| < \varepsilon. \tag{1.18}$$

Since η_1 and η_2 are continuous in $[0, \infty)$, and $\eta_1(0) = \eta_2(0) = 0$, there exists a $\delta_1 > 0$ such that the last two conditions (1.18) are fulfilled if $E_0 < \delta_1$. By denoting

$$\varepsilon_1 = \min\left(\frac{\varepsilon^2}{2}, \delta_1\right),$$

we deduce that the inequalities (1.18) are satisfied if

$$E_0 = \frac{v_0^2}{2} + G(x_0) < \varepsilon_1. \tag{1.19}$$

But G is continuous in $[0, \infty)$ and $G(x) \geqslant G(0) = 0$. Hence, there is a $\delta_2 > 0$ such that $G(x_0) < \dfrac{\varepsilon_1}{2}$ for $|x_0| < \delta_2$. By putting now

$$\delta = \min(\sqrt{\varepsilon_1}, \delta_2),$$

we see that (1.19) is satisfied if

$$|x_0| < \delta, \qquad |v_0| < \delta,$$

and hence the zero solution is stable. On the other hand, by theorem 6, the conditions

$$\lim_{t\to\infty} |x(t)| = 0, \qquad \lim_{t\to\infty} |v(t)| = 0$$

are fulfilled by any solution of (1.4), and hence the zero solution is asymptotically stable for $t \to \infty$.

§ 2. THE TOPOLOGICAL METHOD

a) General properties of the phase trajectories

As has already been mentioned in the preceding section, the equation of motion (1.2) is equivalent to the system of differential equations (1.3). Since the time t does not appear explicitly in this system, we may take x as a new independent variable and v as a function of x. Then the dependence of v on x can be deduced by integrating the corresponding differential equation resulting from (1.3)

$$\frac{dv}{dx} = -\frac{f(v) + g(x)}{v}, \tag{2.1}$$

with the initial condition

$$v(x_0) = v_0. \tag{2.2}$$

The plane of the variables x, v is called the *phase plane*, and the graphs of the solutions of (2.1) in this plane are called *phase trajectories*. We can imagine that each phase trajectory is described by the motion of a point $P(x, v)$, whose coordinates at time t are given by the solution $x(t)$, $v(t)$ of system (1.3); this point is called a *phase point* or a *representative point*.

The points (x, v) of the phase plane in which both right-hand sides of the system

$$\dot{x} = X(x, v), \qquad \dot{v} = V(x, v) \tag{2.3}$$

vanish are said to be *singular points*. The other points of the phase plane are called *regular points*. Any singular point $S(x_s, v_s)$ represents in fact a phase trajectory reduced to a single point and corresponding to the solution $x(t) \equiv x_s$, $v(t) \equiv v_s$ of system (2.3).

In the particular case of system (1.3), the only singular point is the origin of the phase plane, since by H_3 we have simultaneously $v = 0$ and $f(v) + g(x) = 0$ if and only if $x = v = 0$.

We shall now demonstrate some of the main properties of the phase trajectories of system (1.3) or, equivalently, of Eq. (2.1).

Theorem 1. *Equation* (2.1) *has a unique solution that satisfies the initial condition* (2.2) *and is continuous with its first derivative for all* $v \neq 0$.

Indeed, for any $v = 0$, the right-hand side of (2.1) is a continuous function of v and x satisfying a Lipschitz condition within any finite region of the phase plane. Thereby the stated theorem is proved.

Theorem 2. *At the points where the trajectory crosses the x-axis, the tangent to the trajectory is parallel to the v-axis, and the concavity of the trajectory is directed toward the origin.*

Indeed, we derive from (2.1)

$$\frac{dx}{dv} = -\frac{v}{f(v) + g(x)} \tag{2.4}$$

and

$$\frac{d^2x}{dv^2} = -\frac{f(v) + g(x) - v[f'(v) + g'(x)\,dx/dv]}{[f(v) + g(x)]^2}. \tag{2.5}$$

Hence, by making use of H_3, we have

$$\left.\frac{dx}{dv}\right|_{v=0} = 0\,, \qquad \left.\frac{d^2x}{dv^2}\right|_{v=0} = -\frac{1}{g(x)}. \tag{2.6}$$

This completes the proof since $g(x)$ and x have the same sign.

Theorem 3. *If the motion is oscillatory, the trajectory alternately crosses the positive and negative x-axes. Between any two consecutive points of intersection, v is a single-valued function of x.*

The first part of the theorem follows immediately from Lemma 2, § 1, because $\dot{x}$ vanishes when the trajectory crosses the x-axis. From (1.2) we find the relation between the extreme values of the displacement x and the corresponding values of the acceleration,

$$\ddot{x} = -g(x), \tag{2.7}$$

a relation which is independent of f.

Between two consecutive points of intersection of the trajectory with the x-axis the velocity has the same sign. Therefore, a straight line parallel to the v-axis cannot cross the corresponding arc of the trajectory in two points because, otherwise, x ought to be increasing ($v > 0$) at one of these points and decreasing ($v < 0$) at the other. Hence the velocity v is a single-valued function of x between any two consecutive points of intersection of the trajectory with the x-axis.

Theorem 4. *The extreme values of the velocity, hence the zeros of the acceleration, too, correspond to the points at which the trajectory crosses the curve*

$$f(v) + g(x) = 0. \tag{2.8}$$

At these points the tangent to the trajectory is parallel to the x-axis and the concavity of the trajectory is directed toward the origin.

The first part of the theorem results immediately by putting $\ddot{x} = 0$ into (1.2). By differentiating (2.1) with respect to x we obtain further

$$\frac{d^2v}{dx^2} = -\frac{[vf'(v) - f(v) - g(x)]\dfrac{dv}{dx} + vg'(x)}{v^2}. \tag{2.9}$$

Since v and $\dot{v}$ are never zero together, we deduce from (2.1), (2.8), and (2.9) that

$$\left.\frac{dv}{dx}\right|_{\dot{v}=0} = 0 \quad , \quad \left.\frac{d^2v}{dx^2}\right|_{\dot{v}=0} = -\frac{g'(x)}{v}, \tag{2.10}$$

and this completes the proof because $g'(x) > 0$.

Theorem 5. *The extreme values of the acceleration correspond to the points in which the trajectory crosses the curve*

$$f'(v)[f(v) + g(x)] - vg'(x) = 0. \tag{2.11}$$

Indeed, by differentiating (1.2) with respect to t, we obtain

$$\dddot{x} = -\dot{v}f'(v) - vg'(x) = f'(v)[f(v) + g(x)] - vg'(x), \tag{2.12}$$

wherefrom the previous statement follows at once.

In the important particular case when $f'(0) = 0$, the whole x-axis belongs to the curve (2.11), and hence the displacement and the acceleration simultaneously take extreme values, which are related by (2.7).

Since f and g are assumed to be only *piecewise* continuously differentiable, the curve (2.11) may have discontinuities in the phase plane, corresponding to the angular points of the elastic and damping characteristics.

Theorem 6. *If f and g satisfy hypotheses $H_1 - H_4$, then the phase point approaches the origin (the singular point) as $t \to \infty$.*

This theorem is a direct consequence of Theorem 6, § 1.

By making use of the theorems demonstrated in this section we can fully visualize the possible forms of the trajectories for the systems under consideration. We distinguish three cases:

– The trajectory crosses the x-axis in at most one regular point; the motion is aperiodic (Fig. 7*a*, *b*, *c*).

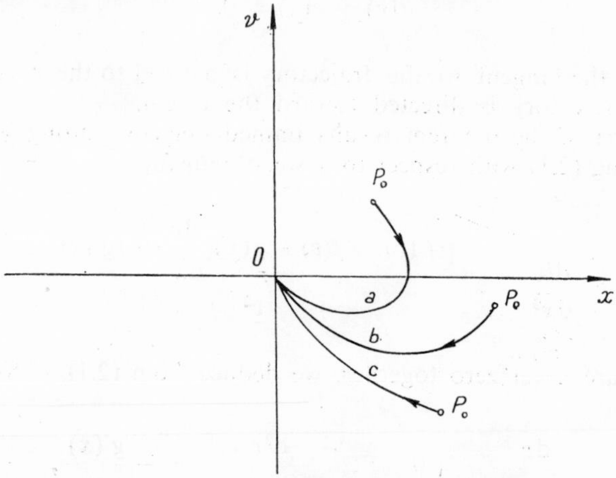

Fig. 7

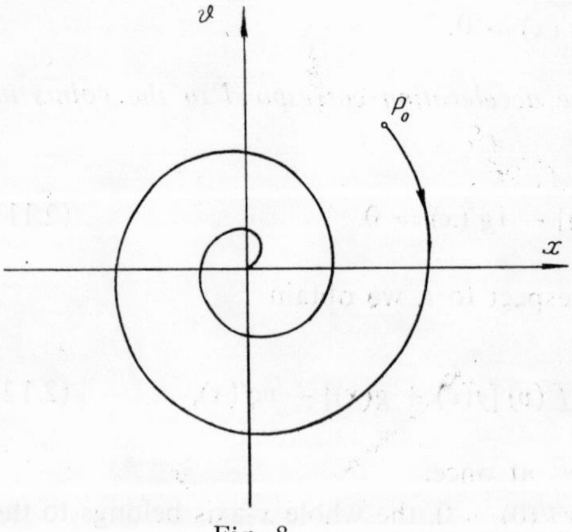

Fig. 8

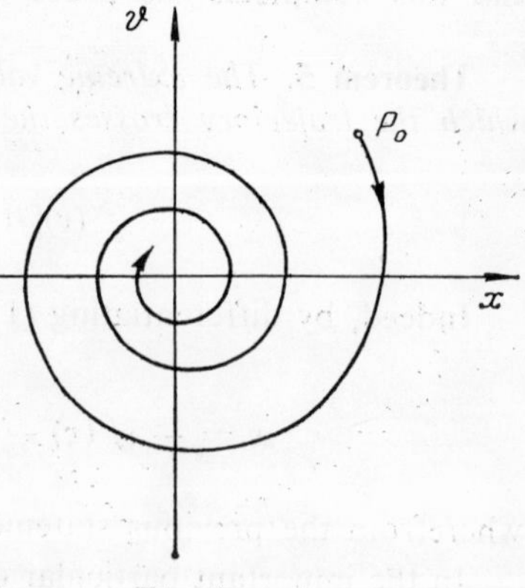

Fig. 9

– The trajectory crosses the x-axis at two or more points, whose number is, however, finite; the motion is at first oscillatory, then aperiodic (Fig. 8).

– The trajectory crosses the x-axis infinitely many times (Fig. 9); the motion is oscillatory in character, but may be aperiodic up to some finite time.

The influence of the relative values of $f(v)$ and $g(x)$ on the character of the motion has been minutely analyzed by ZIEMBA [151], under the hypothesis that f and g are analytic and odd. Most of his results may be extended under our more general assumptions $H_1 - H_4$, a generalization we shall not, however, attempt here.

We finally remark that the phase trajectories drawn in Fig. 7*a*, *b*, *c* correspond qualitatively to the graphs of $x(t)$ in Fig. 4 *a*, *b*, *c*, and the phase trajectories drawn in Figs 8 and 9 correspond to the graphs of $x(t)$ in Figs 5 and 6, respectively.

b) The behavior of the phase trajectories near the singular point

Studying the character of the trajectories in the neighborhood of a singular point allows us to obtain important qualitative information about the motions taking place near the equilibrium position.

In the subsequent analysis we shall adapt to system (1.3) the results obtained by PERRON [144], FROMMER [61], FORSTER [58], LONN [105], and WINTNER [105] on the isolated singular points of the systems of class C^1 (see also the monographs by SANSONE and CONTI [177], chap. V, NEMYTSKY and STEPANOV [131], Chapter II, § 4, and CESARI [32], Chapter III, § 9).

Since in our case the origin is the only singular point, it is adequate to develop the functions f and g into Maclaurin series with remainders of second order in v and x, respectively:

$$f(v) = 2hv + \varphi(v), \tag{2.13}$$

$$g(x) = \omega^2 x + \psi(x). \tag{2.14}$$

From the hypotheses $H_1 - H_5$ we infer

$$2h = f'(0) \geqslant 0, \qquad \omega^2 = g'(0) > 0, \tag{2.15}$$

$$\varphi(0) = \psi(0) = 0, \tag{2.16}$$

$$\lim_{v \to 0} \frac{\varphi(v)}{v} = \varphi'(0) = 0, \qquad \lim_{x \to 0} \frac{\psi(x)}{x} = \psi'(0) = 0, \tag{2.17}$$

$$\varphi'(v) \geqslant -2h, \qquad \psi'(x) \geqslant -\omega^2, \tag{2.18}$$

$$-2h < \lim_{|v| \to \infty} \frac{\varphi(v)}{v} < \infty, \qquad \omega^2 < \lim_{|x| \to \infty} \frac{\psi(x)}{x} < \infty. \tag{2.19}$$

Further, by introducing (2.13) and (2.14) into (1.3), we obtain

$$\dot{x} = v, \qquad \dot{v} = \omega^2 x + \psi(x) + 2hv + \varphi(v). \tag{2.20}$$

To study the behavior of the trajectories near the origin it is convenient to introduce polar coordinates ρ, θ, related to x and v by

$$x = \rho \cos \theta\,, \qquad v = -\,\omega y = -\,\omega\rho \sin \theta\,. \tag{2.21}$$

By taking x, y instead of x, v as unknown functions, the system of differential equations (2.20) becomes

$$\dot{x} = -\,\omega y\,, \qquad \dot{y} = \omega x - 2hy + \frac{1}{\omega}\varphi(-\,\omega y) + \frac{1}{\omega}\psi(x). \tag{2.22}$$

The introduction of the factor ω into $(2.21)_2$ is necessary in order to respect the physical dimensions.[1] The minus sign in the same relation is used for making $\theta(t)$ an increasing function of t. Indeed, in the upper half-plane we have $v > 0$, hence $x(t)$ is increasing and the motion takes place from left to right; in the lower half-plane we have $v < 0$, hence $x(t)$ is decreasing and the motion proceeds from right to left (Fig. 10).

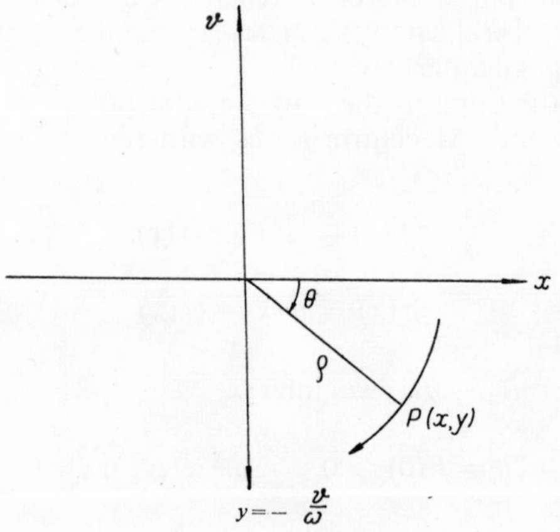

Fig. 10

By now putting (2.21) into (2.22) we obtain after some calculation

$$\dot{\rho} = \rho[-\,2h \sin^2\theta + \Delta(\rho\,, \theta) \sin \theta]\,, \tag{2.23}$$

$$\dot{\theta} = \omega - h \sin 2\theta + \Delta(\rho, \theta) \cos \theta\,, \tag{2.24}$$

where

$$\Delta(\rho, \theta) \equiv \frac{\varphi(-\,\omega\rho \sin \theta) + \psi(\rho \cos \theta)}{\omega\rho}\,. \tag{2.25}$$

[1] The constant ω may be obviously replaced by any other positive constant having the dimension T^{-1}.

Since

$$\rho = \sqrt{x^2 + v^2/\omega^2}\,, \tag{2.26}$$

we deduce from (2.17) that

$$\lim_{\rho \to 0} \Delta(\rho\,, \theta) = 0 \text{ uniformly in } \theta\,. \tag{2.27}$$

In other words, given any $\varepsilon > 0$, there is a $\rho_0 > 0$ such that $|\Delta(\rho, \theta)| < \varepsilon$ for $0 \leqslant \rho \leqslant \rho_0$ and any $0 \leqslant \theta \leqslant 2\pi$.

To more easily understand the trajectory behavior near the origin, we begin by demonstrating some preliminary propositions.

Lemma 1. *If f and g satisfy hypotheses $H_1 - H_4$, then $\lim_{t \to \infty} \rho(t) = 0$, and $\lim_{t\to\infty} \theta(t)$ does exist, finite or infinite.*

Proof. The first part of this lemma follows immediately from Theorem 6 and Eq. (2.26). Let us show now that $\lim_{t\to\infty} \theta(t)$ exists. To make things clearer we consider the representation of the motion in a plane in which ρ and θ are taken as Cartesian coordinates (Fig. 11).

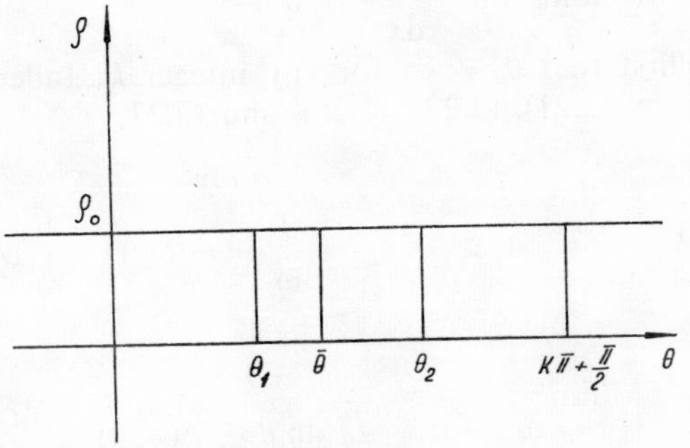

Fig. 11

Let us consider first the case when $\theta(t)$ is bounded for $t_0 \leqslant t < \infty$. We will demonstrate by contradiction that $\theta(t)$ can have but one limiting value as $t \to \infty$. Indeed, let θ_1 and θ_2 be two such values with, say, $\theta_1 < \theta_2$. Then, since the roots of the equation $\omega - h \sin 2\theta = 0$ are isolated, it follows from (2.24) that there exists a $\bar{\theta}$ with $\theta_1 < \bar{\theta} < \theta_2$ such that $\dot{\theta} \neq 0$ on the segment $\theta = \bar{\theta}$, $0 \leqslant \rho \leqslant \rho_0$ if ρ_0 is sufficiently small. On the other hand, since $\lim_{t\to\infty} \rho(t) = 0$, there is a $\bar{t}$ such that for $t > \bar{t}$ the trajectory lies in the strip $0 \leqslant \rho \leqslant \rho_0$. But since θ_1 and θ_2 are limiting values of $\theta(t)$ for $t \to \infty$, there are values of $t > \bar{t}$ for which $\theta(t)$ is as close as we please to θ_1 or θ_2. Hence the trajectory must intersect the segment $\theta = \bar{\theta}$,

$0 \leqslant \rho \leqslant \rho_0$, from left to right and from right to left as well, which contradicts $\dot{\theta} \neq 0$ for this segment. We conclude that $\theta(t)$ has only a limiting value, say θ_0, as $t \to \infty$. Moreover, since $\theta(t)$ was assumed bounded for $t \to \infty$, θ_0 must be finite.

Let us consider now the case when $\theta(t)$ is not bounded for $t \to \infty$. Then, given any positive integer k, there exists at least one t_k such that $\theta(t_k) > k\pi + \dfrac{\pi}{2}$. By (2.24) we have $\dot{\theta}\left(k\pi + \dfrac{\pi}{2}\right) = \omega > 0$, and hence the trajectory always crosses the lines $\theta = k\pi + \dfrac{\pi}{2}$ from left to right. We conclude that, given any positive integer k, we have $\theta(t) > k\pi + \dfrac{\pi}{2}$ for all $t \geqslant t_k$, and hence $\lim\limits_{t\to\infty} \theta(t) = \infty$.

Lemma 2. *If for a given trajectory of system* (2.22) *we have* $\lim\limits_{t\to\infty} \rho(t) = 0$ *and* $\lim\limits_{t\to\infty} \theta(t) = \theta_0$ *with* θ_0 *finite, then the tangent to the trajectory approaches the ray* $\theta = \theta_0$ *as* $t \to \infty$, *and* θ_0 *is a solution of the equation*

$$\omega - h \sin 2\theta = 0\,. \tag{2.28}$$

Proof. The existence of a limit of the tangent to the trajectory is equivalent to the existence of the limit $\lim\limits_{t\to\infty} \dfrac{\mathrm{d}y}{\mathrm{d}x} = \lim\limits_{t\to\infty} \dfrac{\dot{y}}{\dot{x}}$.

We shall prove first that $\theta_0 \neq k\pi$ for any integer k. Indeed, $\theta_0 = k\pi$ with k integer would imply by (2.21), (2.22), (2.25), and (2.27)

$$\begin{aligned}\lim_{t\to\infty} \frac{\dot{x}}{\dot{y}} &= \lim_{t\to\infty} \frac{-\,\omega y}{\omega x - 2hy + \dfrac{1}{\omega}\varphi(-\,\omega y) + \dfrac{1}{\omega}\psi(x)} \\ &= \lim_{t\to\infty} \frac{-\,\omega \sin\theta}{\omega\cos\theta - 2h\sin\theta + \Delta(\rho,\theta)} = 0\,.\end{aligned} \tag{2.29}$$

But, since $\lim\limits_{t\to\infty} x(t) = \lim\limits_{t\to\infty} y(t) = 0$, the existence of the limit (2.29) would further imply by l'Hôpital's rule the existence of the limit

$$\lim_{t\to\infty} \frac{x}{y} = \lim_{t\to\infty} \cot\theta = \lim_{\theta\to\theta_0} \cot\theta\,,$$

and this is impossible for $\theta_0 = k\pi$. We conclude that $\theta_0 \neq k\pi$, and from (2.21), (2.22), (2.25), and (2.27), we derive by a similar calculation

$$\lim_{t\to\infty} \frac{\dot{y}}{\dot{x}} = \lim_{t\to\infty} \frac{\omega\cos\theta - 2h\sin\theta + \Delta(\rho,\theta)}{-\,\omega\sin\theta} = \frac{2h}{\omega} - \cot\theta_0\,. \tag{2.30}$$

Now, by the same l'Hôpital's rule, we deduce that there exist the limit

$$\lim_{t\to\infty} \frac{y}{x} = \lim_{t\to\infty} \tan\theta = \lim_{\theta\to\theta_0} \tan\theta\,, \tag{2.31}$$

which is equal to (2.30). Hence we have $\theta_0 \neq k\pi + \frac{\pi}{2}$ and

$$\tan\theta_0 = \frac{2h}{\omega} - \cot\theta_0,$$

or

$$\omega\tan^2\theta_0 - 2h\tan\theta_0 + \omega = 0\,, \tag{2.32}$$

wherefrom it follows that

$$\sin 2\theta_0 = \frac{2\tan\theta_0}{1-\tan^2\theta_0} = \frac{\omega}{h},$$

which completes the proof. From the above reasoning we conclude also that $\theta_0 \neq \frac{k\pi}{2}$, and hence the tangent to the trajectory cannot tend to one of the coordinate axes.

Equation (2.28) is called the *characteristic equation*, and directions $\theta = \theta_0$, with θ_0 a solution of (2.28), are said to be *exceptional directions*.[1)]

Lemma 3. *If the angle* $\theta_1 \leqslant \theta \leqslant \theta_2$ *contains no exceptional directions, then there exists a* $\rho_0 > 0$ *such that any trajectory passing through points of the sector* $\theta_1 \leqslant \theta \leqslant \theta_2$, $0 \leqslant \rho \leqslant \rho_0$, *leaves this sector in a finite time.*

Proof. Since for $\theta_1 \leqslant \theta \leqslant \theta_2$ we have $\omega - h\sin 2\theta \neq 0$, we can find by virtue of (2.27) a $\rho_0 > 0$ such that

$$\dot\theta = \omega - h\sin 2\theta + \Delta(\rho,\theta)\cos\theta \neq 0$$

for all points of the sector $0 \leqslant p \leqslant \rho_0$, $\theta_1 \leqslant \theta \leqslant \theta_2$. This implies that θ is a monotonic function of t along any arc of trajectory lying within this sector. We conclude that any trajectory passing through points of the sector must leave it in a finite time because, otherwise, we ought to have $\lim_{t\to\infty}\theta(t) < \infty$ and, by lemma 1, also $\lim_{t\to\infty}\theta(t) = \theta_0$ with θ_0 finite, hence $\theta = \theta_0$ would be an exceptional direction, thus contradicting the hypothesis.

1) For a more general definition of the exceptional directions, which reduces, however, in our case, to the definition given above, see, e.g., FROMMER [61], or SANSONE and CONTI [177], p. 182 and p. 216.

Lemma 4. *For any point P of the trajectory let* α, $-\pi < \alpha < \pi$, *be the angle between the radius vector* $\overrightarrow{OP}$ *and the positive tangent at P to trajectory, that is, the tangent oriented in the direction of motion. Then, if* $\theta \neq k\pi$ *for any integer k, we have*

$$\tan \alpha = \rho \frac{d\theta}{d\rho} = \frac{\omega - h \sin 2\theta + \Delta(\rho, \theta) \cos \theta}{- 2h \sin^2\theta + \Delta(\rho, \theta) \sin \theta} . \tag{2.33}$$

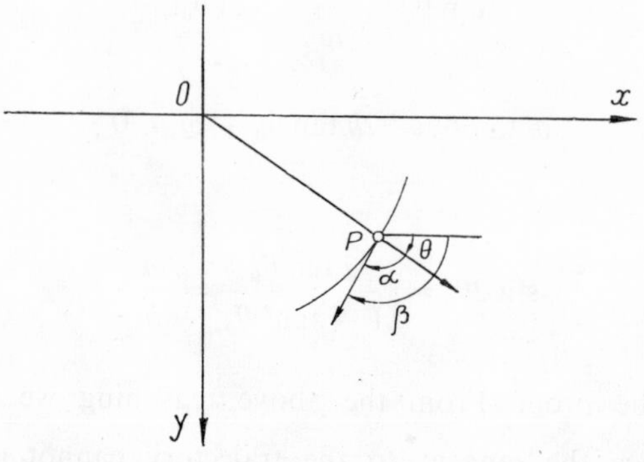

Fig. 12

Proof. With the notations in Fig. 12 we have

$$\tan \alpha = \tan (\beta - \theta) = \frac{\dfrac{dy}{dx} - \tan \theta}{1 + \dfrac{dy}{dx} \tan \theta} . \tag{2.34}$$

Further, we derive from (2.21)

$$\frac{dy}{dx} = \frac{\sin \theta \, d\rho + \rho \cos \theta \, d\theta}{\cos \theta \, d\rho - \rho \sin \theta \, d\theta},$$

and by substituting this expression into (2.34) we get the first part of (2.33). The second part of (2.33) follows immediately by (2.23) and (2.24).

If the origin is the only singular point of the system

$$\dot{x} = ax + by + X(x, y), \tag{2.35}$$

$$\dot{y} = cx + dy + Y(x, y),$$

and if

$$\lim_{\rho \to 0} \frac{X(x, y)}{\rho} = \lim_{\rho \to 0} \frac{Y(x, y)}{\rho} = 0, \tag{2.36}$$

then the behavior of the trajectories near the origin may be always reduced to one of some standard configurations, after which the singular point is called focus, node with one tangent, node with two tangents, center, center-focus, stellar node, or saddle point (see WINTNER [193] or SANSONE and CONTI [177], Chapter V). In the particular case of system (2.22) we shall see below that the singular point can be only of one of the first three types mentioned above. We begin by giving the definitions of these three kinds of singular points. Let $C(0, r)$ denote the set of all points of the phase plane with $0 \leqslant \rho \leqslant r$.

Definitions

1°. *The origin is said to be an* attractor *for system* (2.35), *provided there is an* $r > 0$ *such that* $\lim_{t\to\infty} \rho(t) = 0$ *for all trajectories passing through points of* $C(0, r)$.

2°. *The origin is said to be a* stable focus *for system* (2.35) *if it is an attractor, and if* $\lim_{t\to\infty} \theta(t) = \infty$ *for all trajectories tending to* 0 *(Fig. 13).*

3°. *The origin is called* node with one tangent *for system* (2.35) *if it is an attractor, and if the tangents to all trajectories approaching* 0 *tend to one or the other of two rays issuing from* 0 *in opposite directions (Fig. 14).*

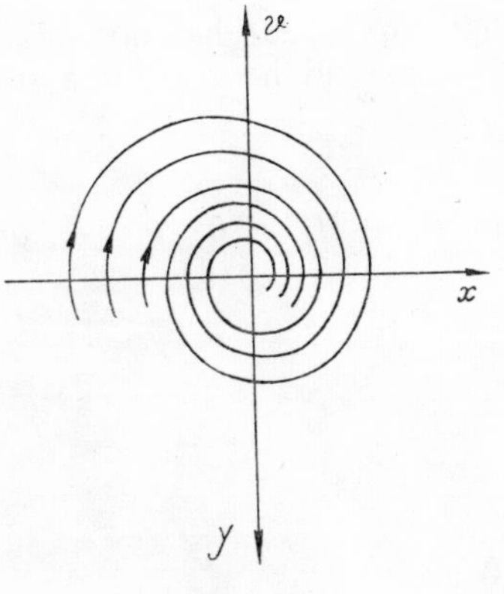

Fig. 13

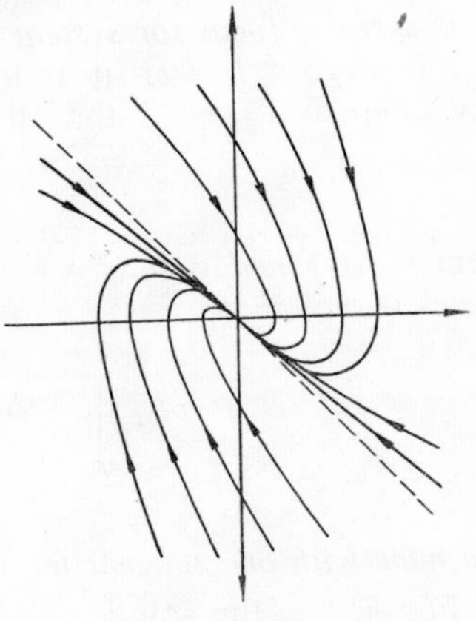

Fig. 14

4°. *The origin is called* node with two tangents *for system* (2.35) *if it is an attractor, if the tangents to two trajectories approaching* 0 *tend to one or the other of two rays issuing from* 0 *in opposite directions, and if the tangents to all other trajectories approaching* 0 *tend to one or the other of another two rays issuing from* 0 *in opposite directions (Fig. 15).*

We will study now the character of the singular point for the system (2.22).

Theorem 7. *If f and g satisfy hypotheses $H_1 - H_4$, then the origin is an attractor for system (2.22).*

This theorem is an immediate consequence of the first part of lemma 1. Moreover, in our case, $\lim_{t\to\infty} \rho(t) = 0$ holds for every trajectory, irrespective of the initial conditions, not only for those trajectories passing through points close enough to the origin.

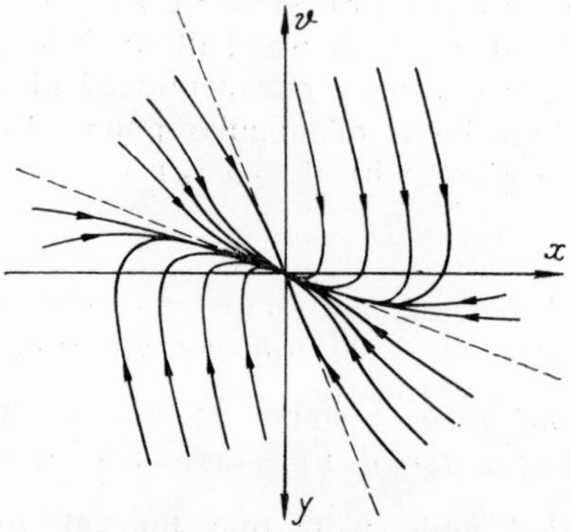

Fig. 15

Theorem 8. *If f and g satisfy hypotheses $H_1 - H_4$, and if $0 \leqslant h < \omega$, then the origin is a stable focus for system (2.22).*

Indeed, from $0 \leqslant h < \omega$ it follows that Eq. (2.28) has no solutions. This implies by Lemmas 1 and 2 that $\lim_{t\to\infty} \theta(t) = \infty$, and hence 0 is a stable focus for (2.22).

Theorem 9. *If f and g satisfy hypotheses $H_1 - H_4$, if $0 < h = \omega$, and if there exists an $\alpha > 0$ such that*

$$\lim_{\rho\to 0} \frac{\Delta(\rho, \theta)}{\rho^\alpha} = 0, \tag{2.37}$$

then 0 is a node with one tangent for system (2.22).

Proof. For $h = \omega$ the solutions of (2.28) are

$$\theta_0 = k\pi + \frac{\pi}{4} \quad \text{with } k \text{ an integer.} \tag{2.38}$$

Let us show that the rays $\theta = \frac{\pi}{4}$ and $\theta = \frac{3\pi}{4}$, oriented toward the origin, have the property of the two rays in the definition of a node with one tangent.

We choose an arbitrary $\bar{\theta}$ such that $0 < \bar{\theta} < \frac{\pi}{4}$ (Fig. 16). By (2.27), there is a $\rho_0 > 0$ such that

$$|\Delta(\rho, \theta)| < \min\{\omega(1 - \sin 2\bar{\theta}), \quad 2\omega \sin^2\bar{\theta}\} \tag{2.39}$$

for $0 \leqslant \rho \leqslant \rho_0$ and any $0 \leqslant \theta \leqslant 2\pi$.

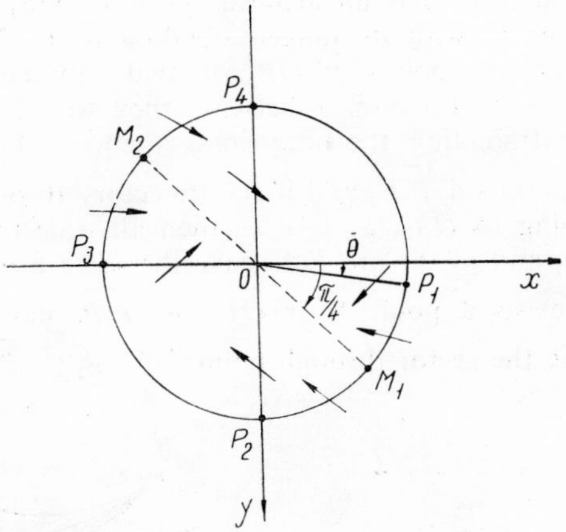

Fig. 16

We consider now the trajectories passing through points of $C(0, \rho_0)$. For $-\frac{\pi}{2} \leqslant \theta \leqslant \bar{\theta}$ or $\frac{\pi}{2} \leqslant \theta \leqslant \pi + \bar{\theta}$ we have $1 - \sin 2\theta \geqslant 1 - \sin 2\bar{\theta}$, so that by (2.24) and (2.39)

$$\dot{\theta} = \omega(1 - \sin 2\theta) + \Delta(\rho, \theta) \cos\theta > 0. \tag{2.40}$$

Since the sectors OP_4P_1 and OP_2P_3 contain no exceptional directions, we conclude by Lemma 3 and (2.40) that the trajectories passing through points of these two sectors must leave them in a finite time in the direction of increasing θ.

Let us consider now the sector $OP_1P_2 \left(\bar{\theta} \leqslant \theta \leqslant \frac{\pi}{2}\right)$ containing a single exceptional direction, namely $\theta = \frac{\pi}{4}$. In view of (2.23), (2.39), and (2.40), we infer that for the points on the arc P_1P_2 $(\rho = \rho_0)$ we have $\dot{\rho} < 0$, and for the points on the rays $\theta = \bar{\theta}$ and $\theta = \frac{\pi}{2}$ we have $\dot{\theta} > 0$. On the other hand, from (2.33) it follows that $\tan \alpha < 0$ for $\theta = \overline{\theta}$ or $\theta = \frac{\pi}{2}$. Therefore, the trajectories cross the boundary of the sector OP_1P_2 as indicated by the arrows in Fig. 16. In this case, by making use of LONN's theorem [105], it may be shown that, if condition (2.37) is satisfied, there exists at least one trajectory entering the sector OP_1P_2 and approaching O. We conclude by lemma 2 that the tangent to this trajectory tends to the exceptional direction $\theta = \frac{\pi}{4}$ as $t \to \infty$.

Let us notice now that, if M is an arbitrary point on OP_1, and if the trajectory through M approaches O with its tangent tending to OM_1 as $t \to \infty$, then all trajectories passing through points of OP_1 situated between O and M have the same property (Fig. 17*a*) because, otherwise, they would cross the trajectory through M, thus contradicting the uniqueness theorem. For the same reason, if M is an arbitrary point on $\widehat{P_1P_2}$ and if the trajectory through M approaches O with its tangent tending to OM_1 as $t \to \infty$, then all trajectories passing through points of the ray OP_1 and of the arc P_1M have the same property (Fig. 17*b*). We conclude that there exists a point S on OP_1 or $\widehat{P_1P_2}$ having the property that all trajectories entering the sector through points of $OP_1 \cup \widehat{P_1P_2}$, situated between

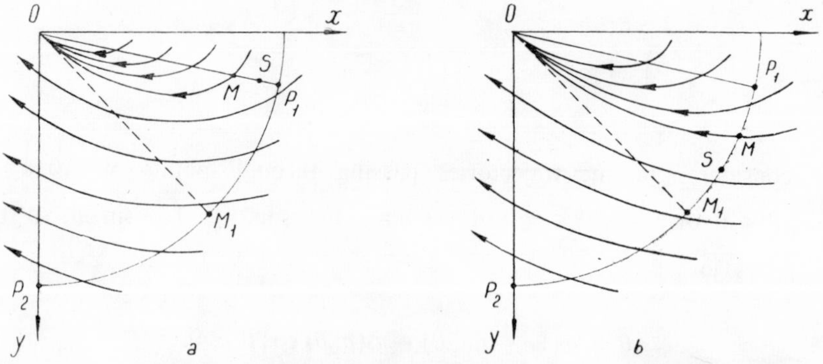

Fig. 17

O and S, approach O with their tangents tending to OM_1 as $t \to \infty$, while the trajectories entering the sector through other points of $OP_1 \cup \widehat{P_1P_2}$ leave the sector in a finite time, by crossing the ray OP_2.

By repeating the same reasoning for the sector OP_3P_2, we infer that the trajectories have near the origin a configuration which is characteristic of a node with one tangent (Fig. 14).

Theorem 10. *If f and g satisfy hypotheses $H_1 - H_4$, and if $0 < \omega < h$, then 0 is a node with two tangents for system (2.22).*

Proof. For $0 < \omega < h$, Eq. (2.28) has the solutions

$$\theta_0 = \frac{k\pi}{2} + \frac{(-1)^k}{2} \arcsin \frac{\omega}{h} \tag{2.41}$$

with k an integer. Let us show that in this case the rays $\theta = \theta_1 = \arcsin \dfrac{\omega}{h}$ and $\theta = \pi + \theta_1$, $\theta = \theta_2 = \pi - \arcsin \dfrac{\omega}{h}$ and $\theta = \pi + \theta_2$, oriented toward the origin, have the properties of the two pairs of rays in the definition of a node with two tangents.

We choose an arbitrary $\bar{\theta}$ such that $0 < \bar{\theta} < \theta_1$ (Fig. 18). By virtue of (2.27) there exists a $\rho_0 > 0$ such that

$$|\Delta(\rho, \theta)| < \min(\omega - h \sin 2\bar{\theta},\ 2h \sin^2 \bar{\theta},\ h - \omega) \tag{2.42}$$

for $0 \leqslant \rho \leqslant \rho_0$ and any $0 \leqslant \theta \leqslant 2\pi$.

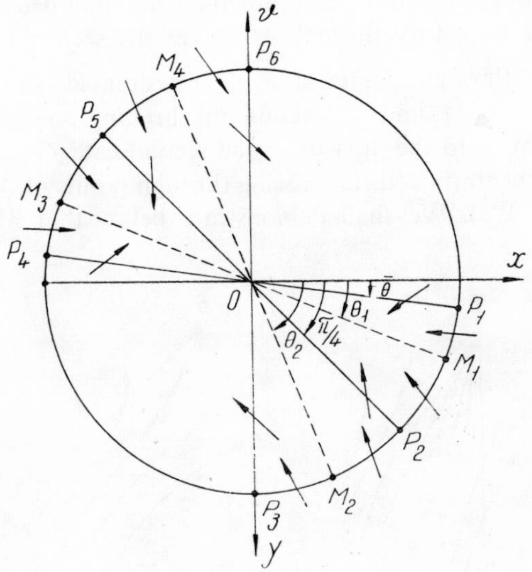

Fig. 18

We consider now the trajectories passing through points of $C(0, \rho_0)$. For $-\dfrac{\pi}{2} \leqslant \theta \leqslant \bar{\theta}$ and $\dfrac{\pi}{2} \leqslant \theta \leqslant \pi + \bar{\theta}$ we have $\omega - h \sin 2\theta \geqslant \omega - h \sin 2\bar{\theta}$, so that, by (2.42), we obtain from (2.24)

$$\dot{\theta} = \omega - h \sin 2\theta + \Delta(\rho, \theta) \cos \theta > 0. \tag{2.43}$$

Since the sectors OP_6P_1 and OP_3P_4 contain no exceptional directions, we conclude from lemma 3 and (2.40) that the trajectories passing through points of these two sectors must leave them in a finite time in the direction of increasing θ.

Further we deduce from (2.23), (2.24), and (2.42), that $\dot{\rho} < 0$ on the arc P_1P_3, $\dot{\theta} > 0$ on the rays OP_1 and OP_3, and $\tan \alpha > 0$ on the ray OP_2. Moreover, we have from (2.33) $\tan \alpha < 0$ on OP_1 and OP_3, and $\tan \alpha > 0$ on OP_2. Therefore, the trajectories cross the boundaries of the sectors OP_1P_2 and OP_2P_3 as indicated by the arrows in Fig. 18. We see now that the trajectories passing through points of the sector OP_1P_2 can no longer leave this sector, and hence approach 0 with their tangents tending to the exceptional direction $\theta = \theta_1$ as $t \to \infty$.

Let us examine now the sector OP_2P_3. The trajectories can enter this sector only through $\overset{\frown}{P_2P_3}$, after which they either leave the sector, by intersecting the rays OP_2 or OP_3 in a finite time, or approach the origin with their tangents tending to the exceptional direction OM_2 as $t \to \infty$. Let us notice now that, if M is an arbitrary point on $\overset{\frown}{P_2P_3}$, and if the trajectory through M leaves the sector through OP_2, then all trajectories passing through points of $\overset{\frown}{P_2P_3}$, situated between P_2 and M, must have the same property because, otherwise, they would intersect the trajectory through M, thus contradicting the uniqueness theorem. For the same reason, if the trajectory through M leaves the sector through OP_3, then all trajectories passing through points of $\overset{\frown}{P_2P_3}$ and situated between M and P_3 have the same property. Let M' and M'' denote the limiting points of the sets of points belonging, respectively, to the first or to the second category as M moves on P_2P_3 from P_2 to P_3. Then all trajectories passing through points of $M'M''$ must approach O as $t \to \infty$ (Fig. 19*a*). We shall demonstrate below that M' and M'' coincide,

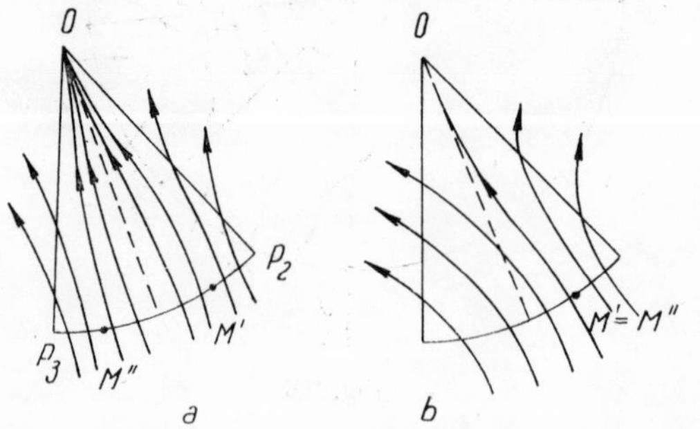

Fig. 19

i.e., there exists a single trajectory which, after entering the sector OP_2P_3, approaches O with its tangent tending to the exceptional direction OM_2 as $t \to \infty$ (Fig. 19*b*).

The proof proceeds by *reductio ad absurdum*. Let us assume that there exist two trajectories of system (2.22) approaching O and with tangents tending to the ray $\theta = \theta_2$ as $t \to \infty$. For large t the two trajectories may be represented by the equations $y = y_1(x)$ and $y = y_2(x)$. We have then

$$\lim_{x\to 0} y_1(x) = \lim_{x\to 0} y_2(x) = 0, \tag{2.44}$$

$$\lim_{x\to 0} \frac{y_1(x)}{x} = y_1'(0) = \lim_{x\to 0} \frac{y_2(x)}{x} = y_2'(0) = \tan\theta_2. \tag{2.45}$$

From the uniqueness theorem it follows that $y_1(x) \neq y_2(x)$ for any x. In order to make a choice we assume that $y_1(x) > y_2(x) > x > 0$ for any $x \neq 0$ in the sector OP_2P_3. We denote $y(x) \equiv y_1(x) - y_2(x)$ and derive from (2.45)

$$y'(0) = \lim_{x\to 0} \frac{y(x)}{x} = \lim_{x\to 0} \frac{y_1(x) - y_2(x)}{x} = 0. \tag{2.46}$$

By (2.22) we have also

$$\frac{dy(x)}{dx} = -\frac{\omega^2 x - 2h\omega y_1(x) + \varphi(-\omega y_1(x)) + \psi(x)}{\omega^2 y_1(x)}$$

$$+\frac{\omega^2 x - 2h\omega y_2(x) + \varphi(-\omega y_2(x)) + \psi(x)}{\omega^2 y_2(x)}$$

or

$$\frac{dy(x)}{dx} = \frac{\omega^2 x + \psi(x) + \varphi(-\omega y_2(x))}{\omega^2 y_1(x) y_2(x)}\, y(x) - \frac{\varphi(-\omega y_1(x)) - \varphi(-\omega y_2(x))}{\omega^2 y_1(x) y_2(x)}\, y_2(x). \tag{2.47}$$

By applying to φ the mean value theorem, we obtain

$$\varphi(-\omega y_1(x)) - \varphi(-\omega y_2(x)) = -\,\omega y(x)\varphi'(-\omega\bar{y}(x))$$

with $y_1(x) < \bar{y}(x) < y_2(x)$, and then (2.47) may be rewritten as

$$\frac{dy(x)}{dx} = \frac{y(x)}{x} \cdot \frac{x^2}{y_1(x)\, y_2(x)} \left\{ \omega^2 + \frac{\psi(x)}{x} + \frac{\varphi(-\omega y_2(x))}{x} + \varphi'(-\omega\bar{y}(x))\frac{y_2(x)}{x} \right\}. \tag{2.48}$$

But from (2.45) and (2.17) it follows that

$$\lim_{x\to 0} \frac{x^2}{y_1(x)\, y_2(x)} = \cot^2\theta_2 < 1,$$

$$\lim_{x\to 0} \frac{\varphi(-\omega y_2(x))}{x} = \lim_{x\to 0} \varphi'(-\omega\bar{y}(x)) = \varphi'(0) = 0,$$

and hence (2.48) takes the form

$$\frac{dy(x)}{dx} = \frac{y(x)}{x}[1 + \varepsilon(x)] \cot^2 \theta_2 \tag{2.49}$$

with $\varepsilon(x) \to 0$ as $x \to 0$. We then have for sufficiently small x

$$\frac{dy(x)}{dx} < \gamma \frac{y(x)}{x}$$

with $\cot^2\theta_2 < \gamma < 1$. By writing this relation in the form

$$\frac{dy(x)}{y(x)} < \gamma \frac{dx}{x}$$

and integrating it from x_0 to x with $0 < x < x_0$, i.e., in the direction of the negative x-axis, we obtain

$$\ln \frac{y(x)}{y(x_0)} > \gamma \ln \frac{x}{x_0},$$

hence

$$\frac{y(x)}{x} > \frac{y(x_0)}{x_0^\gamma} x^{\gamma-1},$$

wherefrom it follows that $y(x)/x \to \infty$ as $x \to 0$, which contradicts (2.46). Therefore, there exists only one trajectory that, after entering the sector P_2OP_3, approaches O with its tangent tending to the exceptional direction OM_2.

By repeating for the sectors OP_4P_5 and OP_5P_6 the same reasoning as for the sectors OP_1P_2 and OP_2P_3, respectively, we deduce that, near the origin, the trajectories have the behavior sketched in Fig. 15, and hence the singular point is a node with two tangents.

Remarks

1°. Theorems 7—10 show that, if f and g satisfy hypotheses $H_1 - H_4$, then the singular point is always an attractor for the system (2.22), and in particular a focus if $0 \leqslant h \leqslant \omega$, a node with one tangent if $0 < h = \omega$, or a node with two tangents if $0 < \omega < h$. Moreover, this implies that the behavior of the solutions of (2.22) near the singular point is the same as that of the solutions of the linear system

$$\dot{x} = -\omega y, \qquad \dot{y} = \omega x - 2hy. \tag{2.50}$$

the only exception being the case $h = 0$, $\omega > 0$, when 0 is a focus for (2.22), but it is a center for the reduced system (2.50). This important result, which holds under still more general hypotheses than those adopted here, is due to POINCARÉ.

2°. The theorems above were demonstrated under the assumption that f is continuously differentiable for $v = 0$. It may be shown, however, that these theorems still hold if, for $v = 0$, f possesses different one-sided derivatives, i.e., $f'_-(0) = h_1$, $f'_+(0) = h_2$ with $h_1 \neq h_2$.

3°. If f and g satisfy the hypotheses $H_1 - H_4$, theorem 7 shows that there are no closed trajectories (cycles or limit cycles) of system (2.22), and hence of system (1.3). In other words, Eq. (1.2) does not admit of periodic solutions. The source of this property may be found in the demonstration of lemma 5, § 1, where assumptions H_3 and H_4 played an essential role. In order to show this in a direct way, we shall prove the following theorem due to BENDIXON [13]:

Theorem 11. *If f and g satisfy hypotheses $H_1 - H_4$, then there exist no closed trajectories of system (1.3).*

Let us notice first that from H_3 and H_4 it follows that $f'(v)$ cannot identically vanish on a whole segment of the v-axis. Now, if there were a closed trajectory Γ of (1.3), bounding a finite region Δ, then, by applying Green's integral transformation, we ought to have

$$0 = \int_\Gamma (v\dot{x} - x\dot{v})\,\mathrm{d}t = \int_\Gamma \{[f(v) + g(x)]\,\mathrm{d}x - v\,\mathrm{d}v\} = -\iint_\Delta f'(v)\,\mathrm{d}x\,\mathrm{d}v,$$

and this would contradict the conditions $f'(v) \geqslant 0$, $f'(v) \not\equiv 0$ in Δ.

In the following chapter we shall consider conservative systems, i.e., systems for which $f(v) \equiv 0$. Since for such systems condition H_3 is no longer fulfilled, we shall reexamine there the behavior of the trajectories near the singular point and the existence of periodic solutions.

§ 3. APPROXIMATE DETERMINATION OF THE INTERVAL OF TIME CORRESPONDING TO A KNOWN ARC OF THE PHASE TRAJECTORY

The exact solution $x(t)$ of Eq. (1.2) may be found only in very few cases. However, in many cases of practical interest, it is still possible to determine first the phase trajectory and then to calculate approximately the time intervals corresponding to various arcs of the trajectory, thus obtaining a rather complete description of the motion. With this in mind, we shall assume in the following that we know an arc of the trajectory, joining the phase points $P'(x', v')$ and $P''(x'', v'')$, and we shall indicate several methods for the approximate determination of the time interval Δt, in which the representative point covers the arc $P'P''$.

If the equation of the trajectory is known in one of the explicit forms

$$v = v(x) \qquad \text{for} \qquad x \in [x', x''], \tag{3.1}$$

or

$$x = x(v) \qquad \text{for} \qquad v \in [v', v''], \tag{3.2}$$

then, by integrating (1.4), we obtain, respectively,

$$\Delta t = \int_{x'}^{x''} \frac{\mathrm{d}x}{v(x)}, \tag{3.3}$$

and

$$\Delta t = -\int_{v'}^{v''} \frac{\mathrm{d}v}{f(v) + g(x(v))}. \tag{3.4}$$

We shall first expound two analytical methods for the evaluation of the integrals (3.3) and (3.4), and then we shall set forth a graphical method, which is particularly convenient for the determination of Δt when the corresponding arc of the trajectory was obtained also by graphical integration of Eq. (2.1).

a) SIMPSON's method

If $v(x) \neq 0$ for $x \in [x', x'']$, then the approximate calculation of the integral (3.3) can be done without difficulty, by applying one of the known methods of numerical integration, for instance SIMPSON's method or the method of LEGENDRE's polynomials. If $v(x)$ vanishes for some values of $x \in [x', x'']$, but we know Eq. (3.2), we may use the integral (3.4) to calculate the time intervals corresponding to some subintervals of $[x', x'']$, chosen so as to include the points where $v(x) = 0$. It may happen, however, that we know only Eq. (3.1) and this equation cannot be solved for x. For such cases we indicate below, after KAUDERER [82], a procedure permitting the removal of the singularities of the integrand in (3.3).

Since the zeros of $v(x)$ are discrete we may always assume that the integrand in (3.3) vanishes only at one or both of the ends of $[x', x'']$ and that $v(x) \neq 0$ for $x \in (x', x'')$. Suppose, e.g., that $v(x') = v(x'') = 0$. This situation always occurs when we want to evaluate the time interval between two extreme values of $x(t)$, i.e., the duration of a semioscillation. By Theorem 3, § 2, the velocity is a single-valued function of x for $x \in [x', x'']$, and the motion takes place within either of the half-planes bounded by the x axis. In order to make a choice, assume that the representative point of the motion moves from P' to P'', hence in the upper half-plane, where $v(x) \geqslant 0$. According to the same theorem we have $x' < 0 < x''$.

We write now $\Delta t = \Delta t' + \Delta t''$, where

$$\Delta t' = \int_{x'}^{(x'+x'')/2} \frac{\mathrm{d}x}{v(x)}, \qquad \Delta t'' = \int_{(x'+x'')/2}^{x''} \frac{\mathrm{d}x}{v(x)}. \tag{3.5}$$

We begin by calculating $\Delta t'$ by means of SIMPSON's formula for four subintervals. To remove the singularity from the lower limit of integration we introduce a new integration variable $\zeta > 0$ by the relation

$$x = \zeta^2 + x', \qquad x' < x < \frac{x' + x''}{2}, \tag{3.6}$$

thus obtaining

$$\Delta t' = 2 \int_0^{\sqrt{(x''-x')/2}} \Phi(\zeta)\, d\zeta, \tag{3.7}$$

where

$$\Phi(\zeta) \equiv \frac{\zeta}{v(\zeta^2 + x')}. \tag{3.8}$$

The value $\Phi(0)$ can be calculated by passing to the limit in (3.8) for $\zeta \to 0$. By l'Hôpital's rule and (2.1), we obtain successively

$$\lim_{\zeta\to 0} [\Phi(\zeta)]^2 = \lim_{\zeta\to 0} \frac{\zeta^2}{[v(\zeta^2 + x')]^2} = \frac{1}{2} \lim_{\zeta\to 0} \frac{1}{\left[v \dfrac{dv}{dx} \right]_{x=\zeta^2+x'}}$$

$$= -\frac{1}{2} \lim_{\zeta\to 0} \frac{1}{g(\zeta^2 + x')} = -\frac{1}{2g(x')}.$$

But $\Phi(\zeta) \geqslant 0$, and hence

$$\Phi(0) = \lim_{\zeta\to 0} \Phi(\zeta) = \frac{1}{\sqrt{-2g(x')}}. \tag{3.9}$$

By applying now SIMPSON's formula for four subintervals (see, e.g., NICOLESCU et al. [133]) to calculate the integral (3.7), we get

$$\Delta t' \approx \frac{x'' - x'}{12} \left\{ \frac{1}{\sqrt{-(x'' - x')\, g(x')}} + \frac{1}{v_1} + \frac{1}{v_4} + \frac{3}{v_9} + \frac{1}{v_{16}} \right\}, \tag{3.10}$$

where

$$v_m = v(x' + \frac{m}{32}(x'' - x')). \tag{3.11}$$

Substituting $x = x'' - \zeta^2$ into $(3.5)_2$ yields by an analogous calculation

$$\Delta t'' \approx \frac{x'' - x'}{12} \left\{ \frac{1}{v_{16}} + \frac{3}{v_{23}} + \frac{1}{v_{28}} + \frac{1}{v_{31}} + \frac{1}{\sqrt{(x''-x')\, g(x'')}} \right\}. \tag{3.12}$$

Finally, adding (3.10) and (3.12) gives

$$\Delta t \approx \frac{x'' - x'}{12} \left\{ \frac{1}{\sqrt{-(x'' - x')\, g(x')}} + \frac{1}{v_1} + \frac{1}{v_4} + \frac{3}{v_9} + \frac{2}{v_{16}} + \frac{3}{v_{23}} + \frac{1}{v_{28}} + \frac{1}{v_{31}} + \frac{1}{\sqrt{(x'' - x')\, g(x'')}} \right\}. \tag{3.13}$$

Formula (3.13) allows the determination of the duration Δt of the semioscillation corresponding to the arc of trajectory joining the points $P'(x', 0)$ and $P''(x'', 0)$ with $x' < 0 < x''$. In case the velocity does not vanish on the arc of the trajectory being considered, i.e., $v(x) \neq 0$ for $x \in [x', x'']$, then the integrand in (3.7) has no singularities and the direct application of SIMPSON's formula yields

$$\Delta t \approx \frac{x'' - x'}{12} \left(\frac{1}{v_1} + \frac{1}{v_4} + \frac{3}{v_9} + \frac{2}{v_{16}} + \frac{3}{v_{23}} + \frac{1}{v_{28}} + \frac{1}{v_{31}} \right). \tag{3.14}$$

b) Method of LEGENDRE's polynomials

The use of this method is adequate when a very precise determination of the time intervals is required. For a thorough exposition of the method see,e.g., HOBSON's book [73]; we indicate here only the way in which this method is applied.

We assume that $v(x)$ does not vanish for $x \in [x', x'']$. Otherwise, the singularity could be removed as shown in the preceding section. By introducing the new variable

$$x = \frac{x'' - x'}{2} \xi + \frac{x' + x''}{2} \tag{3.15}$$

the integral (3.3) becomes

$$\Delta t = \frac{2}{x'' - x'} \int_{-1}^{1} \Phi(\xi)\, d\xi \tag{3.16}$$

with

$$\Phi(\xi) \equiv \frac{1}{v\left(\dfrac{x'' - x'}{2} \xi + \dfrac{x' + x''}{2}\right)}. \tag{3.17}$$

If we make use now of LEGENDRE's polynomial of degree n, $P_n(\xi)$, as interpolation polynomial in the integral (3.16), we obtain the evaluation

$$\Delta t \approx \frac{2}{x'' - x'} \sum_{r=1}^{n} A_r \Phi(a_r), \tag{3.18}$$

where a_r, $r = 1, 2, \ldots, n$ are the zeros of $P_n(\xi)$ in the interval $[-1, 1]$, and the coefficients A_r are given by

$$A_r = \frac{1}{P_n'(a_r)} \int_{-1}^{1} \frac{P_n(\xi) d\xi}{\xi - a_r}. \tag{3.19}$$

It may be proved that formula (3.18) gives exact values for the integral (3.16) provided $\Phi(\xi)$ is a polynomial of degree $2n - 1$ or lower. If $\Phi(\xi)$ is an arbitrary continuous function, then the higher the degree of the interpolation polynomial $P_n(\xi)$, the better the evaluation obtained for the integral (3.16). On the other hand, we see from (3.19) that the coefficients A_r do not depend on Φ, so that they were calculated once for all for various values of n. We indicate below, after HOBSON [73], the values with five exact decimals of a_r and A_r, corresponding to $n = 3$, 5, and 7.

$$n = 3: \quad a_1 = -a_3 = 0.77460, \qquad a_2 = 0,$$

$$A_1 = A_3 = \frac{5}{9}, \qquad A_2 = \frac{8}{9}.$$

$$n = 5: \quad a_1 = -a_5 = 0.90618, \qquad a_2 = -a_4 = 0.53847,$$

$$A_1 = A_5 = 0.23793, \qquad A_2 = A_4 = 0.47863,$$

$$A_3 = 0.56889, \qquad a_3 = 0.$$

$$n = 7: \quad a_1 = -a_7 = 0.94911, \qquad a_2 = -a_6 = 0.74153,$$

$$a_3 = -a_5 = 0.40585, \qquad a_4 = 0,$$

$$A_1 = A_7 = 0.12948, \qquad A_2 = A_6 = 0.27970,$$

$$A_3 = A_5 = 0.38183, \qquad A_4 = 0.41796.$$

The use of more than five decimals for a_r and A_r, or of LEGENDRE's polynomials of degree higher than 7, is of no practical interest for our problem.

c) Graphical method

Suppose now that the trajectory was determined by some graphical method (see § 4), and that we want to calculate approximately the interval of time in which the phase point covers a sufficiently small arc $P'P''$ of the trajectory (Fig. 20).

The bisector of the angle $P'OP''$ cuts the trajectory at P. Let $\widehat{POx} = \theta$, $\widehat{P'OP''} = \Delta\theta$, $OP = \rho$, $OP' = \rho + \Delta\rho'$, $OP'' = \rho - \Delta\rho''$. We consider first the case when the arc $P'P''$ does not intersect the coordinate axes $\left(\theta \neq \dfrac{k\pi}{2} \text{ for any integer } k\right)$. The circle with center at O and radius ρ crosses the lines through P' and P'' parallel to Oy at M' and M'', respectively. Let $\Delta\beta' = \widehat{P'OM'}$, $\Delta\beta'' = \widehat{P''OM''}$.

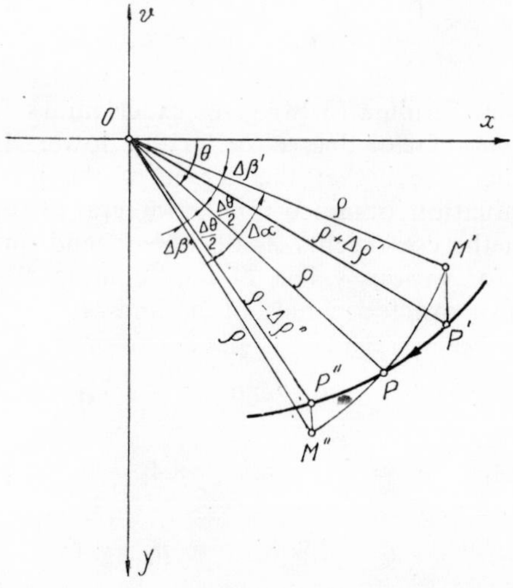

Fig. 20

By applying the law of sines to the triangle $OM'P'$, and taking into account that $\widehat{OP'M'} = \dfrac{\pi}{2} - \theta + \dfrac{\Delta\theta}{2}$, $\widehat{OM'P'} = \dfrac{\pi}{2} + \theta - \dfrac{\Delta\theta}{2} - \Delta\beta'$, we obtain

$$\frac{\rho - \Delta\rho'}{\rho} = \frac{\cos\left(\theta - \dfrac{\Delta\theta}{2} - \Delta\beta'\right)}{\cos\left(\theta - \dfrac{\Delta\theta}{2}\right)} = \cos\Delta\beta' + \tan\left(\theta - \frac{\Delta\theta}{2}\right)\sin\Delta\beta'. \tag{3.20}$$

If $\Delta\theta$, $\Delta\beta'$, $\Delta\beta''$ are small, we have

$$\cos\Delta\beta' \approx 1, \qquad \sin\Delta\beta' \approx \Delta\beta', \qquad \tan\left(\theta - \frac{\Delta\theta}{2}\right) \approx \tan\theta \tag{3.21}$$

and then from (3.20) it follows that

$$\frac{\rho - \Delta\rho'}{\rho} \approx 1 + \Delta\beta' \cdot \tan\theta. \tag{3.22}$$

Analogously, we have from the triangle $OM'P'$

$$\frac{\rho + \Delta\rho''}{\rho} \approx 1 - \Delta\beta'' \cdot \tan\theta. \tag{3.23}$$

By putting now $\Delta\beta' + \Delta\beta'' = \Delta\beta$, $\Delta\rho' + \Delta\rho'' = \Delta\rho$, we deduce from the last three relations that

$$\frac{\Delta\rho}{\rho}\cot\theta \approx -\Delta\beta. \tag{3.24}$$

On the other hand, by differentiating $(2.21)_1$ and considering $(2.21)_2$, we obtain

$$\cos\theta\,\mathrm{d}\rho - \rho\sin\theta\,\mathrm{d}\theta = -\omega\rho\sin\theta\,\mathrm{d}t, \tag{3.25}$$

wherefrom, by passing to finite differences, it follows that

$$\omega\Delta t \approx \Delta\theta - \frac{\Delta\rho}{\rho}\cot\theta. \tag{3.26}$$

Finally, by substituting (3.24) into (3.26) and taking into consideration that $\Delta\alpha = \Delta\theta + \Delta\beta$, we find

$$\omega\Delta t \approx \Delta\alpha. \tag{3.27}$$

Relation (3.27) affords a direct procedure for the determination of Δt by measuring $\Delta\alpha$ and dividing it by ω. This operation may be performed for any arc of the trajectory not crossing the coordinate axes.

Assume now that the point P lies on the y-axis. We have then $\cot\theta = 0$, and from (3.26) it follows that

$$\omega\Delta t \approx \Delta\theta, \tag{3.28}$$

a relation which determines Δt.

Finally, if P is situated on Ox, then Eq. (3.26) no longer has a meaning. However, by differentiating $(2.21)_2$ with respect to t, we obtain

$$\dot{v} = -\omega\dot{\rho}\sin\theta - \omega\rho\dot{\theta}\cos\theta,$$

that is $\dot{v} = -\omega x\dot{\theta}$ since now $\sin\theta = 0$. By comparing the last equation with $(1.4)_2$, in which we put $v = 0$, we infer

$$\dot{\theta} = \frac{g(x)}{\omega x},$$

wherefrom it follows that

$$\Delta t \approx \frac{\omega x}{g(x)} \Delta\theta. \tag{3.29}$$

This relation gives again Δt provided we replace x by the abscissa of the point P, at which the trajectory crosses the x-axis.

Formulas (3.27), (3.28), and (3.29) allow the approximate determination of the interval of time corresponding to a known arc of trajectory. Obviously, the smaller the arc PP', the less the error.

§ 4. GRAPHICAL AND GRAPHIC-ANALYTICAL METHODS

When nonlinear vibration problems are being studied, the exact integration of the equation of motion can be performed but rarely. This is why the use of approximate methods (graphical, graphic-analytical, analytical, or numerical) plays an important role in solving such problems.

We shall begin with the graphical and graphic-analytical methods, which are closer to our previous considerations concerning the phase plane.

In this section we continue to make use of the variables x and $y = -v/\omega$, both of which have the same physical dimension and allow, therefore, the use of the same units on both axes of coordinates. From (2.1) it follows that the differential equation of motion takes in these variables the form

$$\frac{\mathrm{d}y}{\mathrm{d}x} = -\frac{f(-\omega y) + g(x)}{\omega^2 y}. \tag{4.1}$$

The graphical methods are based on the determination of the tangent line to the trajectory at an arbitrary given point. Starting then from the point $P_0(x_0, y_0)$, corresponding to the initial conditions $x(t_0) = x_0$, $y(t_0) = y_0 = -v_0/\omega$, one traces the tangent line to the trajectory at P_0, on which one chooses a close point P_1. One draws further the tangent line at P_1, on which one chooses another point P_2, close to P_1, and so on. After thus plotting the approximate trajectory, the intervals of time $\Delta t_1, \Delta t_2, \ldots$, corresponding to the arcs $P_0P_1, P_1P_2, \ldots$ of the trajectory, may be determined by the graphical method indicated in § 3c.

The first two graphical methods we shall set forth below — LIÉNARD's method and the method of isoclines — are applicable, respectively, when f or g are linear functions of their arguments. The third method – SCHÄFFER's method – may still be applied when both f and g are nonlinear functions. At the end of this section we shall explain the so-called "delta method." This is a graphic-analytical method permitting the simultaneous determination of the phase trajectory and of the time intervals corresponding to various arcs of it. The delta method may also be applied in the case of forced vibrations.

a) LIÉNARD's method

This method may be applied when the elastic characteristic of the oscillator is linear, i.e.,

$$g(x) = \omega^2 x. \tag{4.2}$$

Then, by (4.1), we obtain for the slope of the tangent to the trajectory

$$\frac{\mathrm{d}y}{\mathrm{d}x} = -\frac{x + \dfrac{f(-\omega y)}{\omega^2}}{y}. \tag{4.3}$$

In order to trace the tangent to the trajectory at a given point $P(x, y)$ we proceed as follows. We plot in the plane xOy the curve of equation $x = -f(-\omega y)/\omega^2$, which represents an affine transformation of the damping characteristic (Fig. 21). The parallel through P to Ox cuts this curve at N, and the parallel through N to Oy intersects Ox at S. Then the perpendicular at P to SP represents the tangent to the trajectory at P.

Indeed, from Fig. 21, we see that the characteristic points of the construction described have the coordinates

$$N\left(-\frac{f(-\omega y)}{\omega^2},\, x\right), \qquad S\left(-\frac{f(-\omega y)}{\omega^2},\, 0\right),$$

and the line MP has the slope

$$m_{SP} = \frac{y}{x + \dfrac{f(-\omega y)}{\omega^2}},$$

which is inverse and has opposite sign to that of the tangent to the trajectory, given by (4.3).

Consider as an application the linear oscillator with dry friction, for which

$$g(x) = \omega^2 x, \qquad f(v) = R \operatorname{sgn} v = R\frac{v}{|v|}. \tag{4.4}$$

In this case we have

$$f(-\omega y) = R\,\frac{y}{|y|},$$

and hence the curve $x = -f(-\omega y)/\omega^2$ reduces to two half-lines through the points $S_1\left(\frac{R}{\omega^2}, 0\right)$ and $S_2\left(-\frac{R}{\omega^2}, 0\right)$, parallel to Oy (Fig. 22). It may be readily

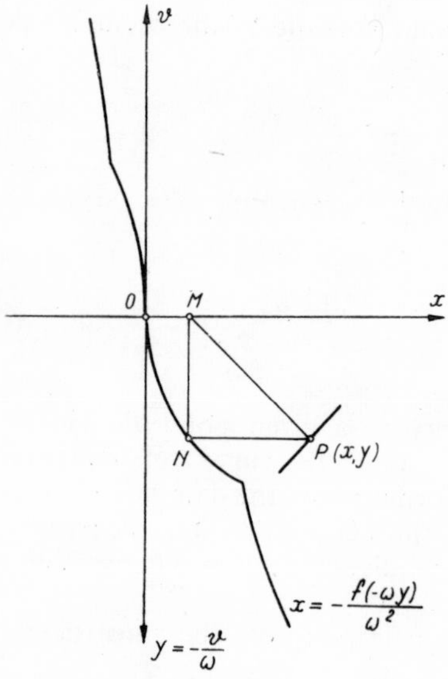

Fig. 21

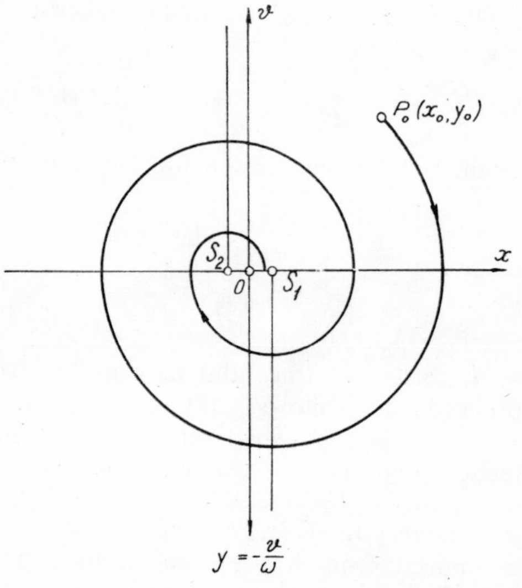

Fig. 22

seen that now the point S in the construction of LIÉNARD's method coincides with S_1 for all points of the trajectory situated in the half-plane $y > 0$, and with S_2 for all points of the trajectory situated in the half-plane $y < 0$. Since the tangent to the trajectory at P is always perpendicular to S_1P, or S_2P, we see that the trajectory consists of a sequence of semicircles, those in the upper half-plane having center at S_1, and those in the lower half-plane having center at S_2. The radius of each semicircle is $2R/\omega^2$ less than the radius of the preceding one. When this radius becomes smaller than R/ω^2, the motion comes to an end. It may be shown that the number n of complete semioscillations up to the end of the motion satisfies the relation

$$\frac{1}{2}\left(\frac{\omega^2 d}{R} - 1\right) < n < \frac{1}{2}\left(\frac{\omega^2 d}{R} + 1\right), \tag{4.5}$$

where d is the distance of P_0 to the origin.

b) The method of isoclines

An isocline is a curve in the phase plane having the property that the tangents to the trajectories through its points cross the x-axis at a constant angle. If the isoclines are drawn, then the tangent to the trajectory at any point $P(x, y)$ may be immediately traced, because we know the angle between this tangent and Ox, which is given by the isocline passing through P.

The isoclines may be easily drawn when the damping is linear, i.e., $f(v) = 2hv$. We have then from (4.1)

$$\frac{dy}{dx} = \frac{2h}{\omega} - \frac{g(x)}{\omega^2 y}. \tag{4.6}$$

If we require now that the angle between the tangent to the trajectory and Ox have a constant value, say α, we obtain from (4.6) the equation of the corresponding isocline

$$y = \frac{g(x)}{\omega^2(2h/\omega - \tan \alpha)}. \tag{4.7}$$

We remark that in our case the isoclines result from affine transformations (multiplication of the ordinates by constant numbers) of the elastic characteristic of the oscillator, i.e., of the curve of equation $y = g(x)$.

Consider as an application the case when

$$\frac{2h}{\omega} = 1, \qquad g(x) = \omega^2 \left(x + \frac{x^3}{a^2}\right), \tag{4.8}$$

where $a = 2$ units of length. The corresponding isoclines, which have the equation

$$y = \frac{1}{1 - \tan \alpha}\left(x + \frac{x^3}{4}\right), \tag{4.9}$$

as well as the trajectory passing through the point $P_0(2, 0)$ obtained by the method of isoclines, are plotted in Fig. 23.

c) SCHÄFER's method [178]

As we have already mentioned, this method may still be applied if both characteristics are nonlinear.

In order to determine the tangent to the trajectory at an arbitrary point $P(x, y)$, one first plots in the phase plane the curves C_1 and C_2 having, respectively, the equations $y = -f(\omega x)/\omega^2$ and $y = g(x)/\omega^2$, which result by affine transformations of the two characteristics (Fig. 24). One traces further the bisector of the coordinate axes of equation $y = -x$.

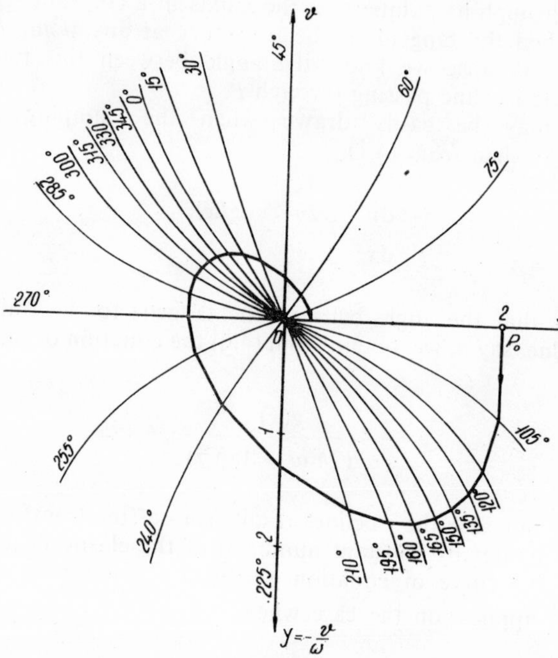

Fig. 23

The parallel from P to Ox crosses the bisector at R, and the parallel from P to Oy cuts the curve C_2 at Q. The parallel through R to Oy intersects the curve C_1 at S and the parallel from S to Oy cuts Oy at V. Then the tangent at P to the phase trajectory is parallel to VT.

Indeed, we deduce from the above construction that the coordinates of the characteristic points are

$$R(-y, y), \qquad S(-y, -f(-\omega y)/\omega^2), \qquad V(0, -f(-\omega y)/\omega^2),$$
$$Q(x, g(x)/\omega^2), \qquad T(-y, g(x)/\omega^2),$$

and hence VT has the slope

$$m_{VT} = -\frac{f(-\omega y) + g(x)}{\omega^2 y},$$

which is inverse and has opposite sign to that of the tangent to the trajectory, given by (4.1).

d) Delta method

Unlike the graphical methods expounded above, by means of which the phase trajectory is obtained as a sequence of line segments the delta method approximates the trajectory by a sequence of circular arcs.

By (2.14), Eq. (4.3) may be written

$$\frac{dy}{dx} = -\frac{x + \delta(x, y)}{y}, \tag{4.10}$$

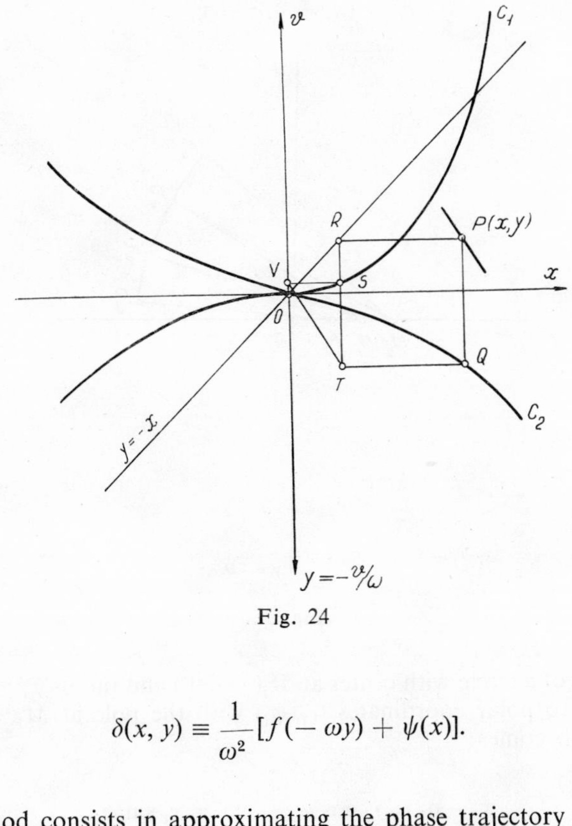

Fig. 24

where

$$\delta(x, y) \equiv \frac{1}{\omega^2}[f(-\omega y) + \psi(x)]. \tag{4.11}$$

The delta method consists in approximating the phase trajectory in the neighborhood of an arbitrary point $P_i(x_i, y_i)$ of it, by the integral curve of Eq. (4.10), in which, however, the function $\delta(x, y)$ is replaced by its value at P_i, i.e.,

$$\delta_i = \delta(x_i, y_i) = \frac{1}{\omega^2}\{f(-\omega y_i) + \psi(x_i)\}. \tag{4.12}$$

By integrating the equation

$$\frac{dy}{dx} = -\frac{x + \delta_i}{y} \tag{4.13}$$

with the initial condition $y(x_i) = y_i$, we obtain

$$(x + \delta_i)^2 + y^2 = r_i^2, \tag{4.14}$$

where

$$r_i^2 = (x_i + \delta_i)^2 + y_i^2, \tag{4.15}$$

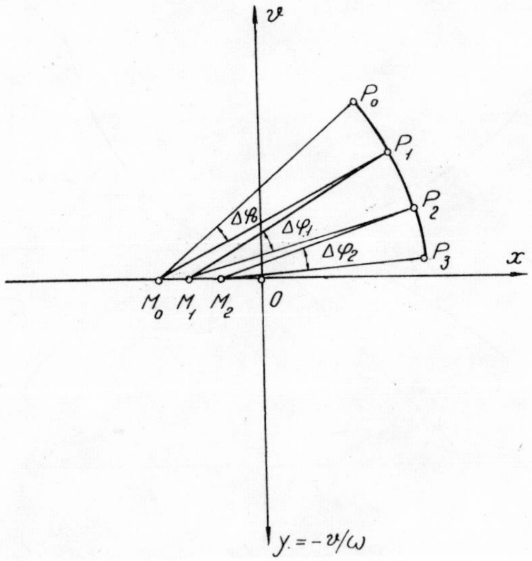

Fig. 25

i.e., the equation of a circle with center at $M_i(-\delta_i, 0)$ and radius $r_i = M_iP_i$ (Fig. 25). By passing now to polar coordinates (r_i, φ_i) with the pole at M_i, the equation of the circle (4.14) becomes

$$x + \delta_i = r_i \cos \varphi_i, \qquad y = r_i \sin \varphi_i, \tag{4.16}$$

and we have successively

$$dt = \frac{dx}{v} = -\frac{1}{\omega}\frac{dx}{y} = \frac{r_i \sin \varphi_i \, d\varphi_i}{\omega \, r_i \sin \varphi_i} = \frac{d\varphi_i}{\omega}. \tag{4.17}$$

Therefore, the interval of time corresponding to the arc P_iP_{i+1} approximating the trajectory is obtained from the corresponding polar angle $\Delta\varphi_i$, measured at M_i, by the quite simple formula

$$\Delta t_i = \frac{\Delta\varphi_i}{\omega}. \tag{4.18}$$

From the above considerations the rule for plotting the trajectory follows immediately. Starting from $P_0(x, y_0)$, calculate $\delta_0 = \delta(x_0, y_0)$ by (4.12). Draw an arc of circle with center at $M_0(-\delta_0, 0)$ and radius M_0P_0, and choose another point P_1 close to P_0 on this arc. Determine graphically the coordinates (x_1, y_1) of P_1 and calculate $\delta_1 = \delta(x_1, y_1)$, and so on.

Concomitantly with plotting the trajectory we may determine the intervals of time $\Delta t_0, \Delta t_1, \ldots$, corresponding to the arcs $P_0P_1, P_1P_2, \ldots$, by measuring the polar angles $\Delta\varphi_0, \Delta\varphi_1, \ldots$, and dividing them by ω. This also permits the use of the delta method in the case of forced vibrations and for more complicated nonlinearities. Indeed, if the equation of motion has the form

$$\ddot{x} + f(x, \dot{x}, t) = 0, \tag{4.19}$$

with the initial conditions

$$x(t_0) = x_0, \qquad \dot{x}(t_0) = v_0, \tag{4.20}$$

then the phase trajectory in the phase plane xOy, with $y = -\dot{x}/\omega$, is given by the differential equation

$$\frac{dy}{dx} = -\frac{x + \delta(x, y, t)}{y}, \tag{4.21}$$

with the initial condition

$$y(x_0) = -v_0/\omega, \tag{4.22}$$

where ω is an arbitrary positive constant having the dimension T^{-1}, and

$$\delta(x, y, t) \equiv \frac{1}{\omega^2} f(x, -\omega y, t) - x. \tag{4.23}$$

The construction and the reasoning above still hold, with the only difference that now

$$\delta_i = \delta(x_i, y_i, t_i). \tag{4.24}$$

The times t_i are to be determined by the same recurrence formula

$$t_{i+1} = t_i + \Delta t_i$$

as in the case of free vibrations.

The delta method was introduced by JACOBSEN [77] and BULAND [21], and was developed by SKOWRONSKI and ZIEMBA [182], who also analyzed the possibility of controlling the error of the method.

§ 5. ANALYTICAL METHODS

We set forth in this section a few analytical methods that allow the approximate determination of the free oscillations of the so-called *weakly nonlinear systems*. The equations of motion of these systems contain a parameter ε, which is supposed to be "small", and become linear for $\varepsilon = 0$. For instance, the equation of motion of an autonomous weakly nonlinear system has the form

$$\ddot{x} + \omega^2 x = \varepsilon f(x, \dot{x}), \tag{5.1}$$

where ε is a small parameter. For $\varepsilon = 0$, this equation has the general solution

$$x_0(t) = a \cos(\omega t + \gamma), \tag{5.2}$$

where a and γ are constants to be determined by the initial conditions.

The analytical methods try to find an approximate solution of Eq. (5.1) of the form

$$x(t) = x_0(t) + \bar{x}(\varepsilon, t), \tag{5.3}$$

which tends to (5.2) as $\varepsilon \to 0$, i.e.,

$$\lim_{\varepsilon \to 0} \bar{x}(\varepsilon, t) = 0 \quad \text{uniformly in } t. \tag{5.4}$$

These methods are also called *small-parameter methods*. In a generalized form, which will be considered in chapter IV, they may be applied to study forced vibrations as well.

a) Perturbation method

The perturbation method aims at getting a *periodic* solution of Eq. (5.1) in the form of a power series in ε. This method, introduced about 1830 by POISSON, was at first applied formally, without any theoretical justification. Nevertheless, it has been successfully used to obtain some effective solutions, especially in celestial mechanics. By the end of the last century, the method had been improved, as far as calculation is concerned, by GYLDEN, LINDSTEDT, BOHLIN, and others. However, the main contribution to the perturbation method is due to POINCARÉ [146], who elaborated in 1892 its theoretical grounds and made possible its systematic application to various problems of nonlinear oscillations. We shall discuss in this section a variant of the perturbation method proposed by LINDSTEDT [104].

By Theorem 11, § 2, the autonomous dissipative systems considered in this book have no periodic oscillations. Consequently, we confine ourselves to applying the perturbation method to conservative oscillating systems with weak nonlinearity. We consider in particular the equation of motion

$$\ddot{x} + \omega^2 x = -\varepsilon\psi_1(x) \tag{5.5}$$

where $\psi_1(x)$ is a polynomial or power series in x with infinite radius of convergence.

Let us suppose that Eq. (5.5) has a periodic solution $x(t)$ of some period $T(\varepsilon) = \omega + 0(\varepsilon)$. We cannot try a solution of the form

$$x(t) = x_0(t) + \varepsilon x_1(t) + \varepsilon^2 x_2(t) + \ldots, \tag{5.6}$$

because some function $x_k(t)$ could be aperiodic. Such a situation occurs, e.g., for the expansion of the periodic function $\sin(1 + \varepsilon)t$:

$$\sin(1 + \varepsilon)\, t = \sin t + \varepsilon\, t \cos t - \frac{\varepsilon^2 t^2}{2} \sin t + \ldots,$$

whose coefficients are not periodic. Terms like t cost, in which the time t appears as "amplitude" are called *secular terms*. It is obvious that the existence of such terms, which grow beyond any bound as $t \to \infty$, destroys the periodicity of the expansion when only a finite number of its terms are considered which is usually the case.

This difficulty may be avoided by developing the period $T(\varepsilon)$ in a power series of ε,

$$T(\varepsilon) = \frac{2\pi}{\omega}(1 + \varepsilon h_1 + \varepsilon^2 h_2 + \ldots), \tag{5.7}$$

and introducing into Eq. (5.5) the new independent variable

$$\tau = \frac{2\pi t}{T} = \frac{\omega t}{1 + \varepsilon h_1 + \varepsilon^2 h_2 + \ldots}. \tag{5.8}$$

Then, Eq. (5.5) becomes

$$\frac{d^2 x}{d\tau^2} + x(1 + \varepsilon h_1 + \varepsilon^2 h_2 + \ldots)^2 = -\frac{\varepsilon}{\omega^2}\psi_1(x)(1 + \varepsilon h_1 + \varepsilon^2 h_2 + \ldots)^2, \tag{5.9}$$

and has a periodic solution $x(\tau)$ of constant period 2π. Consequently, we may assume for $x(\tau)$ in (5.9) the power series

$$x(\tau) = x_0(\tau) + \varepsilon x_1(\tau) + \varepsilon^2 x_2(\tau) + \dots, \tag{5.10}$$

whose coefficients $x_k(\tau)$ have to be periodic functions of τ of period 2π.

Suppose the initial conditions are

$$x(0) = a, \qquad \dot{x}(0) = 0. \tag{5.11}$$

We satisfy the second condition by requiring that

$$\left.\frac{dx_k}{d\tau}\right|_{\tau=0} = 0 \quad \text{for} \quad k = 0, 1, 2, \dots \tag{5.12}$$

By substituting (5.10) into (5.11) and taking into account that

$$(1 + \varepsilon h_1 + \varepsilon^2 h_2 + \dots)^2 = 1 + 2\varepsilon h_1 + \varepsilon^2(2h_2 + h_1^2) + \dots,$$

and

$$\psi_1(x) = \psi_1(x_0) + (\varepsilon h_1 + \varepsilon^2 h_2 + \dots)\,\psi_1'(x_0) + \frac{1}{2}(\varepsilon h_1 + \varepsilon^2 h_2 + \dots)^2\,\psi''(x_0) + \dots$$

$$= \psi_1(x_0) + \varepsilon x_1 \psi_1'(x_0) + \varepsilon^2\left[x_2\psi_1'(x_0) + x_1^2\frac{\psi_1''(x_0)}{2}\right] + \dots,$$

we obtain, by equating coefficients of like powers of ε, the set of recursive linear differential equations

$$\left.\begin{aligned}
&\frac{d^2x_0}{d\tau^2} + x_0 = 0,\\
&\frac{d^2x_1}{d\tau^2} + x_1 = -\,2h_1x_0 + \varphi_1(\tau),\\
&\frac{d^2x_2}{d\tau^2} + x_2 = -\,2h_2x_0 + \varphi_2(\tau),\\
&\dots\dots\dots\dots\dots\\
&\frac{d^2x_m}{d\tau^2} + x_m = -\,2h_mx_0 + \varphi_m(\tau),\\
&\dots\dots\dots\dots\dots
\end{aligned}\right\} \tag{5.13}$$

where

$$\varphi_1(\tau) \equiv -\frac{1}{\omega^2}\psi_1(x_0),$$
$$\varphi_2(\tau) \equiv -2h_1x_1 - h_1^2x_0^2 - \frac{1}{\omega^2}[x_1\psi_1'(x_0) + 2h_1\psi_1(x_0)] \tag{5.14}$$

and $\varphi_m(\tau)$ is a known function of $x_0, x_1, \ldots, x_{m-1}$.

The general solution of the first equation in (5.13) satisfying (5.12) is

$$x_0(\tau) = M\cos\tau, \tag{5.15}$$

where M is an arbitrary constant to be determined in the next step.

To integrate the second of Eqs. (5.13) we develop the even function

$$\varphi_1(\tau) = -\frac{1}{\omega^2}\psi_1(M\cos\tau) \tag{5.16}$$

in a Fourier series. We obtain

$$\varphi_1(\tau) = C_{01} + \sum_{k=1}^{\infty} C_{k1}\cos k\tau \tag{5.17}$$

and Eq. $(5.13)_2$ becomes

$$\frac{d^2x_1}{d\tau^2} + x_1 = -2h_1M\cos\tau + C_{01} + \sum_{k=1}^{\infty} C_{k1}\cos k\tau. \tag{5.18}$$

The solution of this equation does not contain secular terms provided that there are no terms containing $\cos\tau$ and $\sin\tau$ in the right-hand side. This condition gives $-2h_1M - C_{11} = 0$, whence

$$h_1 = -\frac{C_{11}}{2M}. \tag{5.19}$$

The general solution of (5.18) then is

$$x_1(\tau) = C_{01} + \sum_{k=2}^{\infty}\frac{C_{k1}}{1-k^2}\cos k\tau + M_1\cos\tau + N_1\sin\tau. \tag{5.20}$$

From (5.12) we deduce that $N_1 = 0$ and, since M_1 remains arbitrary, we may take for the sake of simplicity $M_1 = 0$. We now have

$$x_1(\tau) = C_{01} + \sum_{k=2}^{\infty}\frac{C_{k1}}{1-k^2}\cos k\tau. \tag{5.21}$$

Introducing (5.15) and (5.21) into $(5.14)_2$ yields $\varphi_2(\tau)$, which is also an even function, and hence can be expanded in a cosine series, and so on.

Assume that we have determined in this way the functions $x_k(\tau)$, and the constants h_k, $k = 1, 2, \ldots, m-1$, and that we may calculate $\varphi_m(\tau)$, which is even. By expanding $\varphi_m(\tau)$ in a Fourier series, we obtain

$$\varphi_m(\tau) = C_{0m} + \sum_{k=1}^{\infty} C_{km} \cos k\tau \tag{5.22}$$

and the mth equation in (5.13) becomes

$$\frac{d^2 x_m}{d\tau^2} + x_m = -2h_m M \cos\tau + C_{0m} + \sum_{k=1}^{\infty} C_{km} \cos k\tau. \tag{5.23}$$

Requiring the periodicity of $x_m(\tau)$ gives

$$h_m = \frac{C_{1m}}{2M},$$

and we deduce as before, by (5.12), that $x_m(\tau)$ may be taken in the form

$$x_m(\tau) = C_{0m} + \sum_{k=2}^{\infty} \frac{C_{km}}{1-k^2} \cos k\tau.$$

We may thus successively determine the functions $x_k(\tau)$ and the constants h_k. The solution of (5.9) is then (5.10), where τ is given by (5.8) and the period T by (5.7). Finally, the constant M may be calculated from the first condition (5.11).

It is apparent from the above that there are no special difficulties in applying the perturbation method up to any step. However, the second and following steps do not qualitatively change the approximate solution; they only introduce small quantitative corrections of order ε^2 or higher, which, usually, do not justify the amount of calculation involved. We shall see this better from the examples given in the next chapter.

POINCARÉ has shown by an example that, in general, the series (5.10) obtained by the method of LINDSTEDT may not converge. In the variant of the perturbation method devised by him [1], the solution is obtained once again as a power series in ε, which uniformly converges to $x(t)$ if ε and the initial amplitude $|a|$ are sufficiently small, but which may contain secular terms. POINCARÉ was interested in astronomical problems, in which the presence of secular terms is harmless, because of the relatively slow motions of the planets. However, for the study

[1] The method of POINCARÉ will be explained in § 16, in connection with its application to the study of forced nonlinear vibrations.

of nonlinear vibrations with comparatively high frequencies, a casting-out method of the secular terms, as the one of LINDSTEDT discussed above, seems to be better suited. Moreover, since the expansion (5.10) is practically limited to its first one or two terms, one is mainly interested in the asymptotic behavior for $\varepsilon \to 0$ of this truncated expansion, and the possible divergence of the whole series is generally immaterial.

b) Method of KRYLOV and BOGOLYUBOV

The method of KRYLOV and BOGOLYUBOV and the analogous method previously developed by VAN DER POL [147] follow the basic idea of the *method of variation of constants* of LAGRANGE.

For $\varepsilon = 0$, the general solution of the equation

$$\ddot{x} + \omega^2 x = \varepsilon f(x, \dot{x}) \tag{5.24}$$

may be written in the form

$$x = A \cos \omega t + B \sin \omega t, \tag{5.25}$$

where A, B are constants to be determined by the initial conditions.

In order to calculate an approximate solution of Eq. (5.24) for $\varepsilon \neq 0$ but small, VAN DER POL proposed retaining the same form (5.25) of the solution, but considering the quantities A and B as "slowly varying" functions of time to be determined. By using this method, VAN DER POL obtained a series of important results concerning stationary solutions, the behavior of the solutions as t increases, and so on. However, this method was initially applied on purely intuitive grounds. Many problems, such as the theoretical justification of the method, its field of applicability, and the way of obtaining approximations of higher order, were solved as late as 1937 by KRYLOV and BOGOLYUBOV. These authors also modified the form of the first approximation by taking instead of (5.26) the solution

$$x = a \cos (\omega t + \gamma),$$

with "slightly varying" amplitude and phase. We shall expound in the following an improved variant of the method of KRYLOV and BOGOLYUBOV, developed by BOGOLYUBOV and MITROPOLSKY [16]. We confine ourselves as before to considering the equation

$$\ddot{x} + \omega^2 x = - \varepsilon[f_1(\dot{x}) + \psi_1(x)]. \tag{5.26}$$

For $\varepsilon = 0$, the general solution of this equation is

$$x = a \cos \alpha, \qquad \alpha = \omega t + \gamma \tag{5.27}$$

and has constant amplitude and constant rate of the total phase, i.e.,

$$\dot{a} = 0, \qquad \dot{\alpha} = \omega.$$

For $\varepsilon \neq 0$, the quantities $\dot{a}$ and $\dot{\alpha}$ generally depend on a and ε, and Eq. $(5.27)_1$ must be completed by terms depending on a, α, and ε. Consequently, we seek an approximate solution of Eq. (5.26) of the form

$$x = a \cos \alpha + \varepsilon u_1(a, \alpha) + \varepsilon^2 u_2(a, \alpha) + \ldots + \varepsilon^m u_m(a, \alpha), \tag{5.28}$$

which is called the approximation of the $(m + 1)^{th}$ order. $u_k(a, \alpha)$ are supposed to be periodic functions of α with period 2π, and a, α, as functions of t, have to satisfy the differential equations with separable variables

$$\dot{a} = \varepsilon A_1(a) + \varepsilon^2 A_2(a) + \ldots + \varepsilon^m A_m(a), \tag{5.29}$$

$$\dot{\alpha} = \omega + \varepsilon B_1(a) + \varepsilon^2 B_2(a) + \ldots + \varepsilon^m B_m(a). \tag{5.30}$$

We first try to determine the functions $u_k(a, \alpha)$, $A_k(a)$, and $B_k(a)$, $k = 1, \ldots, m$, so that Eq. (5.26) be satisfied to within an error of $(m + 1)^{th}$ order in ε. As in the method of LINDSTEDT, the recursive determination of these functions does not give rise to any qualitative difficulties. However, the calculation is usually performed only for $m = 1$ or 2, because the contribution of higher-order terms is generally unimportant and because the formulas obtained for higher approximations are very intricate.

The practical applicability of the method is determined in the first place not by the convergence of the expansion (5.28) as $m \to \infty$, but by the asymptotic properties of the approximate solution as $\varepsilon \to 0$, for a given m. Therefore, the expressions leading to secular terms may be cast out at each step, even if this procedure could determine, as in the method of LINDSTEDT, the divergence of the expansion (5.28) as $m \to \infty$.

After determining the functions u_k, A_k, and B_k, and assuming that the initial conditions are given in the form

$$a(t_0) = a_0, \qquad \alpha(t_0) = \alpha_0, \tag{5.31}$$

we may determine the solution by means of two integrations. By integrating (5.29) we deduce that

$$t(a) = t_0 + \int_{a_0}^{a} \frac{\mathrm{d}a}{\varepsilon A_1(a) + \varepsilon^2 A_2(a) + \ldots + \varepsilon^m A_m(a)} \, . \tag{5.32}$$

Solving this equation with respect to a yields

$$a = a(t). \tag{5.33}$$

Substituting (5.33) into (5.30) and integrating, it follows that

$$\alpha(t) = \alpha_0 + \omega(t - t_0) + \int_{t_0}^{t} [\varepsilon B_1(a) + \varepsilon^2 B_2(a) + \ldots + \varepsilon^m B_m(a)]\mathrm{d}t. \tag{5.34}$$

Finally, by introducing (5.33) and (5.34) into (5.28), we obtain the approximate solution $x(t)$.

Before explaining the method of determining the functions $u_k(a, t)$, $A_k(a)$, and $B_k(a)$, we note that there is a certain freedom in the choice of these functions. It may be shown ([10], p. 38) that this arbitrariness may be used to eliminate the fundamental harmonic (the terms containing $\sin \alpha$ and $\cos \alpha$) from the Fourier expansions of the functions $u_k(a, \alpha)$, by imposing the conditions

$$\int_0^{2\pi} u_k(a, \alpha) \cos \alpha \, d\alpha = 0, \qquad \int_0^{2\pi} u_k(a, \alpha) \sin \alpha \, d\alpha = 0, \tag{5.35}$$

for $k = 1, \ldots, m$. On the other hand, it will be clear from the following that these conditions have to be satisfied, for, otherwise, some u_k would contain secular terms.

It can also be shown ([10], p. 38) that, by neglecting terms of $(m+1)^{\text{th}}$ order in the expansions for $\dot{a}$ and $\dot{\alpha}$, one introduces an error of mth order into (5.82), and hence it would be useless to retain the term of mth order in this last expansion. Consequently, we take $m = 2$ in (5.28) – (5.30), i.e.,

$$x = a \cos \alpha + \varepsilon u_1(a, \alpha) + \varepsilon^2 u_2(a, \alpha), \tag{5.36}$$

$$\dot{a} = \varepsilon A_1(a) + \varepsilon^2 A_2(a), \tag{5.37}$$

$$\dot{\alpha} = \omega + \varepsilon B_1(a) + \varepsilon^2 B_2(a), \tag{5.38}$$

but we shall neglect $u_2(a, \alpha)$ in the expression of x at a further stage. We shall thus obtain the equations of the first two approximations, which in general give entirely satisfactory results. The terms of third or higher order in ε will be systematically omitted in the following formulas.

By differentiating Eq. (5.36) twice with respect to t, we obtain

$$\left.\begin{aligned}
\dot{x} &= \dot{a}\left(\cos\alpha + \varepsilon\frac{\partial u_1}{\partial a} + \varepsilon^2\frac{\partial u_2}{\partial a}\right) + \dot{\alpha}\left(-a\sin\alpha + \varepsilon\frac{\partial u_1}{\partial \alpha} + \varepsilon^2\frac{\partial u_2}{\partial \alpha}\right),\\
\ddot{x} &= \ddot{a}\left(\cos\alpha + \varepsilon\frac{\partial u_1}{\partial a} + \varepsilon^2\frac{\partial u_2}{\partial a}\right) + \ddot{\alpha}\left(-a\sin\alpha + \varepsilon\frac{\partial u_1}{\partial \alpha} + \varepsilon^2\frac{\partial u_2}{\partial \alpha}\right) +\\
&\quad + \dot{a}^2\left(\varepsilon\frac{\partial^2 u_1}{\partial a^2} + \varepsilon^2\frac{\partial^2 u_2}{\partial a^2}\right) + 2\dot{a}\dot{\alpha}\left(-\sin\alpha + \varepsilon\frac{\partial^2 u_1}{\partial a\,\partial\alpha} + \alpha^2\frac{\partial^2 u_2}{\partial a\,\partial\alpha}\right) +\\
&\quad + \dot{\alpha}^2\left(-a\cos\alpha + \varepsilon\frac{\partial^2 u_1}{\partial \alpha^2} + \varepsilon^2\frac{\partial^2 u_2}{\partial \alpha^2}\right).
\end{aligned}\right\} \tag{5.39}$$

Next, we deduce from (5.37) and (5.38) that

$$\ddot{a} = \left(\varepsilon \frac{dA_1}{da} + \varepsilon^2 \frac{dA_2}{da}\right)(\varepsilon A_1 + \varepsilon^2 A_2) = \varepsilon^2 A_1 \frac{dA_1}{da},$$

$$\ddot{\alpha} = \left(\varepsilon \frac{dB_1}{da} + \varepsilon^2 \frac{dB_2}{da}\right)(\varepsilon A_1 + \varepsilon^2 A_2) = \varepsilon^2 A_1 \frac{dB_1}{da},$$

$$\dot{a}^2 = (\varepsilon A + \varepsilon^2 A_2)^2 = \varepsilon^2 A_1^2,$$

$$\dot{a}\dot{\alpha} = (\varepsilon A_1 + \varepsilon^2 A_2)(\omega + \varepsilon B_1 + \varepsilon^2 B_2) = \varepsilon \omega A_1 + \varepsilon^2(\omega A_2 + A_1 B_1).$$

$$\dot{\alpha}^2 = (\omega + \varepsilon B_1 + \varepsilon^2 B_2)^2 = \omega^2 + 2\varepsilon\omega B_1 + \varepsilon^2(B_1^2 + 2\omega B_2).$$

Introducing these expressions, as well as (5.37) and (5.38), into (5.39) yields

$$\dot{x} = -\omega a \sin\alpha + \varepsilon\left(A_1 \cos\alpha - aB_1 \sin\alpha + \omega \frac{\partial u_1}{\partial \alpha}\right), \tag{5.40}$$

$$\begin{aligned} \ddot{x} = & -a\omega^2 \cos\alpha + \varepsilon\left(-2\omega A_1 \sin\alpha - 2\omega a B_1 \cos\alpha + \omega^2 \frac{\partial^2 u_1}{\partial \alpha^2}\right) + \\ & + \varepsilon^2 \left[\left(A_1 \frac{dA_1}{da} - aB_1^2 - 2\omega a B_2\right)\cos\alpha - \left(2\omega A_2 + \alpha A_1 B_1 + \right.\right. \\ & \left.\left. + aA_1 \frac{dB_1}{da}\right)\sin\alpha + 2\omega A_1 \frac{\partial^2 u_1}{\partial a \partial \alpha} + 2\omega B_1 \frac{\partial^2 u_1}{\partial \alpha^2} + \omega^2 \frac{\partial^2 u_2}{\partial \alpha^2}\right]. \end{aligned} \tag{5.41}$$

On the other hand, by taking into consideration (5.36) and (5.40), it follows that

$$\varepsilon\psi(x) = \varepsilon\psi_1(a\cos\alpha) + \varepsilon^2 u_1 \psi_1'(a\cos\alpha), \tag{5.42}$$

$$\varepsilon f(\dot{x}) = \varepsilon f_1(-\omega a \sin\alpha) + \varepsilon^2\left(A_1 \cos\alpha - aB_1 \sin\alpha + \omega \frac{\partial u_1}{\partial \alpha}\right) f_1'(-\omega a \sin\alpha). \tag{5.43}$$

Finally, by substituting (5.36) and (5.41) — (5.43) into (5.26) and equating coefficients of ε and ε^2, we obtain the conditions that have to be satisfied if (5.36) should give the solution of the differential equation (5.26) to within an error of third order in ε:

$$\omega^2\left(\frac{\partial^2 u_1}{\partial \alpha^2} + u_1\right) = -\psi_1(a\cos\alpha) - f_1(-\omega\, a\sin\alpha) + 2\omega A_1 \sin\alpha + \\ + 2\omega\, a\, B_1 \cos\alpha, \tag{5.44}$$

$$\omega^2\left(\frac{\partial^2 u_2}{\partial \alpha^2} + u_2\right) = -u_1\psi_1'(a\cos\alpha) - (A_1\cos\alpha - aB_1\sin\alpha + \\ + \omega\frac{\partial u_1}{\partial \alpha}\Big) f_1'(-\omega\, a\sin\alpha) + \left(aB_1^2 + 2\omega\, a\, B_2 - A_1\frac{dA_1}{da}\right)\cos\alpha + \\ + \left(2\omega A_2 + 2A_1B_1 + aA_1\frac{dB_1}{da}\right)\sin\alpha - 2\omega A_1\frac{\partial^2 u_1}{\partial a\,\partial\alpha} - 2\omega B_1\frac{\partial^2 u_1}{\partial \alpha^2} \tag{5.45}$$

By developing now the function $\psi_1(a\cos\alpha) + f_1(-\omega\, a\sin\alpha)$ in a Fourier series with respect to α, we have

$$\psi_1(a\cos\alpha) + f_1(-\omega\, a\sin\alpha) = C_0(a) + \sum_{k=1}^{\infty}[C_k(a)\cos k\alpha + D_k(a)\sin k\alpha], \tag{5.46}$$

where

$$C_0(a) = \frac{1}{2\pi}\int_0^{2\pi}[\psi_1(a\cos\alpha) + f_1(-\omega\, a\sin\alpha)]\,d\alpha,$$

$$C_k(a) = \frac{1}{\pi}\int_0^{2\pi}[\psi_1(a\cos\alpha) + f_1(-\omega\, a\sin\alpha)]\cos k\alpha\,d\alpha,$$

$$D_k(a) = \frac{1}{\pi}\int_0^{2\pi}[\psi_1(a\cos\alpha) + f_1(-\omega\, a\sin\alpha)]\sin k\alpha\,d\alpha, \quad k = 1, 2, \ldots$$

The solution $u_1(a, \alpha)$ of Eq. (5.44) is a periodic function of α only if the right-hand side of the equation does not contain the fundamental harmonic of period 2π. Introducing (5.46) into (5.44) and equating to zero the coefficients of $\sin\alpha$ and $\cos\alpha$ in the right-hand side gives

$$A_1(a) = \frac{D_1(a)}{2\omega}, \qquad B_1(a) = \frac{C_1(a)}{2\omega a},$$

and Eq. (5.44) becomes

$$\frac{\partial^2 u_1}{\partial \alpha^2} + u_1 = -\frac{C_0(a)}{\omega^2} - \frac{1}{\omega^2} \sum_{k=2}^{\infty} [C_k(a) \cos k\alpha + D_k(a) \sin k\alpha]. \tag{5.47}$$

Since, by (5.35), the solution $u_1(a, \alpha)$ of this equation may not contain $\sin \alpha$ and $\cos \alpha$, its expression is readily found to be

$$u_1(a, \alpha) = -\frac{C_0(a)}{\omega^2} + \frac{1}{\omega^2} \sum_{k=2}^{\infty} \frac{C_k(a) \cos k\alpha + D_k(a) \sin k\alpha}{k^2 - 1}.$$

The functions $A_2(a)$, $B_2(a)$, and $u_2(a, \alpha)$ may be similarly obtained from (5.45). However, as we have already mentioned, $u_2(a, \alpha)$ does not contribute to the second approximation. Therefore, we content ourselves with calculating $A_2(a)$ and $B_2(a)$ on requiring that the terms containing $\sin \alpha$ and $\cos \alpha$ in the right-hand side of Eq. (5.45) vanish. It follows that

$$A_2(a) = -\frac{1}{2\omega}\left(2 A_1 B_1 + a A_1 \frac{\mathrm{d}B_1}{\mathrm{d}a}\right) + \frac{1}{2\pi\omega} \int_0^{2\pi} \Big[u_1 \psi'(a \cos \alpha) +$$

$$+ \left(A_1 \cos \alpha - a B_1 \sin \alpha + \omega \frac{\partial u_1}{\partial \alpha}\right) f_1'(-\omega a \sin \alpha) \Big] \sin \alpha \, \mathrm{d}\alpha,$$

$$B_2(a) = -\frac{1}{2\omega}\left(B_1^2 - \frac{A_1}{a} \frac{\mathrm{d}A_1}{\mathrm{d}a}\right) + \frac{1}{2\pi\omega a} \int_0^{8\pi} \Big[u_1 \psi_1'(a \cos \alpha) +$$

$$+ \left(A_1 \cos \alpha - a B_1 \sin \alpha + \omega \frac{\partial u_1}{\partial \alpha}\right) f_1'(-\omega a \sin \alpha) \Big] \cos \alpha \, \mathrm{d}\alpha \cdot$$

By conveniently dividing the integration interval and using the substitutions $\sin \alpha = t$ and $\cos \alpha = t$, respectively, it may be shown that the relations

$$\int_0^{2\pi} F(\cos \alpha) \sin \alpha \, \mathrm{d}\alpha = 0, \qquad \int_0^{2\pi} F(\sin \alpha) \cos \alpha \, \mathrm{d}\alpha = 0 \tag{5.48}$$

hold for every function which is integrable on the interval $[-1, 1]$. These relations may be used in order to simplify the expressions of $A_1(a)$ and $B_1(a)$ (see Eq. (5.52) below).

To facilitate further reference we summarize here the formulas which are necessary for the determination of the first two approximations by the method of KRYLOV and BOGOLYUBOV. Other possible simplifications of these formulas, valid only in particular cases, will be considered in §§ 8 and 11.

— *First approximation*

$$x_{\mathrm{I}} = a \cos \alpha, \tag{5.49}$$

$$t(a) = t_0 + \frac{1}{\varepsilon} \int_{a_0}^{a} \frac{\mathrm{d}a}{A_1(a)}, \tag{5.50}$$

$$\alpha(t) = \alpha_0 + \omega(t - t_0) + \varepsilon \int_{t_0}^{t} B_1(a)\, \mathrm{d}t, \tag{5.51}$$

$$A_1(a) = \frac{1}{2\pi\omega} \int_0^{2\pi} f_1(-\omega a \sin \alpha) \sin \alpha \, \mathrm{d}\alpha, \qquad B_1(a) = \frac{1}{2\pi\omega a} \int_0^{2\pi} \psi_1(a \cos \alpha) \cos \alpha \, \mathrm{d}\alpha. \tag{5.52}$$

— *Second approximation*

$$x_{\mathrm{II}} = a \cos \alpha + \varepsilon u_1(a, \alpha), \tag{5.53}$$

$$t(a) = t_0 + \frac{1}{\varepsilon} \int_{a_0}^{a} \frac{\mathrm{d}a}{A_1(a) + \varepsilon A_2(a)}, \tag{5.54}$$

$$\alpha(t) = \alpha_0 + \omega(t - t_0) + \varepsilon \int_{t_0}^{t} [B_1(a) + \varepsilon B_2(a)]\, \mathrm{d}t, \tag{5.55}$$

$$u_1(a, \alpha) = -\frac{C_0(a)}{\omega^2} + \frac{1}{\omega^2} \sum_{k=2}^{\infty} \frac{C_k(a) \cos k\alpha + D_k(a) \sin k\alpha}{k^2 - 1}, \tag{5.56}$$

$$\left.\begin{aligned} C_k(a) &= \frac{1}{\pi} \int_0^{2\pi} [\psi_1(a \cos \alpha) + f_1(-\omega a \sin \alpha)] \cos k\alpha \, \mathrm{d}\alpha, \\ D_k(a) &= \frac{1}{\pi} \int_0^{2\pi} [\psi_1(a \cos \alpha) + f_1(-\omega a \sin \alpha)] \sin k\alpha \, \mathrm{d}\alpha, \quad k = 1, 2, \ldots, \\ C_0(a) &= \frac{1}{2\pi} \int_0^{2\pi} [\psi_1(a \cos \alpha) + f_1(-\omega a \sin \alpha)] \, \mathrm{d}\alpha, \end{aligned}\right\} \tag{5.57}$$

$$A_1(a) = D_1(a)/2\omega, \qquad B_1(a) = C_1(a)/2\omega a, \tag{5.58}$$

$$\left.\begin{aligned} A_2(a) &= -\frac{1}{2\omega}\left(2A_1B_1 + aA_1\frac{\mathrm{d}B_1}{\mathrm{d}a}\right) + \frac{1}{2\pi\omega}\int_0^{2\pi}\Big[u_1\,\psi_1'(a\cos\alpha) + \\ &+ \left(A_1\cos\alpha - a\,B_1\sin\alpha + \omega\frac{\partial u_1}{\partial\alpha}\right)f_1'(-\omega\,a\sin\alpha)\Big]\sin\alpha\,\mathrm{d}\alpha, \\ B_2(a) &= -\frac{1}{2\omega}\left(B_1^2 - \frac{A_1}{a}\frac{\mathrm{d}A_1}{\mathrm{d}a}\right) + \frac{1}{2\pi\omega}\int_0^{2\pi}\Big[u_1\psi_1'(a\cos\alpha) + \\ &+ \left(A_1\cos\alpha + \omega\frac{\partial u_1}{\partial\alpha} - aB_1\sin\alpha\right)f_1'(-\omega\,a\sin\alpha)\Big]\cos\alpha\,\mathrm{d}\alpha. \end{aligned}\right\} \tag{5.59}$$

We shall see in the next two chapters, that the method of KRYLOV and BOGOLYUBOV is a powerful tool for the determination of stationary and transitory nonlinear vibrations. The application of a generalized form of this method to the study of forced vibrations will be discussed in Chapter IV.

c) Method of equivalent linearization

The method of equivalent linearization, devised by KRYLOV and BOGOLYUBOV also tries to replace the differential equation of motion of a weakly nonlinear system, e.g.,

$$\ddot{x} + \varepsilon[f_1(\dot{x}) + \psi_1(x)] + \omega^2 x = 0, \tag{5.60}$$

by a linear differential equation,

$$\ddot{x} + 2\bar{h}\dot{x} + \bar{\omega}^2 x = 0, \tag{5.61}$$

whose coefficients are chosen in such a way that the solutions of these two equations differ by terms of second or higher order in ε.

Although the solution of Eq. (5.61) may be calculated directly by the method of KRYLOV and BOGOLYUBOV, without previously determining the coefficients of this equation, there are circumstances under which the equivalent linearization is expedient. Such a situation frequently occurs when a nonlinear oscillator is coupled with other linear oscillators. For instance, the free vibrations of the sprung mass corresponding to one axle of a vehicle may be described by Eq. (5.60), if the elastic and damping characteristics of the suspension are weakly nonlinear, and if the interactions of these vibrations with those corresponding to other degrees of freedom of the vehicle may be neglected. In fact, as has been already mentioned in the introduction, such a simplified mathematical model may be used only as a first approximation. In order to describe the complicated vibration of a vehicle, it is necessary to solve a system of simultaneous linear and nonlinear differential equations and, to this end, the preliminary linearization of the nonlinear equations of the system may be very efficient.

The method of equivalent linearization is closely connected with the calculation of the first approximation by the method of KRYLOV and BOGOLYUBOV. Indeed, we have seen that the solution (5.49) obtained at the first stage of approximation satisfies Eq. (5.60) to within an error of second order in ε. It is then sufficient to determine the parameters $\bar{h}$ and $\bar{\omega}$ by requiring that the solution (5.49) satisfy Eq. (5.61) to within the same error.

Let us first notice that we can derive the expressions of $\dot{x}$ and $\ddot{x}$ corresponding to (5.49) by putting $u_1(a, \alpha) = u_2(a, \alpha) = 0$ into (5.40) and (5.41), respectively. We then have

$$\dot{x} = -\omega a \sin \alpha + \varepsilon(A_1 \cos \alpha - aB_1 \sin \alpha),$$

$$\ddot{x} = -a\omega^2 \cos \alpha + \varepsilon(-2\omega A_1 \sin \alpha - 2\omega a B_1 \cos \alpha).$$

By introducing these expressions into (5.61), neglecting terms of second order in ε, and equating to zero the coefficients of $\sin \alpha$ and $\cos \alpha$, we obtain the following system of algebraic equations:

$$a(\bar{\omega}^2 - \omega^2) + 2\varepsilon(\bar{h}A_1 - aB_1) = 0, \tag{5.62}$$

$$a\bar{h}(\omega + \varepsilon B_1) + \omega\varepsilon A_1 = 0. \tag{5.63}$$

From (5.63) we see that

$$\bar{h} = \frac{\omega\varepsilon A_1}{a(\omega + \varepsilon B_1)} \approx -\frac{\varepsilon A_1}{a}\left(1 - \frac{\varepsilon}{\omega} B_1\right)$$

and, if we neglect the term of second order,

$$\bar{h}(a) = -\frac{\varepsilon A_1(a)}{a} = -\frac{\varepsilon}{2\pi\omega a}\int_0^{2\pi} f_1(-\omega a \sin \alpha) \sin \alpha \, d\alpha. \tag{5.64}$$

Finally, by substituting (5.64) into (5.62) and again neglecting the term of second order, we obtain

$$\bar{\omega}^2 = \omega^2 + 2\varepsilon B_1(a) = \omega^2 + \frac{\varepsilon}{\pi\omega a}\int_0^{2\pi} \psi_1(a \cos \alpha) \cos \alpha \, d\alpha. \tag{5.65}$$

It is readily seen from (5.64) and (5.65) that the parameters of the linearized equation depend on the amplitude a. One may first take for a in (5.64) and (5.65) the initial value of the amplitude, a_0. If the interval of time considered is too long, the linearization has to be piecewise repeated, by taking as a value for a in (5.64) and (5.65) for each subinterval of time the value of the amplitude obtained at the end of the preceding subinterval by integrating the linearized equation of motion (5.61).

Chapter II Free vibrations of conservative systems

§ 6. GENERAL PROPERTIES OF THE VIBRATIONS OF CONSERVATIVE SYSTEMS

An oscillating system is said to be *conservative* if its total energy is time-independent.

Taking into consideration the hypothesis H_3, we derive from (1.6) the following theorem.

Theorem 1. *A system whose equation of motion has the form (1.2) is conservative if and only if* $f(\dot{x}) \equiv 0$.

Consequently, the equation of motion of conservative systems is of the form

$$\ddot{x} + g(x) = 0, \tag{6.1}$$

and the energy equation reduces in their case to

$$E(t) \equiv \frac{v^2}{2} + G(x) = E_0, \tag{6.2}$$

where $v^2/2$ is the kinetic energy per unit mass, $G(x) \equiv \int_0^x g(x)\,dx$ is the potential energy per unit mass, and

$$E_0 = \frac{v_0^2}{2} + G(x_0) \tag{6.3}$$

is the initial total energy of the system per unit mass.

Conservative systems do not fully satisfy the hypotheses adopted in the introduction of this book, because the condition $\dot{x} f(\dot{x}) > 0$ for $\dot{x} \neq 0$ is no longer fulfilled. Consequently, the results obtained by using the first part of H_3, such as Theorem 6, § 1 and Theorems 6 — 11, § 2, do not apply to conservative systems. On the other hand, most of the results derived by the topological method in § 2a still hold for conservative systems, so that we find it useful to recollect here these results, by appropriately amplifying them.

The only singular point and at the same time the only equilibrium point of the system is the origin ($x = v = 0$), for, by H_3, the restoring force per unit mass, $g(x)$, vanishes only for $x = 0$. This equilibrium is always *stable* in our case, because, as pointed out in § 1, the potential energy $G(x)$ has an absolute minimum for $x = 0$, i.e.,

$$G(x) \geqslant G(0) = 0 \tag{6.4}$$

for all $x \in (-\infty, \infty)$. By taking into consideration Eq. (2.14),

$$g(x) = \omega^2 x + \psi(x), \tag{6.5}$$

where $\omega^2 = g'(0) > 0$, we may write (6.2) in the form

$$\omega^2 x^2 + v^2 - 2[E_0 - \Psi(x)] = 0, \tag{6.6}$$

where

$$\Psi(x) \equiv \int_0^x \psi(x)\,\mathrm{d}x \leqslant E_0. \tag{6.7}$$

From (6.2) it follows that the phase trajectories of conservative systems are closed and that the total energy is constant along each phase trajectory. Moreover, we derive from (6.6) that the trajectories are as close to the ellipses

$$\omega^2 x^2 + v^2 - 2E_0 = 0 \tag{6.8}$$

as we like, provided that $|x|$ is small enough, and hence 0 is a *center*. This property differentiates the conservative systems from the dissipative systems studied in § 2b, for which the origin is an attractor.

The phase trajectory crosses the x-axis at the points $M_1(x_1, 0)$ and $M_2(x_2, 0)$, where x_1 and x_2 are the solutions of eq. $G(x) = E_0$ (see Fig. 26) [1]. By theorem 2, § 2, the trajectory has at these points a vertical tangent, its concavity is directed toward the origin, and we have

$$\left.\frac{\mathrm{d}^2 x}{\mathrm{d}v^2}\right|_{v=0} = -\frac{1}{g(x_{1,2})}. \tag{6.9}$$

By Theorem 3, § 2, it follows that x_1 and x_2 have opposite signs (as also readily seen from Fig. 26), and that the velocity is a single-valued function of x when the representative point of the motion is moving from M_1 to M_2, or from M_2 to M_1. We have namely from (6.2)

$$v = \pm \sqrt{2[E_0 - G(x)]}, \tag{6.10}$$

where the plus and minus signs correspond to the motion of the representative point in the upper and lower half-planes, respectively.

[1] The equation $G(x) - E_0 = 0$ has exactly two solutions, because, by (1.5), the function $G(x)$ is strictly increasing for $x > 0$, strictly decreasing for $x < 0$, takes a negative value for $x = 0$, and approaches $+\infty$ as $x \to \pm\infty$.

The extreme values of the velocity, hence also the zeros of the acceleration, correspond to the points $N_1(0, v_1)$ and $N_2(0, v_2)$ at which the trajectory crosses the v-axis (Fig. 27). Indeed, putting $\ddot{x} = 0$ into (6.1) yields $g(x) = 0$, and the unique solution of this equation is $x = 0$. Then, the ordinates of N_1 and N_2 are directly obtained from (6.2):

$$v_{1,2} = \pm \sqrt{E_0}. \tag{6.11}$$

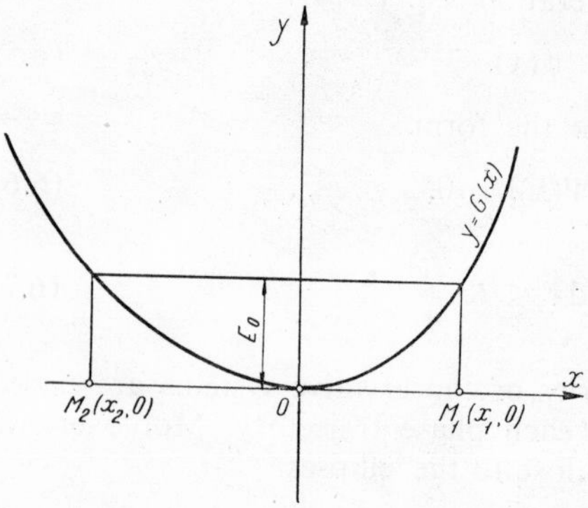

Fig. 26

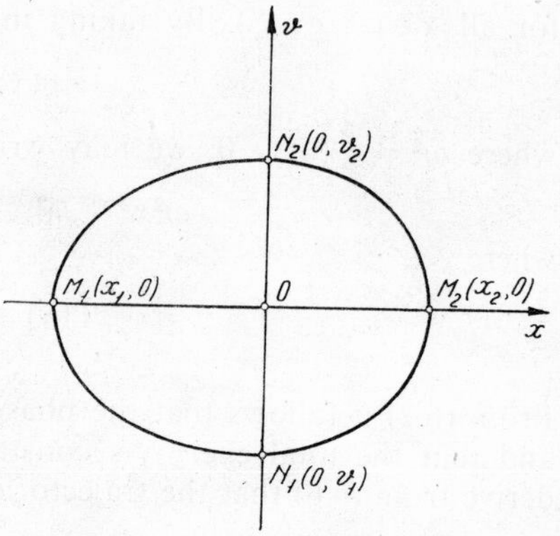

Fig. 27

According to Theorem 4, § 2, the tangent to the trajectory at the points N_1 and N_2 is parallel to the x-axis, the concavity of the trajectory is directed toward the origin, and we have by $(2.10)_1$ and (6.5)

$$\left.\frac{\mathrm{d}^2 v}{\mathrm{d}x^2}\right|_{v=v_{1,2}} = -\frac{\omega^2}{v_{1,2}}. \tag{6.12}$$

Finally, by Theorem 5, § 2, we deduce that the extreme values of the acceleration, a_1 and a_2, correspond to the points M_1 and M_2 at which the trajectory crosses the x-axis and the displacement takes extreme values. Indeed, Eq. (2.11) reduces now to $vg'(x) = 0$, and this equation is satisfied only for $v = 0$, because $g'(x) > 0$. On the other hand, we infer from (6.1) that the extreme values of the acceleration are connected to those of the displacement by the simple relation

$$a_{1,2} = -g(x_{1,2}). \tag{6.13}$$

From the above it follows that the motion in the phase plane is *periodic* and takes place along the curve $M_1N_2M_2N_1M_1$. The time in which the representative point covers once the curve is called a *period*.

By using (3.3) and taking into consideration (6.10), we obtain the expression of the period

$$T = \int_{x_1}^{x_2} \frac{dx}{\sqrt{2[E_0 - G(x)]}} - \int_{x_2}^{x_1} \frac{dx}{\sqrt{2[E_0 - G(x)]}},$$

or

$$T = \sqrt{2} \int_{x_1}^{x_2} \frac{dx}{\sqrt{E_0 - G(x)}} \cdot \tag{6.14}$$

We also see that the times in which the representative point covers the arcs of trajectory $M_1N_2M_2$ and $M_2N_1M_1$ are equal to each other.

§ 7. VARIOUS TYPES OF ELASTIC CHARACTERISTICS AND THEIR INFLUENCE ON THE PERIOD OF THE VIBRATION

Definition 1. *An elastic characteristic is said to be* hardening, linear, or softening, *according as the ratio* $g(x)/x$ *increases, is constant, or decreases as* $|x|$ *increases.*

This definition, which is due to KLOTTER, is very general, since it may also be applied to discontinuous elastic characteristics. We can easily find a geometrical interpretation of the definition. Let M be a current point on the graph of $g(x)$ and φ the angle between the x-axis and the ray OM, measured from the positive x-axis if $x > 0$, and from the negative x-axis if $x < 0$. Then φ increases, is constant, or decreases as $|x|$ increases, according as the elastic characteristic is hardening (Fig. 28*a*), linear (Fig. 28*b*), or softening (Fig. 28*c*). It may, of course, happen that the elastic characteristic be piecewise hardening or softening (Fig. 28*d*).

By hypothesis H_2, the function $g(x)$ is continuously differentiable on the whole x-axis, with the possible exception of a finite number of points, in which it still possesses one-sided derivatives. Therefore, definition 1 may be replaced in our case by the following simpler one:

Definition 2. *An elastic characteristic is called* hardening, linear *or* softening, *according as* $[g(x)/x]'/x$ *is positive, zero, or negative.*

In the technical literature an elastic characteristic is sometimes called *progressive*, *linear*, or *regressive*, according as the angle α formed by the tangent to the trajectory with the positive x-axis for $x > 0$, and with the negative x-axis for $x < 0$, increases, is constant, or decreases, as $|x|$ increases. In case $g(x)$ is twice differentiable, this definition reduces to:

Definition 3. *An elastic characteristic is called* progressive, linear, *or* regressive, *according as* $g''(x)/x$ *is positive, zero, or negative.*

From this definition it follows that, if a characteristic is progressive, then its concavity is directed toward $+y$ for $x > 0$ and toward $-y$ for $x < 0$; if the characteristic is regressive, then its concavity has opposite sense. We also see that one may replace $g(x)$ by $\psi(x)$ in all definitions above, since, by (6.5), we have

$$\left[\frac{g(x)}{x}\right]' = \left[\omega^2 + \frac{\psi(x)}{x}\right]' = \left[\frac{\psi(x)}{x}\right]', \quad g''(x) = \psi''(x). \tag{7.1}$$

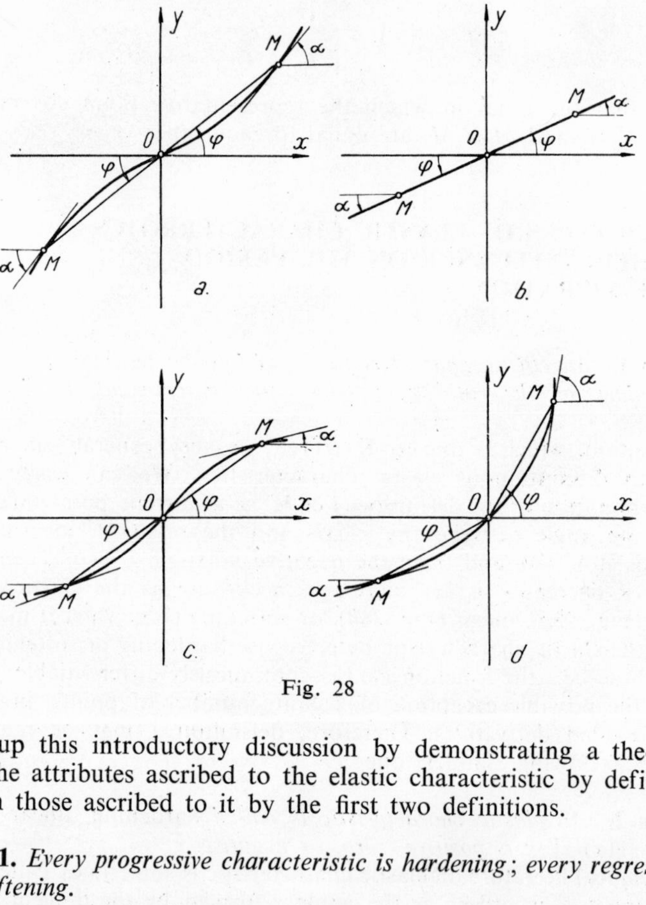

Fig. 28

We end up this introductory discussion by demonstrating a theorem which shows that the attributes ascribed to the elastic characteristic by definition 3 are stronger than those ascribed to it by the first two definitions.

Theorem 1. *Every progressive characteristic is hardening; every regressive characteristic is softening.*

We will prove only the first part of this statement; the proof of the last part is analogous. We have to show that, if $g''(x)$ and x have the same sign, then $[g(x)/x]'$ and x also have the same sign.

Let first assume that $0 < x_1 < x_2$ (Fig. 29). Since $g(0) = 0$, by the mean-value theorem of LAGRANGE, we have

$$\frac{g(x_1)}{x_1} = \frac{g(x_1) - g(0)}{x_1 - 0} = g'(\bar{x}_1), \text{ with } \bar{x}_1 \in (0, x_1),$$

and

$$\frac{g(x_2)-g(x_1)}{x_2-x_1} = g'(\bar{x}_2), \text{ with } \bar{x}_2 \in (x_1, x_2).$$

Since, by hypothesis, $g''(x) > 0$ for $x > 0$, it follows that $g'(x)$ is strictly increasing, and hence $g'(\bar{x}_1) < g'(\bar{x}_2)$. From the last two equations, we obtain after some transformations:

$$\frac{g(x_1)}{x_1} < \frac{g(x_2)}{x_2}.$$

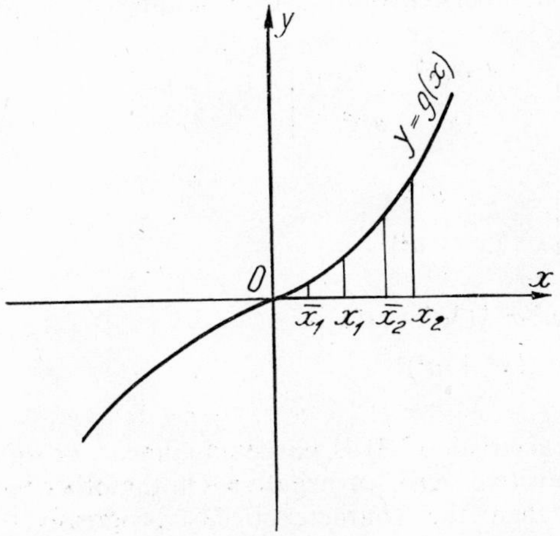

Fig. 29

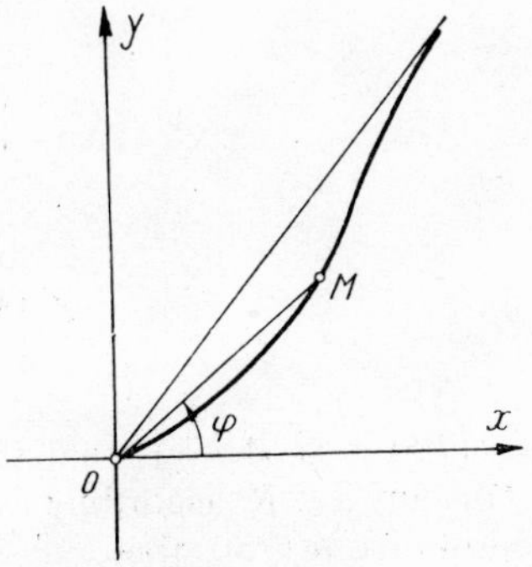

Fig. 30

Since the positive numbers x_1 and x_2 have been chosen arbitrarily we conclude that $g(x)/x$ is strictly increasing, and hence $[g(x)/x]' > 0$ for $x > 0$. The same reasoning may be repeated for $x < 0$.

The converse theorem is not true. For instance, the elastic characteristic with inclined asymptote drawn in Fig. 30 is hardening, but it is not progressive for all x.

In practical applications it is sometimes necessary to find an approximate analytical expression for an experimentally determined elastic characteristic. To this end it is interesting to devise some standard analytical expressions for $g(x)$, which satisfy our hypotheses $H_1 - H_5$ and which contain a sufficient number of free parameters to be determined from the condition of optimum approximation to a given characteristic.

The progressive characteristics may be more easily expressed in an analytical form because every polynomial with positive coefficients and containing only odd powers of x, e.g.,

$$g(x) = a_1x + a_2x^3 + \ldots + a_nx^{2n+1}, \tag{7.2}$$

with $a_k > 0$ for $k = 1, \ldots, n$, provides a progressive characteristic, which fulfills the hypotheses $H_1 - H_5$.

In the following, we will deal with two other analytical expressions of $g(x)$, which depend on three parameters, and which have the advantage of providing elastic characteristics of various types under the appropriate choice of the parameters.

Let us first take

$$y = g(x) = \omega^2 x \left(1 + \frac{kx^2}{x^2 + a^2}\right), \tag{7.3}$$

where k and a are two arbitrary real numbers. From (7.3) we obtain

$$g'(x) = \omega^2 \left[1 + \frac{kx^2(x^2 + 3a^2)}{(x^2 + a^2)^2}\right], \tag{7.4}$$

$$\left[\frac{g(x)}{x}\right]' = \frac{2\omega^2 a^2 kx}{(x^2 + a^2)^2}, \tag{7.5}$$

$$g''(x) = \frac{2\omega^2 a^2 k(3a^2 - x^2)}{(x^2 + a^2)^3}. \tag{7.6}$$

From (7.5) it is obvious that the characteristic (7.3) is hardening, linear, or softening for any $x \in R$, according as k is positive, zero, or negative. On the other hand, it follows from (7.6) that, if $k > 0$, then the characteristic is progressive for $|x| < a\sqrt{3}$ and regressive for $|x| > a\sqrt{3}$, and, if $k < 0$, then the characteristic is regressive for $|x| < a\sqrt{3}$, and progressive for $|x| > a\sqrt{3}$. This example shows once more that the converse of Theorem 1 is not true.

From (7.3) we see that the condition $g(x)/x > 0$ is satisfied for all $x = 0$ only if $k > -1$. By further requiring that $g'(x) > 0$ and taking into account that $g(x)$ takes its minimum value for $k < 0$ and $x = a\sqrt{3}$, we obtain from (7.4)

$$g'_{\min} = g'(a\sqrt{3}) = \omega^2\left(1 + \frac{9k}{8}\right) > 0,$$

and hence k has to satisfy the more restrictive condition $k > -8/9$. The elastic characteristic has an inclined asymptote of equation

$$y = \omega^2(1 + k)\, x,$$

which passes through the origin. By introducing the nondimensional variables $\eta = y/\omega^2 a$ and $\xi = x/a$, the equation of the characteristic becomes

$$\eta(\xi) = \xi \left(1 + \frac{k\xi^2}{\xi^2 + 1}\right), \tag{7.7}$$

and that of the asymptote takes the form

$$\eta = (1 + k)\,\xi.$$

The family of curves (7.7) has been drawn in Fig. 31 for $\xi > 0$ and $k = -8/9$, $\pm 1/2$, $\pm 1/4$, 0, and 1. Since $\eta(-\xi) = -\eta(\xi)$, the graphs of the curves for $\xi < 0$ may be obtained at once by symmetry with respect to the origin. The points of inflection lie on the straight line of equation $\xi = \sqrt{3}$; the change in the direction of concavity is, however, hardly visible unless k is large enough.

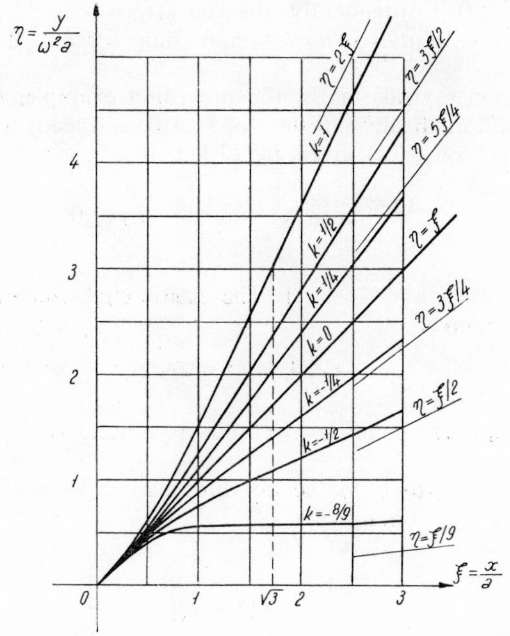

Fig. 31

Let us consider now as a second example

$$y = g(x) = \omega^2 x \left(1 + \frac{kx^2}{x^2 + a|x| + a^2}\right), \tag{7.8}$$

where a and k are real numbers, and $a > 0$. Although Eq. (7.8) strikingly resembles Eq. (7.3), it has, as we shall see below, the advantage of providing elastic characteristics which are of the same type for all x. From (7.8) we have

$$g'(x) = \omega^2 \left[1 + \frac{kx^2(x^2 + 2a|x| + 3a^2)}{(x^2 + a|x| + a^2)^2} \right], \tag{7.9}$$

$$\left[\frac{g(x)}{x} \right]' = \frac{\omega^2 akx(|x| + 2a)}{(x^2 + a|x| + a^2)^2}, \tag{7.10}$$

$$g''(x) = \frac{6\omega^2 a^3 kx(a + |x|)}{(x^2 + a|x| + a^2)^3}. \tag{7.11}$$

We see from (7.10) and (7.11) that $[g(x)/x]'/x$ and $g''(x)/x$ are positive for $k > 0$, and negative for $k < 0$. Consequently, the characteristic is progressive for $k > 0$, and regressive for $k < 0$; it is *a fortiori* hardening for $k > 0$, and softening for $k < 0$.

The condition $g(x)/x$ yields, as in the preceding example, $k > -1$; the condition $g'(x) > 0$ is automatically satisfied for $k > 0$, and leads for $k < 0$, by taking into account the sign of $g''(x)$ given by (7.11), to

$$g'_{\min} = \lim_{x \to \pm\infty} g'(x) = \omega^2(1 + k) > 0,$$

hence again to the restriction $k > -1$. The elastic characteristic has the inclined asymptotes of equations:

$$y = \omega^2[(1 + k)\, x \mp ak],$$

where the minus sign corresponds to the right asymptote and the plus sign to the left one.

By introducing now into (7.8) the nondimensional variables η and ξ defined in the first example, we obtain

$$\eta(\xi) = \xi \left(1 + \frac{k\xi^2}{\xi^2 + |\xi| + 1} \right), \tag{7.12}$$

and the equations of the asymptotes become

$$\eta = (1 + k)\xi \mp k.$$

The family of curves (7.12) has been drawn in Fig. 32 for $\xi > 0$ and $k = 0, \pm 1/4, \pm 1/2, \pm 1$. For $\xi < 0$ the graphs may be drawn by symmetry with respect to the origin, since $\eta(-\xi) = -\eta(\xi)$.

All results obtained in this section may obviously also be applied to classify and study the properties of damping characteristics, by simply substituting $f(v)$, $\varphi(v)$, and $2h$ for $g(x)$, $\psi(x)$, and ω^2, respectively.

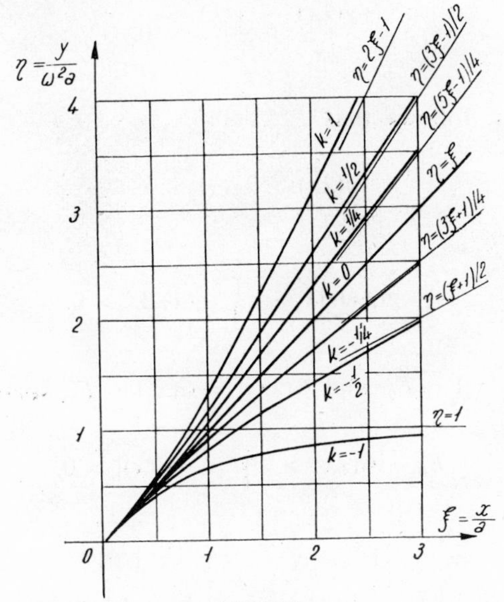

Fig. 32

We conclude this section by demonstrating a theorem which points out how the type of the elastic characteristic influences the dependence of the vibration period on amplitude (see also KAUDERER [82], p. 216).

Theorem 2. *As the total energy E_0 of the oscillation increases, the oscillation period of a conservative system decreases, is constant, or increases, according as the elastic characteristic is hardening, linear, or softening.*

We observe that we may replace in the preceding statement the total energy by the difference $x_2 - x_1$ of the extreme values of the displacement, which is a monotonically increasing function of E_0. The independence of the oscillation period of a linear conservative system on amplitude is well known from the classical theory of vibrations. We will prove only the statement concerning the hardening characteristic; the proof for the softening characteristic proceeds in the same way.

Let x_1, x_2, and x_1', x_2' be the solutions of the equations $G(x) = E_0$ and $G(x) = E_0'$, respectively. If $E_0 < E_0'$, then, from the properties of $G(x)$, we infer $x_1' < x_1 < 0 < x_2 < x_2'$ (see Fig. 26). We will demonstrate that, if the elastic characteristic is hardening for $x \in [x_1', x_2']$, then the periods T and T' of the oscillations corresponding to the total energies E_0 and E_0' satisfy the relation

$$T = \sqrt{2}\int_{x_1}^{x_2} \frac{\mathrm{d}x}{\sqrt{E_0 - G(x)}} > \sqrt{2}\int_{x_1'}^{x_2'} \frac{\mathrm{d}x}{\sqrt{E_0' - G(x)}} = T'. \tag{7.13}$$

We denote $x_2'/x_2 = \lambda > 1$. Since the characteristic is hardening, we have for every $x \in (0, x_2]$

$$\frac{g(\lambda x)}{\lambda x} > \frac{g(x)}{x} > 0,$$

and, since $x > 0$, it follows that

$$g(\lambda x) > \lambda g(x) > 0,$$

whence

$$\int_x^{x_2} g(\lambda x)\,\mathrm{d}x > \lambda \int_x^{x_2} g(x)\,\mathrm{d}x > 0.$$

But $g(x) = G'(x)$, $G(x_2) = E_0$, and $G(\lambda x_2) = G(x_2') = E_0'$, so that the last relation becomes

$$E_0' - G(\lambda x) > \lambda^2[E_0 - G(x)] > 0,$$

and gives

$$\frac{1}{\sqrt{E_0 - G(x)}} > \lambda \frac{1}{\sqrt{E_0' - G(\lambda x)}}.$$

Next, integrating between the limits 0 and x_2 yields

$$\int_0^{x_2} \frac{\mathrm{d}x}{\sqrt{E_0 - G(x)}} > \lambda \int_0^{x_2} \frac{\mathrm{d}x}{\sqrt{E_0' - G(\lambda x)}}.$$

Finally, by replacing λx by x into the right-hand side, we find

$$\int_0^{x_2} \frac{\mathrm{d}x}{\sqrt{E_0 - G(x)}} > \int_0^{x_2'} \frac{\mathrm{d}x}{\sqrt{E_0' - G(x)}}.$$

In a similar way it may be shown that

$$\int_{x_1}^{0} \frac{\mathrm{d}x}{\sqrt{E_0 - G(x)}} > \int_{x_1}^{0} \frac{\mathrm{d}x}{\sqrt{E_0' - G(x)}}.$$

By adding the last two inequalities we obtain (7.13) and the proof is completed.

From this theorem it follows that, if the elastic characteristic is nonlinear and of the same type on the whole range of x in which the vibration takes place, then the period depends on amplitude. There are, however, nonlinear characteristics for which the period is independent of amplitude, provided that they are not of the same type on the whole range of x in which the vibration takes place.

§ 8. APPLICATION OF THE METHOD OF KRYLOV AND BOGOLYUBOV TO WEAKLY NONLINEAR CONSERVATIVE SYSTEMS

The equation of motion of the weakly nonlinear conservative systems results from (5.26), by putting $f_1(\dot{x}) \equiv 0$, as

$$\ddot{x} + \omega^2 x + \varepsilon\psi_1(x) = 0. \tag{8.1}$$

For these systems the formulas giving the first two approximations by the method of KRYLOV and BOGOLYUBOV may be much simplified. By taking into account that $\psi_1(a \cos \alpha)$ is an even function of α, we obtain from (5.57)

$$D_k(a) = 0, \tag{8.2}$$

$$C_k(a) = \frac{1}{\pi}\int_0^{2\pi} \psi_1(a \cos \alpha) \cos k\alpha \, d\alpha, \qquad k = 1, 2, \ldots, \tag{8.3}$$

$$C_0(a) = \frac{1}{2\pi}\int_0^{2\pi} \psi_1(a \cos \alpha) \, d\alpha, \tag{8.4}$$

and, by (5.58), we have

$$A_1(a) = 0, \tag{8.5}$$

$$B_1(a) = \frac{1}{2\pi\omega a}\int_0^{\pi} \psi_1(a \cos \alpha) \cos \alpha \, d\alpha. \tag{8.6}$$

Substituting (8.2) — (8.4) into (5.56) yields

$$u_1(a, \alpha) = -\frac{C_0(a)}{\omega^2} + \frac{1}{\omega^2}\sum_{k=2}^{\infty} \frac{C_k(a) \cos k\alpha}{k^2 - 1} \tag{8.7}$$

and, since $u_1(a, \alpha)$ is an even function of α, it follows from (5.59) that

$$A_2(a) = 0, \tag{8.8}$$

$$B_2(a) = -\frac{1}{2\omega} B_1^2(a) + \frac{1}{2\pi\omega}\int_0^{2\pi} u_1(a, \alpha)\psi_1'(a\cos\alpha)\cos\alpha\, d\alpha =$$

$$= -\frac{1}{2\omega} B_1^2(a) + \frac{1}{2\pi\omega a^3}\Bigg[-C_0(a)\int_0^{2\pi}\psi_1'(a\cos\alpha)\cos\alpha\, d\alpha +$$

$$+\sum_{k=2}^{\infty}\frac{C_k(a)}{k^2-1}\int_0^{2\pi}\psi_1'(a\cos\alpha)\cos\alpha\cos k\alpha\, d\alpha\Bigg].$$

By differentiating (8.3) and (8.4) with respect to a, we obtain

$$C_k'(a) = \frac{1}{\pi}\int_0^{2\pi}\psi_1'(a\cos\alpha)\cos\alpha\cos k\alpha\, d\alpha,$$

$$C_0'(a) = \frac{1}{2\pi}\int_0^{2\pi}\psi_1'(a\cos\alpha)\cos\alpha\, d\alpha,$$

and hence the expression of $B_2(a)$ becomes

$$B_2(a) = -\frac{1}{2\omega} B_1^2(a) + \frac{1}{2\omega^3 a}\left[\sum_{k=2}^{\infty}\frac{C_k(a)\, C_k'(a)}{k^2-1} - 2C_0(a)\, C_0'(a)\right]. \tag{8.9}$$

Taking into account (8.5) and (8.8), we deduce from (5.37) for the first as well as for the second approximation that $\dot{a} = 0$, and hence

$$a = a_0 = \text{const.} \tag{8.10}$$

Therefore, the amplitude of the fundamental harmonic $a(t)$ is stationary, as in the case of linear systems. The nonlinear character of the motion becomes, however, apparent, through the dependence of the vibration frequency on amplitude, and through the correction term $u_1(a, \alpha)$ in the expression of $x_{II}(a, \alpha)$.

Indeed, we deduce from (5.51) at the first approximation

$$\alpha(t) = \alpha_0 + \omega_I(a_0).(t - t_0), \tag{8.11}$$

where

$$\omega_I(a_0) = \omega + \varepsilon B_1(a_0), \tag{8.12}$$

and from (5.55) at the second approximation

$$\alpha(t) = \alpha_0 + \omega_{\text{II}}(a_0).(t - t_0), \tag{8.13}$$

where

$$\omega_{\text{II}}(a_0) = \omega + \varepsilon B_1(a_0) + \varepsilon^2 B_2(a_0). \tag{8.14}$$

Finally, the first approximation gives for the solution

$$x_{\text{I}} = a_0 \cos \alpha, \tag{8.15}$$

where $\alpha(t)$ is to be replaced by (8.11), and the second approximation yields

$$x_{\text{II}} = a_0 \cos \alpha + \varepsilon u_1(a_0, \alpha), \tag{8.16}$$

where $\alpha(t)$ has the expression (8.13) and $u_1(a, \alpha)$ is given by (8.7). The constants α_0 and a_0, occurring in both approximations, are determined by the initial conditions.

The extreme values of the displacement may now be easily obtained from (8.16). Indeed, by differentiating (8.16) with respect to t, neglecting terms of second order in ε, and considering (8.7), (8.13), and (8.14), we obtain

$$\frac{dx_{\text{II}}}{dt} = -\omega a_0 \sin \alpha - \varepsilon \left[a_0 B_1(a_0) \sin \alpha + \frac{1}{\omega} \sum_{k=2}^{\infty} \frac{k C_k(a_0) \sin k\alpha}{k^2 - 1} \right].$$

Since $\sin k\alpha$ vanishes together with $\sin \alpha$, x_{II} takes its extreme values for $\alpha = 0$ and $\alpha = \pi$. By substituting these values of α into (8.7) and (8.16), we obtain

$$x_{\text{II}}(a_0, 0) = a_0 - \frac{\varepsilon C_0(a_0)}{\omega^2} + \frac{\varepsilon}{\omega^2} \sum_{k=2}^{\infty} \frac{C_k(a_0)}{k^2 - 1}, \tag{8.17}$$

$$x_{\text{II}}(a_0, \pi) = -a_0 - \frac{\varepsilon C_0(a_0)}{\omega^2} + \frac{\varepsilon}{\omega^2} \sum_{k=2}^{\infty} \frac{(-1)^k C_k(a_0)}{k^2 - 1}. \tag{8.18}$$

§ 9. THE OSCILLATOR WITH CUBIC ELASTIC RESTORING FORCE

The oscillator with cubic elastic restoring force occupies an important place in the theory of nonlinear systems, since it is the simplest oscillator displaying specific nonlinear properties. On the other hand, it provides a first approximation for the behavior of a much larger class of oscillators. Indeed, for sufficiently small values of $|x|$, the cubic characteristic may approximate as well as we please an elastic characteristic given by an arbitrary analytic function of x. Finally, another argument in favor of a detailed study of this oscillator is the possibility it gives of comparing an exact solution with the approximate solutions obtained by various analytical methods.

a) The exact solution

Let us consider the equation of motion of a conservative oscillator with cubic elastic restoring force, written in the form

$$\ddot{x} + \omega^2 x(1 + \beta x^2) = 0, \tag{9.1}$$

where β is a constant of dimension L^{-1}. We now have

$$g(x) = \omega^2 x(1 + \beta x^2), \tag{9.2}$$

and hence the potential energy per unit mass is

$$G(x) = \int_0^x g(x)\,\mathrm{d}x = \frac{\omega^2 x^2}{2}\left(1 + \frac{\beta}{2}x^2\right). \tag{9.3}$$

By requiring that $g'(x) > 0$ for all x, we obtain from (9.2) $\beta > 0$. Then $g''(x) = 6\,\omega^2\beta x$ has the same sign as x, and hence the systems studied have progressive elastic characteristic (see § 7).

Assuming now the initial conditions

$$x(0) = x_0 > 0, \qquad v(0) = 0, \tag{9.4}$$

we obtain from (6.2) and (6.3) the energy equation

$$\frac{v^2}{2} + \frac{\omega^2 x^2}{2}\left(1 + \frac{\beta}{2}x^2\right) = E_0, \tag{9.5}$$

where

$$E_0 = \frac{\omega^2 x_0^2}{2}\left(1 + \frac{\beta}{2}x_0^2\right). \tag{9.6}$$

From (9.5) and (9.6) we deduce the velocity as function of the displacement

$$v(x) = \pm\,\omega\sqrt{(x_0^2 - x^2)\left[1 + \frac{\beta}{2}(x_0^2 + x^2)\right]}, \tag{9.7}$$

from which the extreme values of the velocity

$$v_{1,2} = \pm\, \omega x_0 \sqrt{1 + \frac{\beta}{2} x_0^2}$$

result for $x = 0$. The extreme values of the displacement may be obtained by putting $v = 0$ into (9.7); we find, as expected, $x_{1,2} = \mp x_0$. From (9.5) we see that the phase trajectories are symmetric with respect to both coordinate axes.

Next, it follows by (3.3) and (9.7) that the time necessary for the representative point to move in the lower half-plane from $M_2(x_0, 0)$ to the point of abscissa x is

$$t(x) = -\frac{1}{\omega} \int_{x_0}^{x} \frac{dx}{\sqrt{(x_0^2 - x^2)\left[1 + \frac{\beta}{2}(x_0^2 + x^2)\right]}} \tag{9.8}$$

By interchanging the limits of integration and putting $x = x_0 \xi$, where ξ is a new variable, Eq. (9.8) becomes

$$t(x) = \frac{\sqrt{2}}{\omega x_0 \beta} \int_{\frac{x}{x_0}}^{1} \frac{d\xi}{\sqrt{(1 - \xi^2)\left(\frac{2 + \beta x_0^2}{\beta x_0^2} + \xi^2\right)}}\,. \tag{9.9}$$

This expression may be further transformed by means of the elliptic integral of the first kind [1], $F(\varphi; k)$. We then obtain

$$t(x) = \frac{F\left(\arccos \frac{x}{x_0}; k\right)}{\omega \sqrt{1 + \beta x_0^2}}\,, \tag{9.10}$$

where

$$k = \sqrt{\frac{\beta x_0^2}{2(1 + \beta x_0^2)}}\,. \tag{9.11}$$

Finally, by inverting the function (9.10), we find the exact solution of the equation of motion

$$x = x_0 \operatorname{cn}\left(\omega t \sqrt{1 + \beta x_0^2};\quad k\right), \tag{9.12}$$

where cn is JACOBI's elliptic function, called the elliptic cosine.

[1] As regards the properties of the elliptic integrals used in this section see, e.g., JAHNKE, EMDE, LÖSCH [79], pp. 44—48.

We also notice that the quarter of the vibration period may be calculated by putting $x = 0$ into (9.12), and hence the period is given by

$$T(x_0) = \frac{4}{\omega\sqrt{1 + \beta x_0^2}} F\left(\frac{\pi}{2}; k\right) = \frac{4}{\omega\sqrt{1 + \beta x_0^2}} K(k), \tag{9.13}$$

where $K(k)$ is the complete elliptic integral of the first kind. From (9.11) and (9.13) we may obtain the dependence of the period on the dimensionless parameter

$$\lambda = \beta x_0^2, \tag{9.14}$$

for sufficiently small values of this parameter. Since

$$K(k) = \frac{\pi}{2}\left(1 + \frac{k^2}{4} + \frac{9k^4}{64} + \frac{25k^6}{256} + \cdots\right),$$

$$k^2 = \frac{\beta x_0^2}{2(1 + \beta x_0^2)} = \frac{\lambda}{2(1 + \lambda)} = \frac{\lambda}{2}(1 - \lambda + \lambda^2 - \lambda^3 + \ldots),$$

$$\frac{1}{\sqrt{1 + \beta x_0^2}} = \frac{1}{\sqrt{1 + \lambda}} = 1 - \frac{\lambda}{2} + \frac{3\lambda^2}{8} - \frac{5\lambda^3}{16} + \ldots,$$

we obtain after some calculation

$$T(\lambda) = \frac{2\pi}{\omega}\left(1 - \frac{3\lambda}{8} + \frac{57\lambda^2}{256} - \frac{315\lambda^3}{2048} + \ldots\right). \tag{9.15}$$

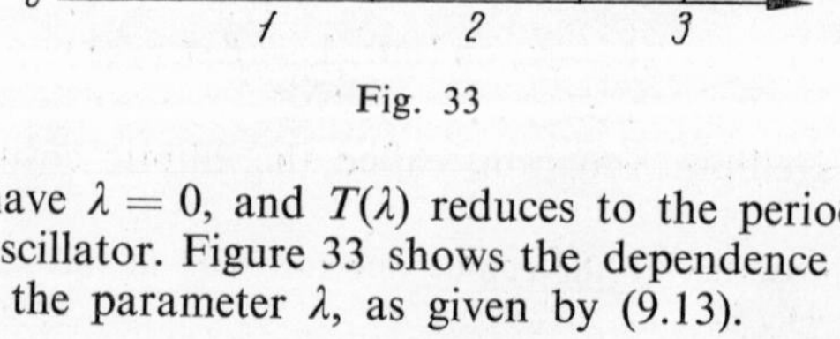

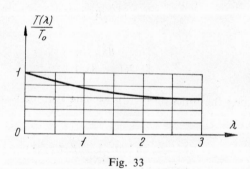

Fig. 33

For $\beta = 0$, we have $\lambda = 0$, and $T(\lambda)$ reduces to the period $T_0 = 2\pi/\omega$ of the linear conservative oscillator. Figure 33 shows the dependence of the dimensionless quantity $T(\lambda)/T_0$ on the parameter λ, as given by (9.13).

Returning now to the solution (9.12) and using the series expansion of the elliptic cosine,

$$\operatorname{cn}(u; k) = \frac{2\pi}{kK}\left(\frac{\sqrt{q}}{1+q}\cos\frac{\pi u}{2K} + \frac{q^{3/2}}{1+q^3}\cos\frac{3\pi u}{2K} + \ldots\right),$$

where

$$q(k) = \mathrm{e}^{-\frac{\pi K(k')}{K(k)}}, \qquad k' = \sqrt{1-k^2}, \tag{9.16}$$

we obtain

$$x = x_0 \cdot \frac{2\pi}{kK}\left(\frac{\sqrt{q}}{1+q}\cos\alpha + \frac{q^{3/2}}{1+q^3}\cos 3\alpha + \ldots\right), \tag{9.17}$$

where

$$\alpha = \frac{\pi\omega t\sqrt{1+\beta x_0^2}}{2K} = \frac{2\pi}{T}t, \tag{9.18}$$

denotes the total phase of the motion. Equations (9.15) and (9.17) may be very useful, especially when comparing the exact solution to approximate solutions obtained by other methods.

b) Use of the perturbation method

We will apply now the perturbation method to determine an approximate solution of Eq. (9.1) satisfying the initial conditions (9.4), under the assumption that the elastic nonlinearity is weak. As shown in § 5a, we try a solution of the form

$$x(\tau) = M\cos\tau + \varepsilon x_1(\tau) + \varepsilon^2 x_2(\tau) + \varepsilon^3 x_3(\tau) + \ldots, \tag{9.19}$$

where

$$\tau = \frac{2\pi}{T}, \tag{9.20}$$

$$T = \frac{2\pi}{\omega}(1 + \varepsilon h_1 + \varepsilon^2 h_2 + \varepsilon^3 h_3 + \ldots), \tag{9.21}$$

M is an arbitrary constant to be determined from the initial conditions, and ε is a small parameter that will be taken equal to the dimensionless quantity βM^2.

By taking τ as new independent variable and considering (9.20) and (9.21), Eq. (9.1) becomes

$$\frac{d^2x}{d\tau^2} + x(1 + \varepsilon h_1 + \varepsilon^2 h_2 + \ldots)^2 = -\frac{\varepsilon x^3}{M^2}(1 + \varepsilon h_1 + \varepsilon^2 h_2 + \ldots)^2.$$

Substituting (9.19) into this equation and equating coefficients of like powers of ε, we obtain for the first three approximations:

$$\frac{d^2x_1}{d\tau^2} + x_1 = -2h_1 M \cos \tau - M \cos^3 \tau, \tag{9.22}$$

$$\frac{d^2x_2}{d\tau^2} + x^2 = -2h_2 M \cos \tau - h_1(2x_1 + h_1 M \cos \tau) -$$

$$- (3x_1 + 2h_1 M \cos \tau) \cos^2 \tau, \tag{9.23}$$

$$\frac{d^2x_3}{d\tau^2} + x_3 = -2h_3 M \cos \tau - 2h_1 x_2 - x_1(2h_2 + h_1^2) -$$

$$- 2h_1 h_2 M \cos \tau - M(2h_2 + h_1^2) \cos^3\tau +$$

$$+ 3(2h_1 x_1 + x_2) \cos^2 \tau + \frac{3}{M} x_1^2 \cos \tau. \tag{9.24}$$

Equation (9.22) may be also written in the form

$$\frac{dx_1}{d\tau^2} + x_1 = -M\left(2h_1 + \frac{3}{4}\right) \cos \tau - \frac{1}{4} \cos 3\tau,$$

from which, by (5.19) and (5.21), it follows that

$$h_1 = -\frac{3}{8} \tag{9.25}$$

and

$$x_1(\tau) = \frac{1}{32} \cos 3\tau. \tag{9.26}$$

By introducing the last two relations into (9.23), we obtain

$$\frac{d^2x_2}{d\tau^2} + x_2 = M\left(-2h_2 + \frac{51}{128}\right)\cos\tau + \frac{21M}{128}\cos 3\tau - \frac{3M}{128}\cos 5\tau,$$

and we deduce in the same way as above that

$$h_2 = \frac{51}{256}, \tag{9.27}$$

$$x_2(\tau) = -\frac{21M}{1024}\cos 3\tau + \frac{M}{1024}\cos 5\tau. \tag{9.28}$$

Finally, by substituting (9.25)—(9.28) into (9.24), we have

$$\frac{d^2x_3}{d\tau^2} + x_3 = -M\left(2h_3 + \frac{915}{4096}\right)\cos\tau - \frac{417M}{4096}\cos 3\tau + \frac{129M}{4096}\cos 5\tau - \frac{3M}{2048}\cos 7\tau,$$

from which it follows that

$$h_3 = -\frac{915}{8192}, \tag{9.29}$$

$$x_3(\tau) = \frac{417M}{32768}\cos 3\tau - \frac{43M}{32768}\cos 5\tau + \frac{M}{32768}\cos 7\tau. \tag{9.30}$$

Taking into consideration (9.26), (9.28), and (9.30), we obtain from (9.19) the approximate solution satisfying Eq. (9.1) to within an error of third order in ε:

$$x(\tau)/M = \cos\tau + \frac{\varepsilon}{32}\left(1 - \frac{21}{32}\varepsilon + \frac{417}{1024}\varepsilon^2\right)\cos 3\tau + \frac{\varepsilon^2}{1024}\left(1 - \frac{43}{32}\varepsilon\right)\cos 5\tau + \frac{\varepsilon^3}{32768}\cos 7\tau. \tag{9.31}$$

Next, by introducing (9.25), (9.27), and (9.29) into (9.21), we derive the corresponding approximation for the period

$$T = \frac{2\pi}{\omega}\left(1 - \frac{3\varepsilon}{8} + \frac{51\varepsilon^2}{256} - \frac{915\varepsilon^3}{8192}\right). \tag{9.32}$$

The constant M may be calculated by using (9.31) and the initial condition $(9.4)_1$. We have

$$\frac{x_0}{M} = 1 + \frac{\varepsilon}{32} - \frac{5\varepsilon^2}{254} + \frac{375\varepsilon^3}{32768}, \tag{9.33}$$

and, since $\varepsilon = \beta M^2$, we obtain the algebraic equation

$$M + \frac{\beta}{32} M^3 - \frac{5\beta^2}{256} M^5 + \frac{375\beta^3}{32768} M^7 = x_0, \tag{9.34}$$

which determines M when β and x_0 are known. In order to compare the approximate expression (9.32) of the period to the exact expression (9.15), we first notice that

$$x_0/M = \frac{(\lambda/\beta)^{1/2}}{(\varepsilon/\beta)^{1/2}} = \left(\frac{\lambda}{\varepsilon}\right)^{1/2}$$

By putting this expression into (9.33) and squaring the equation obtained, we have

$$\lambda = \varepsilon + \frac{\varepsilon^2}{16} - \frac{39\varepsilon^3}{1024} + \dots \tag{9.35}$$

Finally, substituting (9.35) into (9.15) gives

$$T(\varepsilon) = \frac{2\pi}{\omega}\left(1 - \frac{3\varepsilon}{8} + \frac{51\varepsilon^2}{256} - \frac{915\varepsilon^3}{8192} + \dots\right), \tag{9.36}$$

and hence the approximate expression (9.32) of the period, as obtained by the perturbation method, coincides with the exact expression (9.36), as expected, to within terms of third order in ε.

c) Use of the method of Krylov and Bogolyubov

We will apply in this section the method of Krylov and Bogolyubov, under the form developed for conservative systems in § 8, to the equation of motion (9.1). Introducing the dimensionless parameter $\varepsilon = \beta a_0^2$, where a_0 is the *constant* amplitude of the fundamental harmonic, we obtain

$$\ddot{x} + \omega^2 x\left(1 + \frac{\varepsilon}{a_0^2} x^2\right) = 0. \tag{9.37}$$

In our case $\psi(x) = \omega^2 x^3/a_0^2$, and hence

$$\psi_1(a_0 \cos \alpha) = \frac{\omega^2 a^3}{4a_0} (\cos 3\alpha + 3 \cos \alpha), \tag{9.38}$$

from which it follows that

$$C_1(a_0) = \frac{3\omega^2 a_0}{4}, \qquad C_3(a_0) = \frac{\omega^2 a_0}{4}, \qquad C_3'(a_0) = \frac{3\omega^2}{4}, \tag{9.39}$$

and all other $C_k(a_0)$ vanish. Next, we have from (8.6), (8.7), and (8.9), by making use of (9.38) and (9.39),

$$B_1(a_0) = \frac{C_1(a_0)}{2\omega a_0} = \frac{3\omega}{8}, \tag{9.40}$$

$$u_1(a_0, \alpha) = \frac{a_0}{32} \cos 3\alpha, \tag{9.41}$$

$$B_2(a_0) = -\frac{1}{2\omega} B_1^2(a_0) + \frac{1}{16\omega^3 a_0} C_3(a_0)\, C_3'(a_0) = -\frac{15\omega}{254}. \tag{9.42}$$

Substituting these relations into (8.11) — (8.16), we obtain the first approximation

$$x_{\mathrm{I}} = a_0 \cos \alpha, \tag{9.43}$$

where

$$\alpha(t) = \alpha_0 + \omega_{\mathrm{I}}(a_0)\,.(t - t_0), \tag{9.44}$$

$$\frac{\omega_{\mathrm{I}}(a_0)}{\omega} = 1 + \frac{3\varepsilon}{8}, \tag{9.45}$$

and the second approximation

$$x_{\mathrm{II}} = a_0 \left(\cos \alpha + \frac{\varepsilon}{32} \cos 3\alpha\right), \tag{9.46}$$

where

$$\alpha(t) = \alpha_0 + \omega_{\mathrm{II}}(a_0)\,.(t - t_0), \tag{9.47}$$

$$\frac{\omega_{\mathrm{II}}(a_0)}{\omega} = 1 + \frac{3\varepsilon}{8} - \frac{15\varepsilon^2}{256}. \tag{9.48}$$

The constants α_0 and a_0 are to be determined from the initial conditions.

The second approximation for $x(t)$, given by the method of KRYLOV and BOGOLYUBOV, is equal to the sum of the approximations of zeroth and first order obtained by the perturbation method. As for the period, we have from (9.48) to within terms of third order in ε:

$$T = \frac{2\pi}{\omega}\left(1 - \frac{3\varepsilon}{8} + \frac{51\varepsilon^2}{256}\right).$$

and this expression is equal to the sum of the approximations of zeroth, first, and second order obtained by the perturbation method. The difference in the number of steps between the solution and its period occurs because in the method of KRYLOV and BOGOLYUBOV one neglects at the nth step the terms of nth order in the expression of x and of $(n+1)^{\text{th}}$ order in that of T.

d) Equivalent linearization

We close the study of the conservative oscillator with cubic elastic restoring force by applying to this oscillator the method of equivalent linearization explained in § 5c.

Putting $f_1(\dot{x}) = 0$ and $\varepsilon\psi_1(x) = \beta\omega^2 x^3$ into (5.64) and (5.65) yields

$$\bar{h}(a) = 0, \tag{9.49}$$

$$\bar{\omega}^2(a) = \omega^2\left(1 + \frac{3}{4}\beta a^2\right), \tag{9.50}$$

and hence the equivalent linear equation is

$$\ddot{x} + \omega^2\left(1 + \frac{3}{4}\beta a^2\right)x = 0, \tag{9.51}$$

where a denotes the amplitude. The general solution of this equation is:

$$x(t) = a\cos(\bar{\omega}t + \gamma), \tag{9.52}$$

where the constants a and γ are to be determined by the initial conditions.

§ 10. CONSERVATIVE SYSTEMS WITH AMPLITUDE-INDEPENDENT PERIOD OF VIBRATION

We have seen in § 7 that, if the elastic characteristic is of the same type (hardening or softening) on the whole range of x in which the vibration takes place, then the period depends on amplitude. In some practical problems it is, however, advisable to use nonlinear systems with amplitude-independent period of vibration. Such systems may be obtained by using different elastic characteristics for different ranges of x.

Before solving the problem formulated above, we will study a more general one, namely the determination of an elastic characteristic corresponding to a given dependence of the period on amplitude.

Let (6.1) be the equation of motion and denote as in § 6 by x_1 and x_2, $x_1 < 0 < x_2$, the roots of the equation

$$G(x) = E_0. \tag{10.1}$$

By taking into account (6.14), we may write the vibration period in the form

$$T(E_0) = T_1(E_0) + T_2(E_0), \tag{10.2}$$

where

$$T_1(E_0) = \sqrt{2} \int_{x_1}^{0} \frac{dx}{\sqrt{E_0 - G(x)}}, \tag{10.3}$$

$$T_2(E_0) = \sqrt{2} \int_{0}^{x_2} \frac{dx}{\sqrt{E_0 - G(x)}}. \tag{10.4}$$

As shown in § 1, the function $G(x)$ is strictly decreasing on $(-\infty, 0]$ and strictly increasing on $[0, \infty)$. Let $x = \eta_1(G)$ and $x = \eta_2(G)$ be the inverse functions of $G = G(x)$ for $x \in (-\infty, 0]$ and $x \in [0, \infty)$, respectively. By substituting $x = \eta_1(G)$ into (10.3) and taking into consideration that $G(x_1) = E_0$, and hence $x_1 = \eta_1(E_0)$, we obtain

$$T_1(E_0) = -\sqrt{2} \int_0^{E_0} \frac{\eta_1'(G)\, dG}{\sqrt{E_0 - G}}. \tag{10.5}$$

Equation (10.5) is an integral equation of Abel type for the unknown function $\eta_1(G)$, whose solution is

$$\eta_1(G) = -\frac{1}{\pi\sqrt{2}} \int_0^G \frac{T_1(E_0)}{\sqrt{G - E_0}}\, dE_0. \tag{10.6}$$

Similarly, by substituting $x = \eta_2(G)$ into (10.4), we obtain

$$\eta_2(G) = \frac{1}{\pi\sqrt{2}} \int_0^G \frac{T_2(E_0)}{\sqrt{G - E_0}}\, dE_0. \tag{10.7}$$

Finally, by substracting (10.6) from (10.7) and considering (10.2), we have

$$\eta_2(G) - \eta_1(G) = \frac{1}{\pi\sqrt{2}} \int_0^G \frac{T(E_0)}{\sqrt{G - E_0}} \,\mathrm{d}E_0. \tag{10.8}$$

We notice that, if the period is given as function of the total energy, or, what comes to the same thing as a function of amplitude, then this dependence does not fully determine the elastic characteristic, since only the difference $\eta_2(G) - \eta_1(G)$ occurs in (10.8), and this allows one to impose supplementary conditions.

For instance, if we require that the elastic characteristic be symmetric with respect to the origin and correspond to the given dependence $T = T(E_0)$, then we have

$$g(-x) = -g(x), \qquad G(-x) = G(x), \qquad \eta_2(G) = -\eta_1(G) \tag{10.9}$$

and we obtain from (10.8) the unique solution

$$\eta_2(G) = -\eta_1(G) = \frac{1}{2\pi\sqrt{2}} \int_0^G \frac{T(E_0)}{\sqrt{G - E_0}} \,\mathrm{d}E_0. \tag{10.10}$$

By introducing the given expression of $T(E_0)$ into (10.10) and calculating the integral, we obtain $\eta_2(G)$ and $\eta_1(G)$, and then, by inversion, $G(x)$. We see also that the expression of the elastic characteristic may be directly derived from $\eta_2(G)$ and $\eta_1(G)$ by using the equation

$$g(x) = G'(x) = \pm \frac{1}{\eta_2'(G)}, \tag{10.11}$$

where the plus and minus signs correspond to $x > 0$ and $x < 0$, respectively. For further considerations concerning the calculation of the integral (10.10) by means of series expansions and for the application of the method to horology problems we refer to KAUDERER [60], § 45.

Let us consider now the problem of determining the elastic characteristic corresponding to an amplitude-independent period given by

$$T(E_0) = \frac{2\pi}{\omega}, \tag{10.12}$$

where, as in § 6, $\omega^2 = g'(0)$. The degree of arbitrariness allowed by (10.8) may be used now, by assuming that the elastic characteristic $g(x)$ is given for $x < 0$, and hence the function $x = \eta_1(G)$ is known. Substituting (10.12) into (10.8) yields for $x > 0$

$$x = \eta_2(G) = \eta_1(G) + \frac{\sqrt{2}}{\omega}\int_0^G \frac{dE_0}{\sqrt{G - E_0}}\eta_1(G) + \frac{2\sqrt{2G}}{\omega}. \tag{10.13}$$

By inverting this function, we obtain $G(x)$ and, by differentiating, we immediately find $g(x)$ for $x > 0$.

The function $g(x)$ obtained in this way has, in general, different analytical expressions for $x > 0$ and $x < 0$, respectively. However, as the following example shows, there are elastic characteristics that have the same analytical form in their whole range of definition, although the corresponding periods are independent of amplitude. Let

$$g(x) = 2a\omega^2\left(\sqrt{\frac{a}{a - x}} - 1\right), \qquad a > 0, \tag{10.14}$$

be the elastic characteristic defined for $x \in (-\infty, a)$. It is readily seen that $g(0) = 0$, $g'(0) = \omega^2$, and $g'(x) > 0$ for all $x \in (-\infty, a)$. By integrating, we find from (10.14)

$$G(x) = \int_0^x g(x)\,dx = 4\omega^2a^2\left(1 - \frac{x}{2a} - \sqrt{1 - \frac{x}{a}}\right). \tag{10.15}$$

Inverting this function yields for $x < 0$

$$x = \eta_1(G) = -\frac{G}{2\omega^2a} - \frac{\sqrt{2G}}{\omega}, \tag{10.16}$$

and for $x > 0$

$$x = \eta_2(G) = -\frac{G}{2\omega^2a} + \frac{\sqrt{2G}}{\omega}. \tag{10.17}$$

By introducing now $\eta_1(G)$ from (10.16) into (10.13), we obtain for $\eta_2(G)$ the very expression (10.17). Therefore, the period of the conservative system with elastic characteristic (10.14) is independent of amplitude for all $x \in (-\infty, a)$.

From (10.14) it follows also that

$$g''(x) = \frac{3\omega^2a}{(a - x)^2}\sqrt{\frac{a}{a - x}} > 0,$$

and hence the elastic characteristic is regressive for $x \in (-\infty, 0)$, and progressive for $x \in (0, a)$. If one introduces the dimensionless variables $\eta = g(x)/a\omega^2$ and

$\xi = x/a$ into (10.14), then the equation of the elastic characteristic assumes the form

$$\eta(\xi) = 2\left(\frac{1}{\sqrt{1-\xi}} - 1\right).$$

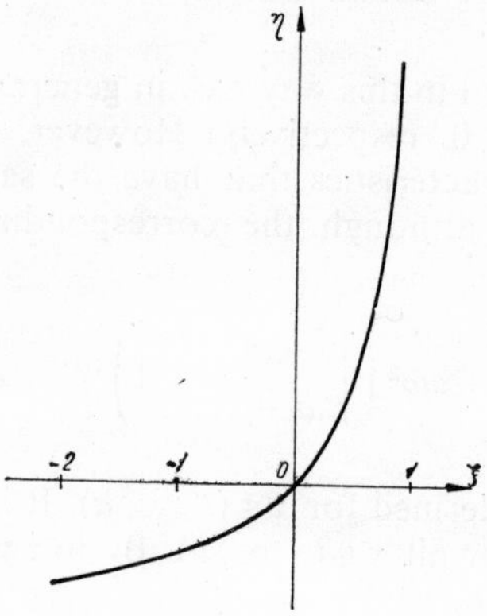

Fig. 34

Figure 34 shows the graph of this function.

Chapter III Free vibrations of dissipative systems

We have defined the conservative systems as systems whose total energy is time-independent. However, real systems always exchange energy with the surrounding medium. Conservative systems provide only a first approximation of real systems, which is the more accurate, the weaker the energy exchanges with the surroundings.

Systems whose total energy is time-dependent are called *nonconservative*. Such systems give in general a better description of real oscillators.

When adopting hypothesis H_3, we have $vf(v) > 0$ for $v \neq 0$, and hence it follows from (1.6) that

$$\frac{dE}{dt} = -vf(v) < 0 \qquad \text{for} \quad v \neq 0,$$

i.e., the total energy is a monotonically decreasing function of time. Nonconservative systems having this property are said to be *dissipative*. In order to remain within the limits proposed in the introduction for the object of our study, we will leave aside other nonconservative systems, e.g., the *self-excited* systems, and we shall thoroughly treat the dissipative systems, which have an exceptional importance for the description of mechanical oscillators. This chapter will be entirely devoted to the discussion of free vibrations of dissipative systems.

§ 11. APPLICATION OF THE METHOD OF KRYLOV AND BOGOLYUBOV TO SYSTEMS WITH LINEAR ELASTIC CHARACTERISTIC AND WEAK DISSIPATION

As regards the application of the method of KRYLOV and BOGOLYUBOV, systems with linear elastic characteristic and weak dissipation are in some sense complementary to the weakly nonlinear conservative systems studied in § 8, since now $\psi_1(x) \equiv 0$ and $f_1(\dot{x}) \not\equiv 0$. This time, however, $f_1(-\omega a \sin \alpha)$ is not necessarily an odd function, although it depends only on $\sin \alpha$. That is why, in order to obtain some simplifications of the formulas, we shall start by expanding in a cosine Fourier series the even function $f_1(a \cos \alpha)$:

$$f_1(a \cos \alpha) = \sum_{k=0}^{\infty} F_k(a) \cos k\,\alpha. \tag{11.1}$$

By replacing α by $\alpha + \pi/2$ and a by ωa in this formula, we have

$$f_1(-\,\omega a \sin \alpha) = \sum_{k=0}^{\infty} F_k(\omega a) \cos k\left(\alpha + \frac{\pi}{2}\right). \tag{11.2}$$

A comparison with (5.46) yields

$$C_k(a) = F_k(\omega a) \cos \frac{k\pi}{2}, \qquad D_k(a) = -\,F_k(\omega a) \sin \frac{k\pi}{2} \tag{11.3}$$

and then it follows from (5.58) that

$$A_1(a) = -\,\frac{F_1(\omega a)}{2\omega}, \qquad B_1(a) = 0. \tag{11.4}$$

Equations (5.49) – (5.52) now become

$$x_{\mathrm{I}} = a \cos \alpha, \tag{11.5}$$

$$t(a) = t_0 - \frac{2\omega}{\varepsilon} \int_{a_0}^{a} \frac{da}{F_1(\omega a)}, \tag{11.6}$$

$$\alpha(t) = \alpha_0 + \omega(t - t_0). \tag{11.7}$$

We notice that, at the first stage of approximation, the frequency is independent of amplitude and damping.

Let us consider now the second approximation. From (5.56) we obtain

$$u_1(a, \alpha) = -\,\frac{F_0(\omega a)}{\omega^2} + \frac{1}{\omega^2} \sum_{k=2}^{\infty} \frac{F_k(\omega a) \cos k\left(\alpha + \dfrac{\pi}{2}\right)}{k^2 - 1}. \tag{11.8}$$

Substituting now (11.2), (11.4), and (11.8) into (5.59) gives

$$A_2(a) = \frac{A_1(a)}{2\pi\omega} \int_0^{2\pi} f_1'(-\,\omega a \sin \alpha) \sin \alpha \cos \alpha \, d\alpha -$$

$$-\,\frac{1}{2\pi\omega^2} \sum_{k=2}^{\infty} \frac{kF_k(\omega a)}{k^2 - 1} \int_0^{2\pi} f_1'(-\,\omega a \sin \alpha) \cos^2 \alpha \, d\alpha -$$

$$-\,\frac{1}{2\pi\omega^2 a} \sum_{k=2}^{\infty} \frac{kF_k(\omega a)}{k^2 - 1} \int_0^{2\pi} f_1'(-\,\omega a \sin \alpha) \sin k\left(\alpha + \frac{\pi}{2}\right) \cos \alpha \, d\alpha.$$

Replacing α by $\alpha - \pi/2$ and taking into account that every function which depends only on $\cos\alpha$ is an even function of α, we have

$$\int_0^{2\pi} f_1'(-\omega a \sin\alpha) \sin k\left(\alpha + \frac{\pi}{2}\right) \sin\alpha \, d\alpha =$$

$$= -\int_{\frac{\pi}{2}}^{2\pi + \frac{\pi}{2}} f_1'(\omega a \cos\alpha) \cos\alpha \sin k\alpha \, d\alpha = 0,$$

and, for $k = 1$, also

$$\int_0^{2\pi} f_1'(-\omega a \sin\alpha) \cos\alpha \sin\alpha \, d\alpha = 0,$$

and hence

$$A_2(a) = 0. \tag{11.9}$$

Next, by changing α with $\alpha - \pi/2$, integrating by parts, and considering (11.1), we successively find

$$\int_0^{2\pi} f_1'(-\omega a \sin\alpha) \sin k\left(\alpha + \frac{\pi}{2}\right) \cos\alpha \, d\alpha =$$

$$= \int_{\frac{\pi}{2}}^{2\pi + \frac{\pi}{2}} f_1'(\omega a \cos\alpha) \sin k\alpha \sin\alpha \, d\alpha =$$

$$= -\frac{1}{\omega a}\int_{\frac{\pi}{2}}^{2\pi + \frac{\pi}{2}} \sin k\alpha \, df_1(\omega a \cos\alpha) =$$

$$= \frac{k}{\omega a}\int_{\frac{\pi}{2}}^{2\pi + \frac{\pi}{2}} f_1(\omega a \cos\alpha) \cos k\alpha \, d\alpha = \frac{\pi k F_k(\omega a)}{\omega a},$$

and, for $k = 1$, also

$$\int_0^{2\pi} f_1'(-\omega a \sin\alpha) \cos^2\alpha \, d\alpha = \frac{\pi F_1(\omega a)}{\omega a}.$$

Hence

$$B_2(a) = \frac{F_1(\omega a)}{8\omega^3 a} \frac{dF_1(\omega a)}{da} - \frac{F_1^2(\omega a)}{4\omega^3 a^2} - \frac{1}{2\omega^3 a^2} \sum_{k=2}^{\infty} \frac{k^2 F_k^2(\omega a)}{k^2 - 1}. \tag{11.10}$$

Finally, by introducing (11.8) – (11.10) into (5.53) – (5.55), we obtain for the second approximation:

$$x_{\mathrm{II}} = a \cos \alpha + \frac{\varepsilon}{\omega^2} \sum_{\substack{k=0 \\ k \neq 0}}^{\infty} \frac{F_k(\omega a) \cos k\left(\alpha + \dfrac{\pi}{2}\right)}{k^2 - 1}, \tag{11.11}$$

$$t(a) = t_0 - \frac{2\omega}{\varepsilon} \int_{a_0}^{a} \frac{da}{F_1(\omega a)}, \tag{11.12}$$

$$\alpha(t) = \alpha_0 + \omega(t - t_0) + \varepsilon^2 \int_{t_0}^{t} B_2(a)\, dt. \tag{11.13}$$

We see that the first two approximations give the same dependence of the amplitude on time. On the other hand, the frequency becomes a function of a at the second stage of approximation through the term of second order in ε.

§ 12. DISSIPATIVE SYSTEMS WITH DRY FRICTION

As a first example of a dissipative system we shall consider the oscillator with *dry friction*, which is also called an oscillator with *Coulomb damping*.

COULOMB idealized the friction force between two surfaces in direct contact with each other by taking the magnitude of this force as constant and its direction opposite to that of the motion. Under this assumption, the governing equation of motion may be written in the form

$$\ddot{x} + R \operatorname{sgn} \dot{x} + g(x) = 0, \tag{12.1}$$

where R is a possitive constant equal to the product between the coefficient of friction and the gravity acceleration, $g(x)$ is the elastic restoring force per unit mass, and

$$\operatorname{sgn} \dot{x} = \frac{|\dot{x}|}{\dot{x}} = \begin{cases} 1 & \text{for} \quad \dot{x} > 0, \\ -1 & \text{for} \quad \dot{x} < 0. \end{cases} \tag{12.2}$$

For $\dot{x} = 0$ the friction force may take any value in the interval $[-R, R]$, and is equal at all times to the elastic restoring force. Denoting by x' and x'', with $x' < 0 < x''$, the roots of the equation

$$|g(x)| = R, \tag{12.3}$$

we see that, for $x \in [x', x'']$ and $\dot{x} = 0$, the system is in neutral equilibrium under the action of the friction force and of the elastic restoring force.

By multiplying Eq. (12.1) by dx and integrating between the limits 0 and x, we obtain the energy equation

$$\frac{v^2}{2} + Rx \operatorname{sgn} v + G(x) = C. \tag{12.4}$$

The equation of motion (12.1), as well as the energy equation (12.4) assume different forms for $v > 0$ and $v < 0$, because of the term $R \operatorname{sgn} v$. Consequently, the constant of integration C takes different values for the successive arcs of trajectory in the upper and lower half-planes. The value of this constant is determined at the beginning of the motion by the initial conditions, and, subsequently, by requiring continuity of the displacement at the points where the trajectory crosses the x-axis.

Assuming the initial conditions

$$x(0) = x_1 < x', \qquad v(0) = 0, \tag{12.5}$$

the velocity is positive during the first half-oscillation, and then, by (12.4) and (12.5), we obtain

$$C = Rx_1 + G(x_1), \tag{12.6}$$

$$v = \sqrt{2}\sqrt{G(x_1) - G(x) + R(x_1 - x)}. \tag{12.7}$$

The next extreme value of the displacement, $x_2 > 0$, is determined by equating to zero the expression (12.7) of the velocity, and hence is given by the equation

$$G(x_1) - G(x_2) = -R(x_1 - x_2). \tag{12.8}$$

During the second half-oscillation the velocity is negative and then, by (12.4) and (12.5), it follows that

$$C = -Rx_2 + G(x_2), \tag{12.9}$$

$$v = -\sqrt{2}\sqrt{G(x_2) - G(x) - R(x_2 - x)}. \tag{12.10}$$

By equating to zero the expression (12.10) of the velocity, we find that the next extreme value of the displacement, $x_3 < 0$, is given by the equation

$$G(x_2) - G(x_3) = R(x_2 - x_3), \tag{12.11}$$

and so on. The oscillatory process continues until, for some extreme displacement, say x_n, the conditions $v = 0$ and $|g(x_n)| \leqslant R$ are simultaneously satisfied. Then, as has been already mentioned, the system reaches a state of neutral equilibrium.

The extreme values of the displacement may be determined either by solving numerically Eqs. (12.8), (12.11), etc., or by a plain graphical method, which will be explained in the following. Let M_1 be the point of abscissa x_1 on the graph of the potential energy $G(x)$ (see Fig. 35). The straight line of slope $-R$ through M_1 cuts the graph of $G(x)$, as shown by Eq. (12.8), in a point of abscissa x_2, say M_2. Through M_2 draw a straight line of slope R which cuts the graph, as shown by Eq. (12.11), at a point M_3 of abscissa x_3, and so on. The motion ceases when some extreme value of the displacement lies in the equilibrium interval $[x', x'']$.

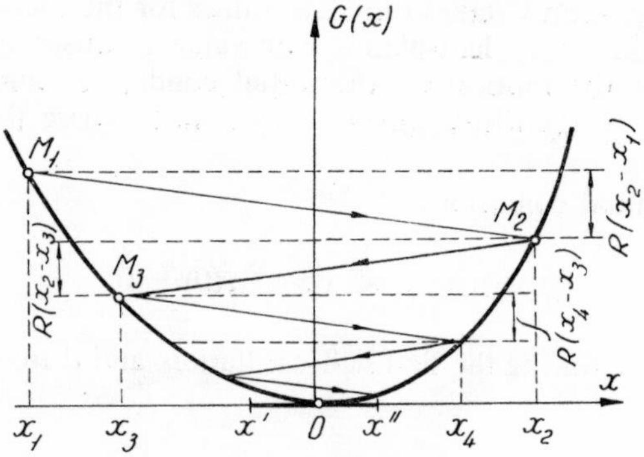

Fig. 35

After successively determining the extreme values of the displacement, we may calculate the duration of the half-oscillations by means of the formulas

$$T_1 = \frac{1}{\sqrt{2}} \int_{x_1}^{x_2} \frac{dx}{\sqrt{G(x_1) - G(x) + R(x_1 - x)}}, \tag{12.12}$$

$$T_2 = \frac{1}{\sqrt{2}} \int_{x_3}^{x_2} \frac{dx}{\sqrt{G(x_2) - G(x) - R(x_2 - x)}}, \tag{12.13}$$

$$T_3 = \frac{1}{\sqrt{2}} \int_{x_3}^{x_4} \frac{dx}{\sqrt{G(x_3) - G(x) + R(x_3 - x)}}, \tag{12.14}$$

and so on.

We have seen in § 4a how the method of LIÉNARD may be used in order to graphically obtain the phase trajectory of an oscillator with dry friction and linear elastic restoring force. Let us now treat analytically the same example. We have

$$g(x) = \omega^2 x, \qquad G(x) = \frac{\omega^2 x^2}{2}, \qquad -x' = x'' = \frac{R}{\omega^2}, \tag{12.15}$$

and the phase trajectory (12.4) consists of a sequence of half-ellipses, each of which corresponding to a half-oscillation. Equation (12.8) now becomes

$$(x_1 - x_2)\left(\frac{\omega^2}{2}x_1 + \frac{\omega^2}{2}x_2 + R\right) = 0,$$

and gives

$$x_2 = -x_1 - \frac{2R}{\omega^2} > 0. \tag{12.16}$$

Analogously, we obtain from (12.11)

$$x_3 = -x_2 + \frac{2R}{\omega^2} < 0, \tag{12.17}$$

and we see by induction that two consecutive extreme values of the displacement are related by

$$|x_{k+1}| - |x_k| = -\frac{2R}{\omega^2}, \tag{12.18}$$

This last equation shows that the sequence of the absolute values of the extreme displacements is a decreasing arithmetic progression with the common difference $-2R/\omega^2$.

By introducing now $(12.15)_2$ into (12.12) we obtain the interval of time corresponding to the first half-oscillation

$$T_1 = \frac{1}{\sqrt{2}}\int_{x_1}^{x_2} \frac{\mathrm{d}x}{\sqrt{\frac{\omega^2}{2}(x_1^2 - x^2) + R(x_1 - x)}},$$

which, considering (12.16), may be successively transformed to give

$$T_1 = \frac{1}{\omega}\int_{x_1}^{x_2} \frac{\mathrm{d}x}{\sqrt{(x - x_1)(x_2 - x)}} =$$

$$= -\frac{1}{\omega}\arcsin \frac{x_1 + x_2 - 2x}{x_2 - x_1}\bigg|_{x_1}^{x_2} = \frac{\pi}{\omega}. \tag{12.19}$$

By repeating the above calculation for any half-oscillation one always finds the same result. Therefore, the duration of each half-oscillation is equal to the half-period of the simple harmonic oscillator. However, the oscillator with dry friction needs more time than the harmonic one to move from a position of extreme displacement to the next position of zero displacement than from this last position to the next position of extreme displacement. For example, we have for the first half-oscillation

$$T_1' = \frac{1}{\omega}\int_{x_1}^{0} \frac{dx}{\sqrt{(x-x_1)(x_2-x)}} = \frac{\pi}{2\omega} + \frac{1}{\omega}\arcsin\frac{x_1+x_2}{x_1-x_2},$$

$$T_1'' = \frac{1}{\omega}\int_{0}^{x_2} \frac{dx}{\sqrt{(x-x_1)(x_2-x)}} = \frac{\pi}{2\omega} - \frac{1}{\omega}\arcsin\frac{x_1+x_2}{x_1-x_2},$$

$$T_1' - T_1'' = \frac{2}{\omega}\arcsin\frac{2R/\omega^2}{-x_1-2R/\omega^2} > 0.$$

We finally remark that the equation of motion of the oscillator with dry friction may be directly integrated for each half-oscillation, and then the solutions obtained may be matched by requiring the continuity of the displacement and velocity. Thus, assuming the initial conditions

$$x(0) = x_1 < 0, \qquad \dot{x}(0) = 0,$$

the equation of motion for the first half-oscillation, $\ddot{x} + \omega^2 x = -R$, has the solution

$$x(t) = -\frac{R}{\omega^2} + \left(x_1 + \frac{R}{\omega^2}\right)\cos\omega t, \tag{12.20}$$

By denoting $z = x + R/\omega^2$, $z_1 = x_1 + R/\omega^2$, Eq. (12.20) becomes

$$z(t) = z_1 \cos\omega t, \qquad 0 \leqslant t \leqslant t_1. \tag{12.21}$$

The second half-oscillation begins at time $T_1 = \pi/\omega$, when the conditions

$$x(T_1) = x_2 = -x_1 - R/\omega^2 > 0, \qquad \dot{x}(T_1) = 0$$

have to be fulfilled. The solution of the equation

$$\ddot{x} + \omega^2 x = R,$$

satisfying these initial conditions is

$$x(t) = \frac{R}{\omega^2} + \left(x_2 - \frac{R}{\omega^2}\right)\cos\omega t, \tag{12.22}$$

By denoting $z = x - R/\omega^2$, $z_2 = x_2 - R/\omega^2$, Eq. (12.22) may be also written

$$z(t) = z_2 \cos \omega t, \qquad \pi/\omega \leqslant t \leqslant 2\pi/\omega, \tag{12.23}$$

and so on. It is apparent from the above that the motion consists of a sequence of harmonic half-oscillations with respect to a horizontal axis which has to be

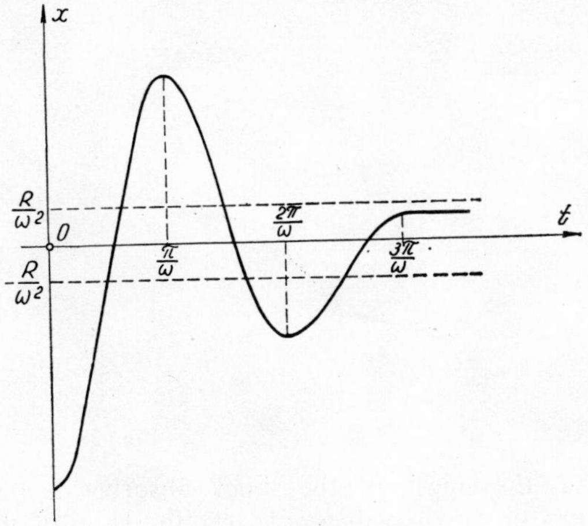

Fig. 36

alternately translated up and down by $2R/\omega^2$ from one half-oscillation to the next one (Fig. 36). The oscillator stops moving when the graph has a horizontal tangent at a point lying in the equilibrium strip $|x| \leqslant R/\omega^2$.

§ 13. SYSTEMS WITH QUADRATIC DAMPING

Systems with quadratic damping appropriately describe those real systems for which the damping of the oscillations is produced by a turbulent liquid flow inside the damper.

A typical example of an oscillator with quadratic damping is the suspension of a vehicle equipped with hydraulic shock absorbers. By neglecting the elasticity of the tires and the coupling between the vibrations of the front and rear axles, the vibrations of the vehicle that are symmetrical with respect to the longitudinal axis may be studied on the single-degree-of-freedom system illustrated in Fig. 37. Suspensions with nonlinear elastic and damping characteristics are frequently utilized, because nonlinearity limits displacements and velocities, reduces the extreme values of the acceleration, and leads to a more uniform dynamic loading of the suspension. The damping nonlinearity is usually achieved by using hydraulic or hydropneumatic shock absorbers. Figure 38 schematically shows the damping characteristic of a hydraulic shock absorber. We have denoted by v the velocity of the sprung mass with respect to the axle and by $f(v)$ the damping force per unit sprung mass. The motion of the piston is called *extension stroke* or *compression*

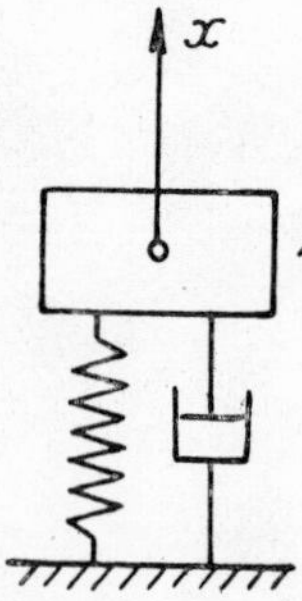

Fig. 37

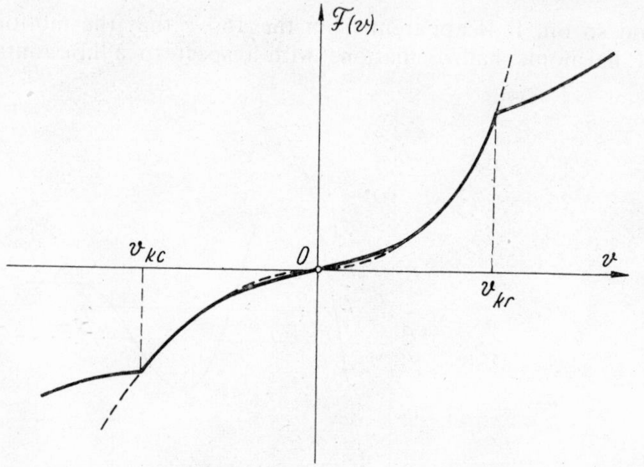

Fig. 38

stroke, according as the length of the shock absorber is increasing ($v > 0$) or decreasing ($v < 0$). For small values of $|v|$, the flow of the oil in the shock absorber is laminar and the damping characteristic is linear. As $|v|$ increases, the flow becomes turbulent and the characteristic takes a parabolic form. Finally, when the velocity reaches a certain "critical" value, the pressure inside the shock absorber becomes large enough to bring about the opening of the relief valves, and this modifies the further dependence of the damping force on velocity.[1] In Fig. 38 the critical value of the velocity has been denoted by v_e for the extension stroke and by v_c for the compression one.

For $v \in [v_c, v_e]$ the damping characteristic may be roughly approximated by two parabolas tangent at the origin to the v-axis (the dashed curves in Fig. 38), which may be symmetrical or unsymmetrical to each other with respect to the origin. We shall begin by considering such a simplified characteristic consisting of two parabolas. The more general case of a damping characteristic that may be piecewise approximated by straight lines and parabolic arcs will be examined in the next section.

a) The oscillator with quadratic damping and nonlinear elastic restoring force

The oscillator with symmetric quadratic damping characteristic has been minutely studied (see, e.g., PRASIL [151], KLOTTER [88], RICHARDSON [168], KAUDERER [82], § 50 b) because of its important applications and because it belongs to the

[1] For other details concerning the flow regime and the design of a shock absorber when its damping characteristic is prescribed see DINCĂ & TEODOSIU [46].

few nonlinear systems whose trajectories in the phase plane may be found by merely calculating one indefinite integral, as in the case of the oscillator with dry friction.

The equation of motion of this oscillator is

$$\ddot{x} + a_e \dot{x} \, |\dot{x}| + g(x) = 0, \tag{13.1}$$

where a_e is a positive constant.[1] By denoting $\dot{x} = v$, we have $|\dot{x}| = v \operatorname{sgn} v$, and hence Eq. (13.1) may be also written

$$v \frac{\mathrm{d}v}{\mathrm{d}x} + a_e v^2 \operatorname{sgn} v + g(x) = 0,$$

or

$$\frac{\mathrm{d}v^2}{\mathrm{d}x} + 2a_e v^2 \operatorname{sgn} v + 2g(x) = 0. \tag{13.2}$$

By taking v^2 as an unknown function, we can immediately integrate this linear differential equation to obtain

$$v = \sqrt{2}\,\mathrm{e}^{-a_e x} \sqrt{C - V_+(x)} \qquad \text{for} \quad v \geqslant 0, \tag{13.3}$$

$$v = -\sqrt{2}\,\mathrm{e}^{a_e x} \sqrt{C - V_-(x)} \qquad \text{for} \quad v \leqslant 0, \tag{13.4}$$

with the notations

$$V_+(x) \equiv \int_0^x g(x)\,\mathrm{e}^{2a_e x}\,\mathrm{d}x, \qquad V_-(x) \equiv \int_0^x g(x)\,\mathrm{e}^{-2a_e x}\,\mathrm{d}x. \tag{13.5}$$

The constant C is to be determined for the first half-oscillation by the initial conditions, and for the subsequent half-oscillations by requiring the continuity of x at the points where the velocity vanishes.

By assuming initial conditions of the form

$$x(0) = x_1 < 0, \qquad v(0) = 0, \tag{13.6}$$

we see that the velocity must be positive during the first half-oscillation. Hence, by (13.3) and (13.6), we have

$$v(x) = \sqrt{2}\,\mathrm{e}^{-a_e x} \sqrt{V_+(x_1) - V_+(x)}.$$

[1] The subscript "e" means extension. Since the damping characteristic is symmetrical with respect to the origin, we have to use the same damping coefficient a_e for the extension as well as for the compression strokes.

The next extreme value of the displacement, say $x_2 > 0$, results by equating this expression to zero i.e., by solving the equation

$$V_+(x_1) = V_+(x_2). \tag{13.7}$$

During the second half-oscillation the velocity is negative, and hence, by (13.4), it follows that

$$v(x) = -\sqrt{2}\, e^{a_e x} \sqrt{V_-(x_2) - V_-(x)}\,. \tag{13.8}$$

The next extreme value of the displacement, say $x_3 < 0$, may be calculated by equating to zero the expression (13.8), i.e., by solving the equation

$$V_-(x_2) = V_-(x_3). \tag{13.9}$$

In the same way, we obtain for the next half-oscillation ($v > 0$):

$$v(x) = \sqrt{2}\, e^{-a_e x} \sqrt{V_+(x_3) - V_+(x)}, \tag{13.10}$$

$$V_+(x_3) = V_+(x_4), \tag{13.11}$$

and so on.

The extreme values of the displacement may be determined either by numerically solving Eqs. (13.7), (13.9), (13.11), etc, or by using the following graphical method (see also Fig. 39). Plot the graphs of the functions $V_+(x)$ and $V_-(x)$ as

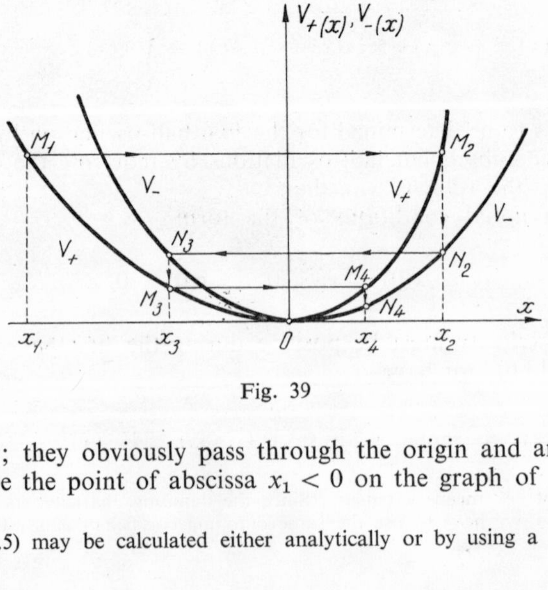

Fig. 39

given by (13.5)[1]; they obviously pass through the origin and are tangent to the x-axis. Let M_1 be the point of abscissa $x_1 < 0$ on the graph of V_+. The straight

[1] Integrals (13.5) may be calculated either analytically or by using a numerical method.

line through M_1 parallel to Ox cuts the graph of V_+ once again at a point of abscissa $x_2 > 0$; call it M_2. The perpendicular through M_2 on Ox intersects the graph of V_- in a point, say N_2. The straight line through N_2 parallel to Ox cuts the graph of V_- a second time in a point of abscissa $x_3 < 0$; call it N_3. Drop a perpendicular from N_3 on the x-axis intersecting the graph of V_+ at a point M_3. The straight line through M_3 parallel to Ox cuts the graph of V_+ once again at a point M_4 of abscissa $x_4 > 0$, and so on. The justification of this graphical method directly follows from considering Eqs. (13.7), (13.9), (13.11), and the like.

After determining the dependence of v on x, one may calculate the duration of the half-oscillations by using one of the numerical methods discussed in § 3a and b.

For the application of the method explained above in case the elastic characteristic is linear or parabolic we refer to KAUDERER [82], § 50. Attention is also called to a paper by SZPUNAR and BOGUSZ [188], in which the oscillator whose elastic and damping characteristics are both quadratic and symmetric with respect to the origin is thoroughly studied.

As we shall immediately see, the results concerning the oscillator with quadratic damping and linear elastic restoring force may be considerably amplified, even if the damping characteristic is asymmetric.

b) The oscillator with asymmetric quadratic damping and linear elastic restoring force

Most of the shock absorbers utilized for suspensions of road vehicles have asymmetric characteristics. By neglecting the laminar character of the flow for small velocities and considering only under-critical regimes, the damping force produced by a hydraulic shock absorber may be approximated by the function

$$P(v) = \begin{cases} -k_e v^2 & \text{for} \quad v \geqslant 0, \\ k_c v^2 & \text{for} \quad v \leqslant 0, \end{cases} \tag{13.12}$$

where the positive numbers k_e and k_c correspond to the extension and compression strokes, respectively. For road vehicles, the ratio

$$\alpha = k_e/k_c \tag{13.13}$$

is generally greater than 1 (cf. Fig. 38).

Suppose the front (or the rear) suspension of a road vehicle consists of two springs of stiffness c and of two hydraulic shock absorbers of damping characteristic (13.12). Let m be the part of the sprung mass corresponding to this axle and let us denote

$$a_e = \frac{2k_e}{m}, \qquad a_c = \frac{2k_c}{m}, \qquad \omega^2 = \frac{2c}{m}. \tag{13.14}$$

Under the simplifying assumptions formulated in the introduction, the free vibrations of the mass m may be described by the equations

$$\begin{aligned} \ddot{x} + a_e\dot{x}^2 + \omega^2 x = 0 \quad &\text{for} \quad \dot{x} \geqslant 0. \\ \ddot{x} - a_c\dot{x}^2 + \omega^2 x = 0 \quad &\text{for} \quad \dot{x} \leqslant 0, \end{aligned} \tag{13.15}$$

which may be also written in the form

$$\begin{aligned} \frac{\mathrm{d}v^2}{\mathrm{d}x} + 2a_e v^2 + 2\omega^2 x = 0 \qquad &\text{for} \quad v \geqslant 0, \\ \frac{\mathrm{d}v^2}{\mathrm{d}x} - 2a_c v^2 + 2\omega^2 x = 0 \qquad &\text{for} \quad v \leqslant 0. \end{aligned} \tag{13.16}$$

Assume the initial conditions

$$x(0) = x_1 < 0, \qquad v(0) = 0. \tag{13.17}$$

Since the differential equations (13.16) are linear with respect to v^2, they may be immediately integrated to obtain

$$\begin{aligned} v(x) &= \mathrm{e}^{-a_e x} \sqrt{C - \frac{\omega^2}{2a_e^2} \mathrm{e}^{2a_e x} (2a_e x - 1)}\,, \qquad v \geqslant 0, \\ v(x) &= -\mathrm{e}^{a_c x} \sqrt{C + \frac{\omega^2}{2a_c^2} \mathrm{e}^{-2a_c x} (2a_c x + 1)}\,, \qquad v \leqslant 0, \end{aligned} \tag{13.18}$$

The constant C is to be determined for the first half-oscillation by the initial conditions (13.17), and for the subsequent half-oscillations by requiring the continuity of x at the points x_n where x takes extreme values, hence at which

$$v(x_n) = 0, \qquad n = 1, 2, \ldots \tag{13.19}$$

Assuming that x_n has been recursively determined by (13.18) and (13.19), we have for the nth half-oscillation, i.e., for $x \in [x_n, x_{n+1}]$:

$$v(x) = \frac{\omega\sqrt{2}}{2a_e} \mathrm{e}^{-a_e x} \sqrt{\mathrm{e}^{2a_e x_n}(2a_e x_n - 1) - \mathrm{e}^{2a_e x}(2a_e x - 1)} \geqslant 0, \quad x_n < 0, \tag{13.20}$$

or

$$v(x) = -\frac{\omega\sqrt{2}}{2a_c} \mathrm{e}^{a_c x} \sqrt{\mathrm{e}^{-2a_c x}(2a_c x + 1) - \mathrm{e}^{-2a_c x}(2a_c x_n + 1)} \leqslant 0, \quad x_n > 0, \tag{13.21}$$

according to whether n is odd or even.

The complete determination of the solution $x(t)$ would require the calculation of the integral

$$t(x) = \int_{x_1}^{x} \frac{dx}{v(x)} \tag{13.22}$$

and then the solution of this equation for x. However, since the integral (13.22) can be calculated only numerically and only for constant values of x, this direct approach is not applicable practically. Therefore, we shall confine ourselves to indicating a method allowing a rapid and rather exact calculation of the sequence of extreme values of the displacement and of the corresponding intervals of time.

Let us introduce in (13.20) and (13.21), respectively, the new independent variables

$$\xi = -2a_e x, \qquad \xi' = 2a_c x, \tag{13.23}$$

which are related by

$$\xi = -\frac{a_e}{a_c}\xi' = -\alpha\xi'. \tag{13.24}$$

Equations (13.20), (13.21) may now be written as

$$v(\xi) = \frac{\omega\sqrt{2}}{2a_e}\sqrt{1 + \xi - e^{\xi - \xi_n}(1 + \xi_n)} \geqslant 0, \qquad \xi_n = -2a_e x_n > 0,$$

or (13.25)

$$v(\xi) = -\frac{\omega\sqrt{2}}{2a_c}\sqrt{1 + \xi' - e^{\xi' - \xi'_n}(1 + \xi'_n)} \leqslant 0, \qquad \xi'_n = 2a_c x_n > 0.$$

for $\xi \in [\xi_n, \xi_{n+1}]$, according to whether n is odd or even. These equations may be used successively to determine the extreme values of the displacement. Thus, by $(13.17)_1$, we have $\xi_1 = -2a_e x_1$, and ξ_2 is determined by the condition $v(\xi_2) = 0$, i.e., by the equation

$$1 + \xi_2 - e^{\xi_2 - \xi_1}(1 + \xi_1) = 0.$$

Then, by (13.24), it follows that $\xi'_2 = -\xi_2/\alpha$ and we may calculate ξ'_3 from the condition $v(\xi'_3) = 0$, i.e., from the equation

$$1 + \xi'_3 - e^{\xi'_3 - \xi'_2}(1 + \xi'_2) = 0.$$

Next we have $\xi_3 = -\alpha\xi'_3$ and so on. Finally, we can immediately calculate by (13.23) the extreme values of the displacement, which, as we see, do not depend on ω.

The advantage of using Eqs. (13.25) instead of (13.20) and (13.21) is that the reduced extreme values of the displacement, ξ_n and ξ'_n, are successively obtained as solutions of the same transcendental equation, namely

$$1 + u_{n+1} - e^{u_{n+1}-u_n}(1 + u_n) = 0, \tag{13.26}$$

where u_n is to be replaced by ξ_n or ξ'_n, according to whether n is odd or even.

Equation (13.26) may be also written in the form

$$\frac{e^{u_n}}{1 + u_n} = \frac{e^{u_{n+1}}}{1 + u_{n+1}}. \tag{13.27}$$

Since u_{n+1} is negative, u_{n+1} approaches -1 as $u_n \to \infty$. By (13.23), this implies $|x_2| < 1/(2a_e)$ if $x_1 < 0$, and $|x_2| < 1/(2a_c)$ if $x_1 > 0$, irrespective of the initial magnitude of the displacement. We see that quadratic damping imposes a limitation upon the magnitude of the displacement, which is the more severe, the larger the initial value of the displacement.

We will now explain a method permitting approximate calculation of the root u_{n+1} of the transcendental equation (13.27) and giving good results for

$$0 \leqslant u_n \leqslant 6, \tag{13.28}$$

i.e., for most of the cases of practical interest.

By expanding e^u in a power series, we obtain

$$\frac{e^u}{1+u} = 1 + \frac{u^2}{2} \cdot \frac{1 + \dfrac{u}{3} + \dfrac{u^2}{12} + \dfrac{u^3}{60} + \dfrac{u^4}{360} + \dots}{1+u}. \tag{13.29}$$

We see that the fractional function in the right-hand side of (13.29) is discontinuous at $u = -1$, and hence a direct expansion of this fraction in a power series would lead to slow convergence, especially in the neigborhood of this value of u. In order to obtain a more rapidly convergent expansion it proves advantageous to write Eq. (13.29) in the form

$$\frac{e^u}{1+u} = 1 + \frac{u^2}{2S(u)}, \tag{13.30}$$

where

$$S(u) = \frac{1+u}{1 + \dfrac{u}{3} + \dfrac{u^2}{12} + \dfrac{u^3}{60} + \dfrac{u^4}{360} + \dots},$$

whence

$$S(u) = 1 + \frac{2u}{3} - \frac{11u^2}{36} + \frac{4u^3}{135} + \frac{11u^4}{6480} + \cdots \tag{13.31}$$

This last expansion is convergent for any finite u and has the value

$$S(u) = \frac{u^2(1+u)}{2(e^u - 1 - u)} \quad \text{for } u \neq 0, \qquad S(0) = 4. \tag{13.32}$$

With the aid of (13.32) we may determine the error done when truncating the expansion (13.31).

If u_n takes values in the interval (13.28), the solution of Eq. (13.27) may be calculated with a satisfactory accuracy by retaining only the first four terms of the expansion (13.31). Therefore, we write Eq. (13.27) in the form

$$\frac{u_n^2}{1 + \dfrac{2u_n}{3} - \dfrac{11u_n^2}{36} + \dfrac{4u_n^3}{135}} = \frac{u_{n+1}^2}{1 + \dfrac{2u_{n+1}}{3} - \dfrac{11u_{n+1}^2}{36} + \dfrac{4u_{n+1}^3}{135}}.$$

By multiplying both sides of this equation by the least common denominator, we obtain

$$(u_n - u_{n+1})\left(u_n + u_{n+1} + \frac{2}{3}u_n u_{n+1} - \frac{4}{135}u_n^2 u_{n+1}^2\right) = 0,$$

and, as $u_n \neq u_{n+1}$, it follows that

$$u_{n+1} = -\frac{u_n}{1 + \dfrac{2}{3}u_n - \dfrac{4}{135}u_n^2 u_{n+1}}.$$

This equation may be solved by successive approximations. We first put $u_{n+1}^{(0)} = 0$ into the right-hand side to obtain

$$u_{n+1}^{(1)} = -\frac{u_n}{1 + \dfrac{2}{3}u_n}. \tag{13.33}$$

Next, by replacing u_{n+1} with $u_{n+1}^{(1)}$, we derive the second approximation

$$u_{n+1}^{(2)} = -\frac{u_n}{1 + \frac{2}{3}u_n - \frac{4}{135}u_n^2 u_{n+1}^{(1)}}. \tag{13.34}$$

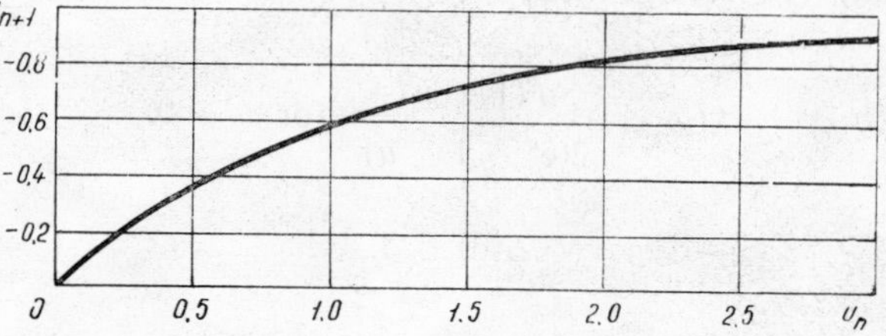

Fig. 40

The values of u_{n+1} obtained by numerically solving the transcendental equation (13.27) to three decimal places are listed in Table 1 for $0 \leqslant u_n \leqslant 6$. It is readily seen that (13.34) gives the root of Eq. (13.27) to an error smaller than 1% for $0 \leqslant u_n \leqslant 3.5$, and equal to 4% for $u_n = 6$. Formula (13.33) may be also used to an error of 1% but only for $0 \leqslant u_n \leqslant 1$. Figure 40 shows the variation of u_{n+1} with u_n; it should be noted that u_{n+1} asymptotically approaches -1 as $u_n \to \infty$.

Table 1

u_n	u_{n+1}	u_n	u_{n+1}	u_n	u_{n+1}
0.000	0.000	0.700	—0.475	1.800	—0.789
0.050	—0.048	0.750	—0.497	1.900	—0.806
0.100	—0.094	0.800	—0.518	2.000	—0.821
0.150	—0.136	0.850	—0.538	2.200	—0.849
0.200	—0.176	0.900	—0.558	2.400	—0.871
0.250	—0.214	0.950	—0.576	2.600	—0.890
0.300	—0.250	1.000	—0.594	2.800	—0.907
0.350	—0.284	1.100	—0.626	3.000	—0.921
0.400	—0.316	1.200	—0.656	3.500	—0.948
0.450	—0.346	1.300	—0.684	4.000	—0.965
0.500	—0.375	1.400	—0.709	4.500	—0.977
0.550	—0.402	1.500	—0.732	5.000	—0.985
0.600	—0.427	1.600	—0.752	5.500	—0.991
0.650	—0.452	1.700	—0.771	6.000	—0.994

In order to better illustrate the limiting action of the quadratic damping on the displacement, the extreme values of the displacement x_n, $n = 1, 2, \ldots, 8$, have been plotted in Fig. 41 against the initial displacement, $x_1 < 0$, for $a_e = 0.20\ \text{cm}^{-1}$, $a_c = 0.05\ \text{cm}^{-1}$. The limiting values of x_n, $n = 1, 2, \ldots, 8$, as $x_1 \to -\infty$ have been written near the corresponding horizontal asymptotes.

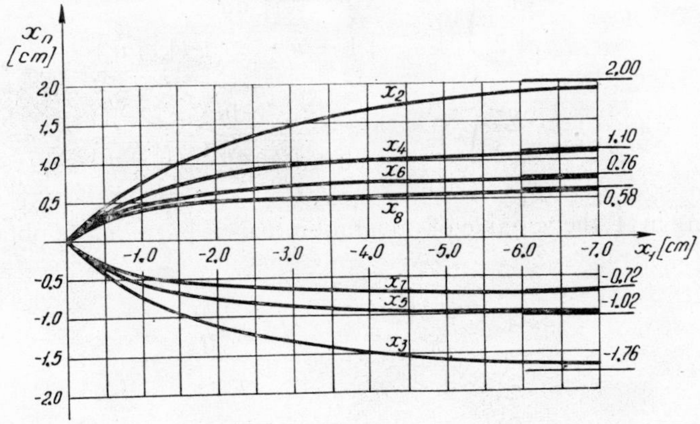

Fig. 41

Let us now determine the approximate duration of the half-oscillations. From (13.22) it follows that the nth half-oscillation lasts a time

$$T_n = \int_{x_n}^{x_{n+1}} \frac{dx}{v(x)}.$$

By changing the variable of integration according to (13.23) and taking into account (13.25), we obtain

$$T_n = A(u_n) . T_0, \tag{13.35}$$

where $T_0 = \pi/\omega$ is the period of the simple harmonic vibration,

$$A(u_n) = -\frac{1}{\pi\sqrt{2}} \int_{u_n}^{u_{n+1}} \frac{du}{\sqrt{1 + u - e^{u-u_n}(1 + u_n)}}, \tag{13.36}$$

and u_n is to be replaced by $\xi_n = -2a_e x_n$ or by $\xi'_n = 2a_c x_n$, according as n is odd or even. As has been already shown, u_n uniquely determines u_{n+1}, so that $A(u_n)$ ultimately depends only on u_n. Since the radicand in the right-hand side of (13.36) vanishes for $u = u_n$ and $u = u_{n+1}$, we can use the procedure explained in § 3a to remove these singularities and to calculate approximately the integral by means of SIMPSON's formula. With this in mind, we write

$$A(u_n) = A' + A'',$$

where

$$A' = -\frac{1}{\pi\sqrt{2}}\int_{u_n}^{\frac{u_n+u_{n+1}}{2}} \frac{\mathrm{d}u}{\sqrt{1+u-\mathrm{e}^{u-u_n}(1+u_n)}},$$

$$A'' = -\frac{1}{\pi\sqrt{2}}\int_{\frac{u_n+u_{n+1}}{2}}^{u_{n+1}} \frac{\mathrm{d}u}{\sqrt{1+u-\mathrm{e}^{u-u_n}(1+u_n)}}.$$

By changing in A' the variable of integration to $\zeta = \sqrt{u_n - u}$, we obtain

$$A' = \frac{\sqrt{2}}{\pi}\int_0^{\sqrt{\frac{u_n-u_{n+1}}{2}}} f(\zeta)\,\mathrm{d}\zeta,$$

where

$$f(\zeta) \equiv \frac{\zeta}{\sqrt{1+u_n-\zeta^2-(1+u_n)\,\mathrm{e}^{-\zeta^2}}}.$$

To determine the value of this function at the lower limit of integration, we first calculate the limit

$$\lim_{\zeta\to 0}[f(\zeta)]^2 = \lim_{\zeta\to 0}\frac{\zeta^2}{1+u_n-\zeta^2-(1+u_n)\,\mathrm{e}^{-\zeta^2}}.$$

By l'HÔPITAL's rule, it follows that

$$\lim_{\zeta\to 0}[f(\zeta)]^2 = \lim_{\zeta\to 0}\frac{1}{-1+(1+u_n)\,\mathrm{e}^{-\zeta^2}} = \frac{1}{u_n}$$

and hence

$$f(0) = \lim_{\zeta\to 0} f(\zeta) = \frac{1}{\sqrt{u_n}}.$$

We now apply SIMPSON's formula for four subintervals to obtain the approximate value of A':

$$A' \approx \frac{\sqrt{2}}{12\pi}\sqrt{\frac{u_n-u_{n+1}}{2}}\left[f(0) + 4f\left(\sqrt{\frac{u_n-u_{n+1}}{2}}\right) + 2f\left(\frac{1}{2}\sqrt{\frac{u_n-u_{n+1}}{2}}\right) + \right.$$
$$\left. + 4f\left(\frac{3}{4}\sqrt{\frac{u_n-u_{n+1}}{2}}\right) + f\left(\sqrt{\frac{u_n-u_{n+1}}{2}}\right)\right],$$

which, by making use of the expressions previously obtained for $f(\zeta)$ and $f(0)$, becomes

$$A' \approx \frac{\sqrt{2}}{24\pi}(u_n - u_{n+1})\left(\sqrt{\frac{2}{u_n - u_{n+1}}}\cdot\frac{1}{\sqrt{u_n}} + \frac{1}{\varphi_1} + \frac{1}{\varphi_4} + \frac{3}{\varphi_9} + \frac{1}{\varphi_{16}}\right),$$

where

$$\varphi_m = \sqrt{1 + u_n - \frac{m}{32}(u_n - u_{n+1}) + (1 + u_n)\,\mathrm{e}^{-\frac{m}{32}(u_n - u_{n+1})}}\,.$$

In the same way, by substituting $\zeta = \sqrt{u - u_{n-1}}$ into the integral giving A'', we find

$$A'' \approx \frac{\sqrt{2}}{24\pi}(u_n - u_{n+1})\left(\sqrt{\frac{2}{u_n - u_{n+1}}}\cdot\frac{1}{\sqrt{-u_{n+1}}} + \frac{1}{\varphi_{31}} + \frac{1}{\varphi_{28}} + \frac{3}{\varphi_{23}} + \frac{1}{\varphi_{16}}\right).$$

Finally, summing up the results obtained for A' and A'' yields

$$A(u_n) \approx \frac{\sqrt{2}}{24\pi}(u_n - u_{n+1})\left[\sqrt{\frac{2}{u_n - u_{n+1}}}\left(\frac{1}{\sqrt{u_n}} + \frac{1}{\sqrt{-u_{n+1}}}\right) + \frac{1}{\varphi_1} + \frac{1}{\varphi_4} + \frac{3}{\varphi_9} + \frac{2}{\varphi_{16}} + \frac{3}{\varphi_{23}} + \frac{1}{\varphi_{28}} + \frac{1}{\varphi_{31}}\right]. \tag{13.37}$$

Equation (13.37) allows one to determine the duration of the nth half-oscillation for various values of u_n, provided that u_{n+1} has been previously calculated, e.g., by the approximate method explained above. Table 2 indicates the values of the correction factor $A(u_n) = T_n/T_0$ for $0 \leqslant u_n \leqslant 6$, calculated by the approximate formula (13.37). As has been already mentioned, if $\alpha > 1$, then $u_n < 1$ for any initial conditions and for $n \geqslant 2$. Inspection of Table 2 reveals that in this case the correction of T_0 is less than 3%. We may, therefore, consider that the duration of the half-oscillations is practically equal to the half-period of the simple harmonic oscillator, with the only exception of the first half-oscillation whose duration may be considerably longer than T_0 (for $u_n = 6$ longer by 34%).

As a first numerical example we take $a_e = 0.24$ cm^{-1}, $a_c = 0.08$ cm^{-1}, $\alpha = 3$, $x_1 = -3$ cm. The recurrent determination of x_n and T_n/T_0, $n = 1, 2, \ldots$, is performed by making use of Tables 1 and 2 and applying a linear interpolation technique. The results obtained are listed in table 3, where the arrows indicate the successive determination of ξ_n and ξ_n'.

Table 2

u_n	0	1	2	3	4	5	6
T_n/T_0	1.000	1.026	1.077	1.137	1.203	1.272	1.341

Thus, for $x_1 = -3$ cm, we find $\xi_1 = -2\, a_e\, x_1 = 1.440$, and then it follows from table 1 that $\xi_2 = -0.717$. By (13.24), the corresponding value of ξ' is $\xi_2' = -\xi_2/\alpha = 0.239$; then, by making use again of table 1, we find $\xi_3' = -0.206$, and (13.24) yields $\xi_3 = -\alpha\xi_3' = 0.618$, and so on. After completing the first two lines of table 3, the extreme values of the displacement, x_n, are calculated by the formulas $x_n = -\xi_n/(2\, a_e)$ or $x_n = \xi_n'/(2\, a_c)$, according as n is odd or even. Then, the reduced durations of the half-oscillations are easily found by setting in table 2: $u_n = \xi_n$ if n is odd, and $u_n = \xi_n'$ if n is even. Figure 42 illustrates the

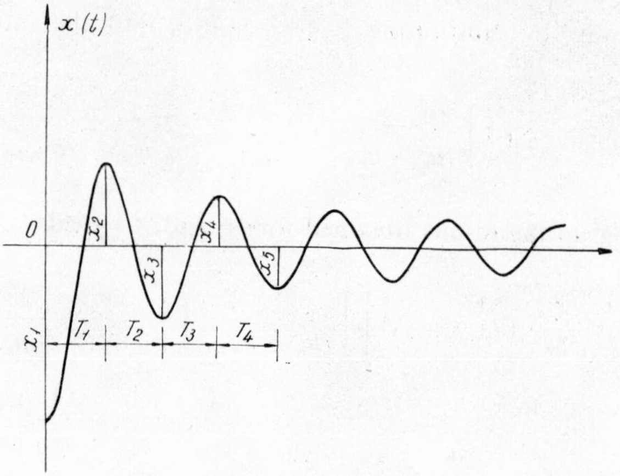

Fig. 42

approximate graph of the displacement obtained by joining the points of coordinates $\left(\sum_{k=1}^{n} T_k, x_n\right)$ by sinusoidal arcs.

Table 3

n	1	2	3	4	5	6	7	8	9	10
ξ_n	1.440	−0.717	0.618	−0.436	0.396	−0.314	0.294	−0.246	0.231	−0.201
ξ_n'	−0.480	0.239	−0.206	0.145	−0.132	0.105	−0.098	0.082	−0.077	0.067
x_n	−3.00	1.49	−1.29	0.91	−0.83	0.66	−0.61	0.51	−0.48	0.42
$\frac{T_n}{T_0}$	1.048	1.006	1.016	1.004	1.010	1.003	1.008	1.002	1.006	1.002

When the damping characteristic is *symmetric* with respect to the origin, all results obtained in this section still hold by merely setting $\alpha = 1$ and hence $\xi_n = \xi'_n$ for any n. As an example of this kind we take $a_e = a_c = 0.24 \text{ cm}^{-1}$, $x_1 = -3$ cm. The corresponding calculation is indicated in Table 4. As expected, the decay of the oscillations is now more rapid, since the damping force produced during the compression stroke of the shock absorber is greater than in the preceding example.

Table 4

n	1	2	3	4	5	6	7	8	9	10
ξ_n	1.440	−0.717	0.482	−0.365	0.294	−0.246	0.211	−0.184	0.163	−0.146
ξ'_n	−1.440	0.717	−0.482	0.365	−0.294	0.246	−0.211	0.184	−0.163	0.146
x_n	−3.00	1.49	−1.01	0.76	−0.61	0.51	−0.44	0.38	−0.34	0.30
$\frac{T_n}{T_0}$	1.048	1.019	1.013	1.010	1.008	1.006	1.005	1.005	1.004	1.004

In applications concerning road vehicles, it is very important to know the extreme values of the acceleration, which determine the dynamic loading transmitted to the chassis and the body of the vehicle. To find these values, we first differentiate $(13.15)_1$ by taking into account (13.20), and obtain

$$\ddot{x} = -a_e\dot{x}^2 - \omega^2 x =$$

$$= -\frac{\omega^2}{2a_e}[1 + (2a_e x_n - 1)\, e^{2a_e(x_n - x)}], \qquad x_n < 0, \tag{13.38}$$

from which it follows that

$$\dddot{x} = \omega^2 \dot{x}(2a_e x_n - 1)\, e^{2a_e(x_n - x)}. \tag{13.39}$$

Consequently, $\dddot{x}$ vanishes together with $\dot{x}$, i.e., for $x = x_n$. Then, by putting $\dot{x} = 0$ and $x = x_n$ into $(13.38)_1$, we deduce

$$\ddot{x}_{\text{extr}} = -\omega^2 x_n = -\omega^2 x_{\text{extr}}. \tag{13.40}$$

The same reasoning may be repeated for $x_n > 0$ by starting from $(13.15)_2$ and (13.21), and leads to the same result (13.40). We see that the extreme values of the acceleration occur at the same time as those of the displacement, as in the case of simple harmonic motions. Equation (13.40) permits the easy calculation of the extreme values of the acceleration immediately after determining the extreme values of the displacement by the approximation method discussed above.

To close this section, we briefly review a linearizing procedure used sometimes in the study of vehicle vibrations.

When studying the bouncing and pitching motion of a road vehicle, the real damping characteristic of the suspension is frequently replaced by a linear one (see, e.g., ROTENBERG [174] and DERBAREMDIKER [43]). For instance, when the vehicle is considered as a single-degree-of-freedom system, one generally makes use of the equation of motion

$$\ddot{x} + 2h\dot{x} + \omega^2 x = 0,$$

whose well-known solution for $h < \omega$ is

$$x = A\mathrm{e}^{-ht} \sin(\omega_0 t + \gamma),$$

where

$$\omega_0 = \sqrt{\omega^2 - h^2} = \omega\sqrt{1 - \zeta^2}, \qquad \zeta = \frac{h}{\omega}.$$

As a parameter characterizing the damping of this "equivalent" linear system, one takes the so-called *damping ratio*[1)], ζ, as defined by the last relation above.

In the case of a linear viscous damping, the damping ratio ζ, and the *constant* ratio p of two consecutive extreme displacements of the same sign are related by the equation

$$\zeta = \frac{1}{\sqrt{1 + \dfrac{4\pi^2}{\ln^2 p}}}. \tag{13.41}$$

Some writers use this relation in order to define a linear oscillator "equivalent" with the nonlinear one. To this end, the ratio p is experimentally determined by exciting free vibrations of the vehicle, and then the equivalent damping ratio is calculated by Eq. (13.41) as if the damping characteristic were linear.

This kind of linearization is, of course, very expedient, but has some evident inconveniences. When designing suspensions with nonlinear damping, e.g., suspen-

1) In English literature this quantity is also termed *fraction of critical damping* (see, e.g., BLAKE [15]), whereas in Russian literature it is sometimes called *coefficient of aperiodicity* (see, e.g., ROTENBERG [174]), since it vanishes for the simple harmonic motion, equals 1 when the damping is critical, and is greater than 1 for aperiodic, i.e., nonoscillatory motions.

sions equipped with hydraulic shock absorbers, the parameters characterizing the damping are usually chosen by comparison with other suspensions that have given satisfactory results as regards the comfort and the dynamic loading of the vehicle [1]. When the method of linearization described above is applied, one compares the damping ratios of the equivalent linear systems instead of directly comparing the parameters defining the nonlinear damping. This repeated linearization may obviously lead to the loss of exactly those features that make a suspension better than others. On the other hand, as we have seen above, for systems with nonlinear damping, the ratio between two consecutive extreme displacements of the same sign is no longer constant, and hence ζ cannot be uniquely determined in this case by Eq. (13.41). For systems with symmetric or asymmetric quadratic damping such disadvantages may be eliminated, by first determining the parameters a_e, a_c, and ω from the oscillogram of the free vibrations of the vehicle, and then by directly comparing the values obtained for various suspensions.

In order to calculate the parameters a_e and a_c it is necessary to measure on the oscillogram the first three extreme values of the displacement, i.e., x_1, x_2, and x_3. Then, supposing that $x_1 < 0$, we know the ratios

$$\frac{\xi_2}{\xi_1} = \frac{-2a_e x_2}{-2a_e x_1} = \frac{x_2}{x_1} = K_{21},$$

$$\frac{\xi_3'}{\xi_2'} = \frac{2a_c x_3}{2a_c x_2} = \frac{x_3}{x_2} = K_{23}.$$

It now suffices to draw in Fig. 40 the straight lines passing through the origin and having the slopes K_{21} and K_{32}. The abscissas of the intersecting points of these lines with the graph will then be ξ_1 and ξ_2', respectively. Finally, the parameters a_e and a_c immediately result from the relations

$$a_e = \frac{\xi_1}{-2x_1}, \qquad a_c = \frac{\xi_2'}{2x_2}.$$

c) Use of the method of KRYLOV and BOGOLYUBOV

We have seen in § 11 how the formulas of the first and second approximations given by the method of KRYLOV and BOGOLYUBOV may be simplified when studying systems with linear elastic restoring force and weak damping.

To get a feeling for the kind of approximation obtained by this method, we first consider the *linear oscillator with weak viscous damping*, whose equation of motion is

$$\ddot{x} + 2h\dot{x} + \omega^2 x = 0. \tag{13.42}$$

[1] A direct procedure, based on the optimization of the damping characteristic of a nonlinear system acted on by random vibrations, will be discussed in Part III, Chapter III.

By introducing the dimensionless small parameter

$$\varepsilon = \frac{h}{\omega}, \tag{13.43}$$

and using the notations of § 11, we obtain $f_1(\dot{x}) = 2h\,\dot{x}$ and $f_1(\omega a \cos\alpha) = 2\omega^2 a \cos\alpha$. We then have

$$F_k(\omega a) = \frac{1}{\pi}\int_0^{2\pi} f_1(\omega a \cos\alpha)\cos k\alpha\, \mathrm{d}\alpha = \frac{2\omega^2 a}{\pi}\int_0^{2\pi} \cos\alpha \cos k\alpha\, \mathrm{d}\alpha,$$

from which it follows that

$$F_1(\omega a) = 2\omega^2 a, \qquad F_k(\omega a) = 0, \qquad k = 0, 2, 3, \ldots \tag{13.44}$$

By (11.11), we obtain for the second approximation

$$x = a\cos\alpha.$$

Substituting $(13.44)_1$ into (11.12) gives

$$t(a) = -\frac{1}{\omega\varepsilon}\int_{a_0}^{a}\frac{\mathrm{d}a}{a} = -\frac{1}{h}\ln\frac{a}{a_0},$$

and, by inversion,

$$a(t) = a_0 \mathrm{e}^{-ht}.$$

On the other hand, it results from (11.10) and (13.44) that

$$B_2(a) = -\frac{\omega}{2},$$

and then (11.13) with $t_0 = 0$ yields

$$\alpha(t) = \alpha_0 + \omega t\left(1 - \frac{h^2}{2\omega^2}\right).$$

Consequently, the second approximation is

$$x(t) = a_0\mathrm{e}^{-ht}\cos\left[\omega t\left(1 - \frac{h^2}{2\omega^2}\right) + \alpha_0\right]. \tag{13.45}$$

By comparing this approximate solution to the exact one,

$$x(t) = a_0\mathrm{e}^{-ht}\cos\left(\omega t\sqrt{1 - \frac{h^2}{\omega^2}} + \alpha_0\right),$$

we see that the second approximation gives the exact expression of the amplitude. As for the frequency, its approximation given by (13.45) may be obtained by developing the radical in the exact expression in a power series of h^2/ω^2 and retaining only the first two terms of the expansion.

As a second example we consider the *oscillator with linear elastic restoring force and weak asymmetric quadratic damping*, whose equation of motion is

$$\begin{aligned} \ddot{x} + a_e\dot{x}^2 + \omega^2 x = 0 \quad \text{for} \quad \dot{x} \geqslant 0, \\ \ddot{x} - a_c\dot{x}^2 + \omega^2 x = 0 \quad \text{for} \quad \dot{x} < 0. \end{aligned} \tag{13.46}$$

By introducing the dimensionless small parameter

$$\varepsilon = a_e a_0, \tag{13.47}$$

we have

$$f_1(\dot{x}) = \begin{cases} \dfrac{1}{a_0}\dot{x}^2 & \text{for} \quad \dot{x} \geqslant 0, \\ -\dfrac{a_c}{a_0 a_e}\dot{x}^2 & \text{for} \quad \dot{x} < 0, \end{cases}$$

from which, by considering the amplitude a of the fundamental harmonic as positive, it follows that

$$f_1(\omega a \cos\alpha) = \begin{cases} \dfrac{\omega^2 a^2}{a_0}\cos^2\alpha & \text{for} \quad \cos\alpha \geqslant 0, \\ -\dfrac{\omega^2 a^2 a_c}{a_0 a_e}\cos^2\alpha & \text{for} \quad \cos\alpha < 0. \end{cases}$$

We now successively obtain

$$F_k(\omega a) = \frac{\omega^2 a^2}{\pi a_0}\left[\int_0^{\frac{\pi}{2}} \cos^2\alpha \cos k\alpha \, d\alpha - \frac{a_c}{a_e}\int_{\frac{\pi}{2}}^{\frac{3\pi}{2}} \cos^2\alpha \cos k\alpha \, d\alpha + \right.$$

$$\left. + \int_{\frac{3\pi}{2}}^{2\pi} \cos^2\alpha \cos k\alpha \, d\alpha \right] = \frac{\omega^2 a^2}{2\pi a_0}\left(1 + \frac{a_c}{a_e}\right)\left\{\frac{1}{k}\left(\sin\frac{k\pi}{2} - \right.\right.$$

$$\left. - \sin\frac{3k\pi}{2}\right) + \frac{1}{2(k+2)}\left[\sin\frac{(k+2)\pi}{2} - \sin\frac{3(k+2)\pi}{2}\right] +$$

$$\left. + \frac{1}{2(k-2)}\left[\sin\frac{(k-2)\pi}{2} - \sin\frac{3(k-2)\pi}{2}\right]\right\} =$$

$$= \frac{\omega^2 a^2}{\pi a_0}\left(1 + \frac{a_c}{a_e}\right)\left[-\frac{1}{k} + \frac{1}{2(k+2)} + \frac{1}{2(k-2)}\right] \sin\frac{k\pi}{2} \cos k\pi,$$

and hence

$$F_1(\omega a) = \frac{4\omega^2 a^2}{3\pi a_0}\left(1 + \frac{a_c}{a_e}\right), \tag{13.48}$$

$$\left.\begin{aligned} F_{2p+1}(\omega a) &= \frac{4\omega^2 a^2}{\pi a_0}\left(1 + \frac{a_c}{a_e}\right)\cdot\frac{(-1)^{p+1}}{(2p-1)(2p+1)(2p+3)}, \\ F_{2p}(\omega a) &= 0, \qquad p = 1, 2, \ldots \end{aligned}\right\} \tag{13.49}$$

Since

$$\cos(2p+1)\left(\alpha + \frac{\pi}{2}\right) = -\sin[(2p+1)\alpha + p\pi] = (-1)^{p+1}\sin(2p+1)\alpha,$$

we have from (11.11)

$$x = a\cos\alpha + \frac{4a^2(a_e + a_c)}{\pi}\sum_{p=1}^{\infty}\frac{\sin(2p+1)\alpha}{(2p-1)\,2p\,(2p+1)(2p+2)(2p+3)} =$$

$$= a\cos\alpha + \frac{4a^2(a_e + a_c)}{\pi}\left(\sin 3\alpha + \frac{1}{21}\sin 5\alpha + \ldots\right). \tag{13.50}$$

On the other hand, substituting (13.48) and (13.49) into (11.10) yields

$$B_2(a) = -\frac{\omega a^2 C}{\pi^2 a_0^2}\left(1 + \frac{a_c}{a_e}\right)^2,$$

where

$$C = \sum_{p=1}^{\infty}\frac{1}{[(2p+1)^2 - 1][(2p+1)^2 - 4]^2} = \frac{1}{25} + \frac{1}{1323} + \frac{1}{12150} + \ldots$$

$$= 0.0407\ldots.$$

We then have from (11.12) with $t_0 = 0$

$$t(a) = -\frac{3\pi}{2\omega(a_e + a_c)}\int_{a_0}^{a}\frac{da}{a^2} = \frac{3\pi}{2\omega(a_e + a_c)}\left(\frac{1}{a} - \frac{1}{a_0}\right),$$

and, by inversion,

$$a(t) = \frac{a_0}{1 + \dfrac{2\omega(a_e + a_c)a_0}{3\pi}t}. \tag{13.51}$$

Finally, we obtain by (11.13)

$$\alpha(t) = \alpha_0 + \omega t - \frac{\omega C(a_e + a_c)^2}{\pi^2} \int_0^t [a(t)]^2 \, dt =$$

$$= \alpha_0 + \omega t - \frac{\omega C(a_e + a_c)^2 a_0^2}{\pi^2} \int_0^t \frac{dt}{\left[1 + \dfrac{2\omega(a_e + a_c)a_0}{3\pi} t\right]^2},$$

which gives

$$\alpha(t) = \alpha_0 + \omega t + \frac{3C(a_e + a_c)a_0}{2\pi} \left[\frac{1}{1 + \dfrac{2\omega(a_e + a_c)a_0}{3\pi} t} - 1\right]. \tag{13.52}$$

Therefore, the second approximation is given by (13.50), where $a(t)$ and $\alpha(t)$ have to be replaced by (13.51) and (13.52), respectively.[1)]

To prove once more the strength of the method of KRYLOV and BOGOLYUBOV let us compare the *first* approximation given by this method, namely

$$x(t) = \frac{a_0}{1 + \dfrac{2\omega(a_e + a_c)\, a_0}{3\pi} t} \cos(\omega t + \alpha_0), \tag{13.53}$$

to the exact solution obtained in the preceding section.

Taking $\alpha_0 = \pi$, we derive from (13.53) the expressions of the first two extreme displacements of the same sign, by setting $t = 0$ for $x_1 < 0$, and $t = 2\pi/\omega$ for $x_3 < 0$:

$$x_1 = -a_0,$$

$$x_3 = -\frac{a_0}{1 + \dfrac{4(a_e + a_c)a_0}{3}} = \frac{x_1}{1 - \dfrac{4(a_e + a_c)x_1}{3}}. \tag{13.54}$$

On the other hand, we have by (13.33)

$$\xi_2 = -\frac{\xi_1}{1 + \dfrac{2}{3}\xi_1}, \qquad \xi_3' = -\frac{\xi_2'}{1 + \dfrac{2}{3}\xi_2'}.$$

1) By putting $a_e = a_c$ into these formulas, we recover the results obtained by BOGOLYUBOV and MITROPOLSKY [16] for the system with symmetric quadratic damping.

Since $\xi_1 = -2a_e x_1$, $\xi_2 = -2a_e x_2$, $\xi_2' = 2a_c x_2$, $\xi_3' = 2a_c x_3$, it follows that

$$x_2 = -\frac{x_1}{1 - \dfrac{4a_e}{3} x_1}, \qquad x_3 = -\frac{x_2}{1 + \dfrac{4a_c}{3} x_2},$$

and, by eliminating x_2, we find the same relation (13.54) between x_3 and x_1. We have seen in the preceding section that Eq. (13.33) yields the ratio $|x_{n+1}/x_n|$ to an error of 1% if $2a_c|x_n| < 1$ and $2a_e|x_n| < 1$. It appears, therefore, that Eq. (13.54) obtained by the method of KRYLOV and BOGOLYUBOV gives the ratio $|x_{n+2}/x_n|$ to an error of 2%, provided $|x_n|$ satisfies the same conditions. Moreover, we conclude that the method of KRYLOV and BOGOLYUBOV provides a very good first approximation in the case of a comparatively strong nonlinearity, too, on condition that the initial displacement and velocity be small enough.

As a last example, consider the *oscillator with linear elastic restoring force and symmetric parabolic damping* (with linear and quadratic terms), whose equation of motion is

$$\ddot{x} + 2h\dot{x} + a_e\dot{x}|\dot{x}| + \omega^2 x = 0. \tag{13.55}$$

By introducing the small dimensionless parameter

$$\varepsilon = \frac{h}{\omega}, \tag{13.56}$$

we have

$$f_1(\omega a \cos\alpha) = 2\omega^2 a \cos\alpha + \frac{\omega^3 a_e a^2}{h} \cos\alpha\,|\cos\alpha|\,. \tag{13.57}$$

To the first approximation we obtain

$$F_1(\omega a) = \frac{1}{\pi}\int_0^{2\pi} f_1(\omega a\cos\alpha)\cos\alpha\, d\alpha = \frac{2\omega^2 a}{\pi}\int_0^{2\pi}\cos^2\alpha\, d\alpha +$$

$$+\frac{\omega^3 a_e a^2}{\pi h}\left[\int_0^{\frac{\pi}{2}}\cos^3\alpha\, d\alpha - \int_{\frac{\pi}{2}}^{\frac{3\pi}{2}}\cos^3\alpha\, d\alpha + \int_{\frac{3\pi}{2}}^{2\pi}\cos^3\alpha\, d\alpha\right],$$

whence

$$F_1(\omega a) = 2\omega^2 a\left(1 + \frac{4\omega a_e a}{3\pi h}\right). \tag{13.58}$$

From (11.6) with $t_0 = 0$ it follows that

$$t(a) = -\frac{1}{\omega\varepsilon}\int_{a_0}^{a}\frac{da}{a\left(1+\dfrac{4\omega a_e a}{3\pi h}\right)} = -\frac{1}{h}\ln\frac{a\left(1+\dfrac{4\omega a_e a_0}{3\pi h}\right)}{a_0\left(1+\dfrac{4\omega a_e a}{3\pi h}\right)},$$

and, by inversion,

$$a(t) = \frac{a_0 e^{-ht}}{1+\dfrac{4\omega a_e a_0}{3\pi h}(1-e^{-ht})}. \tag{13.59}$$

Finally, by substituting (13.59) and (11.7) into (11.5), we obtain the first approximation for the solution

$$x(t) = \frac{a_0 e^{-ht}}{1+\dfrac{4\omega a_e a_0}{3\pi h}(1-e^{-ht})}\cos(\omega t+\alpha_0). \tag{13.60}$$

d) Equivalent linearization

We close this section by briefly mentioning the results obtained by applying the method of equivalent linearization to some dissipative systems.

For the *oscillator with linear elastic restoring force and asymmetric quadratic damping*, whose equation of motion is (13.46), we have by putting (13.49) into (11.4)

$$A_1(a) = -\frac{F_1(\omega a)}{2\omega} = -\frac{2\omega a^2}{3\pi a_0}\left(1+\frac{a_c}{a_e}\right), \qquad B_1(a) = 0.$$

By taking now into account (13.47), it follows from $(5.64)_1$ and $(5.65)_1$ that the parameters of the equivalent linear system are

$$\bar{h}(a) = \frac{2\omega a(a_e+a_c)}{3\pi}, \qquad \bar{\omega}(a) = \omega. \tag{13.61}$$

For the *oscillator with linear elastic restoring force and symmetric parabolic damping* (with linear and quadratic terms), whose equation of motion is (13.55), we deduce from (11.4) and (13.58) that

$$A_1(a) = -\frac{F_1(a)}{2\omega} = -\omega a\left(1+\frac{4\omega a_e a}{3\pi h}\right), \qquad B_1(a) = 0$$

and, considering (13.56), we obtain by $(5.64)_1$ and $(5.65)_1$ the parameters of the equivalent linear system

$$\bar{h}(a) = h + \frac{4\omega a_e a}{3\pi}, \qquad \bar{\omega}(a) = \omega. \tag{13.62}$$

Finally, for the *oscillator with cubic elastic restoring force and symmetric parabolic damping* (with linear and quadratic terms), whose equation of motion is

$$\ddot{x} + 2h\dot{x} + a_e\dot{x}|\dot{x}| + \omega^2 x(1 + \beta x^2) = 0, \tag{13.63}$$

we may combine the results obtained in the preceding example with those from § 9d, since, as it is readily seen from (5.64) and (5.65), the functions f_1 and ψ_1 separately contribute to $\bar{h}(a)$ and $\bar{\omega}(a)$, respectively. It then follows, by $(13.62)_1$ and (9.50), that the equivalent linear system has the parameters

$$\bar{h}(a) = h + \frac{4\omega a_e a}{3\pi}, \qquad \bar{\omega}^2(a) = \omega^2\left(1 + \frac{3}{4}\beta a^2\right). \tag{13.64}$$

§ 14. THE OSCILLATOR WITH LINEAR ELASTIC RESTORING FORCE AND PIECEWISE LINEAR OR QUADRATIC DAMPING

As we have seen at the beginning of the preceding section, it is sometimes very difficult to represent the real characteristic of a damper through a single analytic expression. In such cases it is, therefore, preferable to approximate piecewise the characteristic by line segments and parabolic arcs. We assume, therefore, that the characteristic is given by equations of the form

$$f(v) = av^2 + 2hv + b, \tag{14.1}$$

and we divide the v-axis into subintervals, by the points at which the coefficients a, h and b change their values; for the sake of convenience we always take the origin as one of the dividing points. The only restriction imposed upon the coefficients of the quadratic function (14.1) are

$$h \geqslant 0, \qquad av \geqslant 0. \tag{14.2}$$

Suppose the intial conditions are

$$x(0) = x_0, \qquad \dot{x}(0) = v_0. \tag{14.3}$$

By considering (14.1), the equation of motion may be written as

$$\ddot{x} + a\dot{x}^2 + 2h\dot{x} + b + \omega^2 x = 0 \qquad \text{for} \qquad \dot{x} \in [v', v''], \tag{14.4}$$

where $[v', v'']$ is any interval for which the coefficients a, h, and b have constant values.

It generally suffices to approximate the characteristic by line segments ($a=b=0$) and by parabolic arcs of quadratic equation without linear term ($h=0$). Therefore, we shall confine ourselves to considering these two cases.

We will show in the following a way of determining the solution of Eq. (14.4) in the phase plane,

$$v = v(x), \tag{14.5}$$

and of calculating the times at which the velocity leaves the interval considered taking one of the values v' or v''. This "limiting" value of the velocity and the corresponding displacement will be denoted by v_l and x_l, respectively. The interval of time in which the velocity varies from v_0 to v_l may be subsequently calculated by one of the approximate methods set forth in § 3. When determining the solution of Eq. (14.4) for some interval, we shall take, of course, as initial values, x_0 and v_0, the "limiting" values x_l and v_l found for the preceding interval.

a) The line segments of the damping characteristic

For a line segment of the damping characteristic, we have $a = b = 0$; the equation of motion (14.4) becomes

$$\ddot{x} + 2h\dot{x} + \omega^2 x = 0 \qquad \text{for} \qquad \dot{x} \in [v', v'']. \tag{14.6}$$

The solution of this equation satisfying the initial conditions (14.3) is in the oscillatory case ($h < \omega$):

$$x = \mathrm{e}^{-ht}\left(x_0 \cos \omega_0 t + \frac{v_0 + hx_0}{\omega_0} \sin \omega_0 t\right), \tag{14.7}$$

where

$$\omega_0 = \sqrt{\omega^2 - h^2}. \tag{14.8}$$

By differentiating with respect to t, we obtain

$$v = \mathrm{e}^{-ht}\left(v_0 \cos \omega_0 t - \frac{hv_0 + \omega^2 x_0}{\omega_0} \sin \omega_0 t\right). \tag{14.9}$$

Equations (14.7) and (14.8) provide a parametric representation of the phase trajectory with the time t as parameter. To find the displacement corresponding to $v = v_l$, we first have to solve the transcendental equation

$$v_l = e^{-ht}\left(v_0 \cos \omega t - \frac{hv_0 + \omega^2 x_0}{\omega_0} \sin \omega_0 t\right), \tag{14.10}$$

and then to introduce the value of t so obtained into (14.7). In particular, if $v_0 = 0$, the solution $t = t_l$ of (14.10) may be directly deduced from the simplified condition

$$\tan \omega_0 t_l = \frac{\omega_0 v_0}{hv_0 + \omega^2 x_0}. \tag{14.11}$$

If v attains an extreme value on the interval $[v', v'']$, then the corresponding time may be calculated from the condition $v = 0$, which leads, after some manipulation, to

$$\tan \omega_0 t = \frac{\omega_0(2hv_0 + \omega^2 x_0)}{(h^2 - \omega_0^2)v_0 + h\omega^2 x_0}. \tag{14.12}$$

Introducing the solution of this equation into (14.9) and (14.7) yields the extreme value of the velocity and the corresponding value of the displacement. As for the extreme values of the acceleration, if any, we can derive two evaluations. First, by eliminating $\ddot{x}$ between (14.7) and the equation obtained from it by differentiating with respect to t, it follows that

$$\dddot{x} = (4h^2 - \omega^2)\,\dot{x} + 2h\omega^2 x.$$

The acceleration takes an extreme value if $\dddot{x} = 0$, hence for

$$\dot{x} = -\frac{2h\omega_0^2}{4h^2 - \omega_0^2}\,x.$$

Introducing this value into (14.6) gives

$$\ddot{x}_{\text{extr}} = -\frac{\omega_0^4 x}{\omega_0^2 - 4h^2}.$$

On the other hand, by putting $\dot{x} = 0$ into (14.6), we see that the extreme values of the displacement, x_{extr}, and the corresponding values of the acceleration are related by

$$\ddot{x} = -\omega_0^2 x_{\text{extr}}.$$

Assuming that the system is underdamped, i.e., $\omega_0 > 2h$, and taking into account that for each half-oscillation $|\ddot{x}| \leqslant |\ddot{x}_{\text{extr}}|$, $|x| \leqslant |x_{\text{extr}}|$, we deduce from the last two equations that

$$1 \leqslant \left| \frac{\ddot{x}_{\text{extr}}}{\omega_0^2 x_{\text{extr}}} \right| \leqslant \frac{\omega_0^2}{\omega_0^2 - 4h^2}.$$

We emphasize that these evaluations are valid only if both the acceleration and the displacement take extreme values in the interval of velocities in which the equation of motion has the form (14.6).

b) The parabolic arcs of the damping characteristic

For a parabolic arc of the damping characteristic, the equation of motion has the form

$$\ddot{x} + a\dot{x}^2 + b + \omega^2 x = 0 \qquad \text{for} \qquad \dot{x} \in [v', v''],$$

in which $a\dot{x} > 0$. Since

$$\ddot{x} = v \frac{\mathrm{d}v}{\mathrm{d}x} = \frac{1}{2} \frac{\mathrm{d}v^2}{\mathrm{d}x}, \tag{14.13}$$

the equation of motion may be written as

$$\frac{\mathrm{d}v^2}{\mathrm{d}x} + 2av^2 + 2(\omega^2 x + b) = 0, \tag{14.14}$$

and the initial condition (14.3) becomes

$$v(x_0) = v_0. \tag{14.15}$$

Since the differential equation (14.14) is linear in v^2 and has constant coefficients, its solution satisfying (14.15) may be immediately found to be

$$v^2 = \frac{\omega^2}{2a^2} \left[1 - 2ax - \frac{2ab}{\omega^2} - \left(1 - 2ax_0 - \frac{2ab}{\omega^2} - \frac{2a^2 v_0^2}{\omega^2} \right) \mathrm{e}^{-2a(x - x_0)} \right]. \tag{14.16}$$

To determine the displacement x_l corresponding to the limiting value $v_l = v'$ or v'' of the velocity, we have to solve Eq. (14.16) for x, with $v = v_l$. Introducing into (14.16) the dimensionless quantities

$$w_l = 1 - 2ax_l - \frac{2ab}{\omega^2} - \frac{2a^2 v_l^2}{\omega^2}, \tag{14.17}$$

$$w_0 = 1 - 2ax_0 - \frac{2ab}{\omega^2} - \frac{2a^2 v_0^2}{\omega^2}, \tag{14.18}$$

we obtain

$$w_l = w_0 e^{-2a(x_l - x_0)}. \tag{14.19}$$

From (14.17) and (14.18) it follows that

$$-2a(x_i - x_0) = w_l - w_0 + \frac{2a^2}{w^2}(v_l^2 - v_0^2),$$

and hence (14.19) becomes

$$\frac{e^{w_l}}{w_l} = \frac{e^{w_0}}{w_0} \cdot e^{\frac{2a^2}{\omega^2}(v_l^2 - v_0^2)}.$$

Finally, by taking the logarithms of both sides, we have

$$\Phi(w_i) = \Phi(w_0) + A, \tag{14.20}$$

with the notations

$$\Phi(w) \equiv w - \ln |w|, \tag{14.21}$$

$$A = \frac{2a^2}{\omega^2}(v_0^2 - v_l^2). \tag{14.22}$$

If the values of x_0, v_0, and v_l are given, we have to solve Eq. (14.20) for w_l, and then to find x_l from (14.17). Equation (14.20) may be conveniently solved by using Tables 5 and 6, in which the values of $\Phi(v)$ have been listed for $w \in [-10, 0]$ and $w \in [0, 10]$, respectively. The solving procedure may be also followed in Fig. 43, where the graph of Φ versus w has been plotted. Calculate first the values of w_0 and A by (14.18) and (14.22), respectively. Next, determine $\Phi(w_0)$ by using Table 5 or 6, calculate $\Phi(w_l)$ from (14.20), and find the corresponding value of w_l, by making use of the same tables. Finally, derive x_l from (14.17).

It is readily seen from (14.19) that w_l has the same sign as w_0. Moreover, since both a and $x_l - x_0$ have the same sign as v, it follows that $|w_l| < |w_0|$. These two conditions uniquely determine the root of Eq. (14.20), as may be also seen from Fig. 43.

If the velocity takes an extreme value in the interval considered, this may be calculated by a rather simple formula. Indeed, by using the notation (14.18), we obtain from (14.13), (14.14), and (14.16)

$$\ddot{x} = \frac{\omega^2}{2a}[w_0 e^{-2a(x - x_0)} - 1]. \tag{14.23}$$

Table 5

w	0	1	2	3	4	5	6	7	8	9
0.0	$+\infty$	2.2026	1.4095	0.9040	0.5163	0.1932	— 0.0892	— 0.3433	— 0.5768	— 0.7936
—1.0	— 1.0000	— 1.1953	— 1.3823	— 1.5624	— 1.7365	— 1.9055	— 2.0700	— 2.2306	— 2.3878	— 2.5419
—2.0	— 2.6931	— 2.8419	— 2.9885	— 3.1329	— 3.2755	— 3.4163	— 3.5555	— 3.6933	— 3.8296	— 3.9647
—3.0	— 4.0986	— 4.2314	— 4.3632	— 4.4939	— 4.6238	— 4.7528	— 4.8809	— 5.0083	— 5.1350	— 5.2610
—4.0	— 5.3863	— 5.5110	— 5.6351	— 5.7586	— 5.8816	— 6.0041	— 6.1261	— 6.2476	— 6.3686	— 6.4892
—5.0	— 6.6094	— 6.7292	— 6.8487	— 6.9677	— 7.0864	— 7.2047	— 7.3228	— 7.4405	— 7.5579	— 7.6750
—6.0	— 7.7918	— 7.9083	— 8.0245	— 8.1405	— 8.2563	— 8.3718	— 8.4871	— 8.6021	— 8.7169	— 8.8315
—7.0	— 8.9459	— 9.0601	— 9.1741	— 9.2879	— 9.4015	— 9.5149	— 9.6281	— 9.7412	— 9.8541	— 9.9669
—8.0	—10.0794	—10.1919	—10.3041	—10.4163	—10.5282	—10.6401	—10.7518	—10.8633	—10.9748	—11.0861
—9.0	—11.1972	—11.3083	—11.4192	—11.5300	—11.6407	—11.7513	—11.8618	—11.9721	—12.0824	—12.1925

Table 6

w	0	1	2	3	4	5	6	7	8	9
0.0	$+\infty$	2.4026	1.8095	1.5040	1.3163	1.1932	1.1108	1.0567	1.0232	1.0054
1.0	1.0000	1.0047	1.0177	1.0376	1.0635	1.0945	1.1300	1.1694	1.2122	1.2581
2.0	1.3069	1.3581	1.4115	1.4671	1.5245	1.5837	1.6445	1.7067	1.7704	1.8353
3.0	1.9014	1.9686	2.0368	2.1061	2.1762	2.2472	2.3191	2.3917	2.4650	2.5390
4.0	2.6137	2.6890	2.7649	2.8414	2.9184	2.9959	3.0739	3.1524	3.2314	3.3108
5.0	3.3906	3.4708	3.5513	3.6323	3.7136	3.7953	3.8772	3.9595	4.0421	4.1250
6.0	4.2082	4.2917	4.3755	4.4595	4.5437	4.6282	4.7129	4.7979	4.8831	4.9685
7.0	5.0541	5.1399	5.2259	5.3121	5.3985	5.4851	5.5719	5.6588	5.7459	5.8331
8.0	5.9206	6.0081	6.0959	6.1837	6.2718	6.3599	6.4482	6.5367	6.6252	6.7139
9.0	6.8028	6.8917	6.9808	7.0700	7.1593	7.2487	7.3382	7.4279	7.5176	7.6075

Equating this expression to zero yields the displacement corresponding to the extreme value of the velocity

$$x = x_0 + \frac{1}{2a} \ln w_0, \qquad w_0 > 0. \tag{14.24}$$

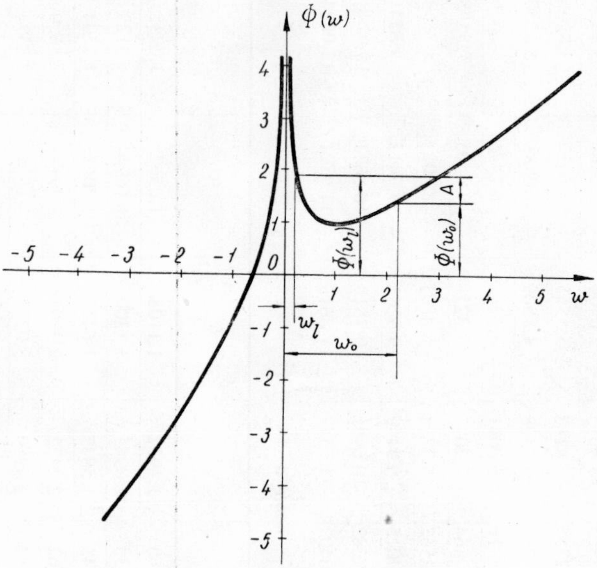

Fig. 43

Finally, by introducing (14.24) into (14.16), we find the square of the extreme velocity

$$v_{\text{extr}}^2 = -\frac{\omega^2}{2a^2}\left(2ax_0 + \ln w_0 + \frac{2ab}{\omega^2}\right), \tag{14.25}$$

which, by (14.18) and (14.21), may be also written as

$$v_{\text{extr}}^2 = v_0^2 + \frac{\omega^2}{2a^2}[\Phi(w_0) - 1]. \tag{14.26}$$

To obtain now the extreme values of the acceleration, we first deduce from (14.23)

$$\dddot{x} = -\omega^2 w_0 \dot{x} e^{-2a(x-x_0)}, \tag{14.27}$$

from which it follows that $\dddot{x}$ vanishes together with $\dot{x}$. Consequently, the acceleration and the displacement simultaneously take extreme values. Moreover, we see from the equation of motion that these extreme values are related by

$$\ddot{x}_{\text{extr}} = -\omega^2 x_{\text{extr}}. \tag{14.28}$$

c) Numerical example

To illustrate the results obtained above, we will give a numerical example concerning the suspension of a truck equipped with hydraulic shock absorbers. The real damping characteristic may be approximated by two line segments and four parabolic arcs, as shown in Fig. 44. The values of the coefficients a, h, and b, corresponding to the six intervals, have been listed in table 7.

We take $\omega = 10$ rad/sec and

$$x_0 = -10 \text{ cm}, \qquad v_0 = 0. \tag{14.29}$$

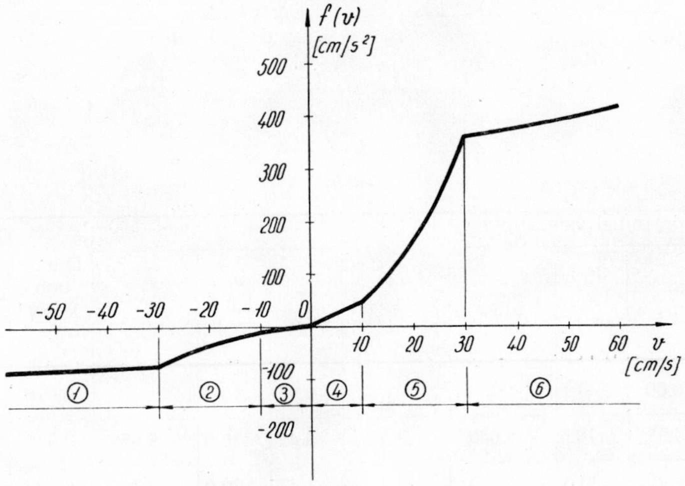

Fig. 44

In table 8 we have indicated the initial and final (limiting) values of the displacement and of the velocity corresponding to each interval. We have also listed the values of some secondary quantities occurring in the calculation, such as w_0, w_l, A, as well as the duration Δt of the motion over each interval. We now show how these values have actually been calculated for the first three intervals.

Interval 4: $v \in [0, 10]$, $x_0 = -10$ cm, $v_0 = 0$, $v_l = 10$ cm/sec. From (14.8) it follows that $\omega_0 = 9.798$ rad/sec. By solving Eq. (14.10), we obtain $\Delta t = t_l = 0.02$ sec, and, by (14.7), we find the corresponding value of the displacement, $x_l = -9.95$ cm.

Interval 5: $v \in [10, 30]$, $x_0 = -9.95$ cm, $v_0 = 10$ cm/sec, $v_l = 30$ cm/sec. From (14.18) and (14.22) we deduce $w_0 = 8.640$ and $A = -2.560$. With the aid of Table 7 we find $\Phi(w_l) = \Phi(8.640) - 2.560 = 3.9236$, and then $w_l = 5.656$, and we obtain by (14.17) the corresponding value of the displacement, $x_l = -9.42$ cm. Finally, by using (3.14), we calculate the approximate duration of the motion over this interval: $\Delta t = 0.025$ sec.

Table 7

Interval	v' [cm/sec]	v'' [cm/sec]	h [s^{-1}]	a [cm^{-1}]	b [cm/sec]
1	—∞	—30	0	—0.004	—68.4
2	−30	−10	0	−0.080	0
3	−10	0	0.4	0	0
4	0	10	2.0	0	0
5	10	30	0	0.400	0
6	30	∞	0	0.020	342.0

Table 8

Velocity interval	Initial values			A	Final values			Duration Δt[sec]	v_{extr} [cm/sec]	Corresponding displacement [cm]
	x_0 [cm]	v_0 [cm/sec]	w_0		x_l [cm]	v_l [cm/sec]	w_l			
4	—10.00	0	—	—	—9.95	10.0	—	0.010	—	—
5	— 9.95	10.0	8.640	—2.560	—9.42	30.0	5.656	0.025	—	—
6	— 9.42	30.0	1.233	0	1.47	30.0	0.797	0.222	62.6	—4.17
5	1.47	30.0	—3.016	2.560	2.46	10.0	—1.288	0.050	—	—
4	2.46	10.0	—	—	2.67	0	—	0.036	—	—
3	2.67	0	—	—	2.47	—10.0	—	0.039	—	—
2	2.47	—10.0	1.382	0	—1.77	—10.0	0.704	0.222	—23.7	0.45
3	— 1.77	—10.0	—	—	—2.02	0	—	0.050	—	—
4	— 2.02	0	—	—	—1.71	10.0	—	0.059	—	—
5	— 1.71	10.0	2.050	0	0.36	10.0	0.391	0.162	14.3	—0.81
4	0.36	10.0	—	—	0.92	0	—	0.107	—	—
3	0.92	0	—	—	—0.81	0	—	0.314	—8.6	0.07
4	— 0.81	0	—	—	0.43	0	—	0.321	6.0	—0.24

Interval 6: $v \in [30, \infty)$, $x_0 = -9.42$ cm, $v_0 = v_l = 30$ cm/sec. From (14.18) and (14.22) it results $w_0 = 1.2328$, $A = 0$, and, with the aid of Table 7, we find $\Phi(w_l) = 1.0242$ and $w_l = 0.797$. Next, by (14.17) and (3.14), it follows that $x_l = 1.47$ cm and $\Delta t = 0.222$ sec. Then, by (14.24) and (14.25), we deduce that in this interval the velocity takes the extreme value $v_{\text{extr}} = 62.6$ cm/sec and that the corresponding value of the displacement is $x = -4.17$ cm, and so on.

The extreme value of the velocity must be provisionally calculated by (14.12), (14.9), or (14.26) for each interval, and then, by comparing the value obtained with v' and v'', it may be seen if the velocity actually takes this extreme value in the interval considered or not, and if the motion leaves the interval to the right or to the left.

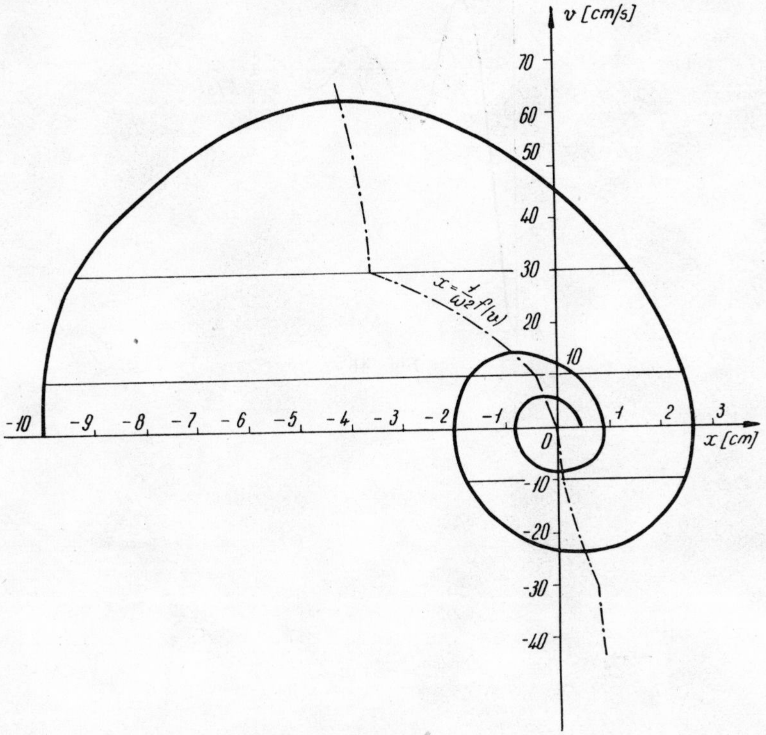

Fig. 45

Figure 45 shows the phase trajectory and the curve of the extreme values of the velocity (dashed line), whose equation (2.8) becomes now

$$x = -\frac{1}{\omega^2} f(x).$$

The trajectory has horizontal tangent at the points where it intersects this curve, and vertical tangent at the points where it cuts the x-axis (cf. theorems 4 and 2,

§ 2). In Fig. 46 we have drawn the approximate graph of $x(t)$; the points of the graph determined by the calculation above have been joined by sinusoidal arcs, with the exception of the parts of the graph corresponding to line segments of the damping characteristic, which have been drawn by using the exact dependence of x on t.

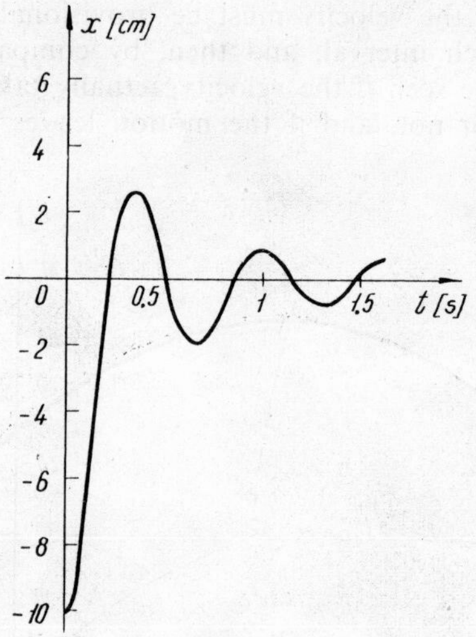

Fig. 46

PART II

Forced vibrations of nonlinear systems acted on by deterministic excitations

Part I has been devoted to free vibrations of nonlinear systems. In Part II we assume the nonlinear oscillators as being acted on by exciting forces that are deterministic functions of t. Denoting such a force by $\mathscr{P}(t)$, the governing equation of motion of the oscillator reads

$$m\ddot{x} + \mathscr{F}(\dot{x}) + \mathscr{G}(x) = \mathscr{P}(t),$$

where $\mathscr{F}(\dot{x})$ is the damping force and $\mathscr{G}(x)$ is the elastic restoring force.

In Chapter IV we shall discuss the qualitative properties of the solutions, with special stress on the existence and stability of periodic solutions, and we shall set forth a few methods permitting the approximate determination of such periodic solutions.

In the following two chapters we shall apply the analytical methods expounded in Chapter IV to the study of periodic oscillations of conservative and dissipative nonlinear systems. We shall also compare the results obtained by various methods, particularly by emphasizing the use of the perturbation method and of the method of KRYLOV and BOGOLYUBOV.

For the sake of completeness, we shall deal with subharmonic vibrations in § 19 and in some other examples, although such vibrations are less important for mechanical than for electrical oscillators.

Chapter IV

General methods of study

§ 15. GENERAL THEOREMS ON THE PROPERTIES OF THE SOLUTIONS

a) Existence, boundedness, and uniqueness of the solutions

Denoting

$$g(x) \equiv \frac{\mathscr{G}(x)}{m}, \qquad f(\dot{x}) \equiv \frac{\mathscr{F}(\dot{x})}{m}, \qquad p(t) \equiv \frac{\mathscr{P}(t)}{m}, \tag{15.1}$$

the equation of forced vibrations of our nonlinear oscillator takes the form

$$\ddot{x} + f(\dot{x}) + g(x) = p(t). \tag{15.2}$$

The initial conditions

$$x(t_0) = x_0, \qquad \dot{x}(t_0) = v_0 \tag{15.3}$$

define a Cauchy problem for this differential equation.

Instead of Eq. (15.2) we could study an equivalent system of two differential equations, by putting $\dot{x} \equiv v$, as we have done in § 1a for the case of free vibrations. However, we prefer to state all theorems about the qualitative properties of the solutions directly for Eq. (15.2), in order to facilitate reference to other standard works.

Theorems 1 and 2 below concern the existence and uniqueness of the solution of Eq. (15.2) satisfying the initial conditions (15.3). More general theorems are stated and proved in standard books on differential equations (see, e.g., NEMYTSKY and STEPANOV [131], Chapter I, CESARI [32] § 1, or HALANAY [64]).

Theorem 1. *If $g(x)$, $f(\dot{x})$, and $p(t)$ are continuous in the intervals $|x - x_0| \leqslant a$, $|\dot{x} - v_0| \leqslant b$, and $|t - t_0| \leqslant c$, respectively, then there exists at least a solution of Eq. (15.2), which is defined in an interval $|t - t_0| \leqslant t_1$, with $0 < t_1 < c$, and satisfies the initial conditions (15.3).*

The number t_1 occurring in the statement above depends on a, b, c, and on the maxima of the functions $|g(x)|$, $|f(\dot{x})|$, and $|p(t)|$ in the intervals $|x - x_0| \leqslant a$, $|\dot{x} - v_0| \leqslant b$, and $|t - t_0| \leqslant c$, respectively. If g, f, and p are continuous for all real values of their arguments, then the following theorem, which is analogous to Theorem 2, § 1, can also be proved.

Theorem 2. *If $g(x), f(\dot{x})$, and $p(t)$ are continuous for $x \in (-\infty, \infty)$, $\dot{x} \in (-\infty, \infty)$, and $t \in [t_0, \infty)$, respectively, then any solution of (15.2) which is bounded together with its derivative for $t \geqslant t_0$ exists in the whole interval $[t_0, \infty)$.*

In order to apply this theorem and to solve some other related problems, it is necessary to study the boundedness of the solutions. In case f, g, and p satisfy our hypotheses H_1 and H_5, the boundedness of x and $\dot{x}$ is assured by the following theorem due to REUTER [125].

Theorem 3. *If $g(x)$ and $f(\dot{x})$ are continuous for $x \in (-\infty, \infty)$ and $\dot{x} \in (-\infty, \infty)$, respectively, if $p(t)$ is continuous and bounded for $t \in [t_0, \infty)$, and if $\lim_{x \to \pm\infty} g(x) = \pm\infty$, $\lim_{\dot{x} \to \pm\infty} f(\dot{x}) = \pm\infty$, then there exist two positive constants,* A and B, *such that for every solution $x(t)$ of Eq. (15.2) there exists a corresponding T_0 such that*

$$|x(t)| \leqslant A, \quad |\dot{x}(t)| \leqslant B \tag{15.4}$$

for $t \geqslant T_0$.

For other more general statements of this kind, which are less restrictive as regards the functions f and g, we refer the reader to the paper by REUTER cited above and to a study by REISSIG [161]. Various forms of Theorem 3, which hold under more general hypotheses concerning one of the functions f and g, in case the other one is linear, have been given by CARTWRIGHT and LITTLEWOOD [24], REUTER [164], MIZOHATA and YAMAGUTI [164], ASCARI [7], REISSIG [159], and LOUD [106, 107]. The last author has proved a statement that will be recorded here in view of its further applications:

Theorem 4. *If $f(\dot{x}) = 2h\dot{x}$, if $g(x)$ is continuous for $x \in (-\infty, \infty)$, $g'(x)$ is piecewise continuous and bounded in any finite interval, if $g(0) = 0$, $g'(x) \geqslant b > 0$, and if $p(t)$ is continuous and bounded for $t \in [t_0, \infty)$, then for any solution $x(t)$ of Eq. (15.2) there exists a corresponding T_0 such that*

$$|x(t)| \leqslant A = \min\left(\frac{P}{b} + \frac{P}{h^2}, \ \frac{P}{b} + \frac{2P}{h\sqrt{b}}\right), \qquad |\dot{x}(t)| \leqslant B = \frac{2P}{h}, \tag{15.4'}$$

for $t \geqslant T_0$, where $P = \sup_{t \in [t_0, \infty)} |p(t)|$.

Since $x(t)$ and $\dot{x}(t)$ are continuous in $[t_0, T_0]$, it follows that they are also bounded in this interval, and hence, by theorem 3, they are bounded in their maximal interval of existence. Then, by using Theorem 2, we deduce:

Theorem 5. *If the hypotheses of Theorem 3 are satisfied, then any solution of Eq. (15.2) exists in the whole interval* $[t_0, \infty)$.

This theorem does not apply to forced vibrations of conservative systems, since for such systems $f(\dot{x}) \equiv 0$, and the condition $\lim_{\dot{x} \to \pm\infty} f(\dot{x}) = \pm\infty$ is no longer fulfilled. However, by imposing supplementary restrictions upon the function $p(t)$, it is still possible to assure the existence of the solution for all t, as shown by the following statement, which is a particular case of a theorem of JOHN [80, 81]:

Theorem 6. *If* $g(x)$ *is continuous and differentiable for* $x \in (-\infty, \infty)$, *if* $g'(x) > 0$, *and if* $\lim_{x \to \pm\infty} g(x) = \pm\infty$, *then any solution of the equation*

$$\ddot{x} + g(x) = P \cos \nu t \tag{15.5}$$

exists for $t \in (-\infty, \infty)$.

We note that this theorem still holds if the right-hand side has the form $Q \sin \nu t$, as may be readily verified by replacing t with $t + \pi/(2\nu)$. From the theorem of JOHN one can also derive some other criteria for the existence of the solution in the whole interval $t \in (-\infty, \infty)$, in case the right-hand side of Eq. (15.5) is an arbitrary function of $\cos \nu t$ or $\sin \nu t$.

Finally, we observe that the uniqueness of the solution of Eq. (15.2) satisfying the initial conditions (15.3) is assured, as in the case of free vibrations, by the hypotheses H_1 and H_2. Indeed, it follows from these hypotheses that the function $p(t) - f(\dot{x}) - g(x)$ is continuous with respect to t and satisfies a Lipschitz condition with respect to the couple of variables x and $\dot{x}$, and hence Eq. (15.2) has at most one solution satisfying the initial conditions (15.3) (see, e.g., SANSONE and CONTI [177], Capter I, ROSEAU [172], § 4.2, and HALANAY [65]). From this statement and Theorems 1 and 4 we deduce the following theorem about the *existence and uniqueness* of the solution of the Cauchy problem (15.2), (15.3):

Theorem 7. *If* $g(x)$, $f(\dot{x})$, *and* $p(t)$ *fulfill the hypotheses* H_1, H_2, *and* H_5, *then Eq. (15.2) has exactly one solution satisfying the initial conditions (15.3). This solution exists in the whole interval* $[t_0, \infty)$.

b) Existence and stability of periodic solutions

The most important and practically the most studied case of Eq. (15.2) is that of $p(t)$ being a periodic function of t.

Before discussing the specific properties of periodic vibrations of nonlinear systems, it is useful to recall the main results concerning the steady-state vibrations of linear oscillators.

We begin with the vibrations performed by a linear conservative oscillator under the action of a simple harmonic exciting force. The governing equation of motion is then

$$\ddot{x} + \omega^2 x = P \cos \nu t \tag{15.6}$$

and, for $\nu \neq \omega$, has the general solution

$$x(t) = a_0 \cos(\omega t + \gamma) + \frac{P}{\omega^2 - \nu^2} \cos \nu t, \tag{15.7}$$

where a_0 and γ are constants to be determined from the initial conditions (15.3). If $\nu = \omega$, Eq. (15.6) has the solution

$$x(t) = a_0 \cos(\omega t + \gamma) + \frac{P}{2\omega} t \sin \omega t,$$

and the phenomenon of *resonance* occurs, since $|x(t)| \to \infty$ as $t \to \infty$.

Studying periodic solutions of Eq. (15.6) is very important, because their classification may be extended to nonlinear systems (see LEVENSON [97], STOKER [184], p. 7, SANSONE and CONTI [177], p. 360). By assuming for the sake of simplicity initial conditions of the form

$$x(0) = a, \qquad \dot{x}(0) = 0,$$

the solution (15.7) becomes

$$x(t) = \left(a - \frac{P}{\omega^2 - \nu^2}\right) \cos \omega t + \frac{P}{\omega^2 - \nu^2} \cos \nu t. \tag{15.8}$$

The periodic solutions resulting from (15.8) may be classified as follows.

1. If $\omega/\nu = [1 - P/(\omega^2 a)]^{-1\,2}$, then the solution (15.8) reduces to

$$x(t) = a \cos \nu t,$$

has the same minimum period $2\pi/\nu$ as the external force, and is said to be a *harmonic*. If this condition is not fulfilled, there are still another three cases in which the solution is periodic, and which are recorded below.

2. If $\omega/\nu = p$, with p an integer greater than 1, the solution (15.8) still has the minimum period $2\pi/\nu$. However, it results from the superposition of two vibrations, one of frequency ν and the other of frequency $p\nu$, which is called a *superharmonic* of the first.

3. If $\omega/\nu = 1/q$, with q an integer greater than 1, the solution (15.8) has the minimum period $2\pi q/\nu$. It results from the superposition of two vibrations, one of frequency ν, and the other with frequency ν/q, which is said to be a *subharmonic* of the first.

4. If $\omega/\nu = p/q$, with p, q integers, greater than 1, and relatively prime, the solution (15.8) has the minimum period $2\pi q/\nu$. It results from the superposition of two vibrations, one of frequency ν and the other of frequency $p\nu/q$, which is called a *supersubharmonic* of the first.

Let us consider now the linear dissipative oscillator, whose equation of motion is

$$\ddot{x} + 2h\dot{x} + \omega^2 x = P \cos \nu t, \qquad h > 0. \tag{15.9}$$

The general solution of the corresponding homogeneous equation depends on the damping ratio, $\zeta = h/\omega$ and may be put in the form

$$x_1(t) = \begin{cases} C_1 e^{-\omega t(\zeta + \sqrt{\zeta^2 - 1})} + C_2 e^{-\omega t(\zeta - \sqrt{\zeta^2 - 1})} & \text{for } \zeta > 1, \\ C_1 e^{-ht}(t + C_2) & \text{for } \zeta = 1, \\ a_0 e^{-ht} \cos(\omega t \sqrt{1 - \zeta^2} + \gamma) & \text{for } 0 < \zeta < 1. \end{cases} \tag{15.10}$$

where C_1, C_2, a, and γ are arbitrary constants. However, in all these cases we have

$$\lim_{t\to\infty} x_1(t) = \lim_{t\to\infty} \dot{x}_1(t) = 0. \tag{15.11}$$

The general solution of Eq. (18.9) may be written as

$$x(t) = x_1(t) + \bar{x}(t), \tag{15.12}$$

where $x_1(t)$ is given by (15.10), and

$$\bar{x}(t) = \frac{P}{\sqrt{(\omega^2 - \nu^2)^2 + 4h^2\nu^2}} \cos(\nu t + \vartheta) \tag{15.13}$$

is a particular solution of Eq. (15.9), whose phase $\vartheta \in [-\pi, 0]$ may be obtained from the equations

$$\sin \vartheta = -\frac{2h\nu}{\sqrt{(\omega^2 - \nu^2)^2 + 4h^2\nu^2}}, \quad \cos \vartheta = \frac{\omega^2 - \nu^2}{\sqrt{(\omega^2 - \nu^2)^2 + 4h^2\nu^2}}. \tag{15.14}$$

The constants C_1 and C_2, or a_0 and γ are determined by the initial conditions (15.3).

From (15.10)–(15.14) we see that Eq. (15.9) has exactly one periodic solution, namely $x(t)$; this solution corresponds to those initial conditions for which $x_1(t) \equiv 0$, that is $x_0 = \bar{x}(t_0)$, $v_0 = \dot{\bar{x}}(t_0)$. The periodic solution has the same minimum period as the external force and is asymptotically stable in the large, i.e., if $x(t)$ is any other solution of Eq. (15.9) corresponding to different initial conditions, we have

$$\lim_{t\to\infty} | x(t) - \bar{x}(t) | = 0, \qquad \lim_{t\to\infty} | \dot{x}(t) - \dot{\bar{x}}(t) | = 0. \tag{15.15}$$

A periodic solution that is asymptotically stable in the large is said to be a *steady state*. If the equation of motion has a steady-state solution, which, by definition, is the only solution with this property, then all other solutions are called *transitory*. In other words, a linear dissipative oscillator under the action of a simple harmonic external force performs a steady-state vibration only for certain initial values of the displacement and of the velocity, x_0 and v_0, which depend on the frequency v and on the phase δ of the external force.

It is always possible to assume two of the four quantities x_0, v_0, ν, and δ as given and to calculate the values of the other two quantities corresponding to the periodic solution. In the reasoning above the external force has been supposed given and the corresponding initial values x_0 and v_0 of the periodic response have been calculated as functions of ν, for $\delta = 0$. Alternatively, we shall sometimes assume that x_0 and v_0 are prescribed and we shall determine the external force that is able to bring about a periodic motion satisfying these initial conditions [1]. When using this approach it is always possible to take the initial conditions in the form

$$x(0) = a, \qquad \dot{x}(0) = 0,$$

by choosing the time origin at one of the velocity zeros of the periodic solution.

Let us consider now the nonlinear equation (15.2) and let us assume that $p(t)$ is a periodic function of t with minimum period T. We will state without proof, as in the preceding section, the main results available in the literature about the periodic solutions of this equation, by adjusting them to the hypotheses adopted by us on the functions $g(x)$, $f(\dot{x})$, and $p(t)$. We shall see that, if the damping is sufficiently strong, or if the elastic characteristic does not depart too far from linearity, then the nonlinear oscillator may perform, as the linear one, exactly one periodic vibration. Moreover, this periodic vibration has the same minimum period as the external force and is asymptotically stable in the large.

We begin with a theorem of REUTER [125] on the existence of periodic solutions.

Theorem 8. *If $g(x)$ and $f(\dot{x})$ are continuous for $x \in (-\infty, \infty)$ and $\dot{x} \in (-\infty, \infty)$, respectively, if their derivatives exist, are piecewise continuous and bounded in each finite interval, if $\lim_{x \to \pm\infty} g(x) = \pm\infty$, $\lim_{\dot{x} \to \pm\infty} f(\dot{x}) = \pm\infty$, and if $p(t)$ is continuous, bounded, and periodic of minimum period T for $t \in [t_0, \infty)$, then there exists at least one periodic solution of Eq. (15.2) in $[t_0, \infty)$ with minimum period T.*

We note that this theorem does not permit one to decide whether there exists exactly one periodic solution of period T or more of such solutions, and, if there is only one, whether it is asymptotically stable in the large. The answer to these questions is given by the following theorem of OPIAL [105], which generalizes and improves a series of results previously obtained for other particular cases of Eq. (15.2).

[1] This point of view will be for instance adopted for the classification of periodic motions of nonlinear oscillators (see below in this section).

Theorem 9. *If $g(x)$, $f(\dot{x})$, and $p(t)$ satisfy the hypotheses of Theorem 7, if the functions $g'(x)$, $g''(x)$, and $f'(\dot{x})$ exist for $|x| \leqslant A$ and $|\dot{x}| \leqslant B$ and fulfill the conditions*

$$g'(x) > 0 \ for \ |x| \leqslant A, \qquad f'(\dot{x}) > 0 \ for \ |\dot{x}| \leqslant B, \tag{15.16}$$

$$2 \inf_{|\dot{x}| \leqslant B} \frac{f'(\dot{x})}{|\dot{x}|} > \sup_{|x| \leqslant A} \frac{|g''(x)|}{g'(x)}, \tag{15.17}$$

where A and B are the constants occurring in (15.4), then Eq. (15.2) has exactly one periodic solution with minimum period T, and this solution is asymptotically stable in the large.

By using this theorem, it may be easily verified that Eq. (15.2) has *a fortiori* exactly one periodic solution of minimum period T. This solution is asymptotically stable in the large if our hypotheses H_1, H_2, H_4, and H_5 are satisfied and if, in addition, $p(t)$ is periodic with minimum period T, and $g''(x)$ exists and satisfies the inequality (15.17). This last condition expresses just the requirement that the damping be strong enough or that the elastic characteristic be sufficiently close to a linear one. To better understand this point, we will give two statements, which may be immediately proved by putting $g(x) = \omega^2 x$ and, respectively, $f(\dot{x}) = 2h\dot{x}$ into Theorem 8.

Corollary 1. *If $f(\dot{x})$ is continuous in $\dot{x} \in (-\infty, \infty)$, if $f'(\dot{x})$ is piecewise continuous, strictly positive for $|\dot{x}| \leqslant B$, and bounded in any finite interval, if $\lim_{\dot{x}\to\pm\infty} f(\dot{x}) = \pm\infty$, and if $p(t)$ is continuous, bounded, and has the minimum period T, for $t \in [t_0, \infty)$, then the equation*

$$\ddot{x} + f(\dot{x}) + \omega^2 x = p(t) \tag{15.18}$$

has exactly one periodic solution with minimum period T, and this solution is asymptotically stable in the large.

Various forms of this corollary, which are valid under more or less restrictive conditions imposed on the function $f(\dot{x})$, have been previously proved by CACCIOPOLI and GHIZZETTI [22], LEVINSON [100], ASCARI [7], and REISSIG [159].

Corollary 2. *If $g(x)$ is continuous, monotonic strictly increasing and is twice continuously differentiable for $|x| \leqslant A$, if $\lim_{x\to\pm\infty} g(x) = \pm\infty$, if $p(t)$ is continuous, bounded and periodic with minimum period T for $t \in [t_0, \infty)$, and if*

$$h > \frac{B}{4} \sup_{|x| \leqslant A} \frac{|g''(x)|}{g'(x)}, \tag{15.19}$$

then the equation

$$\ddot{x} + 2h\dot{x} + g(x) = p(t) \tag{15.20}$$

has exactly one periodic solution of minimum period T, and this solution is asymptotically stable in the large.

Corollary 2 may also be deduced from a theorem of SEIFERT [179] concerning the more general case when the coefficient of $\dot{x}$ in (15.20) depends on x. Various forms of this corollary have been previously proved by CARTWRIGHT and LITTLEWOOD [24], CARTWRIGHT [23], REUTER [164], and BLAIR and LOUD [14], by imposing on h conditions that are more restrictive than (15.19). The case $g(x) = \omega^2 x(1 + \beta x^2)$ has also been treated by LOUD [106].

The following theorem that also assures the existence, uniqueness, and asymptotical stability in the large of the periodic solution of Eq. (15.20), and no longer assumes the existence of $g''(x)$, has been independently proved by ZLÁMAL [194] and LOUD [107].

Theorem 10. *If $g(x)$ is continuous, if $g'(x)$ is piecewise continuous, strictly positive, and bounded for $|x| \leqslant A$, if $p(t)$ is continuous, bounded, and periodic with minimum period T for $t \in [t_0, \infty)$, and if*

$$h^2 > \frac{1}{2} \sup_{|x| \leqslant A} g'(x), \tag{15.21}$$

then Eq. (15.20) has exactly one periodic solution of minimum period T, and this solution is asymptotically stable in the large.

Conditions (15.19) and (15.21) may be considered as giving either a superior bound for the departure of $g(x)$ from linearity or an inferior bound for the damping factor, both of which assure the existence, uniqueness, and asymptotical stability in the large of the periodic solution. In most cases of practical interest, the condition (15.19) proves to be weaker than (15.21) if both of them are applicable, i.e., if $g'(x)$ and $g''(x)$ do exist. To illustrate this point, let us consider DUFFING'S equation with damping

$$\ddot{x} + 2h\dot{x} + \omega^2 x(1 + \beta x^2) = p(t),$$

with $\beta > 0$, and let us use the above conditions as upper limitations for β. Choosing for the parameters of the oscillator and of the external force the values $h = 4$ rad/sec, $\omega = 10$ rad/sec, $P = 100$ cm/sec², we obtain from (15.4′) $A = 11$ cm, $B = 100$ cm/sec, and the conditions (15.19) and (15.21) yield, respectively,

$$\sup_{|x| \leqslant A} \frac{\beta\,|x|}{1 + 3\beta x^2} < \frac{1}{75}\,\text{cm}^{-1}, \tag{15.19′}$$

$$1 + 3\beta A^2 < 0.08. \tag{15.21′}$$

The function $\beta|x|/(1 + 3\beta x^2)$ has the maximum value $\sqrt{\beta/12}$, and hence, for the condition (15.19′) to be satisfied it is sufficient to take $\sqrt{\beta/12} < 1/75\ \text{cm}^{-1}$, whence $\beta < 0.00214\ \text{cm}^{-2}$, whereas the inequality (15.21′) is not fulfilled for any positive value of β.

Inspection of (15.17) reveals that, for comparatively weak damping, the existence of a steady-state solution of the same minimum period as the external force is no longer assured. It is then possible, under certain circumstances for superharmonic, subharmonic, or supersubharmonic vibrations to occur, as in the case of the linear conservative oscillator. We consider as an example the equation of motion

$$\ddot{x} + \varepsilon f_1(\dot{x}) + \omega^2 x + \varepsilon\psi_1(x) = P \cos \nu(t + \delta),$$

with the initial conditions

$$x(0) = a, \qquad \dot{x}(0) = 0.$$

For $\varepsilon = 0$ we recover the equation of motion of the linear conservative oscillator, which, as we have seen above, has periodic solutions if the ratio ω/ν takes one of the values $[1 - P/(\omega^2 a)]^{-1/2}$, p, $1/q$, or p/q, where p and q are integers, greater than 1, and relatively prime, but otherwise arbitrary. It is to be expected that, if ε is sufficiently small, there exist various types of periodic solutions of the above nonlinear equation for certain values of the phase δ depending on ε, and provided that the ratio ω/ν take certain values in the neighborhood of the values $[1 - P/(\omega^2 a)]^{-1/2}$, p, $1/q$, or p/q, which also depend on ε. In §§ 16b and 19b we shall give sufficient conditions for the existence of such solutions.

We indicate now, after LEVENSON [97], how the classification of periodic solutions previously given may be generalized for the nonlinear equation and the initial conditions considered above. A periodic solution of this equation is said to be a *harmonic* or a *superharmonic* (with respect to the external force) if it has the miminum period $2\pi/\nu$, and if the ratio $\omega/\nu(\varepsilon)$ tends to $[1 - P/(\omega^2 a]^{-1/2}$ or to p, respectively, as $\varepsilon \to 0$. A periodic solution of this equation is said to be a *subharmonic* or a *supersubharmonic* (with respect to the external force) if it has the minimum period $2\pi q/\nu$ and if the ratio $\omega/\nu(\varepsilon)$ tends to $1/q$ or to p/q, respectively, as $\varepsilon \to 0$, where p and q are integers, greater than 1, and relatively prime, but otherwise arbitrary.

If in addition $P = 0(\varepsilon)$, which will be frequently assumed in the following, then the nonlinear oscillator can perform periodic oscillations with period $2\pi q/\nu$ only if $\omega - p\nu/q = 0(\varepsilon)$, where now the integers p and q may also take the value 1. This situation is sometimes termed "resonance", although, in contradistinction to the usual resonance occurring in forced vibrations of linear oscillators, the ratio ω/ν may be different from 1, and the amplitude of the vibrations does not tend to infinity as $\omega - p\nu/q \to 0$ or, equivalently, as $\varepsilon \to 0$. In order to emphasize that the words *resonance* and *nonresonance* should be carefully used in the nonlinear case, we shall always write them with quotes. By combining the above definitions, we shall speak of *subharmonic*, *superharmonic*, or *supersubharmonic* "resonances," according as $\omega \approx \nu/q$, $\omega \approx p\nu$, or $\omega \approx p\nu/q$. The very important case $\omega \approx \nu$ will be termed *principal "resonance."*

A particular case of Eq. (15.2), which requires, however, a special treatment, is that of the dry friction. In this case the function $f(\dot{x})$ includes a term like $R \operatorname{sgn} \dot{x}$ and hence is no longer continuous for $\dot{x} = 0$. The essential results obtained by REISSIG [155—158] on forced vibrations of systems with dry friction will be expounded in § 22.

Before closing this section we briefly consider the periodic vibrations of conservative systems. Now $f(\dot{x}) \equiv 0$ and the governing equation becomes

$$\ddot{x} + g(x) = p(t). \tag{15.22}$$

In this case, Theorems 1, 6, and 7 still assure the existence for $t \in [t_0, \infty)$ and the uniqueness of the solution of Eq. (15.22) satisfying the initial conditions (15.3). However, all other theorems proved in this section loose, generally, their validity.

There are considerably less qualitative results available in the literature concerning forced vibrations of conservative systems than of dissipative systems. From the results that may be rightly situated in this book we choose — in accordance with the general treatment of this section — those regarding the existence of a periodic solution of Eq. (15.22), in case $p(t)$ is a periodic function of t.

Theorem 11. (OPIAL [138]). *If* $g(x)$ *is continuous for all* x, *if* $x\,g(x) > 0$ *for* $x \neq 0$, $\lim_{x \to \pm\infty} g(x) = \pm\infty$, *if* $p(t)$ *is continuous and periodic with minimum period* T *for* $t \in [t_0, \infty)$, *and if*

$$T < 4 \liminf_{|x| \to \infty} T(x), \tag{15.23}$$

where

$$T(x) \equiv \left| \int_0^x \frac{ds}{\sqrt{G(x) - G(s)}} \right|, \quad G(x) \equiv \int_0^x g(s)\, ds, \tag{15.24}$$

then Eq. (15.22) *has at least a periodic solution with minimum period* T.

It should be noted that, unlike theorem 8, which holds only for dissipative systems, the above statement warrants the existence of a periodic solution only if, in addition to other hypotheses currently adopted by us about $g(x)$ and $p(t)$, the period T of the external force is sufficiently small. A similar result has been obtained by SEIFERT [180], under the assumption that $p(t)$ is continuous and periodic and its minimum period is small enough, $|p(t)| \leqslant P$ for $t \in [t_0, \infty)$, $g(x)$ is continuous, increasing, $g(0) = 0$, and $|g(x)| > P$ for sufficiently large values of x. However, the superior bound assumed by SEIFERT is not only more complicated than (15.23), but it depends on both $g(x)$ and $p(t)$.

We end herewith our review of the principal qualitative results concerning forced vibrations of nonlinear systems. In the following sections of this chapter we will focus our attention mainly on the quantitative methods for calculating the solutions.

§ 16. PERTURBATION METHOD

In the preceding section we have seen that, if the external force is periodic and satisfies, together with the elastic and damping characteristics, the hypotheses adopted in the introduction, then there exists at least one periodic solution of the equation of motion, which has the same minimum period as the external force. The remaining part of this chapter will be devoted entirely to the exposition of various quantitative methods for the determination of such periodic solutions.

The available methods generally permit finding the periodic solutions to any desired accuracy, but only if the elastic nonlinearity and the damping are weak. Then, the equation of motion may be written in the form

$$\ddot{x} + \omega^2 x = -\varepsilon[\psi_1(x) + f_1(\dot{x})] + p_0(\nu t) + \varepsilon p_1(\nu t), \tag{16.1}$$

where ν is the frequency of the external force, and ε is a small positive and dimensionless parameter.

A strict reasoning would require the introduction into Eq. (16.1) of new dimensionless variables in order to compare numerically all quantities and to define ε directly in terms of dimensionless parameters of the system. On the other hand, such an approach needs special care, because by introducing dimensionless variables one changes the scale, and this can considerably modify the order of magnitude of the original variables. Moreover, the choice of the parameter ε is, as we shall see below, rather arbitrary, since final formulas contain only the functions $\psi(x) = \varepsilon\psi_1(x)$, $f(\dot{x}) = \varepsilon f_1(\dot{x})$, $p(t) = p_0(t) + \varepsilon p_1(t)$, and hence do not depend on ε, but only on the external force and on the characteristics of the system. To avoid these complications, we shall frequently keep the original variables in the equation of motion, by considering ε merely as a coefficient establishing which quantities are "small", even if they have different dimensions.

The first method we will deal with is the perturbation method of POINCARÉ [146], to which we have already referred to in § 5a. In this method, the functions $\psi_1(x)$ and $f_1(\dot{x})$ are supposed to be analytic functions of their arguments, and the periodic solution is sought in the form of a power series in ε, whose coefficients are periodic functions of νt with period 2π. The first term of this expansion, denoted by $\bar{x}(t)$, and called *generating solution*, is the general solution of Eq. (16.1) for $\varepsilon = 0$, i.e.,

$$\ddot{\bar{x}} + \omega^2 \bar{x} = p_0(t). \tag{16.2}$$

To substantiate this method it is necessary to answer the following questions:

1. Is there any periodic solution of Eq. (16.1) that reduces to the generating one for $\varepsilon = 0$?

2. If such a solution does exist, is it an analytic function of ε with nonzero radius of convergence?

3. How should the iterative method be modified in case "resonance" occurs, i.e., if $\omega \approx p\nu$ with p an integer?

The last two questions will be examined below, by separately considering the "resonance" and the "nonresonance" cases. The answer to the first question is given by the following theorem due to POINCARÉ. [1)]

[1)] A proof of this theorem for more general systems than those considered here is given by ROSEAU [172], p. 72.

Theorem. *Consider the equation*

$$\ddot{x} + G(t, x, \dot{x}; \varepsilon) = 0, \tag{16.3}$$

where ε *is a small parameter, and assume that the following conditions are fulfilled:*

1. Equation $\ddot{x} + G(t, x, \dot{x}; 0) = 0$ *has a solution in* $|t - t_0| \leqslant h$ *for some positive* h, *say* $x = \bar{x}(t)$, *which satisfies the initial conditions* $\bar{x}(t_0) = \bar{x}_0$, $\dot{\bar{x}}(t_0) = \bar{v}_0$ *and is called a* generating solution.

2. $G(t, x, \dot{x}; \varepsilon)$ *is continuous with respect to the set of all its arguments and can be expanded in a power series of* $x - \bar{x}(t)$, $\dot{x} - \dot{\bar{x}}(t)$, ε, *which is convergent in* $|x - \bar{x}(t)| \leqslant \rho_1$, $|\dot{x} - \dot{\bar{x}}(t)| \leqslant \rho_2$, $|\varepsilon| \leqslant \rho$ for $|t - t_0| \leqslant h$ *and some positive* ρ_1, ρ_2, ρ.

Under these assumptions:

1. *There exist some positive* ρ_1^*, ρ_2^*, ρ^* *such that, for* $|x_0 - \bar{x}_0| \leqslant \rho_1^*$, $|v_0 - \bar{v}_0| \leqslant \rho_2^*$, $|\varepsilon| \leqslant \rho^*$, *there exists a function* $x(t; x_0, v_0, \varepsilon)$ *which is a solution of* (16.3) *in* $|t - t_0| \leqslant h$, *satisfies the initial conditions*

$$x(t_0; x_0, v_0, \varepsilon) = x_0, \qquad \dot{x}(t_0; x_0, v_0, \varepsilon) = v_0,$$

and reduces to the generating solution for $\varepsilon = 0$, $x_0 = \bar{x}_0$, $v_0 = \bar{v}_0$, i.e.,

$$x(t; \bar{x}_0, \bar{v}_0, 0) \equiv \bar{x}(t), \qquad \dot{x}(t; \bar{x}_0, \bar{v}_0, 0) \equiv \dot{\bar{x}}(t).$$

2. $x(t; x_0, v_0)$ *can be expanded in a power series of* $|x_0 - \bar{x}_0|$, $|v_0 - \bar{v}_0|$, *and* ε, *which is convergent in* $|x_0 - \bar{x}_0| \leqslant \rho_1^*$, $|v_0 - \bar{v}_0| \leqslant \rho_2^*$, $|\varepsilon| \leqslant \rho^*$ for $|t - t_0| \leqslant h$. *For the same values of* x_0, v_0, ε, *and* t, *this expansion can be twice differentiated term by term with respect to* t *and the series obtained converge to the first and second time derivatives of* $x(t; x_0, v_0, \varepsilon)$, *respectively.*

It is readily seen that, if $\psi_1(x)$ and $f_1(x)$ are entire functions of their arguments, and if $p(\nu t)$ is continuous for $t \in [t_0, \infty)$, then the theorem above assures the existence of a solution of Eq. (16.1) that exists for $t \in [t_0, \infty)$ and reduces to the generating solution for $\varepsilon = 0$. Moreover, if the magnitudes of $x_0 - \bar{x}_0$, $v_0 - \bar{v}_0$, and ε are sufficiently small, the solution of Eq. (16.1) can be expanded in a power series with respect to these quantities.

We will now show that, if $p(\nu t)$ is a periodic function of νt with period 2π, and if we denote by x_0, v_0 the initial values corresponding to a periodic solution $x(t; x_0, v_0, \varepsilon)$ of Eq. (16.1), then $x_0 - \bar{x}_0$ and $v_0 - \bar{v}_0$ are analytic functions of ε, and hence, by Poincaré's theorem, the solution is also an analytic function of ε, which may be denoted by $x(t; \varepsilon)$. The proof below follows with some slight modifications the one given by Malkin [113] (see also Minorsky [122], pp. 232–240) and considers separately the "resonance" and the "nonresonance" cases.

a) "Nonresonance" case ($\omega \neq p\nu$ for all p, p integer)

By applying the transformation

$$\tau = \nu t, \tag{16.4}$$

Eq. (16.1) becomes

$$x'' + \eta^2 x = \frac{1}{\nu^2} p_0(\tau) + \varepsilon F(\tau, x, x'), \tag{16.5}$$

where

$$x' \equiv \frac{\mathrm{d}x}{\mathrm{d}\tau}, \qquad x'' \equiv \frac{\mathrm{d}^2 x}{\mathrm{d}\tau^2}, \qquad \eta = \frac{\omega}{\nu}, \tag{16.6}$$

$$F(\tau, x, x') \equiv -\frac{1}{\nu^2} [\psi_1(x) + f_1(\nu x') - p_1(\tau)]. \tag{16.7}$$

By (16.4), the period of the solution sought becomes equal to 2π, and hence is independent of ε. Therefore, we may try a solution in form of a power series of ε whose coefficients are also periodic functions of t with period 2π.[1)]

We have recalled in the preceding section that the *linear* oscillator acted on by a periodic external force may perform exactly one periodic motion, corresponding to certain initial conditions, which depend on the frequency and phase of the external force. The motions corresponding to any other initial conditions are not periodic. It is, therefore, natural to assume that the motion of a *nonlinear* oscillator will be periodic only if the displacement and velocity take certain initial values depending on ε, which will be denoted by $x_0(\varepsilon)$ and $v_0(\varepsilon)$.

Let then $\gamma_1(\varepsilon)$ be the initial value of the difference between the sought solution and the generating one, and $\gamma_2(\varepsilon)$ the initial value of the difference of their time derivatives, i.e.,

$$x_0(\varepsilon) - \bar{x}_0 \equiv \gamma_1(\varepsilon), \qquad v_0(\varepsilon) - \bar{v}_0 \equiv \gamma_2(\varepsilon). \tag{16.8}$$

By POINCARÉ's theorem, Eq. (16.5) has a solution, $x(\tau;\, \gamma_1, \gamma_2, \varepsilon)$, which is analytic with respect to $\gamma_1, \gamma_2, \varepsilon$, for sufficiently small magnitudes of these parameters, and which reduces, for $\varepsilon = 0$, to the generating solution $\bar{x}(\tau)$, that is, to the solution of the equation

$$x'' + \eta^2 x = \frac{1}{\nu^2} p_0(\tau). \tag{16.9}$$

We will now prove that the required periodicity of $x(\tau;\, \gamma_1, \gamma_2, \varepsilon)$ implies the analyticity of $\gamma_1(\varepsilon)$ and $\gamma_2(\varepsilon)$ if ε is small enough.

[1)] Cf. § 5a where the same transformation has been applied in the case of free vibrations.

As $p_0(\tau)$ is periodic with period 2π, it may be expanded in a Fourier series, say

$$p_0(\tau) = a_0 + \sum_{k=1}^{\infty} (a_k \cos k\tau + b_k \sin k\tau). \tag{16.10}$$

Supposing that η is not an integer, the *only* periodic solution of Eq. (16.9) with period 2π is

$$\bar{x}(\tau) = \frac{a_0}{\omega^2} + \sum_{k=1}^{\infty} \left(\frac{a_k}{\omega^2 - k^2\nu^2} \cos k\tau + \frac{b_k}{\omega^2 - k^2\nu^2} \sin k\tau \right). \tag{16.11}$$

By a well-known result from the theory of linear differential equations, we deduce that the solution of (16.5) has to satisfy the integral equation

$$\left.\begin{aligned} x(\tau;\, \gamma_1, \gamma_2, \varepsilon) &= C_1 \sin \eta\tau + C_2 \cos \eta\tau + \bar{x}(\tau) - \\ &- \frac{\varepsilon}{\eta} \int_{\tau_0}^{\tau} F[s, x(s),\; x'(s)] \sin \eta(\tau - s)\, ds, \end{aligned}\right\} \tag{16.12}$$

where $x(s;\, \gamma_1, \gamma_2, \varepsilon)$ has been shortly denoted as argument of F by $x(s)$. Without restricting the generality, we may set $\tau_0 = 0$. It follows then from (16.12), by making use of (16.8), that $C_1 = \gamma_2/\eta$, $C_2 = \gamma_1$, and hence

$$\left.\begin{aligned} x(\tau;\, \gamma_1, \gamma_2, \varepsilon) &= \gamma_1 \cos \eta\tau + \frac{\gamma_2}{\eta} \sin \eta\tau + \bar{x}(\tau) - \\ &- \frac{\varepsilon}{\eta} \int_0^{\tau} F[s, x(s),\, x'(s)] \sin \eta(\tau - s)\, ds, \\ x'(\tau;\, \gamma_1, \gamma_2, \varepsilon) &= -\gamma_1\eta \sin \eta\tau + \gamma_2 \cos \eta\tau + \bar{x}'(\tau) - \\ &- \varepsilon \int_0^{\tau} F[s, x(s),\, x'(s)] \cos \eta(\tau - s)\, ds. \end{aligned}\right\} \tag{16.13}$$

In order for the solution to be periodic it is necessary and sufficient that the displacement and velocity take at time $\tau = 2\pi$ the same values as the initial ones, that is

$$\left.\begin{aligned} x(2\pi;\, \gamma_1, \gamma_2, \varepsilon) - x(0;\, \gamma_1, \gamma_2, \varepsilon) &= 0, \\ x'(2\pi;\, \gamma_1, \gamma_2, \varepsilon) - x'(0;\, \gamma_1, \gamma_2, \varepsilon) &= 0. \end{aligned}\right\} \tag{16.14}$$

Substituting (16.13) into (16.14) yields

$$\left.\begin{aligned} G_1(\gamma_1, \gamma_2, \varepsilon) &\equiv \gamma_1(\cos 2\eta\tau - 1) + \frac{\gamma_2}{\eta} \sin 2\eta\pi - \\ &- \frac{\varepsilon}{\eta} \int_0^{2\pi} F[s, x(s), x'(s)] \sin \eta(2\pi - s)\, ds = 0, \\ G_2(\gamma_1, \gamma_2, \varepsilon) &\equiv -\gamma_1\eta \sin 2\eta\pi + \gamma_2(\cos 2\eta\pi - 1) - \\ &- \varepsilon \int_0^{2\pi} F[s, x(s), x'(s)] \cos \eta(2\pi - s)\, ds = 0. \end{aligned}\right\} \tag{16.15}$$

For $\varepsilon = 0$, it results from (16.8) that $\gamma_1 = \gamma_2 = 0$ and hence conditions (16.15) are identically fulfilled. On the other hand, since η is not an integer, we have

$$\left.\frac{\partial(G_1, G_2)}{\partial(\gamma_1, \gamma_2)}\right|_{\varepsilon=\gamma_1=\gamma_2=0} = (\cos 2\eta\pi - 1)^2 + \sin^2 2\eta\pi \neq 0,$$

By applying the existence theorem of implicit functions, we infer that system (16.15) has exactly one solution that vanishes together with ε. Moreover, as $x(\tau; \gamma_1, \gamma_2, \varepsilon)$ is an analytic function of $\gamma_1, \gamma_2, \varepsilon$, we deduce from (16.15) that $G_1(\gamma_1, \gamma_2, \varepsilon)$, $G_2(\gamma_1, \gamma_2, \varepsilon)$ are analytic functions of their arguments, too, and hence, by virtue of the same theorem of implicit functions, it follows that $\gamma_1(\varepsilon)$ and $\gamma_2(\varepsilon)$ are analytic functions of ε, for ε sufficiently small.

Therefore, the periodic solution $x(\tau; \gamma_1, \gamma_2, \varepsilon)$ can be expanded in a power series in ε, which is convergent for ε sufficiently small and $\tau \in [0, \infty)$. This expansion may be written in the form

$$x(\tau; \varepsilon) = \bar{x}(\tau) + \varepsilon x_1(\tau) + \varepsilon^2 x_2(\tau) + \ldots, \tag{16.16}$$

where $x_k(\tau)$, $k = 1, 2, \ldots$ are periodic functions of τ with period 2π.

The practical calculation of the periodic solutions proceeds as follows. Introducing (16.16) into (16.5) and equating coefficients of like powers of ε leads to a system of recurrent differential equations. Thus, we have at the $(k+1)^{\text{th}}$ step:

$$x_k'' + \eta^2 x_k = F_k(\tau), \tag{16.17}$$

where $F_k(\tau)$ is a periodic function of τ with period 2π, which depends on the functions $\bar{x}(\tau), x_1(\tau), \ldots, x_{k-1}(\tau)$ already determined at the previous stages. In particular,

$$F_1(\tau) \equiv F[\tau, \bar{x}(\tau), \bar{x}'(\tau)]. \tag{16.18}$$

We note that, since η is not an integer, each of the Eqs. (16.17) has exactly one periodic solution, which may be determined in the same way as that used to calculate $\bar{x}(\tau)$ from (16.9). This procedure fails to apply when η is equal or close to an integer; this "resonance" case will be considered in the following.

b) "Resonance" case ($\omega \approx p\nu, p$ an integer)[1)]

If $\omega \approx p\nu$ with p an integer, we may write

$$\eta^2 = p^2 + \varepsilon\bar{\Delta}. \tag{16.19}$$

In this case the perturbation method may be applied only if the coefficients of the harmonic of pth order in the expansion of $p_0(\tau)$ are supposed to vanish together with ε, i.e., $a_p = \varepsilon\bar{a}_p$, $b_p = \varepsilon\bar{b}_p$, because, otherwise, the generating solution could not be a periodic function of τ. By putting

$$\bar{p}_0(\tau) \equiv p_0(\tau) - a_p \cos p\tau - b_p \sin p\tau, \tag{16.20}$$

and considering (16.19), Eq. (16.5) becomes

$$x'' + p^2 x = \frac{1}{\nu^2}\,\bar{p}_0(\tau) + \varepsilon\bar{F}(\tau, x, x'), \tag{16.21}$$

where

$$\bar{F}(\tau, x, x') \equiv F(\tau, x, x') - \frac{1}{\nu^2}\,(x\bar{\Delta} - \bar{a}_p \cos p\,\tau - \bar{b}_p \sin p\,\tau). \tag{16.22}$$

For $\varepsilon = 0$, we have

$$x'' + p^2 x = \frac{1}{\nu^2}\,\bar{p}_0(\tau). \tag{16.23}$$

Unlike Eq. (16.9) in the "nonresonance" case, Eq. (16.23) has not only one periodic solution with period 2π, but infinitely many such solutions, namely

$$\bar{x}(\tau) = M \cos p\tau + N \sin p\tau + \varphi(\tau), \tag{16.24}$$

1) In the form set forth in this section the perturbation method permits to investigate only principal and superharmonic "resonances." However, in a somewhat modified form, called method of successive approximations it may also be used, as we shall see in § 19b, to study subharmonic „resonance".

where M and N are arbitrary constants and

$$\varphi(\tau) = \frac{a_0}{\nu^2} + \frac{1}{\nu^2} \sum_{\substack{k=1 \\ k \neq p}}^{\infty} \left(\frac{a_k}{k^2 - p^2} \cos k\tau + \frac{b_k}{k^2 - p^2} \sin k\tau \right). \tag{16.25}$$

On the other hand, the periodicity conditions (16.15) may now be written, by replacing η with p, as

$$G_1(\gamma_1, \gamma_2, \varepsilon) \equiv \frac{\varepsilon}{p} \int_0^{2\pi} \bar{F}[s, x(s), x'(s)] \sin ps \, ds = 0,$$

$$G_2(\gamma_1, \gamma_2, \varepsilon) = - \varepsilon \int_0^{2\pi} \bar{F}[s, x(s), x'(s)] \cos ps \, ds = 0.$$

These conditions are identically satisfied for $\varepsilon = 0$. Since they have to be fulfilled for any ε sufficiently small, we may divide both sides by ε and $-\varepsilon$, respectively, to obtain

$$\left.\begin{aligned} H_1(\gamma_1, \gamma_2, \varepsilon) &\equiv \int_0^{2\pi} \bar{F}[s, x(s), x'(s)] \sin ps \, ds = 0, \\ H_2(\gamma_1, \gamma_2, \varepsilon) &\equiv \int_0^{2\pi} \bar{F}[s, x(s), x'(s)] \cos ps \, ds = 0. \end{aligned}\right\} \tag{16.26}$$

We first determine the constants M and N by requiring that Eqs. (16.26) be satisfied for $\varepsilon = 0$, hence for $x(s) \equiv \bar{x}(s)$. We suppose, therefore, that there exists a solution (M, N) of the system

$$\left.\begin{aligned} Q(M, N) \equiv \int_0^{2\pi} F[s, &M \cos ps + N \sin ps + \varphi(s), \\ &- Mp \sin ps + Np \cos ps + \varphi'(s)] \sin ps \, ds = 0, \\ R(M, N) \equiv \int_0^{2\pi} F[s, &M \cos ps + N \sin ps + \varphi(s), \\ &- Mp \sin ps + Np \cos ps + \varphi'(s)] \cos ps \, ds = 0, \end{aligned}\right\} \tag{16.27}$$

i.e., that

$$\frac{\partial(Q, R)}{\partial(M, N)} \neq 0. \tag{16.28}$$

We will prove that, under this assumption, system (16.26) may be solved for γ_1 and γ_2 as functions of ε, as in the "nonresonance" case. Indeed, for $\eta = p$ and $\varepsilon = \gamma_1 = \gamma_2 = 0$, we have from (16.16)

$$\frac{\partial x}{\partial \gamma_1} = \cos p\tau, \quad \frac{\partial x}{\partial \gamma_2} = \frac{1}{p}\sin p\tau, \quad \frac{\partial x'}{\partial \gamma_1} = -p\sin p\tau, \quad \frac{\partial x'}{\partial \gamma_2} = \cos p\tau,$$

and then, for the same values of the parameters, it follows from (16.26) that

$$\frac{\partial H_1}{\partial \gamma_1} = \int_0^{2\pi}\left(\frac{\partial F}{\partial x}\cos ps - p\frac{\partial F}{\partial x'}\sin ps\right)\sin ps\, ds = \frac{\partial Q}{\partial M},$$

$$\frac{\partial H_1}{\partial \gamma_2} = \int_0^{2\pi}\left(\frac{1}{p}\frac{\partial F}{\partial x}\sin ps + \frac{\partial F}{\partial x'}\cos ps\right)\sin ps\, ds = \frac{1}{p}\frac{\partial Q}{\partial N},$$

$$\frac{\partial H_2}{\partial \gamma_1} = \int_0^{2\pi}\left(\frac{\partial F}{\partial x}\cos ps - p\frac{\partial F}{\partial x'}\sin ps\right)\cos ps\, ds = \frac{\partial R}{\partial M},$$

$$\frac{\partial H_2}{\partial \gamma_2} = \int_0^{2\pi}\left(\frac{1}{p}\frac{\partial F}{\partial x}\sin ps + \frac{\partial F}{\partial x'}\cos ps\right)\cos ps\, ds = \frac{1}{p}\frac{\partial R}{\partial N},$$

whence

$$\frac{\partial(H_1, H_2)}{\partial(\gamma_1, \gamma_2)} = \frac{1}{p}\,\frac{\partial(Q, R)}{\partial(M, N)} \neq 0.$$

We deduce, as in the "nonresonance" case, that system (16.26) has exactly one solution, $\gamma_1 = \gamma_1(\varepsilon)$, $\gamma_2 = \gamma_2(\varepsilon)$, which vanishes together with ε, and can be developed in a power series in ε that is convergent for ε sufficiently small. Therefore, the periodic solution looked for may be also expanded in a power series of the form

$$x(\tau; \varepsilon) = \bar{x}(\tau) + \varepsilon x_1(\tau) + \varepsilon^2 x_2(\tau) + \ldots, \tag{16.29}$$

where now $\bar{x}(\tau)$ is given by (16.24), and M, N are determined by (16.27). We have seen that in the "nonresonance" case only one periodic generating solution does exist and that the periodic solution of the weakly nonlinear equation may be found in the neighborhood of this generating solution. In the "resonance" case the situation is somewhat different: there are infinitely many (periodic) generating solutions (16.24), but only one of them, namely that satisfying system (16.27), is close to the periodic solution of the weakly nonlinear equation.

Substituting (16.17) into (16.21) and equating coefficients of like powers of ε leads to a set of recursive differential equations of the same form as (16.17). In particular, we have for x_1

$$x_1'' + p^2 x_1 = \overline{F}[\tau, \overline{x}(\tau), \overline{x}'(\tau)]. \tag{16.30}$$

In order for a periodic solution with period 2π of this equation to exist, it is necessary that the right-hand side do not contain the harmonic of pth order, that is

$$\int_0^{2\pi} \overline{F}[s, \overline{x}(s), \overline{x}'(s)] \sin ps \, ds = 0, \qquad \int_0^{2\pi} \overline{F}[s, \overline{x}(s), \overline{x}'(s)] \cos ps \, ds = 0.$$

By using (16.24), it may be seen that these conditions are identical with Eqs. (16.27), which determine the constants M and N.

Next, we write the solution of (16.30) in the form

$$x_1(\tau) = M_1 \cos p\tau + N_1 \sin p\tau + \varphi_1(\tau),$$

where $\varphi_1(\tau)$ is a particular solution of Eq. (16.30), which may be easily obtained by developing the right-hand side in a Fourier series. Then, constants M_1 and N_1 are calculated by equating to zero the coefficients of the pth harmonic in the differential equation for x_2 and so on.

c) Illustration of the method: the linear oscillator with weak viscous damping

To better understand how the perturbation method really works, let us first apply it to a linear oscillator with weak viscous damping,[1] whose equation of motion is

$$\ddot{x} + 2h\dot{x} + \omega^2 x = P \cos \nu t. \tag{16.31}$$

As is well-known, the general exact solution of this equation for $h/\omega < 1$ is

$$x(t) = a_0 e^{-ht} \cos (t \sqrt{\omega^2 - h^2} + \gamma) + \frac{P(\omega^2 - \nu^2)}{(\omega^2 - \nu^2) + 4h^2\nu^2} \cos \nu t$$

$$+ \frac{2h\nu P}{(\omega^2 - \nu^2)^2 + 4h^2\nu^2} \sin \nu t, \tag{16.32}$$

[1] Other applications of the method to nonlinear oscillators will be given in the following two chapters.

where a_0 and γ are constants to be determined from the initial conditions. The only periodic solution of this equation is

$$x(t) = \frac{P(\omega^2 - \nu^2)}{(\omega^2 - \nu^2)^2 + 4h^2\nu^2} \cos \nu t + \frac{2h\nu P}{(\omega^2 - \nu^2)^2 + 4h^2\nu^2} \sin \nu t. \quad (16.33)$$

Since this solution is asymptotically stable in the large, it is a steady state (cf. § 15b).

— "*Nonresonance*" *case*

By applying the transformation $\tau = \nu t$ and denoting

$$\frac{h}{\nu} = \varepsilon \bar{h}, \qquad \frac{\omega^2}{\nu^2} = \eta^2, \quad (16.34)$$

where ε is a small dimensionless parameter, Eq. (16.31) becomes

$$x'' + \eta^2 x = \frac{P}{\nu^2} \cos \tau - 2\varepsilon x'. \quad (16.35)$$

Assume that η is not an integer. Then, for $\varepsilon = 0$, we obtain from (16.35) the generating solution

$$\bar{x}(\tau) = \frac{P}{\nu^2(\eta^2 - 1)} \cos \tau.$$

We seek now a periodic solution of Eq. (16.35) of the form

$$x(\tau; \varepsilon) = \bar{x}(\tau) + \varepsilon x_1(\tau) + \varepsilon^2 x_2(\tau) + \ldots$$

Substituting this series into (16.35) and equating coefficients of like powers of ε, we have

$$x_1'' + \eta^2 x_1 = -2\bar{h}\bar{x}',$$

$$x_2' + \eta^2 x_2 = -2\bar{h}x_1', \text{ etc.}$$

The first of these equations may be rewritten as

$$x_1'' + \eta^2 x_1 = \frac{2P\bar{h}}{\nu^2(\eta^2 - 1)} \sin \tau$$

and has the unique periodic solution

$$x_1(\tau) = \frac{2P\bar{h}}{\nu^2(\eta^2 - 1)^2} \sin \tau.$$

Similarly, we find

$$x_2(\tau) = - \frac{4P\bar{h}^2}{\nu^2(\eta^2 - 1)^3} \cos \tau,$$

$$x_3(\tau) = - \frac{8P\bar{h}^3}{\nu^2(\eta^2 - 1)} \sin \tau, \text{ etc.},$$

and hence the solution, $x(\tau; \varepsilon)$, has the form

$$x(\tau; \varepsilon) = \frac{P}{\nu^2(\eta^2 - 1)} \left[1 - \frac{4\varepsilon^2\bar{h}^2}{(\eta^2 - 1)^2}\right] \cos \tau +$$

$$+ \frac{2P\varepsilon\bar{h}}{\nu^2(\eta^2 - 1)^2} \left[1 - \frac{4\varepsilon^2\bar{h}^2}{(\eta^2 - 1)^2}\right] \sin \tau + 0(\varepsilon^4)$$

or, by (16.34),

$$x\left(t; \frac{h}{\nu}\right) = \frac{P}{\omega^2 - \nu^2} \left[1 - \frac{4h^2\nu^2}{(\omega^2 - \nu^2)^2}\right] \cos \nu t +$$

$$+ \frac{2P\nu h}{(\omega^2 - \nu^2)^2} \left[1 - \frac{4h^2\nu^2}{(\omega^2 - \nu^2)^2}\right] \sin \nu t + 0\left(\frac{h^4}{\nu^4}\right). \tag{16.36}$$

It is readily seen that this approximate solution contains the first four terms from the expansion of the exact solution (16.33) in a power series in h/ν.

— "*Resonance*" *case*

Let us investigate now the behavior of the solution in the neighborhood of the principal "resonance", i.e., for $\eta^2 = 1 + \varepsilon\bar{\Delta}$. In this case we have to assume, as has been shown before, that the amplitude of the external force is of order ε. By putting $P = \varepsilon\bar{P}\nu^2$ into (16.35), we obtain

$$x'' + x = -\varepsilon(x\bar{\Delta} + 2\bar{h}x' - \bar{P} \cos \tau). \tag{16.37}$$

For $\varepsilon = 0$, this equation has the general solution

$$\bar{x}(\tau) = M \cos \tau + N \sin \tau,$$

where M and N are arbitrary constants. We try now a periodic solution of Eq. (16.37) in the form

$$x(\tau; \varepsilon) = \bar{x}(\tau) + \varepsilon x_1(\tau) + \varepsilon^2 x_2(\tau) + \ldots$$

By introducing this expansion into (16.37) and equating like powers of ε, we obtain

$$x_1'' + x_1 = -\bar{x}\bar{\Delta} - 2\bar{h}\bar{x}' + \bar{P}\cos\tau,$$

$$x_2'' + x_2 = -x_1\bar{\Delta} - 2\bar{h}x_1', \text{ etc.}$$

Considering now the expression of $\bar{x}(\tau)$ already found, the first of these equations becomes

$$x_1'' + x = -(M\cos\tau + N\sin\tau)\,\bar{\Delta} + 2\bar{h}\,(M\sin\tau - N\cos\tau) + \bar{P}\cos\tau.$$

The periodicity of x_1 requires the vanishing of the coefficients of $\sin\tau$ and $\cos\tau$ in the right-hand side, which gives

$$M = \frac{\bar{P}\bar{\Delta}}{\bar{\Delta}^2 + 4\bar{h}^2}, \qquad N = \frac{2\bar{P}\bar{h}}{\bar{\Delta}^2 + 4\bar{h}^2}.$$

Since $\bar{h} = \dfrac{h}{\varepsilon\nu}$, $\bar{\Delta} = \dfrac{\omega^2 - \nu^2}{\varepsilon\nu^2}$, $\bar{P} = \dfrac{P}{\varepsilon\nu^2}$, it results

$$x(t) = \frac{P(\omega^2 - \nu^2)}{(\omega^2 - \nu^2)^2 + 4h^2\nu^2}\cos\nu t + \frac{2Ph\nu}{(\omega^2 - \nu^2)^2 + 4h^2\nu^2}\sin\nu t.$$

We see that the approximation of zeroth order in ε now gives just the exact solution (16.33) and, therefore, the convergence of the method is more rapid in the "resonance" than in the "nonresonance" case.

§ 17. METHOD OF KRYLOV AND BOGOLYUBOV

We will expound the method of KRYLOV and BOGOLYUBOV in connection with the equation

$$\ddot{x} + \varepsilon f_1(\dot{x}) + \omega^2 x + \varepsilon\psi_1(\varepsilon) = \varepsilon p_1(\nu t), \tag{17.1}$$

where ε is a small dimensionless parameter, and $p_1(\nu t)$ is a periodic function of νt with period 2π, which may be expanded in a Fourier series and has mean value zero on every period. We suppose that the functions $f(\dot{x}) = \varepsilon f_1(\dot{x})$, $g(x) = \omega^2 x + \varepsilon\psi_1(x)$ satisfy the conditions of theorem 8, § 15, which assure the existence of at least one periodic solution with frequency ν. The remarks from the beginning of § 16 concerning the choice and the use of the small parameter ε still hold.

As in the perturbation method, we distinguish between two cases which will be separately treated: the "resonance" case, when $\omega \approx p\nu/q$, with p and q integers, greater than 1, and relatively prime, and the "nonresonance" case, for other values of ν sufficiently far from "resonance." We give below in a slightly modified form an improved variant of the method of KRYLOV and BOGOLYUBOV, which has been elaborated by BOGOLYUBOV and MITROPOLSKY [16] in 1955.

It is noteworthy that, unlike the perturbation method, which merely allows the determination of periodic solutions, the method of KRYLOV and BOGOLYUBOV may be used to calculate approximately both periodic and nonperiodic solutions. However, we shall confine ourselves to applying this method, like the perturbation method, only to periodic motions.

For the study of transitory vibrations and of the passing-through-"resonance" phenomena in nonlinear systems, as well as for other interesting applications of the method of KRYLOV and BOGOLYUBOV, we refer to the book by MITROPOLSKY [123].

Let us finally remark that Eq. (17.1), which will be considered in this section, is less general than Eq. (16.1), which has been studied by the perturbation method, since now the whole external force is supposed to be of the order $0(\varepsilon)$. In exchange, this restriction permits a considerable simplification of the formulas and seems to lead to a higher accuracy of the approximate solution, as has been shown by LEVENSON [98, 99] for the equation of DUFFING.

a) "Nonresonance" case ($\omega \neq p\nu/q$)

In the "nonresonance" case the method of KRYLOV and BOGOLYUBOV may be applied in a way similar to that used in § 5b to study free vibrations of weakly nonlinear oscillators.

For $\varepsilon = 0$, Eq. (17.1) has the solution $x = a\cos\alpha$, where $\dot{a} = 0$ and $\dot{\alpha} = \omega$. For $\varepsilon \neq 0$, we try a solution of the form

$$x = a\cos\alpha + \varepsilon u_1(a, \alpha, \tau) + \ldots + \varepsilon^m u_m(a, \alpha, \tau), \tag{17.2}$$

where $\tau = \nu t$. The difference between this expression and Eq. (5.28) adopted in the case of free vibrations is that the functions u_k depend now on τ, too, and are periodic functions with period 2π in both α and τ. However, the amplitude and the total phase of the first harmonic are supposed to satisfy the same differential equations (5.29) and (5.30), i.e.,

$$\dot{a} = \varepsilon A_1(a) + \varepsilon^2 A_2(a) + \ldots + \varepsilon^m A_m(a), \tag{17.3}$$

$$\dot{\alpha} = \omega + \varepsilon B_1(a) + \varepsilon^2 B_2(a) + \ldots + \varepsilon^m B_m(a). \tag{17.4}$$

because, far from "resonance," the phase of the solution is not related to that of the external force, and hence this last one does not influence the values of a and α.

As in the case of free vibrations, the asymptotic developments (17.2)–(17.4) contain a certain degree of arbitrariness, which may be used by requiring that [1])

$$\int_0^{2\pi} u_k(a, \alpha, \tau) \sin \alpha \, d\alpha = 0, \qquad \int_0^{2\pi} u_k(a, \alpha, \tau) \cos \alpha \, d\alpha = 0, \tag{17.5}$$

for $k = 1, 2, \ldots, m$. To obtain the formulas of the first two approximations, we set $m = 2$ in (17.2)–(17.4), but we shall neglect $u_2(a, \alpha, \tau)$ in the expression of $x(t)$, because, as has been already mentioned in § 5, the errors of third order in ε, which are allowed in the expressions of $\dot{a}$ and $\dot{\alpha}$ at the second stage of approximation, bring about errors of second order in ε in the expression of $x(t)$.

The calculation proceeds as for free vibrations with the only difference that now the functions u_k depend on t not only through a and α but also through τ. Requiring that (17.2) satisfy Eq. (17.1) to within an error of third order in ε, we obtain now instead of (5.44) and (5.45) the equations

$$\omega^2 \frac{\partial^2 u_1}{\partial \alpha^2} + 2\omega\nu \frac{\partial^2 u_1}{\partial \alpha \partial \tau} + \nu^2 \frac{\partial^2 u_1}{\partial \tau^2} + \omega^2 u_1 = -\psi_1(a \cos \alpha) - f_1(-\omega a \sin \alpha) +$$
$$+ p_1(\tau) + 2\omega A_1 \sin \alpha + 2\omega a B_1 \cos \alpha, \tag{17.6}$$

$$\omega^2 \frac{\partial^2 u_2}{\partial \alpha^2} + 2\omega\nu \frac{\partial^2 u_2}{\partial \alpha \partial \tau} + \nu^2 \frac{\partial^2 u_2}{\partial \tau^2} + \omega^2 u_2 = -u_1 \psi_1'(a \cos \alpha) -$$
$$- \left(A_1 \cos \alpha - a B_1 \sin \alpha + \omega \frac{\partial u_1}{\partial \alpha} + \nu \frac{\partial u_1}{\partial \tau} \right) f_1'(-\omega a \sin \alpha) +$$
$$+ \left(2\omega A_2 + 2A_1 B_1 - a A_1 \frac{dB_1}{da} \right) \sin \alpha + \left(2\omega a B_2 + a B_1^2 - A_1 \frac{dA_1}{da} \right) \cos \alpha -$$
$$- 2 \left(\omega A_1 \frac{\partial^2 u_1}{\partial a \partial \alpha} + \omega B_1 \frac{\partial^2 u_1}{\partial \alpha^2} - \nu A_1 \frac{\partial^2 u_1}{\partial a \partial \tau} - \nu B_1 \frac{\partial^2 u_1}{\partial \alpha \partial \tau} \right). \tag{17.7}$$

By expanding the functions $\psi_1(a \cos\alpha) + f_1(-\omega a \sin \alpha)$ and $p_1(\tau)$ in Fourier series, we have

$$\psi_1(a \cos \alpha) + f_1(-\omega a \sin \alpha) = C_0(a) + \sum_{k=1}^{\infty} [C_k(a) \cos k\alpha + D_k(a) \sin k\alpha], \tag{17.8}$$

$$p_1(\tau) = \sum_{k=1}^{\infty} (P_k \cos k\tau + Q_k \sin k\tau), \tag{17.9}$$

[1]) Equations (17.5) are obviously equivalent to the condition that a be the total amplitude of the fundamental harmonic of the solution.

where

$$\left.\begin{aligned} C_0(a) &= \frac{1}{2\pi}\int_0^{2\pi} [\psi_1(a\cos\alpha) + f_1(-\omega a\sin\alpha)]\,d\alpha, \\ C_k(a) &= \frac{1}{\pi}\int_0^{2\pi} [\psi_1(a\cos\alpha) + f_1(-\omega a\sin\alpha)]\cos k\alpha\,d\alpha, \\ D_k(a) &= \frac{1}{\pi}\int_0^{2\pi} [\psi_1(a\cos\alpha) + f_1(-\omega a\sin\alpha)]\sin k\alpha\,d\alpha, \end{aligned}\right\} \tag{17.10}$$

$$P_k = \frac{1}{\pi}\int_0^{2\pi} p_1(\tau)\cos k\tau\,d\tau, \qquad Q_k = \frac{1}{\pi} p_1(\tau)\sin k\tau\,d\tau, \qquad k = 1, 2, \ldots \tag{17.11}$$

Since, according to our hypothesis, $p_1(\tau)$ has a zero mean value on every period, its Fourier expansion (17.9) does not contain a constant term P_0.

Next, we note that the right-hand side of Eq. (17.6) is equal to the sum of a function depending only on a and α and of another function depending only on τ. Therefore, we may seek a solution of Eq. (17.6) in the form

$$u_1(a, \alpha, \tau) = \bar{u}_1(a, \alpha) + \tilde{u}_1(\tau), \tag{17.12}$$

where $\bar{u}_1(a, \alpha)$ and $\tilde{u}_1(\tau)$ are to be determined from the equations

$$\omega^2 \frac{\partial^2 \bar{u}_1}{\partial \alpha^2} + \omega^2 \bar{u}_1 = -C_0(a) + [2\omega A_1(a) - D_1(a)]\sin\alpha +$$

$$+ [2\omega a B_1(a) - C_1(a)]\cos\alpha - \sum_{k=2}^{\infty} [C_k(a)\cos k\alpha + D_k(a)\sin k\alpha], \tag{17.13}$$

$$\nu^2 \frac{\partial^2 \tilde{u}_1}{\partial \tau^2} + \omega^2 \tilde{u}_1 = \sum_{k=1}^{\infty} (P_k \cos k\tau + Q_k \sin k\tau). \tag{17.14}$$

The periodicity of $u_1(a, \alpha, \tau)$ in α requires the vanishing of the terms involving $\sin\alpha$ and $\cos\alpha$ in the right-hand side of (17.13). By taking into account (5.48), we then have, as in the case of free vibrations,

$$A_1(a) = \frac{D_1(a)}{2\omega} = \frac{1}{2\pi\omega}\int_0^{2\pi} f_1(-\omega a\sin\alpha)\sin\alpha\,d\alpha, \tag{17.15}$$

$$B_1(a) = \frac{C_1(a)}{2\omega a} = \frac{1}{2\pi\omega a}\int_0^{2\pi} \psi_1(a\cos\alpha)\cos\alpha\,d\alpha. \tag{17.16}$$

Particular solutions of Eqs. (17.13) and (17.14) can now be easily found, e.g., by expanding $\bar{u}_1(a, \alpha)$ and $\tilde{u}_1(\tau)$ in Fourier series with respect to α and τ, respectively, and considering (17.5). The result reads

$$\bar{u}_1(a, \alpha) = \frac{1}{\omega^2} \sum_{\substack{k=0 \\ k \neq 1}}^{\infty} \frac{C_k(a) \cos k\alpha + D_k(a) \sin k\alpha}{k^2 - 1}, \tag{17.17}$$

$$\tilde{u}_1(\tau) = \sum_{k=1}^{\infty} \frac{P_k \cos k\tau + Q_k \sin k\tau}{\omega^2 - k^2\nu^2}, \tag{17.18}$$

and, by summation, gives $u_1(a, \alpha, \tau)$.

Let us now examine Eq. (17.7). As has been already mentioned, the function $u_2(a, \alpha, \tau)$ does not contribute to the second approximation for $x(t)$. Therefore, we confine ourselves to determining $A_2(a)$ and $B_2(a)$ by requiring the coefficients of $\sin \alpha$ and $\cos\alpha$ in the right-hand side of (17.7) to vanish; otherwise, these terms, which are solutions of the homogeneous equation, would induce secular terms in the expression of $u_2(a, \alpha, \tau)$. Then, by using (17.12) and taking into consideration that $\bar{u}_1(a, \alpha)$ does not contain terms in $\sin\alpha$ and $\cos\alpha$, and that the expansion (17.18) of $\tilde{u}_1(\tau)$ has no constant term, we obtain from (17.7) and (17.5)

$$A_2(a) = -\frac{1}{2\omega}\left(2A_1B_1 + aA_1\frac{dB_1}{da}\right) + \frac{1}{2\pi\omega}\int_0^{2\pi}\Big[\bar{u}_1\psi_1'(a\cos\alpha) +$$

$$+\left(A_1\cos\alpha - aB_1\sin\alpha + \omega\frac{\partial\bar{u}_1}{\partial\alpha}\right)f_1'(-\omega a\sin\alpha)\Big]\sin\alpha\, d\alpha, \tag{17.19}$$

$$B_2(a) = -\frac{1}{2\omega a}\left(aB_1^2 - A_1\frac{dA_1}{da}\right) + \frac{1}{2\pi\omega a}\int_0^{2\pi}\Big[\bar{u}_1\psi_1'(a\cos\alpha) +$$

$$+\left(A_1\cos\alpha - aB_1\sin\alpha + \omega\frac{\partial\bar{u}_1}{\partial\alpha}\right)f_1'(-\omega a\sin\alpha)\Big]\cos\alpha\, d\alpha. \tag{17.20}$$

The first approximation is given by

$$x(t) = a\cos\alpha. \tag{17.21}$$

The amplitude a and the total phase α of this fundamental harmonic have to be determined from the differential equations with separable variables

$$\dot{a} = \varepsilon A_1(a), \qquad \dot{\alpha} = \omega + \varepsilon B_1(a), \tag{17.22}$$

where $A_1(a)$ and $B_1(a)$ are given by (17.15) and (17.16). By comparing (17.21) and (17.22) with (5.49) and (5.51), it is readily seen that in the "nonresonance" case considered here the first approximation coincides with that obtained for free vibrations. This apparently unexpected result may be, however, easily understood, upon remembering that the amplitude of the external force has been assumed proportional to ε, and, therefore, its influence on the vibration is negligible in the first approximation if no "resonance" occurs.

The second approximation is given by

$$x_{\text{II}}(t) = a \cos \alpha + \varepsilon \bar{u}_1(a, \alpha) + \varepsilon \tilde{u}_1(\tau), \tag{17.23}$$

where $\bar{u}_1(a, \alpha)$ and $\tilde{u}_1(\tau)$ have the expressions (17.12) and (17.13), respectively, and a, α are solutions of the differential equations with separable variables

$$\dot{a} = \varepsilon A_1(a) + \varepsilon^2 A_2(a), \qquad \dot{\alpha} = \omega + \varepsilon B_1(a) + \varepsilon^2 B_2(a), \tag{17.24}$$

with $A_2(a)$ and $B_2(a)$ given by (17.19) and (17.20). By comparing (17.19) and (17.20) to (5.59), we see that $A_2(a)$ and $B_2(a)$ still have the same expressions as for free vibrations. Therefore, the only difference occurring at the second stage of approximation between the "nonresonance" case and the free vibration is the appearance of the term $\tilde{u}_1(\nu t)$ in the expression of $x_{\text{II}}(t)$.

On the other hand, Eqs. (17.23) and (17.24) show that, unlike the perturbation method, the method of KRYLOV and BOGOLYUBOV allows the determination of both periodic and nonperiodic solutions. When studying periodic solutions only, we have to set $\dot{a} = \dot{\alpha} = 0$, and then, by $(17.24)_1$, it follows that the constant value (or values) of the amplitude, say $a = a_0$, may be found from the algebraic equation

$$\varepsilon A_1(a_0) + \varepsilon^2 A_2(a_0) = 0.$$

Next, introducing a_0 into $(17.24)_2$ and integrating with respect to t, gives the total phase

$$\alpha = [\omega + \varepsilon B_1(a_0) + \varepsilon^2 B_2(a_0)]t + \gamma.$$

It is noteworthy that in the "nonresonance" case the difference between the phase of the external force and that of the solution is determined only to within an additive constant. This is also a consequence of our assumption that the external force is $0(\varepsilon)$. Indeed, we have seen above that, if the external force contains a term that is independent of ε, then the phase of the solution is completely determined even in the "nonresonance" case (cf. § 16a).

b) "Resonance" case ($\omega \approx p\nu/q$)

We shall show first how the method of KRYLOV and BOGOLYUBOV may be applied when studying the behavior of the solution *in the neighborhood of a "resonance."* At the end of this section we shall also consider the more complicated case when the passing from "nonresonance" to "resonance" and the "resonance" itself are concomitantly investigated.

Let us first assume that $\omega \approx p\nu/q$, with p and q integers, and relatively prime, and let us denote

$$\omega^2 - \left(\frac{p}{q}\nu\right)^2 = \varepsilon\Delta. \tag{17.25}$$

The equation of motion (17.1) may then be written

$$\ddot{x} + \left(\frac{p}{q}\nu\right)^2 x = -\varepsilon[x\Delta + \psi_1(x) + f_1(\dot{x}) - p_1(\nu t)]. \tag{17.26}$$

We apply as before the transformation $\tau = \nu t$ and try a solution of (17.26) of the form

$$x(t) = a\cos\alpha + \varepsilon u_1(a, \alpha, \tau) + \ldots + \varepsilon^m u_m(a, \alpha, \tau), \tag{17.27}$$

where $u_k(a, \alpha, \tau)$, $k = 1, 2, \ldots$ are periodic functions with period 2π in both α and τ. In the "resonance" regions, the phase difference ϑ between the solution and the external force exerts a strong influence on the amplitude and frequency of the forced vibrations. Therefore, we write

$$\alpha = \frac{p}{q}\nu t + \vartheta = \frac{p}{q}\tau + \vartheta \tag{17.28}$$

and assume that the functions A_k, B_k, $k = 1, 2, \ldots$, in (17.3) and (17.4) depend on both a and ϑ. Since, by (17.25), it follows that $\omega = p\nu/q + 0(\varepsilon)$, we have

$$\dot{a} = \varepsilon A_1(a, \vartheta) + \varepsilon^2 A_2(a, \vartheta) + \ldots + \varepsilon^m A_m(a, \vartheta), \tag{17.29}$$

$$\dot{\alpha} = \frac{p}{q}\nu + \varepsilon B_1(a, \vartheta) + \varepsilon^2 B_2(a, \vartheta) + \ldots + \varepsilon^m B_m(a, \vartheta). \tag{17.30}$$

Substituting $(17.28)_1$ into (17.30) gives

$$\dot{\vartheta} = \varepsilon B_1(a, \vartheta) + \varepsilon^2 B_2(a, \vartheta) + \ldots + \varepsilon^m B_m(a, \vartheta). \tag{17.31}$$

Equations (17.29) and (17.31) will be used for calculating the functions $a(t)$ and $\vartheta(t)$ after the functions $A_k(a, \vartheta)$ and $B_k(a, \vartheta)$ have been found.

By introducing now (17.27) into (17.26), considering (17.29) and (17.30), and equating coefficients of ε and ε^2, we obtain, as in the preceding section,

$$\left(\frac{p}{q}\nu\right)^2 \frac{\partial^2 u_1}{\partial\alpha^2} + 2\frac{p}{q}\nu^2\frac{\partial^2 u_1}{\partial\alpha\partial\tau} + \nu^2\frac{\partial^2 u_1}{\partial\tau^2} + \left(\frac{p}{q}\nu\right)^2 u_1 = -\psi_1(a\cos\alpha) -$$

$$-f_1\left(-\frac{p}{q}\nu a\sin\alpha\right) + p_1(\tau) + 2\frac{p}{q}\nu A_1\sin\alpha + \left(2\frac{p}{q}\nu a B_1 - a\Delta\right)\cos\alpha, \tag{17.32}$$

$$\left(\frac{p}{q}\nu\right)^2\frac{\partial^2 u_2}{\partial\alpha^2}+2\frac{p}{q}\nu^2\frac{\partial^2 u^2}{\partial\alpha\partial\tau}+\nu^2\frac{\partial^2 u_2}{\partial\tau^2}+\left(\frac{p}{q}\nu\right)^2=-u_1\psi_1'(a\cos\alpha)-$$
$$-\left(A_1\cos\alpha-a_1B_1\sin\alpha+\frac{p}{q}\nu\frac{\partial u_1}{\partial\alpha}+\nu\frac{\partial u_1}{\partial\tau}\right)f_1'\left(-\frac{p}{q}\nu a\sin\alpha\right)+$$
$$+\left(2\frac{p}{q}\nu A_2+2A_1B_1+aA_1\frac{\partial B_1}{\partial a}+aB_1\frac{\partial B_1}{\partial\vartheta}\right)\sin\alpha+\left(2\frac{p}{q}\nu aB_2+\right.$$
$$\left.+aB_1^2-A_1\frac{\partial A_1}{\partial a}-B_1\frac{\partial A_1}{\partial\vartheta}\right)\cos\alpha-2\left(\frac{p}{q}\nu A_1\frac{\partial^2 u_1}{\partial a\partial\tau}+\nu A_1\frac{\partial^2 u_1}{\partial a\partial\tau}+\right.$$
$$\left.+\frac{p}{q}\nu B_1\frac{\partial^2 u_1}{\partial\alpha^2}+\nu B_1\frac{\partial^2 u_1}{\partial\alpha\partial\tau}\right). \tag{17.33}$$

Consider first Eq. (17.32). The periodicity of $u_1(a, \alpha, \tau)$ with respect to α requires the vanishing of the terms containing $\sin\alpha$ and $\cos\alpha$ in the right-hand side, because they are particular solutions of the homogeneous equation. Since now, unlike the "nonresonance" case, the variables α and τ are coupled by (17.28), from which it follows that $k\tau = kq(\alpha - \vartheta)/p$, it is possible that some harmonics in the expansion of $p(\tau)$ do contribute to the terms in $\sin\alpha$ and $\cos\alpha$. It is, therefore, necessary to distinguish between the following two cases.

1. $q > 1$. In this case we have $kq/p \neq 1$ for any integers k and p, and hence $p(\tau)$ does not contribute to the terms containing $\sin\alpha$ and $\cos\alpha$ in the right-hand side of Eq. (17.32). We then obtain

$$A_1(a,\vartheta)=A_1(a)=\frac{D_1(a)}{2\frac{p}{q}\nu}=\frac{1}{2\frac{p}{q}\pi\nu}\int_0^{2\pi}f_1\left(-\frac{p}{q}\nu a\sin\alpha\right)\sin\alpha\,d\alpha, \tag{17.34}$$

$$B_1(a,\vartheta)=B_1(a)=\frac{C_1(a)+a\Delta}{2\frac{p}{q}\nu a}=\frac{\Delta}{2\frac{p}{q}\nu}+\frac{1}{2\frac{p}{q}\nu\pi a}\int_0^{2\pi}\psi_1(a\cos\alpha)\cos\alpha\,d\alpha. \tag{17.35}$$

By making now use of (17.5), we can find a particular solution of Eq. (17.32) in the same way as for the "nonresonance" case. The result is

$$\left.\begin{aligned}
u_1(a,\alpha,\tau)&=\bar{u}_1(a,\alpha)+\tilde{u}_1(\tau),\\
\bar{u}_1(a,\alpha)&=\frac{1}{\left(\frac{p}{q}\nu\right)^2}\sum_{\substack{k=0\\k\neq1}}^{\infty}\frac{C_k(a)\cos k\alpha+D_k(a)\sin k\alpha}{k^2-1},\\
\tilde{u}_1(\tau)&=\frac{1}{\nu^2}\sum_{k=1}^{\infty}\frac{P_k\cos k\tau+Q_k\sin k\tau}{\frac{p^2}{q^2}-k^2},
\end{aligned}\right\} \tag{17.36}$$

where $C_k(a)$, $D_k(a)$, P_k, and Q_k may still be calculated by Eqs. (17.10) and (17.11) in which, however, ω is to be replaced by $p\nu/q$.

2. $q = 1$. We now have from $(17.28)_2$

$$\alpha = p\tau + \vartheta, \qquad p\tau = \alpha - \vartheta,$$

and hence the harmonic of pth order in the expansion of $p_1(\tau)$ does contain terms in $\sin\alpha$ and $\cos\alpha$. Since

$$P_p \cos p\tau + Q_p \sin p\tau = (P_p \sin \vartheta + Q_p \cos \vartheta) \sin \alpha + (P_p \cos \vartheta - Q_p \sin \vartheta) \cos \alpha,$$

we may immediately introduce the necessary corrections into Eqs. (17.34)—(17.36), which become

$$A_1(a, \vartheta) = \frac{D_1(a) - P_p \sin \vartheta - Q_p \cos \vartheta}{2p\nu}, \tag{17.37}$$

$$B_1(a, \vartheta) = \frac{C_1(a) + a\Delta - P_p \cos \vartheta + Q_p \sin \vartheta}{2p\nu a}, \tag{17.38}$$

$$\left.\begin{aligned} u_1(a, \alpha, \tau) &= \overline{u}_1(a, \alpha) + \widetilde{u}_1(\tau), \\ \overline{u}_1(a, \alpha) &= \frac{1}{\left(\dfrac{p}{q}\nu\right)^2} \sum_{\substack{k=0 \\ k\neq 1}}^{\infty} \frac{C_k(a) \cos k\alpha + D_k(a) \sin k\alpha}{k^2 - 1}, \\ \widetilde{u}_1(\tau) &= \frac{1}{\nu^2} \sum_{\substack{k=1 \\ k\neq p}}^{\infty} \frac{P_k \cos k\tau + Q_k \sin k\tau}{p^2 - k^2}. \end{aligned}\right\} \tag{17.39}$$

Next, let us examine Eq. (17.33). As in the preceding section, the functions $A_2(a)$ and $B_2(a)$ will be determined by requiring the vanishing of the coefficients of $\sin\alpha$ and $\cos\alpha$ in the right-hand side, by taking into account that both $\overline{u}_1(a, \alpha)$, and $\widetilde{u}_1(\tau)$ may contribute to such terms. However, by (17.36) and (17.39), it follows that $\dfrac{\partial^2 u_1}{\partial a \partial \alpha}$, $\dfrac{\partial^2 u_1}{\partial a \partial \tau}$, $\dfrac{\partial^2 u_1}{\partial \alpha^2}$, and $\dfrac{\partial^2 u_1}{\partial \alpha \partial \tau}$ do not involve $\sin\alpha$ and $\cos\alpha$, and for $q = 1$ neither $\sin p\tau$ nor $\cos p\tau$. Hence, we may write for any q

$$\begin{aligned} A_2(a, \vartheta) = &-\frac{1}{2\dfrac{p}{q}\nu}\left(2A_1B_1 + aA_1 \frac{\partial B_1}{\partial a} + aB_1 \frac{\partial B_1}{\partial \vartheta}\right) + \\ &+ \frac{1}{2\dfrac{p}{q}\pi\nu} \int_0^{2\pi} \Big[U(a, \alpha)\, \psi'(a \cos \alpha) + (A_1 \cos \alpha - aB \sin \alpha + \\ &+ V(a, \alpha) f_1'\left(-\frac{p}{q}\nu\, a \sin \alpha\right)\Big] \sin \alpha d\alpha, \end{aligned} \tag{17.40}$$

$$B_2(a, \vartheta) = -\frac{1}{2\frac{p}{q}\nu a}\left(aB_1^2 - A_1\frac{\partial A_1}{\partial a} - B_1\frac{\partial A_1}{\partial \vartheta}\right) +$$

$$+\frac{1}{2\frac{p}{q}\pi\nu a}\int_0^{2\pi}\left[U(a, \alpha)\,\psi_1'(a\cos\alpha) + (A_1\cos\alpha - aB_1\sin\alpha +\right.$$

$$\left.+V(a, \alpha))\,f_1'\left(-\frac{p}{q}\nu a\sin\alpha\right)\right]\cos\alpha\,d\alpha, \tag{17.41}$$

where

$$\begin{aligned} U(a, \alpha) &\equiv \overline{u}_1(a, \alpha) + \hat{u}_1\left(\frac{q}{p}\alpha - \frac{q}{p}\vartheta\right), \\ V(a, \alpha) &\equiv \frac{p}{q}\nu\frac{\partial \overline{u}_1}{\partial \alpha}(a, \alpha) + \nu\hat{u}_1'\left(\frac{q}{p}\alpha - \frac{q}{p}\vartheta\right), \end{aligned} \tag{17.42}$$

and $\hat{u}_1(\tau)$ denotes the sum of all harmonics of $\tilde{u}_1(\tau)$ whose order is a multiple of p.

Summarizing the above results, we see that *the first approximation* is given by

$$x_{\mathrm{I}}(t) = a\cos\left(\frac{p}{q}\nu t + \vartheta\right); \tag{17.43}$$

a and ϑ have to be obtained from the system of differential equations

$$\dot{a} = \varepsilon A_1(a, \vartheta), \quad \dot{\vartheta} = \varepsilon B_1(a, \vartheta), \tag{17.44}$$

where $A_1(a, \vartheta)$ and $B_1(a, \vartheta)$ are given by (17.34), (17.35) or by (17.37), (17.38), according as $q > 1$ or $q = 1$.

The second approximation is given by

$$x_{\mathrm{II}}(t) = a\cos\left(\frac{p}{q}\nu t + \vartheta\right) + \varepsilon u_1(a, \alpha, \tau); \tag{17.45}$$

the amplitude a and the phase ϑ of the first harmonic have to be calculated from the system of differential equations

$$\dot{a} = \varepsilon A_1(a, \vartheta) + \varepsilon^2 A_2(a, \vartheta), \qquad \dot{\vartheta} = \varepsilon B_1(a, \vartheta) + \varepsilon^2 B_2(a, \vartheta), \tag{17.46}$$

where $A_2(a, \vartheta)$ and $B_2(a, \vartheta)$ can be found from (17.40)–(17.42), and $u_1(a, \alpha, \tau)$ is given by (17.36) for $q > 1$, and by (17.39) for $q = 1$.

In the remaining part of this section we will show how the method of Krylov and Bogolyubov has to be modified when *the behavior of the solution in the neighborhood of a "resonance" and the passing from "nonresonance" to "resonance"* are concomitantly considered. In this case we must give up the assumption of $\omega^2 - (p\nu/q)^2$ being small, and hence Eq. (17.25), too. Therefore, we maintain the form (17.27) of the solution and the expression (17.28) of α, i.e.,

$$x(t) = a \cos \alpha + \varepsilon u_1(a, \alpha, \tau) + \ldots + \varepsilon^m u_m(a, \alpha, \tau), \tag{17.47}$$

$$\alpha = \frac{p}{q} \nu t + \vartheta = \frac{p}{q} \tau + \vartheta, \tag{17.48}$$

but we modify Eqs. (17.29)—(17.31), by rewriting them in the form

$$\dot{a} = \varepsilon A_1(a, \vartheta) + \varepsilon^2 A_2(a, \vartheta) + \ldots + \varepsilon^m A_m(a, \vartheta), \tag{17.49}$$

$$\dot{\alpha} = \omega + \varepsilon B_1(a, \vartheta) + \varepsilon^2 B_2(a, \vartheta) + \ldots + \varepsilon^m B_m(a, \vartheta), \tag{17.50}$$

$$\dot{\vartheta} = \omega - \frac{p}{q} \nu + \varepsilon B_1(a, \vartheta) + \varepsilon^2 B_2(a, \vartheta) + \ldots + \varepsilon^m B_m(a, \vartheta). \tag{17.51}$$

We remark that Eqs. (17.49)—(17.51) generalize the equations adopted in the "nonresonance" case, which may be recovered by assuming that $\dot{a}$ and $\dot{\alpha}$ do not depend on ϑ, as well as the equations adopted in the "resonance" case, which may be reobtained by setting $\omega - p\nu/q = 0(\varepsilon)$ and conveniently redefining the functions $B_1(a)$ and $B_2(a)$.

Substituting now (17.47) into (17.1), considering (17.49), (17.50), and equating coefficients of ε and ε^2, we obtain

$$\omega^2 \frac{\partial^2 u_1}{\partial \alpha^2} + 2\omega\nu \frac{\partial^2 u_1}{\partial \alpha \partial \sigma \tau} + \nu^2 \frac{\partial^2 u_1}{\partial \tau^2} + \omega^2 u_1 = -\psi_1(a \cos \alpha) - f_1(-\omega a \sin \alpha) +$$
$$+ p_1(\tau) + \left[2\omega A_1 + \left(\omega - \frac{p}{q}\nu\right) a \frac{\partial B_1}{\partial \vartheta}\right] \sin \alpha + \left[2\omega a B_1 - \left(\omega - \frac{p}{q}\nu\right) \frac{\partial A_1}{\partial \vartheta}\right] \cos \alpha, \tag{17.52}$$

$$\omega^2 \frac{\partial^2 u_2}{\partial \alpha^2} + 2\omega\nu \frac{\partial^2 u_2}{\partial \alpha \partial \tau} + \nu^2 \frac{\partial^2 u_2}{\partial \tau^2} + \omega^2 u_2 = -u_1 \psi_1'(a \cos \alpha) -$$
$$\left(-A_1 \cos \alpha - a B_1 \sin \alpha + \omega \frac{\partial u_1}{\partial \alpha} + \nu \frac{\partial u_1}{\partial \tau}\right) f_1'(-\omega a \sin\alpha) +$$
$$+ \left[2\omega A_2 + \left(\omega - \frac{p}{q}\nu\right) a \frac{\partial B_2}{\partial \vartheta} + 2A_1 B_1 + a A_1 \frac{\partial B_1}{\partial a} + a B_1 \frac{\partial B_1}{\partial \vartheta}\right] \sin \alpha +$$
$$+ \left[2\omega a B_2 - \left(\omega - \frac{p}{q}\nu\right) \frac{\partial A_2}{\partial \vartheta} + a B_1^2 - A_1 \frac{\partial A_1}{\partial a} - B_1 \frac{\partial A_1}{\partial \vartheta}\right] \cos \alpha -$$
$$- \left[2\omega A_1 \frac{\partial^2 u_1}{\partial a \partial \alpha} + 2\nu A_1 \frac{\partial^2 u_1}{\partial a \partial \tau} + 2\omega B_1 \frac{\partial^2 u_1}{\partial \alpha^2} + 2\nu B_1 \frac{\partial^2 u_1}{\partial \alpha \partial \tau} + \right.$$
$$\left. + \left(\omega - \frac{p}{q}\nu\right) \frac{\partial u_1}{\partial \alpha} \frac{\partial B_1}{\partial \vartheta} + \left(\omega - \frac{p}{q}\nu\right) \frac{\partial u_1}{\partial \alpha} \frac{\partial A_1}{\partial \vartheta}\right]. \tag{17.53}$$

The further treatment of these equations is rather similar to that used for studying the "resonance" case. Therefore, we omit the details of the calculation, by indicating, however, all important results. In order to study Eq. (17.52) we have to consider, as before, two different cases.

1. $q > 1$. The vanishing of the coefficients of $\sin\alpha$ and $\cos\alpha$ in the right-hand side of Eq. (17.52) yields the system of differential equations

$$2\omega a B_1 - \left(\omega - \frac{p}{q}\nu\right)\frac{\partial A_1}{\partial \vartheta} = C_1(a),$$

$$2\omega A_1 + \left(\omega - \frac{p}{q}\nu\right) a\frac{\partial B_1}{\partial \vartheta} = D_1(a),$$

which admits of the particular solution

$$A_1(a, \vartheta) = A_1(a) = \frac{D_1(a)}{2\omega} = \frac{1}{2\pi\omega}\int_0^{2\pi} f_1(-\,\omega a \sin\alpha)\sin\alpha\, d\alpha, \tag{17.54}$$

$$B_1(a, \vartheta) = B_1(a) = \frac{C_1(a)}{2\omega a} = \frac{1}{2\pi\omega a}\int_0^{2\pi} \psi_1(a\cos\alpha)\cos\alpha\, d\alpha. \tag{17.55}$$

Next, by considering (17.5), we deduce from (17.52) that

$$\left.\begin{aligned} u_1(a, \alpha, \tau) &= \bar{u}_1(a, \alpha) + \tilde{u}_1(\tau), \\ \bar{u}_1(a, \alpha) &= \frac{1}{\omega^2}\sum_{\substack{k=0\\k\neq 1}}^{\infty} \frac{C_k(a)\cos k\alpha + D_k(a)\sin k\alpha}{k^2 - 1}, \\ \tilde{u}_1(\tau) &= \sum_{k=1}^{\infty} \frac{P_k\cos k\tau + Q_k\sin k\tau}{\omega^2 - k^2\nu^2}. \end{aligned}\right\} \tag{17.56}$$

where $C_k(a)$, $D_k(a)$, P_k, and Q_k are given by (17.10) and (17.11).

2. $q = 1$. In this case we must take into account as before the contribution of the harmonic of pth order of $p_1(\tau)$ to the terms in $\sin\alpha$ and $\cos\alpha$. Equating to zero the coefficients of these terms gives the system of differential equations

$$2\omega a B_1 - (\omega - p\nu)\frac{\partial A_1}{\partial \vartheta} = C_1(a) - P_p\cos\vartheta + Q_p\sin\vartheta,$$

$$2\omega A_1 + (\omega - p\nu)\, a\frac{\partial B_1}{\partial \vartheta} = D_1(a) - P_p\sin\vartheta - Q_p\cos\vartheta,$$

which has the particular solution

$$A_1(a, \vartheta) = \frac{D_1(a)}{2\omega} - \frac{P_p \sin \vartheta + Q_p \cos \vartheta}{\omega + p\nu}, \tag{17.57}$$

$$B_1(a, \vartheta) = \frac{C_1(a)}{2\omega a} + \frac{Q_p \sin \vartheta - P_p \cos \vartheta}{a(\omega + p\nu)}. \tag{17.58}$$

From (17.52) it follows now that

$$\left.\begin{aligned} u_1(a, \alpha, \tau) &= \bar{u}_1(a, \alpha) + \tilde{u}_1(\tau), \\ \bar{u}_1(a, \alpha) &= \frac{1}{\omega^2} \sum_{\substack{k=0 \\ k\neq 1}}^{\infty} \frac{C_k(a) \cos k\alpha + D_k(a) \sin k\alpha}{k^2 - 1}, \\ \tilde{u}_1(\tau) &= \sum_{\substack{k=1 \\ k\neq p}}^{\infty} \frac{P_k \cos k\tau + Q_k \sin k\tau}{\omega^2 - k^2\nu^2}. \end{aligned}\right\} \tag{17.59}$$

Finally, by equating to zero the coefficients of $\sin\alpha$ and $\cos\alpha$ in the right-hand side of Eq. (17.53), we obtain for all q the system of differential equations

$$2\omega A_2 + \left(\omega - \frac{p}{q}\nu\right) a \frac{\partial B_2}{\partial \vartheta} = -\left(2A_1 B_1 + aA_1 \frac{\partial B_1}{\partial a} + aB_1 \frac{\partial B_1}{\partial \vartheta}\right) +$$
$$+ \frac{1}{\pi} \int_0^{2\pi} [U(a, \alpha)\psi_1'(a \cos \alpha) + (A_1 \cos \alpha - aB_1 \sin\alpha +$$
$$+ V(a, \alpha)) f_1'(-\omega a \sin \alpha)] \sin \alpha \, d\alpha, \tag{17.60}$$

$$2\omega a B_2 - \left(\omega - \frac{p}{q}\nu\right) \frac{\partial A_2}{\partial \vartheta} = -\left(aB_1^2 - A_1 \frac{\partial A_1}{\partial \vartheta} - B_1 \frac{\partial A_1}{\partial \vartheta}\right) +$$
$$+ \frac{1}{\pi} \int_0^{2\pi} [U(a, \alpha)\psi'(a \cos \alpha) + (A_1 \cos\alpha - aB_1 \sin \alpha +$$
$$+ V(a, \alpha)) f_1'(-\omega a \sin \alpha)] \cos \alpha \, d\alpha, \tag{17.61}$$

where we have introduced the following notations, analogous to (17.42):

$$U(a, \alpha) \equiv \bar{u}_1(a, \alpha) + k_1 \left(\frac{q}{p}\alpha - \frac{q}{p}\vartheta\right),$$

$$V(a, \alpha) \equiv \omega \frac{\partial \bar{u}_1}{\partial \alpha}(a, \alpha) + \nu \hat{u}_1' \left(\frac{q}{p}\alpha - \frac{q}{p}\vartheta\right).$$

Summarizing the above results, we see that *the first approximation* is given by

$$x_{\mathrm{I}}(t) = a \cos\left(\frac{p}{q}\nu t + \vartheta\right); \tag{17.62}$$

a and ϑ have to be determined from the system

$$\dot{a} = \varepsilon A_1(a, \vartheta), \quad \dot{\vartheta} = \omega - \frac{p}{q}\nu + \varepsilon B_1(a, \vartheta), \tag{17.63}$$

where $A_1(a, \vartheta)$ and $B_1(a, \vartheta)$ are given by Eqs. (17.54), (17.55) or by (17.57), (17.58), according as $q > 1$ or $q = 1$.

The second approximation is given by

$$x_{\mathrm{II}}(t) = a \cos\left(\frac{p}{q}\nu t + \vartheta\right) + \varepsilon u_1(a, \alpha, \tau); \tag{17.64}$$

a and ϑ have to satisfy the system

$$\dot{a} = \varepsilon A_1(a, \vartheta) + \varepsilon^2 A_2(a, \dot{\vartheta}), \quad \dot{\vartheta} = \omega - \frac{p}{q}\nu + \varepsilon B_1(a, \vartheta) + \varepsilon^2 B_2(a, \vartheta), \tag{17.65}$$

where $A_2(a, \vartheta)$ and $B_2(a, \vartheta)$ are particular solutions of system (17.60), (17.61), and $u_1(a, \alpha, \tau)$ is given by (17.56) for $q > 1$, and by (17.59) for $q = 1$.

To study periodic vibrations we require that $\dot{a} = \dot{\vartheta} = 0$, and then from $(17.65)_2$ it follows that

$$\omega - \frac{p}{q}\nu = 0(\varepsilon). \tag{17.66}$$

This relation shows that periodic vibrations cannot occur but in the neighborhood of a "resonance," that is for values of ω/ν which are sufficiently close to a rational number. This condition is apparently always fulfilled, since the field of rational numbers is dense in the field of real numbers. However, for real systems, the expansions of the functions $f_1(-\omega a \sin \alpha) + \psi_1(a \cos \alpha)$ and $p_1(\tau)$ usually contain only a small number of harmonics, and hence the number of the values p, q for which "resonance" can actually occur is also small; it generally suffices to take as values of p and q a few natural numbers close to 1.

On the other hand, from (17.66) it results that, in order to study periodic vibrations in the "resonance" case, we may directly use the method expounded at the beginning of this section, i.e., the equations (17.34)–(17.36), since Eq. (17.25), on which this method has been grounded, is always satisfied.

c) Periodic vibrations and their stability according to the first approximation

As we have already seen in the case of free vibrations, the first approximation provides a rather complete description of the qualitative phenomena characterizing nonlinear vibrations. Therefore, we shall insist in this section upon the equations of the first approximation and upon their consequences for the behavior of periodic solutions.

For the sake of simplicity, we consider the case when the equation of motion (17.1) may be written in the form

$$\ddot{x} + \varepsilon f_1(\dot{x}) + \omega^2 x + \varepsilon\psi_1(x) = \varepsilon P_1 \cos \nu t.$$

In this case all coefficients P_p and Q_p, with the only exception of P_1, are zero, and, from (17.38), (17.43), and (17.44), we deduce for the principal "resonance" ($p = q = 1$, $\omega \approx \nu$) that

$$\left.\begin{aligned} x_{\mathrm{I}}(t) &= a \cos(\nu t + \vartheta), \\ \dot{a} &= \varepsilon \frac{D_1(a) - P_1 \sin \vartheta}{2\nu}, \\ \dot{\vartheta} &= \varepsilon \frac{C_1(a) + a\Delta - P_1 \cos \vartheta}{2\nu a}. \end{aligned}\right\} \tag{17.67}$$

Here and in all examples studied by means of the method of KRYLOV *and* BOGOLYUBOV, *we assume that* $a > 0$, $P_1 > 0$, *by including the eventual difference of sign between the displacement and the external force in the phase difference* ϑ.

The last two formulas may be given a different and more significant form. To this end let us remember that in the case of free vibrations (cf. § 5b and c), the first approximation may be formally obtained as solution of an "equivalent" linear differential equation, whose coefficients, $\bar{h}(a)$ and $\bar{\omega}(a)$, are given by

$$\bar{h}(a) = -\frac{\varepsilon D_1(a)}{2\omega a}, \qquad \bar{\omega}^2(a) - \omega^2 = \frac{\varepsilon C_1(a)}{a}. \tag{17.68}$$

The above formulas become now

$$\dot{a} = -\frac{2\omega a\bar{h}(a) + \varepsilon P_1 \sin \vartheta}{2\nu},$$

$$\dot{\vartheta} = \frac{\bar{\omega}^2(a) - \omega^2}{2\nu} + \varepsilon \frac{a\Delta - P_1 \cos \vartheta}{2a\nu}.$$

Finally, by putting $\varepsilon P_1 = P$, taking into account that $\omega^2 - \nu^2 = \varepsilon\Delta$ implies $\omega = \nu + 0(\varepsilon)$, and neglecting terms of second order in ε, we obtain

$$\left.\begin{aligned} 2\nu\dot{a} &= -2\nu a\bar{h}(a) - P \sin \vartheta, \\ 2\nu a\dot{\vartheta} &= a[\bar{\omega}^2(a) - \nu^2] - P \cos \vartheta. \end{aligned}\right\} \tag{17.69}$$

For periodic vibrations we have $\dot{a} = \dot{\vartheta} = 0$, and hence Eqs. (17.69) give

$$\sin \vartheta = -\frac{2\nu a\overline{h}(a)}{P}, \qquad \cos \vartheta = \frac{a[\overline{\omega}^2(a) - \nu^2]}{P}, \tag{17.70}$$

$$a^2[\overline{\omega}^2(a) - \nu^2]^2 + 4\nu^2\overline{h}^2(a)\} = P^2. \tag{17.71}$$

It is readily seen that the last two formulas coincide with those determining the amplitude a and the phase ϑ of the steady-state vibration of a linear oscillator with damping coefficient $\overline{h}(a)$ and natural frequency $\overline{\omega}(a)$.

We then arrive at the following rule: The amplitude and the phase of the periodic vibration of a weakly nonlinear oscillator in the neighborhood of the principal "resonance" are equal to the amplitude and the phase of the steady-state vibration of the linear oscillator with governing equation

$$\ddot{x} + 2\overline{h}(a)\,\dot{x} + \overline{\omega}^2(a)\,x = P \cos \nu t, \tag{17.72}$$

where the parameters $\overline{h}(a)$ and $\overline{\omega}(a)$ have to be calculated by equivalent linearization leaving out the forcing term, as has been shown in § 5c.

Let us examine now the stability of the periodic vibration in the first approximation, i.e., for ε sufficiently small. For an elegant and rigorous approach to this problem, see HALANAY [65]. We content ourselves in the following with giving a simplified treatment, which goes along the line of BOGOLYUBOV and MITROPOLSKY [16]. Let

$$x^*(t) = a^*(t) \cos [\nu t + \vartheta^*(t)], \tag{17.73}$$

be a motion of the oscillator close to the periodic vibration, where

$$a^*(t) = a + \delta a(t), \qquad \vartheta^*(t) = \vartheta + \delta\vartheta(t),$$

and where a and ϑ are the *constant* values of the amplitude and phase of the periodic vibration, which have been obtained as solutions of (17.70) for given ν and P.

By replacing $a(t)$ and $\vartheta(t)$ in (17.69) by $a^*(t)$ and $\vartheta^*(t)$, and neglecting in the first approximation the products of the variations, we obtain

$$\left.\begin{aligned} 2\nu \frac{d\delta a}{dt} &= -2\nu[\overline{h}(a) + a\overline{h}'(a)]\,\delta a - P \cos \vartheta\delta\vartheta, \\ 2\nu a \frac{d\delta\vartheta}{dt} &= [\overline{\omega}^2(a) - \nu^2 + 2a\overline{\omega}(a)\overline{\omega}'(a)]\delta a + P \sin \vartheta\delta\vartheta. \end{aligned}\right\} \tag{17.74}$$

We see that $\delta a(t)$ and $\delta\vartheta(t)$ satisfy a system of linear differential equations with constant coefficients. As it is well-known, if the roots λ_1 and λ_2 of the characteristic equation,

$$\begin{aligned}&4\nu^2 a\lambda^2 + 2\nu\lambda\,\{2\nu a[\overline{h}(a) + a\overline{h}'(a)] - P\sin\vartheta\} + \\ &+ [\overline{\omega}^2(a) - \nu^2 + 2a\overline{\omega}(a)\overline{\omega}'(a)]\,P\cos\vartheta - 2\nu\,[\overline{h}(a) + a\overline{h}'(a)]\,P\sin\vartheta = 0,\end{aligned} \tag{17.75}$$

are not equal, the general solution of system (17.74) is

$$\delta a(t) = C_1 e^{\lambda_1 t} + C_2 e^{\lambda_2 t}, \qquad \delta\vartheta(t) = D_1 e^{\lambda_1 t} + D_2 e^{\lambda_2 t}.$$

In order for the periodic vibration to be asymptotically stable in the large, it is sufficient that both λ_1 and λ_2 have negative real parts, because then the conditions

$$\lim_{t\to\infty} \delta a(t) = 0, \qquad \lim_{t\to\infty} \delta\vartheta(t) = 0$$

are satisfied. Therefore, we require that the sum of the roots of Eq. (17.75) be negative and their product positive, to obtain the *stability conditions*

$$2\nu a[\overline{h}(a) + a\overline{h}'(a)] - P\sin\vartheta > 0,$$

$$[\overline{\omega}^2(a) - \nu^2 + 2a\overline{\omega}(a)\,\overline{\omega}'(a)]\,P\cos\vartheta - 2\,\nu[\overline{h}(a) + a\overline{h}'(a)]\,P\sin\vartheta > 0,$$

which, by (17.70), become

$$2\nu\,\frac{d}{da}\,[a^2\overline{h}(a)] > 0, \tag{17.76}$$

$$\begin{aligned}&a\{[\overline{\omega}^2(a) - \nu^2]\,[\overline{\omega}^2(a) - \nu^2 + 2a\overline{\omega}(a)\overline{\omega}'(a)] + \\ &+4\nu^2\overline{h}^2(a)\,[\overline{h}(a) + a\overline{h}'(a)]\} > 0.\end{aligned} \tag{17.77}$$

Both conditions above may be written in a more expressive form. We first have from (5.64):

$$a^2\overline{h}(a) = -\frac{\varepsilon}{2\pi\omega}\int_0^{2\pi} f_1(-\,\omega a\sin\alpha)\,a\sin\alpha\,d\,\alpha.$$

Next, by interpreting the equations of the first approximation,

$$x = a\cos\alpha, \qquad v = -\,\nu a\sin\alpha,$$

as a transformation from Cartesian to polar coordinates in the phase plane $(x, -v/\nu)$, and neglecting terms of second order in ε, we obtain

$$a^2\bar{h}(a) = \frac{\varepsilon}{2\pi\nu} \oint_{\Gamma(a)} f_1(v)\,\mathrm{d}x,$$

where $\Gamma(a)$ is the circle with radius a and center at the origin of the phase plane, and where the direction of the path of integration is clockwise (in the sense of increasing α).[1] Finally, by making use of Green's formula, the last equation becomes

$$a^2\bar{h}(a) = \frac{\varepsilon}{2\pi\nu} \iint_{\Delta(a)} f_1'(v)\,\mathrm{d}x\,\mathrm{d}v,$$

where $\Delta(a)$ is the region inside the circle $\Gamma(a)$. We have assumed in the introduction that $f_1'(v)$ is strictly positive, with the possible exception of a finite number of values of v for which it may vanish; hence $a^2\bar{h}(a)$ is an increasing function of a. It follows that condition (17.76) is satisfied by all dissipative oscillators considered in this book.

Let us examine now the inequality (17.77). We first rewrite eq. (17.71) as

$$F(\nu^2, a) \equiv a^2\{[\bar{\omega}^2(a) - \nu^2]^2 + 4\nu^2\bar{h}^2(a)\} - P^2 = 0. \tag{17.78}$$

This relation defines the amplitude of the periodic vibration as an implicit function of the square of exciting frequency, ν^2. The graph of this function for $P = \text{const}$, which has been represented in a hypothetical form in Fig. 47, is called *resonance curve*.[2] Denoting the solution of Eq. (17.78) by $a = a(\nu^2)$, and taking into account that

$$\frac{\mathrm{d}a}{\mathrm{d}(\nu^2)} = -\frac{\partial F}{\partial(\nu^2)}\bigg/\frac{\partial F}{\partial a},$$

we obtain from (17.78)

$$\frac{\mathrm{d}a}{\mathrm{d}(\nu^2)} = \frac{a[\bar{\omega}^2(a) - 2\bar{h}^2(a) - \nu^2]}{[\bar{\omega}^2(a) - \nu^2]\,[\bar{\omega}^2(a) - \nu^2 + 2\bar{\omega}(a)\,\bar{\omega}'(a)] + 4\nu^2\bar{h}(a)\,[\bar{h}(a) + a\bar{h}'(a)]}.$$

[1] Cf. also the comments concerning Eqs. (2.21) and (2.22).

[2] The curve of response amplitude versus the square of exciting frequency for $P = \text{const}$, i.e., the graph of the function $a = a(\nu^2)$, is generally called a *response curve* [1] or a *frequency characteristic* [72]. The part of this curve close to a "resonance" is said to be a *resonance curve*; this term is widely used in Russian literature (see, e.g., BOGOLYUBOV and MITROPOLSKY [16]).

We now see that the inequality (17.77) is satisfied if and only if $da/d(\nu^2)$ and $\bar{\omega}^2(a) - 2\bar{h}^2(a) - \nu^2$ have the same sign. Therefore, the condition for stability according to the first approximation may be written in the form

$$\left.\begin{aligned} \frac{da}{d(\nu^2)} &> 0 \quad \text{for} \quad \nu^2 < \bar{\omega}^2(a) - 2\bar{h}^2(a), \\ \frac{da}{d(\nu^2)} &< 0 \quad \text{for} \quad \nu^2 > \bar{\omega}^2(a) - 2\bar{h}^2(a). \end{aligned}\right\} \tag{17.79}$$

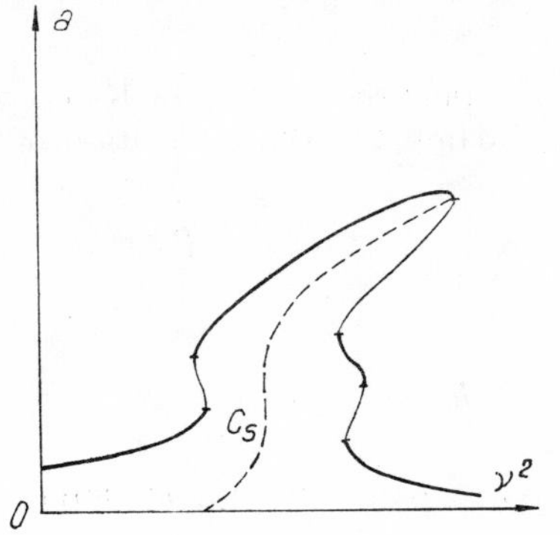

Fig. 47

To obtain a more intuitive interpretation of the relations (17.79) we have represented in Fig. 47 by a broken line the curve C_s of equation $\nu^2 = \bar{\omega}^2(a) - 2\bar{h}^2(a)$, which is sometimes called *skeleton curve* (see, e.g., [16, 82, 122]). We now see that stable periodic vibrations correspond on the left of C_s to arcs of the resonance curve on which a is an increasing function of ν^2, and on the right of C_s to arcs on which a is a decreasing function of ν^2. In Fig. 47 all these arcs are represented by a thickened line. The other arcs of the resonance curve correspond to unstable vibrations. The points at which the resonance curve has a vertical tangent correspond to the passing from a stable to an unstable periodic motion, or vice versa. The existence of unstable solutions, as well as the fact that more than one value of a may correspond to the same value of ν^2, leads to the occurrence of some discontinuities in the variation of the amplitude and phase of the periodic response, when the exciting frequency varies continuously.

We end these comments by a remark concerning the study of periodic vibrations by means of the second approximation. Setting $\dot{a} = \dot{\vartheta} = 0$ in (17.46) yields

$$\varepsilon A_1(a, \vartheta) + \varepsilon^2 A_2(a, \vartheta) = 0, \qquad \varepsilon B_1(a, \vartheta) + \varepsilon^2 B_2(a, \vartheta) = 0.$$

We conclude that for *periodic vibrations*

$$A_1(a, \vartheta) = 0(\varepsilon), \qquad B_1(a, \vartheta) = 0(\varepsilon),$$

and hence we may neglect terms containing $A_1(a, \vartheta)$, $B_1(a, \vartheta)$ in the expressions of $A_2(a, \vartheta)$ and $B_2(a, \vartheta)$, because they introduce only corrections of third order in ε. This observation will be repeatedly used in connection with the applications considered in Chapters V and VI.

d) Illustration of the method: the linear oscillator with weak viscous damping

As an illustration let us apply the method of Krylov and Bogolyubov to the linear oscillator with weak damping, whose equation of motion is

$$\ddot{x} + 2h\dot{x} + \omega^2 x = P \cos \nu t. \tag{17.80}$$

By putting

$$h = \varepsilon h_1, \qquad P = \varepsilon P_1, \tag{17.81}$$

where ε is a small dimensionless parameter, Eq. (17.80) becomes

$$\ddot{x} + \omega^2 x = \varepsilon(-2h_1\dot{x} + P_1 \cos \nu t). \tag{17.82}$$

— "*Nonresonance*" *case*

By assuming $\omega \neq \nu$ and making use of Eqs. (17.21), (17.22), (17.15), and (17.16), where now $f_1(\dot{x}) = 2h_1\dot{x}$, $\psi_1(x) \equiv 0$, we obtain for the first approximation the equations

$$x_{\mathrm{I}}(t) = a \cos \alpha, \qquad A_1(a) = -h_1 a, \qquad B_1(a) = 0,$$

$$\dot{a} = -ha, \qquad \dot{\alpha} = \omega,$$

from which, by integrating the last two equations, it results that

$$x_{\mathrm{I}}(t) = a_0 e^{-ht} \cos(\omega t + \gamma). \tag{17.83}$$

As we have already seen in the general case (cf. § 17a), the external force does not influence this first approximation, since, by assumption, $P = 0(\varepsilon)$.

To obtain the second approximation, we have to apply Eqs. (17.23), (17.24), and (17.17)–(17.20). The result is

$$x_{\text{II}}(t) = a\cos\alpha + \frac{\varepsilon P_1}{\omega^2 - \nu^2}\cos\nu t, \quad A_2(a) = 0, \quad B_2(a) = -\frac{h_1^2}{2\omega},$$

$$\dot{a} = -ha, \qquad \dot{\alpha} = \omega\left(1 - \frac{h^2}{2\omega^2}\right),$$

whence

$$x_{\text{II}}(t) = a_0\,\mathrm{e}^{-ht}\cos\left[\omega\left(1 - \frac{h^2}{2\omega^2}\right)t + \gamma\right] + \frac{P}{\omega^2 - \nu^2}\cos\nu t. \tag{17.84}$$

It is apparent that the second approximation for $x(t)$ contains the first term of the expansion of the exact steady-state solution (16.33) in a power series of ε. Moreover, we now obtain the frequency of the damped part of the exact solution (16.32) up to terms of second order in ε; this can be easily verified by developing the exact expression of the frequency, namely $\sqrt{\omega^2 - h^2}$, in a power series of ε and retaining the first two terms of the expansion.

– *Principal "resonance"*

Suppose now that ν is close to ω and denote $\omega^2 - \nu^2 = \varepsilon\Delta$. By putting $p = q = 1$ into Eqs. (17.43), (17.44), (17.37), and (17.38), we obtain for the first approximation

$$x_{\text{I}}(t) = a\cos(\nu t + \vartheta), \tag{17.85}$$

$$A_1(a, \vartheta) = -h_1 a - \frac{P_1\sin\vartheta}{2\nu}, \qquad B_1(a, \vartheta) = \frac{\Delta}{2\nu} - \frac{P_1\cos\vartheta}{2\nu a},$$

$$\dot{a} = -ha - \frac{P\sin\vartheta}{2\nu}, \qquad \dot{\vartheta} = \frac{\omega^2 - \nu^2}{2\nu} - \frac{P\cos\vartheta}{2\nu a}. \tag{17.86}$$

Theoretically, as we have already mentioned in the general case, it is always possible to calculate $a(t)$ and $\vartheta(t)$ as solutions of system (17.86), and then to derive $x_{\text{I}}(t)$, which will contain both a periodic and a nonperiodic part. However, even in the very simple case above, the integration of system (17.86) is rather difficult, for it comes to approximating the exact solution by the condensed form (17.85), which implies a tedious calculation. In exchange, the periodic vibration can be determined by a straightforward algebraic calculation, by putting $\dot{a} = \dot{\vartheta} = 0$ into (17.86) and solving the equations obtained for a and ϑ. It then follows that

$$a = \frac{P}{\sqrt{(\omega^2 - \nu^2)^2 + 4h^2\nu^2}}, \tag{17.87}$$

$$\sin\vartheta = -\frac{2a\nu h}{P}, \qquad \cos\vartheta = \frac{a(\omega^2 - \nu^2)}{P},$$

$$x_1(t) = \frac{P(\omega^2 - \nu^2)}{(\omega^2 - \nu^2)^2 + 4h^2\nu^2}\cos\nu t + \frac{2h\nu P}{(\omega^2 - \nu^2)^2 + 4h^2\nu^2}\sin\nu t, \tag{17.88}$$

i.e., one recovers the exact expression (16.33) of the steady-state vibration. This result should not be surprising, because now, obviously, $\bar{h}(a) = h$, $\bar{\omega}(a) = \omega$ (cf. the considerations in the preceding section). Furthermore, we see that the stability condition (17.70) now reduces to $h > 0$, and that condition (17.77) is automatically satisfied. For the sake of further comparison with nonlinear systems, we have traced in Fig. 48 the resonance curves of Eq. (17.87) corresponding to various

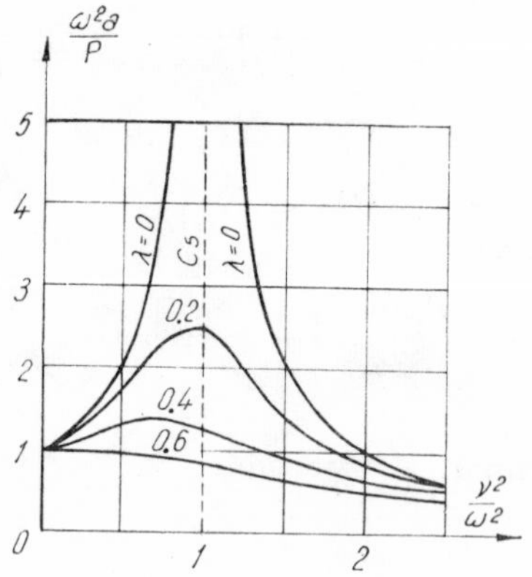

Fig. 48

values of the damping ratio $\zeta = h/\omega$, thus reobtaining the well-known family of response curves of the linear oscillator. The skeleton curve reduces in this case to the vertical line of equation $\omega = \nu$. Resonance proper takes place for $h = 0$ and $\nu \to \omega$, when $a(\nu^2) \to \infty$.

§ 18. OTHER ANALYTICAL METHODS FOR THE APPROXIMATE DETERMINATION OF THE SOLUTIONS

a) The method of finite sums of trigonometric functions

This method may be used for the determination of periodic vibrations excited by periodic external forces. We have already seen that the response of a nonlinear oscillator to a periodic excitation with frequency ν may possibly contain periodic vibrations with frequency ν/q, where q is an integer not less than 1. The actual occurrence of such vibrations, as well as their stability or instability, is conditioned by the characteristics of the oscillator and of the external force.

In any case, if a periodic solution of frequency ν/q does exist, it may be sought in the form of a Fourier series, whose coefficients are determined by requiring the series to satisfy the equation of motion. However, in order to avoid the solving of an infinite system of algebraic equations, it is better to approximate the solution by finite sums of trigonometric functions, i.e.,

$$\bar{x}(t) = a_0 + \sum_{k=1}^{m} \left(a_k \cos \frac{k\nu t}{q} + b_k \sin \frac{k\nu t}{q} \right). \tag{18.1}$$

Substituting this expression into the governing equation and equating the coefficients of all harmonics to zero leads to a set of algebraic equations, whose number is generally greater than that of the coefficients a_k, b_k, and may eventually be even infinite. Therefore, to obtain a compatible system, we limit ourselves to imposing a number of conditions equal to the number of coefficients a_k, b_k to be determined; this procedure yields, obviously, only an approximate solution.

Consider the equation of motion having the general form

$$\ddot{x} + G(x, \dot{x}, \nu t) = 0, \tag{18.2}$$

where $G(x, \dot{x}, \nu t)$ is a periodic function of t with minimum period $2\pi/\nu$, and denote by $E(t)$ the error to within the solution (18.1) satisfies this equation, i.e.,

$$E(t) \equiv \ddot{\bar{x}}(t) + G[\bar{x}(t), \dot{\bar{x}}(t), \nu t]. \tag{18.3}$$

Since $x(t)$ has minimum period $2\pi q/\nu$, it is readily seen from (18.3) that $E(t)$ is also periodic with the same minimum period, and hence it may be expanded in a Fourier series. Equating to zero coefficients of successive harmonics of this expansion gives

$$\int_0^{\frac{2\pi q}{\nu}} E(t) \sin \frac{n\nu t}{q}\, dt = 0, \qquad \int_0^{\frac{2\pi q}{\nu}} E(t) \cos \frac{n\nu t}{q}\, dt = 0, \tag{18.4}$$

where n is an integer. The number of conditions (18.4) we need equals the number of coefficients a_k, b_k in the approximate solution (18.1).

The main advantage of the method of finite sums of trigonometric functions is its logical simplicity. Nevertheless, even the brief exposition above allows considering some disadvantages of the method. First of all, the algebraic equations are nonlinear, and, therefore, the calculation necessary to solve them becomes very complicated if the number of coefficients to be determined is not small enough. On the other hand, the harmonics of superior order in the expansion of $E(t)$ may be neglected only if the magnitudes of the coefficients a_k, b_k decrease rapidly enough as k increases, and this happens only if the parameters of the system satisfy certain conditions. After establishing these conditions one also has to take them into account in the equations determining the coefficients a_k, b_k, too. This whole manipulation may bring about considerable simplifications, but it always requires much attention.

We will now prove that, if the method of finite sums of trigonometric functions is applied to the equation

$$\ddot{x} + \varepsilon f_1(\dot{x}) + \omega^2 x + \varepsilon \psi_1(x) = \varepsilon p_1(\nu t), \tag{18.5}$$

where ε is a small dimensionless parameter, and if one retains in (81.1) only terms with $k = p$, p an integer, then the results obtained for the periodic vibration are identical with those at the first stage of approximation by the method of KRYLOV and BOGOLYUBOV for the neighborhood of the supersubharmonic "resonance" $\omega \approx p\nu/q$.

Assume that $p_1(\nu t)$ can be developed in a Fourier series without constant term:

$$p_1(\nu t) = \sum_{k=1}^{\infty} (P_k \cos k\nu t + Q_k \sin k\nu t), \tag{18.6}$$

where P_k and Q_k are given by (17.11). Suppose also, for the sake of simplicity, that $\psi_1(x)$ is an odd function of x; one may then set $a_0 = 0$ in (18.1).

Since we retain only terms with $k = p$, the approximate solution (18.1) may be written as

$$\bar{x}(t) = a \cos\left(\frac{p}{q} \nu t + \vartheta\right), \tag{18.7}$$

where a and ϑ are two constants related to a_p and b_p by

$$a_p = a \cos \vartheta, \qquad b_p = a \sin \vartheta.$$

The approximate solution (18.7) satisfies Eq. (18.5) to within the error

$$E(t) = \left[\omega^2 - \left(\frac{p}{q}\nu\right)^2\right] a \cos\left(\frac{p}{q}\nu t + \vartheta\right) + \varepsilon\left[\psi_1(a \cos \alpha) + \right.$$

$$\left. + f_1\left(-\frac{p}{q}\nu a \sin \alpha\right) - p_1(\nu t)\right], \tag{18.8}$$

where

$$\alpha(t) = \frac{p}{q} \nu t + \vartheta. \tag{18.9}$$

To determine the free constants a and ϑ we have to use two conditions of the type (18.4). We shall take namely $n = p$ in (18.4), by assuring thus the vanishing of the harmonic of $E(t)$ which has the same order as the approximate solution. It then follows that

$$\int_0^{\frac{2\pi q}{\nu}} E(t) \sin \frac{p\nu t}{q} \, dt = 0, \qquad \int_0^{\frac{2\pi q}{\nu}} E(t) \cos \frac{p\nu t}{q} \, dt = 0. \tag{18.10}$$

In the following we have to distinguish between two cases, as in the method of KRYLOV and BOGOLYUBOV.

1. $q > 1$. In this case, by taking into account (18.6), we have

$$\int_0^{\frac{2\pi q}{\nu}} p_1(\nu t) \sin \frac{p\nu t}{q} \, dt = 0, \qquad \int_0^{\frac{2\pi q}{\nu}} p_1(\nu t) \cos \frac{p\nu t}{q} \, dt = 0,$$

and hence, by substituting (18.8) into (18.10), we obtain

$$-\frac{a\pi q}{\nu}\left[\omega^2 - \left(\frac{p}{q}\nu\right)^2\right] \sin \vartheta + \varepsilon \int_0^{\frac{2\pi q}{\nu}} \left[\psi_1 (a \cos \alpha) + f_1\left(-\frac{p}{q} \nu a \sin \alpha\right)\right] \sin \frac{p\nu t}{q} \, dt = 0,$$

$$\frac{a\pi q}{\nu}\left[\omega^2 - \left(\frac{p}{q}\nu\right)^2\right] \cos \vartheta + \varepsilon \int_0^{\frac{2\pi q}{\nu}} \left[\psi_1 (a \cos \alpha) + f_1\left(-\frac{p}{q} \nu a \sin \alpha\right)\right] \cos \frac{p\nu t}{q} \, dt = 0.$$

Multiplying the first of these equations by $\cos \vartheta$ and the second one by $-\sin \vartheta$ and summing up, and then multiplying the first equation by $-\sin \vartheta$ and the second one by $\cos \vartheta$ and again summing up, we obtain

$$\varepsilon \int_0^{\frac{2\pi q}{\nu}} \left[\psi_1 (a \cos \alpha) + f_1\left(-\frac{p}{q} \nu a \sin \alpha\right)\right] \sin \alpha \, dt = 0,$$

$$\frac{a\pi q}{\nu}\left[\omega^2 - \left(\frac{p}{q}\nu\right)^2\right] + \varepsilon \int_0^{\frac{2\pi q}{\nu}} \left[\psi_1(a \cos \alpha) + f_1\left(-\frac{p}{q}\nu a \sin \alpha\right)\right] \cos \alpha \, dt = 0.$$

Let us apply now the transformation (18.9), by choosing α as variable of integration and taking into account that $d\alpha = (p\nu/q)\,dt$, since ϑ is a constant. Considering (5.48) and dividing out constant factors, we obtain

$$\varepsilon \int_{\vartheta}^{2\pi q + \vartheta} f_1\left(-\frac{p}{q}\nu a \sin\alpha\right) \sin\alpha d\alpha = 0,$$

$$a\pi\left[\omega^2 - \left(\frac{p}{q}\nu\right)^2\right] + \frac{\varepsilon}{p}\int_{\vartheta}^{2\pi p + \vartheta} \psi_1(a\cos\alpha)\cos\alpha\, d\alpha = 0.$$

Next, since all integrands are periodic functions of α with period 2π, we may reduce the interval of integration from $[\vartheta, 2\pi p + \vartheta]$ to $[0, 2\pi]$, on the condition that the result be multiplied by p. It follows that

$$\left.\begin{aligned} &\varepsilon \int_0^{2\pi} f_1\left(-\frac{p}{q}\nu a\sin\alpha\right)\sin\alpha\, d\alpha = 0, \\ &a\pi\left[\omega^2 - \left(\frac{p}{q}\nu\right)^2\right] + \varepsilon\int_0^{2\pi}\psi_1(a\cos\alpha)\cos\alpha\, d\alpha = 0. \end{aligned}\right\} \tag{18.11}$$

Inspection of these equations reveals that, aside from constant factors, they coincide with those obtained in the first approximation by the method of KRYLOV and BOGOLYUBOV for the periodic vibration, i.e.,

$$\varepsilon A_1(a,\vartheta) = 0, \qquad \varepsilon B_1(a,\vartheta) = 0, \tag{18.12}$$

where $A_1(a, \vartheta)$ and $B_1(a, \vartheta)$ are given by (17.34) and (17.35), respectively.

2. $q = 1$. We now have

$$\int_0^{\frac{2\pi}{\nu}} p_1(\nu t)\sin p\nu t\, dt = \frac{\pi Q_p}{\nu}, \qquad \int_0^{\frac{2\pi}{\nu}} p_1(\nu t)\cos p\nu t\, dt = \frac{\pi P_p}{\nu}$$

and, by repeating the same calculation as above, we obtain instead of (18.11) the equations

$$\left.\begin{aligned} &\varepsilon\int_0^{2\pi} f_1\left(-\frac{p}{q}\nu a\sin\alpha\right)\sin\alpha\, d\alpha - \pi P_p\sin\vartheta - \pi Q_p\cos\vartheta = 0, \\ &a\pi(\omega^2 - p^2\nu^2) + \varepsilon\int_0^{2\pi}\psi_1(a\cos\alpha)\cos\alpha\, d\alpha - \pi P_p\cos\vartheta + \pi Q_p\sin\vartheta = 0. \end{aligned}\right\} \tag{18.13}$$

Denoting, as in § 6, by

$$G(x) \equiv \int_0^x g(x)\,\mathrm{d}x$$

the potential energy of the oscillator per unit mass, and considering (18.19), we infer from (6.10) and (6.3) that

$$\dot{x} = \sqrt{2}\,\sqrt{G(x) - G(a)}$$

Separating variables and integrating yields

$$t_0(x) = \frac{1}{\sqrt{2}} \int_a^x \frac{\mathrm{d}x}{\sqrt{G(x) - G(a)}}. \tag{18.20}$$

Since $g(x)$ is odd, $G(x)$ is odd, too, and hence the time $t_0(0)$ over which the oscillator covers the distance from the position of extreme displacement to the equilibrium position is equal to a quarter of the period. Therefore, to the first approximation, the frequency of the vibration is

$$\nu_0 = \frac{\pi}{2t_0(0)}.$$

Substituting now

$$\nu t \approx \nu_0\, t_0\,(x),$$

into (18.17), we obtain

$$\ddot{x} + g_1\,(x) = 0,$$

where $g_1(x) \equiv g(x) - P \cos \nu_0 t_0(x)$, i.e., an equation of the same form as (18.18). The above procedure may be iteratively applied to obtain a sequence of approximate values of the exciting frequency, say $\nu_0, \nu_1, \ldots$, which approaches the exact value of ν corresponding to the prescribed amplitude a. In general, at the nth step, there results:

$$\nu_n = \frac{\pi}{2t_n\,(0)},$$

where

$$\left.\begin{aligned} t_n(x) &\equiv \frac{1}{\sqrt{2}} \int_0^x \frac{\mathrm{d}x}{\sqrt{G_n(x) - G_n(a)}}, \\ G_n(x) &\equiv \int_0^x g_n(x)\,\mathrm{d}x, \qquad g_n(x) \equiv g(x) - P \cos \nu_{n-1}\, t_{n-1}(x). \end{aligned}\right\} \tag{18.21}$$

The convergence is generally very good, and, if P is small enough, even ν_1 gives a fairly accurate approximation for the exciting frequency. RAUSCHER [154] has also indicated how the method should be modified in the case of a damped oscillator.

RAUSCHER's method has the advantage of being also applicable in case the elastic nonlinearity is strong. In exchange, integrals (18.21) cannot be calculated but in a graphical way.

§ 19. SUBHARMONIC VIBRATIONS

At the beginning of this chapter we have seen that a periodic external force with frequency ν acting on a nonlinear oscillator may excite periodic vibrations of frequency ν/q, with q an integer not less than 1, on condition that the damping be small enough. If the equation of motion has the form

$$\ddot{x} + \varepsilon f_1(\dot{x}) + \omega^2 x + \varepsilon \psi_1(x) = \varepsilon P_1 \cos \nu t, \tag{19.1}$$

where ε is a small parameter, then, as shown in § 17, a periodic vibration of frequency ν/q can occur only if the difference $\omega^2 - (\nu/q)^2$ is $0(\varepsilon)$.

In this section we focus our attention on the case of subharmonic vibrations ($q > 1$). We begin by a classification of subharmonic vibrations given by ROSENBERG [173].

If the subharmonic vibration may be represented by the equation

$$x(t) = a_q \cos\left(\frac{\nu}{q}\, t + \vartheta_q\right) + \sum_{k \neq q} a_k \cos\left(\frac{\nu}{k}\, t + \vartheta_k\right), \tag{19.2}$$

where $|a_q| \gg a_k$ for any $k \neq q$, then it is said to be a *strong subharmonic of order* $1/q$. If $a_q \neq 0$ and $a_k = 0$ for any $k \neq q$, the vibration (19.2) is called a *pure subharmonic of order* $1/q$; if, in addition, $\vartheta = 0$, it is said to be a *simple subharmonic of order* $1/q$.

We shall first consider the condition to be satisfied by the nonlinear elastic restoring force of a conservative oscillator of governing equation

$$\ddot{x} + g(x) = P \cos \nu t, \tag{19.3}$$

in order for this equation to admit a simple subharmonic solution

$$x(t) = a \cos \frac{\nu t}{q}. \tag{19.4}$$

This study is particularly interesting, for it proves the actual existence of subharmonic vibrations.

Next, we shall expound a method for the determination of subharmonic vibrations of Eq. (19.1) by successive approximations. Finally, we shall indicate the way one may use the method of KRYLOV and BOGOLYUBOV to study subharmonic vibrations and their stability.

a) Simple subharmonic vibrations of conservative systems

Even if only from a theoretical point of view, it is interesting to prove that there exist equations of the form (19.3) which have *exact* solutions of the form (19.4). This problem has been solved by ROSENBERG [173] and, independently, by KAUDERER [82], with the aid of CHEBYSHEV's polynomials. These polynomials are defined by the relations

$$T_q(x) = \cos (q \arccos x), \quad q = 0, 1, 2, \ldots, \tag{19.5}$$

or, equivalently, by

$$T_q (\cos \alpha) = \cos q \alpha. \tag{19.5'}$$

In order for (19.4) to be a solution of Eq. (19.3), it is necessary that

$$-\frac{\nu^2}{q^2} x + g(x) = P \cos \nu t. \tag{19.6}$$

On the other hand, we successively have from (19.4) and (19.5'):

$$\cos \nu t = \cos q \frac{\nu t}{q} = T_q \left(\cos \frac{\nu t}{q} \right) = T_q \left(\frac{x}{a} \right),$$

and hence, by (19.6), it follows that

$$g(x) = \frac{\nu^2}{q^2} x + PT_q \left(\frac{x}{a} \right). \tag{19.7}$$

We conclude that the equation

$$\ddot{x} + \frac{\nu^2}{q^2} x + PT_q \left(\frac{x}{a} \right) = P \cos \nu t \tag{19.8}$$

has the simple subharmonic solution (19.4) for any integer q greater than 1. By using the definition (19.5'), one may easily find the expressions of CHEBYSHEV's polynomials of various grades. For example

$$T_0(x) = 1, \quad T_1(x) = x, \quad T_2(x) = 2x^2 - 1, \quad T_3(x) = 4x^3 - 3x, \ldots$$

Then, one can determine the relations between the parameters of the system and those of the external force which are necessary for the existence of a simple subharmonic solution. As illustration let us establish under which conditions DUFFING's equation

$$\ddot{x} + \omega^2 x\,(1 + \beta x^2) = P \cos \nu t \tag{19.9}$$

admits the simple subharmonic solution

$$x(t) = a \cos \frac{\nu t}{3}. \tag{19.10}$$

From (19.8) we have for $q = 3$

$$\ddot{x} + \frac{\nu^2}{9}\,x + \frac{Px}{a}\left(\frac{4x^2}{a^2} - 3\right) = P \cos \nu t. \tag{19.11}$$

By comparing this equation to (19.9), we obtain the conditions

$$\omega^2 = \frac{\nu^2}{9} - \frac{3\,P}{a}, \qquad \omega^2 \beta = \frac{4\,P}{a^3},$$

from which, for given ω, β, and P, it results that

$$a = \sqrt[3]{\frac{4P}{\beta \omega^2}}, \qquad \nu^2 = 9\omega^2\left(1 + 3\sqrt[3]{\frac{\beta P^2}{4\omega^4}}\right). \tag{19.12}$$

Inspection of $(19.12)_2$ reveals that, if β and P are $0(\varepsilon)$, then DUFFING's equation admits a simple subharmonic solution of order 1/3 only if $\omega^2 - (\nu/3)^2 = 0(\varepsilon)$, in agreement with our comment at the beginning of this section.

A detailed discussion of various relations existing between the parameters of the elastic characteristic and the quantities a, P, and ν, has been given by KAUDERER [82], § 68. HSU [75] has shown that Eq. (19.8) and its simple subharmonic solution (19.4) may be reduced to the equation of the simple harmonic oscillator and to its steady-state solution under the action of a simple harmonic external force, if one applies a certain transformation of the dependent variable in (19.8).

b) Method of successive approximations

We now set forth a method of successive approximations, which is similar to the perturbation method expounded in § 16, and has been used in various forms by STOKER [184], CARTWRIGHT [23], LUNDQUIST [109], and ROSEAU [172] to study subharmonic vibrations. We follow here with some unimportant changes the exposition of this method given by ROSEAU.

Unlike the method of KRYLOV and BOGOLYUBOV, the method of successive approximations assumes only that the nonlinearity of the oscillator is weak, but it does not require the amplitude of the exciting force to be small. Consequently, the equation of motion may be taken in the form

$$\ddot{x} + \omega^2 x = -\varepsilon\,[\psi_1(x) + f_1(\dot{x})] + P\cos\nu t, \tag{19.13}$$

where ε is a small dimensionless parameter. Let us seek a periodic solution of this equation with frequency ν/q, where q is an integer greater than 1. By applying the transformation

$$\tau = \omega t \tag{19.14}$$

and denoting $x' = \mathrm{d}x/\mathrm{d}\tau$, $x'' = \mathrm{d}^2x/\mathrm{d}\tau^2$, Eq. (19.13) becomes

$$x'' + x = -\frac{\varepsilon}{\omega^2}\,[\psi_1(x) + f_1(\omega x')] + \frac{P}{\omega^2}\cos\frac{\nu}{\omega}\,\tau, \tag{19.15}$$

and the solution looked for has frequency ν/q, i.e., minimum period $2\pi\omega q/\nu$, with respect to τ.

It has been repeatedly pointed out that the equation of motion admits of periodic solutions only for certain initial values of the displacement and velocity, which depend on the frequency and phase of the exciting force. When applying the method of successive approximations, it proves more advantageous to determine the frequency and the phase of the exciting force as functions of the initial values of the displacement and velocity, which are considered as fixed during the calculation. We assume initial conditions of the form

$$x(0) = a, \quad x'(0) = 0. \tag{19.16}$$

This is always possible by taking the time origin at one of the velocity zeros of the periodic motion. Next, denoting by δ the unknown difference of phase between the external force and the periodic solution, we write the equation of motion in the form

$$x'' + x = -\frac{\varepsilon}{\omega^2}\,[\psi_1(x) + f_1(\omega x')] + \frac{P}{\omega^2}\cos\frac{\nu}{\omega}\,(\tau + \delta). \tag{19.17}$$

By a known result in the theory of linear differential equations, we deduce that the solution of (19.17) has to satisfy the integral equation

$$x(\tau) = C_1\cos\tau + C_2\sin\tau + \frac{P}{\omega^2 - \nu^2}\cos\frac{\nu}{\omega}\,(\tau + \delta) -$$

$$-\frac{\varepsilon}{\omega^2}\int_0^{\tau}\{\psi_1[x(s)] + f_1[\omega x'(s)]\}\sin(\tau - s)\,\mathrm{d}s,$$

from which, by considering (19.16), we infer:

$$
\begin{aligned}
x(\tau) &= x_1(\tau) - \frac{\varepsilon}{\omega^2}\int_0^\tau \{\psi_1[x(s)] + f_1[\omega x'(s)]\} \sin(\tau - s)\, ds, \\
x'(\tau) &= x'(\tau) - \frac{\varepsilon}{\omega^2}\int_0^\tau \{\psi_1[x(s)] + f_1[\omega x'(s)]\} \cos(\tau - s)\, ds,
\end{aligned}
\tag{19.18}
$$

where

$$
\begin{aligned}
x_1(\tau) &\equiv a\cos\tau + \frac{P}{\omega^2 - \nu^2}\left[\cos\frac{\nu}{\omega}(\tau + \delta) - \cos\frac{\nu\delta}{\omega}\cos\tau + \right. \\
&\quad \left. + \frac{\nu}{\omega}\sin\frac{\nu\delta}{\omega}\sin\tau\right], \\
x_1'(\tau) &\equiv -a\sin\tau + \frac{P}{\omega^2 - \nu^2}\left[-\frac{\nu}{\omega}\sin\frac{\nu}{\omega}(\tau + \delta) + \right. \\
&\quad \left. + \frac{\nu}{\omega}\sin\frac{\nu\delta}{\omega}\cos\tau + \cos\frac{\nu\delta}{\omega}\sin\tau\right].
\end{aligned}
\tag{19.19}
$$

We now solve system (19.18) iteratively, by taking as first approximation $x(\tau) = x_1(\tau)$, $x'(\tau) = x_1'(\tau)$, and defining the successive approximations by

$$
\begin{aligned}
x_k(\tau) &= x_1(\tau) - \frac{\varepsilon}{\omega^2}\int_0^\tau \{\psi_1[x_{k-1}(s)] + f_1[\omega x_{k-1}'(s)]\} \sin(\tau - s)\, ds, \\
x_k'(\tau) &= x_1'(\tau) - \frac{\varepsilon}{\omega^2}\int_0^\tau \{\psi_1[x_{k-1}(s)] + f_1[\omega x_{k-1}'(s)]\} \cos(\tau - s)\, ds.
\end{aligned}
\tag{19.20}
$$

It may be proved that, if ε is small enough, the sequence $x_k(\tau)$ uniformly approaches a function $x(\tau)$ that satisfies Eq. (19.17) and the initial conditions (19.16).

The function $x_k(\tau)$ will be periodic with period $2\pi\omega q/\nu$ if and only if the displacement and velocity take the same values as the initial ones at time $t = 2\,\pi\omega q/\nu$, i.e.,

$$
x\left(\frac{2\pi q\omega}{\nu}\right) - a = 0, \qquad x'\left(\frac{2\pi q\omega}{\nu}\right) = 0. \tag{19.21}
$$

Substituting (19.18) and (19.19) into (19.21) gives

$$\left(a - \frac{P}{\omega^2 - \nu^2}\cos\frac{\nu\delta}{\omega}\right)\left(1 - \cos\frac{2\pi q\omega}{\nu}\right) - \frac{P\nu}{\omega(\omega^2 - \nu^2)}\sin\frac{\nu\delta}{\omega}\sin\frac{2\pi q\omega}{\nu} +$$

$$+ \frac{\varepsilon}{\omega^2}\int_0^{\frac{2\pi q\omega}{\nu}} \{\psi_1[x(s)] + f_1[\omega x'(s)]\}\sin\left(\frac{2\pi q\omega}{\nu} - s\right) ds = 0,$$

$$\left(a - \frac{P}{\omega^2 - \nu^2}\cos\frac{\nu\delta}{\omega}\right)\sin\frac{2\pi q\omega}{\nu} + \frac{P\nu}{\omega(\omega^2 - \nu^2)}\sin\frac{\nu\delta}{\omega}\left(1 - \cos\frac{2\pi q\omega}{\nu}\right) +$$

$$+ \frac{\varepsilon}{\omega^2}\int_0^{\frac{2\pi q\omega}{\nu}} \{\psi_1[x(s)] + f_1[\omega x'(s)]\}\cos\left(\frac{2\pi q\omega}{\nu} - s\right) ds = 0.$$

We have already seen that the external force may excite a periodic subharmonic vibration of order $1/q$ only if its frequency is close to $q\omega$, i.e.,

$$\frac{q\omega}{\nu} = 1 + 0(\varepsilon).$$

Then, by putting

$$\frac{2\pi q\omega}{\nu} = 2\pi + \varepsilon\eta, \tag{19.22}$$

noting that

$$1 - \cos\frac{2\pi q\omega}{\nu} = 0\,(\varepsilon^2), \qquad \sin\frac{2\pi q\omega}{\nu} = \varepsilon\eta + 0(\varepsilon^3),$$

and taking into account (19.19), we may write the periodicity conditions, after dividing out a factor of ε, in the form

$$\left.\begin{aligned} Q(\delta, \eta) &\equiv \int_0^{2\pi} \{\psi_1[\bar{x}_1'(s)] + f_1[\omega\bar{x}_1'(s)]\}\sin s\, ds + \\ &\quad + \frac{q\eta P}{1 - q^2}\sin q\delta + 0\,(\varepsilon) = 0, \\ R(\delta, \eta) &\equiv \int_0^{2\pi} \{\psi_1[\bar{x}_1(s)] + f_1[\omega\bar{x}_1'(s)]\}\cos s ds + \\ &\quad + \left(a - \frac{P}{1 - q^2}\cos q\delta\right)\eta + 0\,(\varepsilon) = 0. \end{aligned}\right\} \tag{19.23}$$

In the last two equations, we have denoted by $\bar{x}_1(\tau)$ and $\bar{x}_1'(\tau)$ the values of $x_1(\tau)$ and $x_1'(\tau)$ for $\varepsilon = 0$ and $\nu/\omega = q$, that is

$$\left.\begin{aligned} \bar{x}_1(\tau) &\equiv a \cos \tau + \frac{P}{\omega^2(1-q^2)} [\cos q(\tau+\delta) - \\ &\quad - \cos q\delta \cos \tau + q \sin q\delta \sin \tau], \\ \bar{x}_1'(\tau) &\equiv - a \sin \tau + \frac{P}{\omega^2(1-q^2)} [- q \sin q(\tau+\delta) + \\ &\quad + q \sin q\delta \cos \tau + \cos q\, \delta \sin \tau]. \end{aligned}\right\} \tag{19.24}$$

Assume that system (19.23) has for $\varepsilon = 0$ a real solution δ_0, η_0, such that

$$\left.\frac{\partial(Q, R)}{\partial(\delta_0, \eta_0)}\right|_{\varepsilon=0} \neq 0.$$

Then, by applying the existence theorem of implicit functions, we deduce that system (19.23) admits for ε sufficiently small a real solution $\delta(\varepsilon)$, $\eta(\varepsilon)$, which reduces to δ_0, η_0 when $\varepsilon = 0$.

Practically, if ε is small enough, the first approximation, $x_1(\tau)$, given by (19.19) is fairly accurate, and then Eqs. (19.23) yield the required relations between a, η, and δ.

c) The method of KRYLOV and BOGOLYUBOV

In § 17 c we studied periodic vibrations in the neighborhood of the principal "resonance" ($p = q = 1$), as well as their stability or instability according to the first approximation. We will now deal with the same problems in the case of subharmonic vibrations ($p = 1$, $q > 1$). By assuming, as in § 17, that the external force has the form

$$\varepsilon p_1(\nu t) = \varepsilon P_1 \cos \nu t, \tag{19.25}$$

we obtain from Eqs. (17.34)—(17.36), (17.40)—(17.42) for $p = 1$:

$$\left.\begin{aligned} A_1(a, \vartheta) &= \frac{q}{2\pi\nu} \int_0^{2\pi} f_1\left(-\frac{\nu a}{q} \sin \alpha\right) \sin \alpha \, d\alpha, \\ B_1(a, \vartheta) &= \frac{q\Delta}{2\nu} + \frac{q}{2\pi\nu a} \int_0^{2\pi} \psi_1(a \cos \alpha) \cos \alpha \, d\alpha, \end{aligned}\right\} \tag{19.26}$$

$$
\left.\begin{aligned}
u_1(a, \alpha, \tau) &= \bar{u}_1(a, \alpha) + \tilde{u}_1(\tau), \\
\bar{u}_1(a, \alpha) &= \frac{q^2}{\nu^2} \sum_{\substack{k=0 \\ k \neq 1}}^{\infty} \frac{C_k(a) \cos k\alpha + D_k(a) \sin k\alpha}{k^2 - 1}, \\
\tilde{u}_1(\tau) &= \frac{P_1 q^2}{\nu^2(1 - q^2)} \cos \tau,
\end{aligned}\right\} \tag{19.27}
$$

$$
\begin{aligned}
U(a, \alpha) &= \bar{u}_1(a, \alpha) + \frac{P_1 q^2}{\nu^2(1 - q^2)} \cos (q\alpha - q\vartheta), \\
V(a, \alpha) &= \frac{\nu}{q} \frac{\partial \bar{u}_1}{\partial \alpha}(a, \alpha) - \frac{P_1 q^2}{\nu(1 - q^2)} \sin (q\alpha - q\vartheta),
\end{aligned} \tag{19.28}
$$

$$
\left.\begin{aligned}
A_2(a, \vartheta) = &- \frac{q}{2\nu}\left(2A_1B_1 + aA_1\frac{\partial B_1}{\partial a} + aB_1\frac{\partial B_1}{\partial \vartheta}\right) + \\
&+ \frac{q}{2\pi\nu}\int_0^{2\pi}\Big[U(a, \alpha)\, \psi_1'(a \cos \alpha) + \\
&+ (A_1 \cos \alpha - aB_1 \sin \alpha + V(a, \alpha))\, f_1'\Big(-\frac{\nu a}{q} \sin \alpha\Big)\Big] \sin \alpha \, d\alpha, \\
B_2(a, \vartheta) = &- \frac{q}{2\nu a}\left(aB_1^2 - A_1\frac{\partial A_1}{\partial a} - B_1\frac{\partial A_1}{\partial \vartheta}\right) + \\
&+ \frac{q}{2\pi\nu a}\int_0^{2\pi}\Big[U(a, \alpha)\, \psi_1'(a \cos \alpha) + \\
&+ A_1 \cos \alpha - aB_1 \sin \alpha + V(a, \alpha))\, f_1'\Big(-\frac{\nu a}{q} \sin \alpha\Big)\Big] \cos \alpha \, d\alpha.
\end{aligned}\right\} \tag{19.29}
$$

We have already mentioned that for $q > 1$ the influence of the external force does not occur before the second stage of approximation. For periodic subharmonic vibrations, we deduce from (17.46), by setting $\dot{a} = \dot{\vartheta} = 0$, that

$$
\varepsilon A_1(a, \vartheta) + \varepsilon^2 A_2(a, \vartheta) = 0, \quad \varepsilon B_1(a, \vartheta) + \varepsilon^2 B_2(a, \vartheta) = 0. \tag{19.30}
$$

It is now apparent that for periodic vibrations

$$
A_1(a, \vartheta) = 0(\varepsilon), \quad B_1(a, \vartheta) = 0(\varepsilon)
$$

and hence we may neglect terms containing $A_1(a, \vartheta)$, $B_1(a, \vartheta)$ in the expressions of $A_2(a, \vartheta)$, $B_2(a, \vartheta)$, since they introduce into (19.30) only terms of third or higher order in ε. We conclude that the equations of periodic vibrations may be written for $P_1 = \text{const}$ in the form

$$\left.\begin{aligned} M(a, \vartheta, \nu) &\equiv \varepsilon A_1(a, \vartheta) + \frac{\varepsilon^2 q}{2\pi\nu} \int_0^{2\pi} \Big[U(a, \alpha)\, \psi_1' (a \cos \alpha) + \\ &\quad + V(a, \alpha) f_1'\Big(-\frac{\nu a}{q} \sin \alpha \Big)\Big] \sin \alpha \, d\alpha = 0, \\ N(a, \vartheta, \nu) &\equiv \varepsilon B_1(a, \vartheta) + \frac{\varepsilon^2 q}{2\pi\nu a} \int_0^{2\pi} \Big[U(a, \alpha)\, \psi_1'(a \cos \alpha) + \\ &\quad + V(a, \alpha) f_1'\Big(-\frac{\nu a}{q} \sin \alpha \Big)\Big] \cos \alpha \, d\alpha = 0. \end{aligned}\right\} \tag{19.31}$$

By taking into consideration Eqs. (19.28), it is readily seen that system (19.31) can always be solved for $\sin q\vartheta$ and $\cos q\vartheta$ as functions of a and ν, and then, substituting them into the identity $\sin^2 q\vartheta + \cos^2 q\vartheta = 1$, one finds the equation of the resonance curve:

$$F(\nu^2, a) = 0, \tag{19.32}$$

which determines the amplitude of the periodic subharmonic vibration as an implicit function of the square of exciting frequency.

The results obtained by the method of KRYLOV and BOGOLYUBOV cannot be directly compared with those given by the method of successive approximations, because the former assumes $P = 0(\varepsilon)$. On the other hand, as has been shown by LEVENSON [98, 99] for DUFFING's equation, by supposing that the amplitude of the external force is proportional to ε one derives more accurate resonance curves. That is why we will use in the following applications concerning subharmonic vibrations only the method of KRYLOV and BOGOLYUBOV and the method of finite sums of trigonometric functions.

Let us now examine the stability of the periodic subharmonic vibration according to the second approximation. Consider a vibration in the neighborhood of the periodic one, having the form

$$x^*(t) = a^*(t) \cos [\nu t + \vartheta^* (t)] + u_1[a^*(t), \ \nu t + \vartheta^*(t), \nu t],$$

where

$$a^*(t) = a + \delta a(t), \qquad \vartheta^*(t) = \vartheta + \delta\vartheta(t), \tag{19.33}$$

and a, ϑ are the *constant* values of the amplitude and phase of the periodic satisfying Eqs. (19.30) for given ν and P. From the preceding discussion it follows that we may replace the right-hand sides of Eqs. (17.46) by $M(a, \vartheta, \nu)$ and $N(a, \vartheta, \nu)$ if we neglect terms of third order in ε. We then have

$$\dot{a} = M(a, \vartheta, \nu), \qquad \dot{\vartheta} = N(a, \vartheta, \nu). \tag{19.34}$$

Replacing in these equations $a(t)$, $\vartheta(t)$ with $a^*(t)$, $\vartheta^*(t)$, and neglecting the products of the variations, gives

$$\begin{aligned} \frac{d\delta a}{dt} &= \frac{\partial M}{\partial a}\delta a + \frac{\partial M}{\partial \vartheta}\delta\vartheta, \\ \frac{d\delta\vartheta}{dt} &= \frac{\partial N}{\partial a}\delta a + \frac{\partial N}{\partial \vartheta}\delta\vartheta. \end{aligned} \tag{19.35}$$

We see that $\delta a(t)$ and $\delta\vartheta(t)$ satisfy a system of linear differential equations of first order with constant coefficients. If the characteristic equation of the system, i.e.,

$$\lambda^2 - \left(\frac{\partial M}{\partial a} + \frac{\partial N}{\partial \vartheta}\right)\lambda + \frac{\partial M}{\partial a}\frac{\partial N}{\partial \vartheta} - \frac{\partial N}{\partial a}\frac{\partial M}{\partial \vartheta} = 0, \tag{19.36}$$

has two different roots, λ_1 and λ_2, then system (19.35) admits of solutions of the form

$$\delta a(t) = C_1 e^{\lambda_1 t} + C_2 e^{\lambda_2 t}, \qquad \delta\vartheta(t) = D_1 e^{\lambda_1 t} + D_2 e^{\lambda_2 t}.$$

In order for the periodic vibration to be asymptotically stable in the large it is sufficient that both roots of Eq. (19.36) have negative real parts, since then

$$\lim_{t\to\infty} \delta a(t) = 0, \qquad \lim_{t\to\infty} \delta\vartheta(t) = 0.$$

Therefore, we require that the sum of the roots of (19.35) be negative and their product positive, to obtain the *conditions of stability*:

$$\frac{\partial M}{\partial a} + \frac{\partial N}{\partial \vartheta} < 0, \qquad \frac{\partial M}{\partial a}\frac{\partial N}{\partial \vartheta} - \frac{\partial N}{\partial a}\frac{\partial M}{\partial \vartheta} > 0. \tag{19.37}$$

Assuming that condition $(19.37)_1$ is satisfied, we may prove that the points at which the resonance curve (19.32) has a vertical tangent correspond, as in the case of the principal "resonance," to the passing from a stable to an unstable periodic vibration or *vice versa*. Indeed, denoting by $a(\nu)$ and $\vartheta(\nu)$ the roots of system (19.31), we have

$$M[a(\nu), \vartheta(\nu), \nu] \equiv 0, \qquad N[a(\nu), \vartheta(\nu), \nu] \equiv 0.$$

Differentiating these identities yields

$$\frac{\partial M}{\partial a}\frac{\mathrm{d}a}{\mathrm{d}\nu}+\frac{\partial M}{\partial \vartheta}\frac{\mathrm{d}\vartheta}{\mathrm{d}\nu}+\frac{\partial M}{\partial \nu}=0, \qquad \frac{\partial N}{\partial a}\frac{\mathrm{d}a}{\mathrm{d}\nu}+\frac{\partial N}{\partial \vartheta}\frac{\partial \vartheta}{\partial \nu}+\frac{\partial N}{\partial \nu}=0,$$

from which there results:

$$\frac{\mathrm{d}a}{\mathrm{d}\nu}=\frac{\dfrac{\partial M}{\partial \vartheta}\dfrac{\partial N}{\partial \nu}-\dfrac{\partial N}{\partial \vartheta}\dfrac{\partial M}{\partial \nu}}{\dfrac{\partial M}{\partial a}\dfrac{\partial N}{\partial \vartheta}-\dfrac{\partial N}{\partial a}\dfrac{\partial M}{\partial \vartheta}}. \tag{19.38}$$

We now see that the expression in the left-hand side of $(19.37)_2$ changes of sign, and hence the stable or unstable character of the motion also changes at the points where the denominator of the fraction in the right-hand side of (19.38) vanishes, i.e., where the resonance curve has a vertical tangent.

Chapter V — Forced vibrations of conservative systems

§ 20. THE OSCILLATOR WITH CUBIC ELASTIC RESTORING FORCE

It has been already noticed in § 9 that the oscillator with cubic elastic restoring force is the simplest oscillator displaying specific nonlinear properties. On the other hand, the cubic characteristic may satisfactorily approximate for small values of $|x|$ any characteristic given by an analytic odd function of x.

This section will be devoted entirely to the study of forced vibrations of the oscillator with cubic elastic restoring force acted on by a simple harmonic excitation. The governing equation is

$$\ddot{x} + \omega^2 x(1 + \beta x^2) = P \cos \nu t. \tag{20.1}$$

In the main part of this section (§ 20a, b, c) we shall deal with harmonic solutions of Eq. (20.1). In § 20d subharmonic solutions will be considered too. Finally, in § 20e, we shall study the simultaneous action of two periodic external forces with different frequencies, in order to point out the departure from the superposition valid in the linear case.

We assume as before that $g'(x) = \omega^2(1 + 3\beta x^2) > 0$ for any x. This condition is fulfilled either if $\beta > 0$ (as supposed in § 9 for the study of free vibrations), or if $\beta < 0$, but $|x| < 1/\sqrt{-3\beta}$. For the sake of completeness we shall consider both these cases in the following.

Equation (20.1) has been considered for the first time by DUFFING [55] and is called, therefore, DUFFING's equation. In the forties the harmonic and subharmonic solutions of this equation have been thoroughly studied by FRIEDRICHS and STOKER [59], LEVENSON [97], and STOKER [182]. Experimental investigations of periodic solutions of Eq. (20.1) have been undertaken by DUFFING himself, and then by many others, especially by LUDEKE [108].

Exact solutions of DUFFING's equation are known so far only for particular values of the quantities ω, β, and P (see § 19a). Therefore, to study this equation we shall mostly use the methods set forth in the preceding chapter, which allow the approximate calculation of the solution when β is small enough. Before elaborating we will consider once more the order of magnitude of various terms in Eq. (20.1), a problem which has been briefly dealt with at the beginning of § 16.

A reliable method for the determination of the order of magnitude of various physical quantities is to introduce dimensionless variables and parameters, and then to compare numerically the terms occurring in the calculation. Some authors, e.g., LEVENSON [97] and KAUDERER [82], introduce into (20.1) the dimensionless variables

$$\tau = \omega t, \qquad \xi = \frac{\omega^2 x}{P},$$

and arrive at the equation

$$\frac{d^2\xi}{d\tau^2} + \xi(1 + \varepsilon\xi^2) = \cos \Omega\tau, \tag{20.2}$$

where $\varepsilon = \beta P^2/\omega^4$, $\Omega = \nu/\omega$. Then, assuming that ξ is of order of magnitude unity and that $\varepsilon \ll 1$, the order of magnitude of each term is fully determined by its power in ε. This procedure may be applied, β being small, only if P is *not* proportional to β, since, otherwise, the magnitude of ξ could tend to infinity as P, or β tends to zero. In other words, the transformation above is correct when studying the weakly nonlinear equation (20.1) as a deviation for β small but nonzero from the equation of *forced* vibrations of the linear conservative oscillator, i.e., from the equation

$$\ddot{x} + \omega^2 x = P \cos \nu t.$$

However, as has been shown by LEVENSON [99], it is possible to derive a more accurate solution by supposing that P *is* proportional to β, i.e., by studying the weakly nonlinear equation (20.1) as a deviation, for both β and P small but nonzero, from the equation of *free* vibrations of the simple harmonic oscillator, i.e., from the equation

$$\ddot{x} + \omega^2 x = 0.$$

In this case, one may still introduce new dimensionless variables, such as

$$\tau = \omega t, \qquad \xi = x/A,$$

where A is a length of the same order of magnitude as the maximum displacement which may be evaluated by making use of the first relation (15.4′). Then, the equation of motion becomes

$$\frac{d^2\xi}{d\tau^2} + \xi(1 + \varepsilon\xi^2) = \varepsilon P_1 \cos \Omega\tau,$$

where $\varepsilon = \beta A^2$, $\Omega = \nu/\omega$, $\varepsilon P_1 = P/\omega^2 A$; ξ and P_1 are of order of magnitude unity, and the order of magnitude of all other terms is given by the power of ε involved. It may be shown, however, that the final results obtained by this method are independent of A. That is why we will adopt in the following a simplified procedure, which avoids the introduction of a dimensionless displacement but leads, nevertheless, to the same results. We shall only assume that both β and P are proportional to the same small parameter ε. Then, the order of magnitude of all other quantities will be determined by their powers in ε, irrespective of their physical dimensions. Obviously, this method may be applied only when the quantities βA^2 and $P/(\omega^2 A)$ are of the same order of magnitude and are small in comparison with unity. This assumption is to be verified at the end of the calculation, when a more accurate evaluation of A may be easily found by using the approximate solution.

a) Use of the perturbation method

— "*Nonresonance*" *case* ($\omega/\nu \neq p$ for any integer p)

By applying the transformation

$$\tau = \nu t, \tag{20.3}$$

Eq. (20.2) becomes

$$x'' + \eta^2 x(1 + \beta x^2) = \frac{P}{\nu^2} \cos \tau, \tag{20.4}$$

where

$$x'' = \frac{d^2 x}{d\tau^2}, \qquad \eta = \frac{\omega}{\nu}. \tag{20.5}$$

Let us seek a harmonic solution of this equation, i.e., a periodic solution that has the same minimum period as the external force. Assume that β is small and denote

$$\eta^2 \beta = \varepsilon \bar{\beta}, \tag{20.6}$$

where ε is a small parameter. Then the equation of motion may be rewritten as

$$x'' + \eta^2 x = \frac{P}{\nu^2} \cos \tau - \varepsilon \bar{\beta} x^3. \tag{20.7}$$

Since, by assumption, η is not an integer, this equation has for $\varepsilon = 0$ the unique generating solution

$$\bar{x}(\tau) = \frac{P}{\nu^2(\eta^2 - 1)} \cos \tau = \frac{P}{\omega^2 - \nu^2} \cos \tau. \tag{20.8}$$

It has been shown in § 16a that any periodic solution with minimum period 2π of Eq. (20.7) may be expanded in a power series in ε of the form

$$x(\tau;\varepsilon) = \bar{x}(\tau) + x_1(\tau)\,\varepsilon + x_2(\tau)\,\varepsilon^2 + \ldots, \tag{20.9}$$

whose coefficients are periodic functions with minimum period 2π, and which is convergent for ε nonzero but sufficiently small. By substituting (20.9) into (20.7) and equating coefficients of like powers of ε, we obtain

$$x_1'' + \eta^2 x_1 = -\bar{\beta}\bar{x}^3, \tag{20.10}$$

$$x_2'' + \eta^2 x_2 = -3\bar{\beta}x_1\bar{x}^2, \quad \text{etc.} \tag{20.11}$$

Next, introducing (20.8) into (20.10) yields

$$x_1'' + \eta^2 x_1 = -\frac{\bar{\beta}P^3}{4(\omega^2-\nu^2)^3}(\cos 3\tau + 3\cos\tau).$$

Since η is not an integer, the only periodic solution with minimum period 2π of this equation is

$$x_1(\tau) = -\frac{\bar{\beta}P^3}{4(\omega^2-\nu^2)^3}\left(\frac{1}{\eta^2-9}\cos 3\tau + \frac{3}{\eta^2-1}\cos\tau\right). \tag{20.12}$$

Putting (20.8) and (20.12) into (20.11), we may easily calculate $x_2(\tau)$ and so on However, we content ourselves with the first approximation, because the subsequent terms of the expansion do not qualitatively change the solution. From (20.8), (20.9), and (20.12) we deduce that

$$x(\tau;\varepsilon) = \frac{P}{\omega^2-\nu^2}\left\{\left[1-\frac{3\bar{\beta}\nu^2P^2\varepsilon}{4(\omega^2-\nu^2)^3}\right]\cos\tau - \frac{\bar{\beta}\nu^2P^2\varepsilon}{4(\omega^2-\nu^2)^2(\omega^2-9\nu^2)}\cos 3\tau\right\} + 0(\varepsilon^2),$$

or, by reverting to the former variable t and to the parameter β,

$$x(t;\beta) = \frac{P}{\omega^2-\nu^2}\left\{\left[1-\frac{3\beta\omega^2P^2}{4(\omega^2-\nu^2)^3}\right]\cos\nu t - \frac{\beta\omega^2P^2}{4(\omega^2-\nu^2)^2(\omega^2-9\nu^2)}\cos 3\nu t\right\} + 0(\beta^2). \tag{20.13}$$

— *Principal "resonance"* ($\omega \approx \nu$)

Let us now consider the principal "resonance," i.e., the case $\eta \approx 1$. Then, as has been shown in § 16b, we have to assume that the amplitude of the external force is of order ε. By setting

$$\eta^2 = 1 + \varepsilon\bar{\Delta}, \tag{20.14}$$

$$\frac{P}{\nu^2} = \varepsilon\bar{P}, \tag{20.15}$$

Eq. (20.7) assumes the form

$$x'' + x = -\varepsilon\,(x\bar{\Delta} + \bar{\beta}x^3 - \bar{P}\cos\tau) \tag{20.16}$$

and has for $\varepsilon = 0$ the periodic generating solution

$$\bar{x}(\tau) = M\cos\tau + N\sin\tau, \tag{20.17}$$

which depends on two arbitrary constants. For $\varepsilon \neq 0$ we try a solution of the form

$$x(\tau;\varepsilon) = \bar{x}(\tau) + x_1(\tau)\,\varepsilon + x_2(\tau)\,\varepsilon^2 + \ldots \tag{20.18}$$

Substituting this expansion into (20.16) and equating coefficients of like powers of ε, we obtain a set of recursive linear differential equations:

$$\begin{aligned} x_1'' + x_1 &= -\,\bar{x}\,\bar{\Delta} - \bar{\beta}\,\bar{x}^3 + \bar{P}\cos\tau, \\ x_2'' + x_2 &= -\,x_1\bar{\Delta} - 3\,\bar{\beta}\,x_1\bar{x}^2, \quad \text{etc.} \end{aligned} \tag{20.19}$$

Next, introducing $\bar{x}(\tau)$ into $(20.19)_1$ leads to

$$\begin{aligned} x_1'' + x_1 = &-\left(M\bar{\Delta} + \frac{3}{4}\bar{\beta}M^3 + \frac{3}{4}\bar{\beta}\,MN^2 - \bar{P}\right)\cos\tau - \\ &N\left(\bar{\Delta} + \frac{3}{4}\bar{\beta}M^2 + \frac{3}{4}\bar{\beta}N^2\right)\sin\tau - \frac{1}{4}\bar{\beta}M(M^2 - 3N^2)\cos 3\tau - \\ &-\frac{1}{4}\bar{\beta}N\,(3M^2 - N^2)\sin 3\tau. \end{aligned}$$

The periodicity of $x_1(\tau)$ requires the coefficients of $\sin\tau$ and $\cos\tau$ in the right-hand side to vanish, which yields

$$N = 0, \qquad M\bar{\Delta} + \frac{3}{4}\bar{\beta}M^3 - \bar{P} = 0. \tag{20.20}$$

Then, the differential equation reduces to

$$x_1'' + x_1 = -\frac{1}{4}\bar{\beta} M^3 \cos 3\tau$$

and has the general solution

$$x_1(\tau) = M_1 \cos \tau + N_1 \sin \tau + \frac{1}{32}\bar{\beta} M^3 \cos 3\tau, \tag{20.21}$$

where M_1 and N_1 are arbitrary constants to be determined by requiring the periodicity of x_2. Substituting (20.17) and (20.21) into $(20.19)_2$ and equating to zero the coefficients of $\cos \tau$ and $\sin \tau$ in the right-hand side gives

$$N_1 = 0, \qquad M_1\bar{\Delta} + \frac{9}{4}\bar{\beta} M_1 M^2 + \frac{3}{128}\bar{\beta}^2 M^5 = 0. \tag{20.22}$$

It then follows from (20.17), (20.18), and (20.21) that

$$x(\tau; \varepsilon) = (M + \varepsilon M_1) \cos \tau + \frac{\varepsilon}{32}\bar{\beta} M^3 \cos 3\tau + 0(\varepsilon^2). \tag{20.23}$$

Reverting to the variable t and to the parameters β, ω, ν, by means of Eqs. (20.3), $(20.5)_2$, (20.6), (20.14), and (20.15), and taking into account that $\omega^2/\nu^2 = 1 + 0(\varepsilon)$, we may rewrite Eqs. (20.23), $(20.20)_2$, and $(20.22)_2$ as

$$x(t; \varepsilon) = (M + \varepsilon M_1) \cos \nu t + \frac{1}{32}\beta M^3 \cos 3\nu t + 0(\varepsilon^2), \tag{20.24}$$

$$\nu^2 = \omega^2 + \frac{3}{4}\beta\omega^2 M^2 - \frac{P}{M}, \tag{20.25}$$

$$\varepsilon M_1 = -\frac{3\beta^2\omega^2 M^5}{128\left(\omega^2 - \nu^2 + \frac{9}{4}\beta\omega^2 M^2\right)} + 0(\varepsilon^2). \tag{20.26}$$

Given ω, β, ν, and P, we can calculate M from (20.26), and then Eq. (20.24) gives the solution to within an error of second order in ε. The physical meaning of Eq. (20.25), as well as the qualitative properties of the solution (20.24) will be discussed in § 20d, after obtaining similar results by means of other analytical methods. We confine ourselves here to noting that, for certain values of the exciting frequency ν, Eq. (20.25) may yield three different values of the response amplitude M. This phenomenon, which is typical for nonlinear vibrations, is closely related to the stable or unstable character of the periodic response.

b) Use of the method of KRYLOV and BOGOLYUBOV

— "*Nonresonance*" *case* ($\omega \neq p\nu/q$)

We again consider Eq. (20.2) and, assuming that β and P are small, we set

$$\omega^2\beta = \varepsilon\beta_1, \qquad P = \varepsilon P_1, \tag{20.27}$$

where ε is a small parameter. The equation of motion becomes

$$\ddot{x} + \omega^2 x = \varepsilon(-\beta_1 x^3 + P_1 \cos \nu t). \tag{20.28}$$

Putting $f_1(\dot{x}) \equiv 0$, $\psi_1(x) = \beta_1 x^3$ into (17.10) gives

$$C_1(a) = \frac{3}{4}\beta_1 a^3, \qquad C_3(a) = \frac{1}{4}\beta_1 a^3,$$

and all other coefficients $C_k(a)$, $D_k(a)$ vanish. It then follows from (17.15), (17.16), (17.21), and (17.22):

$$A_1(a) = 0, \qquad B_1(a) = \frac{3\beta_1 a^2}{8\omega}, \qquad x_{\mathrm{I}}(t) = a \cos \alpha,$$

$$\dot{a} = 0, \qquad a = a_0, \qquad \dot{\alpha} = \omega\left(1 + \frac{3}{8}\beta a_0^2\right) \equiv \omega_{\mathrm{I}}.$$

Integrating the last equation and substituting a and α into $x_{\mathrm{I}}(t)$ yields the first approximation

$$x_{\mathrm{I}}(t) = a_0 \cos(\omega_{\mathrm{I}} t + \gamma), \tag{20.29}$$

where a_0 and γ are arbitrary constants to be determined by the initial conditions. It is readily seen that this solution coincides with the solution (9.43) obtained for the free vibration.

To find the second approximation we now apply Eqs. (17.23), (17.24), and (17.17)—(17.20), successively obtaining

$$\bar{u}_1(a, \alpha) = \frac{\beta_1 a^3}{32\omega^2}, \qquad \tilde{u}_1(\nu t) = \frac{P_1}{\omega^2 - \nu^2}\cos \nu t,$$

$$A_2(a) = 0, \qquad B_2(a) = -\frac{15\beta_1^2 a^4}{256\,\omega^3},$$

$$\dot{a} = 0, \qquad a = a_0, \qquad \dot{\alpha} = \omega\left(1 + \frac{3}{8}\beta a_0^2 - \frac{15}{256}\beta^2 a_0^4\right) \equiv \omega_{\mathrm{II}},$$

$$x_{\mathrm{II}}(t) = a_0 \cos(\omega_{\mathrm{II}} t + \gamma) + \frac{1}{32}\beta a_0^3 \cos 3(\omega_{\mathrm{II}} t + \gamma) + \frac{P}{\omega^2 - \nu^2}\cos \nu t. \tag{20.30}$$

We remark that the second approximation results from the superposition of the forced vibration of the linear oscillator with the free nonlinear vibration found at the same stage of approximation (cf. Eq. (9.46)). This uncoupling of nonlinear and forcing terms still persists in the second approximation because of our assumption $P = 0(\varepsilon)$. As has been already mentioned at the beginning of § 17, we cannot directly compare in the "nonresonance" case the method of KRYLOV and BOGOLYUBOV with the perturbation method, since the latter assumes that P is independent of ε and may be applied only for periodic vibrations.

– *Principal "resonance"* ($\omega \approx \nu$)

Let us now suppose that ν is close to ω and let us set $\omega^2 - \nu^2 = \varepsilon\Delta$. Since $p = q = 1$, the functions $C_k(a)$ and $D_k(a)$ have the same expressions as in the "nonresonance" case, and hence Eqs. (17.37), (17.38), and (17.44) give as a first approximation

$$A_1(a, \vartheta) = -\frac{P_1 \sin \vartheta}{2\nu}, \qquad B_1(a, \vartheta) = \frac{\Delta}{2\nu} + \frac{3\beta a_1 a^2}{8\nu} - \frac{P_1 \cos \vartheta}{2a\nu},$$

$$\dot{a} = \varepsilon A_1(a, \vartheta), \qquad \dot{\vartheta} = \varepsilon B_1(a, \vartheta).$$

We assume, as in the general theory (cf. § 17c) that a and P_1 are positive, by including the eventual difference of sign between the displacement and the external force in the phase difference ϑ. For the periodic vibration ($\dot{a} = \dot{\vartheta} = 0$) there results:

$$\sin \vartheta = 0, \quad \vartheta = 0 \text{ or } \pi, \quad \cos \vartheta = \pm 1,$$

$$\Delta + \frac{3\beta_1 a^2}{4\nu} \mp \frac{P_1}{a} = 0,$$

from which, by considering (20.27), we obtain

$$\nu^2 = \omega^2 + \frac{3}{4}\beta\omega^2 a^2 \mp \frac{P}{a}. \tag{20.31}$$

Then, from (17.43), we deduce that

$$x_{\text{I}} = a \cos(\nu t + \vartheta), \quad \vartheta = 0 \text{ or } \pi, \tag{20.32}$$

where a is to be determined by Eq. (20.31), in which the minus sign corresponds to $\vartheta = 0$, and the plus sign to $\vartheta = \pi$. It should be noted that periodic vibrations may occur either in phase or in opposition of phase with the exciting force.

To determine the second approximation we use Eqs. (17.39)—(17.41), obtaining

$$u_1(a, \alpha, \tau) = \frac{\beta_1 a^3}{32\nu^2} \cos 3\alpha, \qquad A_2(a, \vartheta) = \frac{P_1 \sin \vartheta}{8\nu^3}\left(\Delta + \frac{9}{4}\beta_1 a^2\right),$$

$$B_2(a, \vartheta) = -\frac{1}{8\nu^3}\left(\Delta + \frac{3}{4}\beta_1 a^2 - \frac{P_1 \cos\vartheta}{a}\right)\left(\Delta + \frac{3}{4}\beta_1 a^2\right) + \frac{3\beta_1^2 a^4}{256\,\nu^3}.$$

For the periodic vibration ($\dot{a} = \dot{\vartheta} = 0$) we have by (17.46)

$$\varepsilon A_1(a, \vartheta) + \varepsilon^2 A_2(a, \vartheta) = 0, \qquad \varepsilon B_1(a, \vartheta) + \varepsilon^2 B_2(a, \vartheta) = 0,$$

whence $\sin\vartheta = 0$, $\vartheta = 0$ or π, $\cos\vartheta = \pm 1$, and

$$\varepsilon\left(\Delta + \frac{3}{4}\beta_1 a^2 \mp \frac{P_1}{a}\right) - \frac{\varepsilon^2}{4\nu^2}\left[\left(\Delta + \frac{3}{4}\beta_1 a^2 \mp \frac{P_1}{a}\right)\left(\Delta + \frac{3}{4}\beta_1 a^2\right) - \frac{3}{32}\beta_1^2 a^4\right] = 0. \tag{20.33}$$

From this relation it follows that

$$\Delta + \frac{3}{4}\beta_1 a^2 \mp \frac{P_1}{a} = 0(\varepsilon)$$

and, substituting this into (20.33) and neglecting terms of third order in ε, we derive

$$\varepsilon\left(\Delta + \frac{3}{4}\beta_1 a^2 \mp \frac{P_1}{a}\right) + \frac{3\varepsilon^2 \beta_1^2 a^4}{128\,\nu^2} = 0. \tag{20.34}$$

Taking into account that $\omega^2/\nu^2 = 1 + \varepsilon\Delta$ and using (20.27), we may rewrite Eq. (20.34) as

$$\omega^2 - \nu^2 + \frac{3}{4}\beta\omega^2 a^2 + \frac{3}{128}\beta^2\omega^2 a^4 \mp \frac{P}{a} = 0. \tag{20.35}$$

Finally, neglecting terms of order $0(\varepsilon)^2$ in eq. (17.43) leads to

$$x_{\text{II}}(t) = a\cos(\nu t + \vartheta) + \frac{1}{32}\beta a^3 \cos 3(\nu t + \vartheta), \qquad \vartheta = 0 \text{ or } \pi, \tag{20.36}$$

where a is to be determined as a root of Eq. (20.35), in which the minus sign corresponds to $\vartheta = 0$, and the plus sign to $\vartheta = \pi$.

Let us compare now the results above with those given by the perturbation method. We first note that, by (20.14), (20.15), and (20.27), there results

$$\beta_1 = \nu^2\bar{\beta}, \qquad \Delta = \nu^2\bar{\Delta}, \qquad P_1 = \nu^2\bar{P}.$$

Putting this into (20.34) gives

$$\varepsilon\left(\bar{\Delta} + \frac{3}{4}\bar{\beta}a^2 \mp \frac{\bar{P}}{a}\right) + \frac{3}{128}\varepsilon^2\bar{\beta}^2a^4 = 0. \tag{20.37}$$

On the other hand, by comparing (20.36) to (20.23), we see that in the method of Krylov and Bogolyubov the total amplitude of the fundamental harmonic has been denoted by a and considered as positive, the sign of this harmonic being included in the phase, whereas in the perturbation method the amplitude of the first harmonic is obtained as an expansion in powers of ε and it may be negative, too, because the phase is taken as zero. Consequently, we have

$$\pm\, a = M + \varepsilon M_1 + 0(\varepsilon^2). \tag{20.38}$$

Introducing this into (20.37) and equating to zero the terms of zeroth and first order in ε, we obtain

$$\bar{\Delta} + \frac{3}{4}\bar{\beta}M^2 - \frac{\bar{P}}{M} = 0,$$

$$\frac{3}{2}\bar{\beta}MM_1 + \frac{\bar{P}M_1}{M^2} + \frac{3}{128}\bar{\beta}^2M^4 = 0.$$

The first of these relations coincides with Eq. $(20.20)_2$, which determines M, and the second relation becomes by considering the first one

$$\frac{9}{4}\bar{\beta}MM_1 + \frac{M_1\bar{\Delta}}{M} + \frac{3}{128}\bar{\beta}^2M^4 = 0,$$

i.e., it reduces to Eq. $(20.22)_2$, which gives M_1. Finally, by substituting (20.38) into (20.36), we find

$$x_{\mathrm{II}}(t) = (M + \varepsilon M_1)\cos \nu t + \frac{1}{32}\bar{\beta}M^3\cos 3\,\nu t + 0(\varepsilon^2),$$

that is exactly the solution (20.24) obtained by the perturbation method.

c) Use of the method of finite sums of trigonometric functions

Like the perturbation method, the method of finite sums of trigonometric functions allows calculating only periodic solutions (cf. § 18 a). We again consider the equation of motion

$$\ddot{x} + \omega^2 x(1 + \beta x^2) = P \cos \nu t \tag{20.39}$$

and try to determine an approximate periodic solution with the same minimum period as the exciting force, by using a finite sum of trigonometric functions of the form (18.1) with $q = 1$.

Since Eq. (20.39) does not change when replacing t by $-t$, we may seek the solution as a sum of cosines. On the other hand, we see that the equation does not change when simultaneously replacing t by $t + \pi/\nu$ and x by $-x$. Consequently, we may retain in the expression (18.1) of the approximate solution only the cosines whose argument is an odd multiple of νt. Furthermore, in order to simplify the calculation, we limit ourselves to considering only the first two nonzero terms of the sum (18.1), i.e.,

$$\overline{x}(t) = x_1 \cos \nu t + x_3 \cos 3\nu t. \tag{20.40}$$

Introducing this into (20.39) yields

$$\begin{aligned}&\left[(\omega^2 - \nu^2)x_1 + \frac{3}{4}\beta\omega^2 x_1 (x_1^2 + x_1 x_3 + 2x_3^2) - P\right] \cos \nu t \\ &+ \left[(\omega^2 - 9\nu^2)x_3 + \frac{1}{4}\beta\omega^2(x_1^3 + 6x_1^2 x_3 + 3x_3^3)\right] \cos 3\nu t \\ &+ \frac{3}{4}\beta\omega^2 x_1 x_3 (x_1 + x_3) \cos 5\nu t + \frac{3}{4}\beta\omega^2 x_1 x_3^2 \cos 7\nu t + \frac{1}{4}\beta\omega^2 x_3^3 \cos 9\nu t = 0.\end{aligned} \tag{20.41}$$

Since only two free coefficients, x_1 and x_3, are available, we can require no more than the vanishing of the coefficients of $\cos \nu t$ and $\cos 3\nu t$ in (20.41), which leads to

$$\begin{aligned}&(\omega^2 - \nu^2)x_1 + \frac{3}{4}\beta\omega^2 x_1(x_1^2 + x_1 x_3 + 2x_3^2) - P = 0, \\ &(\omega^2 - 9\nu^2)\, x_3 + \frac{1}{4}\beta\omega^2(x_1^3 + 6x_1^2 x_3 + 3x_3^3) = 0.\end{aligned} \tag{20.42}$$

We have thus neglected the terms in $\beta x_1^2 x_3$, $\beta x_1 x_3^2$, and βx_3^3 in (20.41), and this is possible only if βx_1^2 and x_3/x_1 are small as compared to unity. By using the same approximation in Eqs. (20.42). we find

$$(\omega^2 - \nu^2)\, x_1 + \frac{3}{4}\, \beta\omega^2 x_1^3 - P = 0, \tag{20.43}$$

$$(\omega^2 - 9\nu^2)\, x_3 + \frac{1}{4}\, \beta\omega^2 x_1^3 = 0. \tag{20.44}$$

In the above calculation β has been supposed small, but no assumption has been made about P. Suppose now that both P and β are of order of magnitude $0(\varepsilon)$, where ε is a small parameter. It then follows from (20.43) that $\omega^2 - \nu^2 = 0(\varepsilon)$, and hence, by (20.44),

$$x_3 = \frac{1}{32}\, \beta x_1^3 + 0(\varepsilon^2). \tag{20.45}$$

Introducing (20.45) into (20.40) gives the approximate solution

$$\bar{x}(t) = x_1 \cos \nu t + \frac{1}{32}\beta x_1^3 \cos 3\nu t + 0(\varepsilon^2). \tag{20.46}$$

The coefficient x_1 of the fundamental harmonic must be calculated from Eq. (20.43), which may be also written as

$$\nu^2 = \omega^2 + \frac{3}{4}\beta\omega^2 x_1^2 - \frac{P}{x_1} + 0(\varepsilon^2). \tag{20.47}$$

By setting now $|x_1| = a$, $x_1 = \pm a$, the last two equations become

$$\bar{x}(t) = \pm \left(a \cos \nu t + \frac{1}{32}\, \beta a^3 \cos 3\nu t \right) + 0(\varepsilon^3),$$

$$\nu^2 = \omega^2 + \frac{3}{4}\, \beta\omega^2 a^2 \mp \frac{P}{a} + 0(\varepsilon^2).$$

It is readily seen that the first of these relations coincides with the second approximation (20.36) given by the method of KRYLOV and BOGOLYUBOV, and the second one is identical with Eq. (20.31) obtained by the same method as the first approximation for the resonance curve.

If a more accurate equation of the resonance curve is desired, one has to retain in Eq. $(20.42)_1$ the term $(3/4)\,\beta\omega^2 x_1^2 x_3$, too, and then, by considering (20.45), there results

$$\nu^2 = \omega^2 + \frac{3}{4}\beta\omega^2 x_1^2 + \frac{3}{128}\beta^2\omega^2 x_1^4 - \frac{P}{x_1} + 0(\varepsilon)^3.$$

Denoting as before $|x_1| = a$, this equation becomes

$$\nu^2 = \omega^2 + \frac{3}{4}\beta\omega^2 a^2 + \frac{3}{128}\beta^2\omega^2 a^4 \mp \frac{P}{a} + 0(\varepsilon^3),$$

and, obviously, coincides with Eq. (20.35) obtained by the method of KRYLOV and BOGOLYUBOV as a second approximation for the resonance curve.

d) Resonance curve. Stability of periodic vibrations. Jump phenomena

Let us now consider in more detail the relation between the amplitude a of the periodic vibration and the exciting frequency ν in the neighborhood of the principal "resonance." We have deduced by three different methods that this relation may be written, to within terms of second order in ε, in the form

$$F\left(\frac{\nu^2}{\omega^2}, a\right) \equiv \frac{\nu^2}{\omega^2} - 1 - \frac{3}{4}\beta a^2 \mp \frac{P}{\omega^2 a} = 0, \tag{20.48}$$

where a and P are assumed to be positive; the minus sign corresponds to $\vartheta = 0$, and the positive sign to $\vartheta = \pi$. The graph of the function

$$\frac{\nu^2}{\omega^2} = 1 + \frac{3}{4}\beta a^2 \mp \frac{P}{\omega^2 a} \tag{20.49}$$

in the plane $(\nu^2/\omega^2, a)$, for $P =$ const, is said to be a *resonance curve*. The curve of equation

$$\frac{\nu^2}{\omega^2} = 1 + \frac{3}{4}\beta a^2, \tag{20.50}$$

which is called *skeleton curve*, correspond to $P = 0$, i.e., to nonlinear free vibrations. In fact, if one replaces ν by $\bar{\omega}(a)$, Eq. (20.50) becomes identical with Eq. (9.50) obtained as a first approximation for the frequency of free vibrations. It is readily seen from (20.49) that a resonance curve and the skeleton curve intercept equal segments on any horizontal line.

Figure 49 shows the resonance curves corresponding to $P_1/\omega^2 = 0.5$ cm, $P_2/\omega^2 = 1$ cm, and $P_3/\omega^2 = 1.5$ cm, for $\beta = 0.02\ \text{cm}^{-2}$ (Fig. 49 *a*) and $\beta = -0.02\ \text{cm}^{-2}$ (Fig. 49 *b*). The skeleton curve is represented by the broken line C_s. Comparing Fig. 49 to Fig. 48 (for $\zeta=0$), we see that the resonance curves of systems with nonlinear restoring forces have a form similar to those of a linear conservative system, but are "swept over" to the right or to the left, depending on whether the elastic characteristic is hardening ($\beta > 0$), or softening ($\beta < 0$). Unlike the case of the linear oscillator, the amplitude a of the nonlinear response is *finite* for any finite values of P and ν, even if $\nu = \omega$.

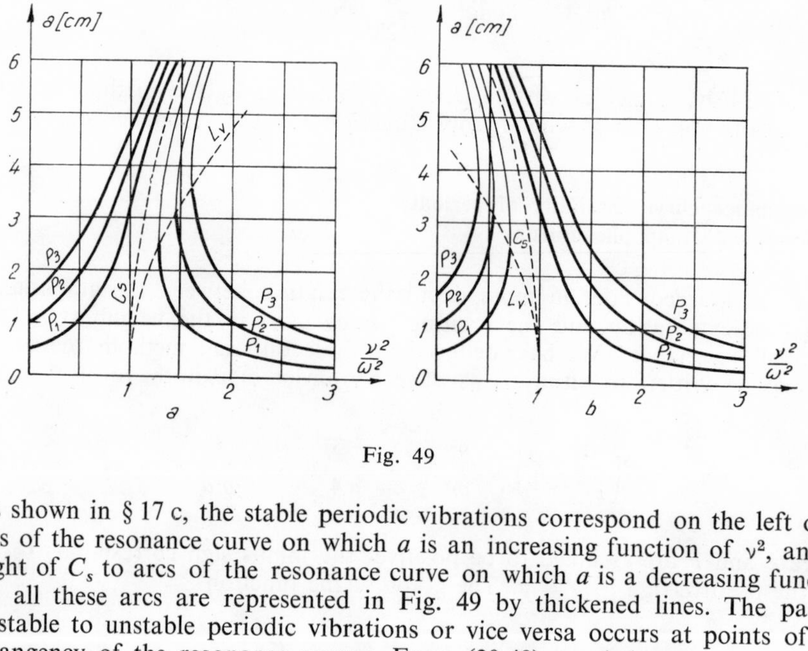

Fig. 49

As shown in § 17 c, the stable periodic vibrations correspond on the left of C_s to arcs of the resonance curve on which a is an increasing function of ν^2, and on the right of C_s to arcs of the resonance curve on which a is a decreasing function of ν^2; all these arcs are represented in Fig. 49 by thickened lines. The passing from stable to unstable periodic vibrations or vice versa occurs at points of vertical tangency of the resonance curves. From (20.48) we deduce:

$$\frac{\mathrm{d}a}{\mathrm{d}(\nu^2/\omega^2)} = -\frac{\dfrac{\partial F}{\partial(\nu^2/\omega^2)}}{\dfrac{\partial F}{\partial a}} = \frac{1}{a\left(\dfrac{3}{2}\beta a^2 \mp \dfrac{P}{\omega^2 a}\right)}. \tag{20.51}$$

At the points where the resonance curve has a vertical tangent the denominator of the fraction in the right-hand side vanishes, and hence

$$\frac{3}{2}\beta a^2 \mp \frac{P}{\omega^2\, a} = 0. \tag{20.52}$$

Eliminating P between Eqs. (20.52) and (20.49), we obtain the equation of the geometric locus of the points where the resonance curves have vertical tangent, namely

$$\frac{\nu^2}{\omega^2} = 1 + \frac{9}{4}\beta a^2. \tag{20.53}$$

The graph of this function is represented in Figs. 49 *a* and 49 *b* by broken lines denoted by L_v. As already mentioned, L_v separates the stable from the unstable conditions.

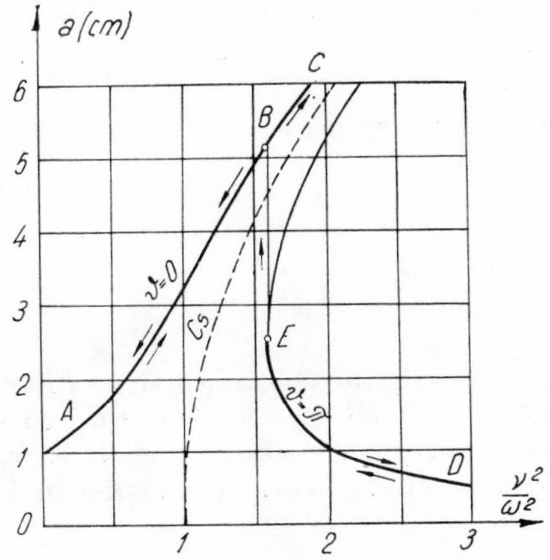

Fig. 50

The existence of points at which the resonance curve has vertical tangent determines the occurrence of discontinuities in the variation of the response amplitude and phase, when either the frequency or the amplitude of the external force varies continuously, the other one being constant.

Let us examine the variation of a with ν, by using the resonance curve illustrated in Fig. 50, corresponding to $P/\omega^2 = 1$ cm, $\beta = 0.04$ cm^{-2}. If the exciting frequency ν gradually increases from 0, the amplitude a increases continuously along the stable branch ABC of the resonance curve, and we have $\vartheta = 0$. Alternatively, if ν decreases from a large value, the periodic vibration is at first in opposition of phase with the exciting force, i.e., $\vartheta = \pi$, and its amplitude continuously increases along DE. At point E, however, this stable zone ends; the amplitude jumps up to the point B, and its phase jumps from π to 0. In case the frequency further decreases, the periodic vibration remains in phase with the external force, and its amplitude continuously decreases along BA.

A similar discontinuity appears in the variation of a when $\nu =$ const and P varies continuously. We shall discuss this situation in more detail when considering the generalized DUFFING's equation with damping (§ 23 d), since damping diversifies the jump phenomena encountered in nonlinear systems.

It is worth noting that the jumps are not instantaneous, but require a few cycles of vibration to establish a steady-state vibration at the new amplitude. In other words, when ν approaches a critical value, the response amplitude increases or decreases at a *finite* rate and transitory vibrations take place till the new steady-state is attained; the character of these transitory vibrations depends on how rapidly ν passes through the critical value (see MITROPOLSKY [162], chap. III, § 6).

We close this section with a few comments concerning the second approximation of the resonance curve. Equation (20.35) obtained at the second stage of approximation may be rewritten as

$$\frac{\nu^2}{\omega^2} = 1 + \frac{3}{4}\beta a^2 + \frac{3}{128}\beta^2 a^4 \mp \frac{P}{\omega^2 a}\cdot \tag{20.54}$$

The resonance curve may be plotted as before, by first tracing the skeleton curve, which now has the equation

$$\frac{\nu^2}{\omega^2} = 1 + \frac{3}{4}\beta a^2 + \frac{3}{128}\beta^2 a^4, \tag{20.55}$$

and then horizontally measuring distances equal to $P/(\omega^2 a)$ on both sides of this curve. Comparing (20.55) to (20.50), we see that the second approximation introduces only a small quantitative correction, which is generally negligible. For instance, in the case of the resonance curve illustrated in Fig. 50 ($\beta = 0.04\ \text{cm}^{-2}$), we obtain for $a = 6$ cm:

$$\frac{3}{4}\beta a^2 = 2.08, \quad \frac{3}{4}\beta a^2 + \frac{3}{128}\beta^2 a^4 = 2.13,$$

i.e., the departure of the skeleton curve from the vertical line of equation $\nu/\omega = 1$ amounts to merely 2.4%. Therefore, we shall usually content ourselves with considering only the first approximation when discussing qualitative phenomena specific to nonlinear vibrations.

e) Subharmonic vibrations

Subharmonic vibrations of various orders of DUFFING's equation have been the object of a great deal of research. We mention here CARTWRIGHT and LITTLEWOOD [24], LEVENSON [97], MINORSKY [121], MC LACHLAN [116, 117], MANDELSTAM and PAPALEXI [114], STOKER [184], KAUDERER [82], BOGOLYUBOV and MITROPOLSKY [16], and HAYASHI [71, 72].

In the following we will study the subharmonic solutions of order 1/3 of the equation

$$\ddot{x} + \omega^2 x(1 + \beta x^2) = P\cos\nu t, \tag{20.56}$$

under the assumption that both β and P are of order $0(\varepsilon)$, where ε is a small parameter. As already mentioned in § 19 c, this hypothesis seems to lead to a better accuracy of the solution in the neighborhood of the "resonance." We shall comparatively apply the method of KRYLOV and BOGOLYUBOV and the method of finite sums of trigonometric functions, in order to better understand the particularities of both methods and to enrich the results obtained.

— *Use of the method of* KRYLOV *and* BOGOLYUBOV

By setting

$$\omega^2\beta = \varepsilon\beta_1, \quad P = \varepsilon P_1, \quad \nu t = \tau \tag{20.57}$$

and assuming that ε is a small parameter, Eq. (20.56) becomes

$$\ddot{x} + \omega^2 x = -\varepsilon(\beta_1 x^3 - P_1 \cos\tau). \tag{20.58}$$

We seek a strong subharmonic solution of order 1/3 of this equation, by using the results derived in § 19 c. Putting $p = 1$, $q = 3$, and

$$\omega^2 - \frac{\nu^2}{9} = \varepsilon\Delta, \tag{20.59}$$

and making use of the notations explained in §§ 17 and 19, we obtain

$$f_1(\dot{x}) \equiv 0, \quad \psi_1(x) = \beta_1 x^3, \quad \alpha = \frac{\nu t}{3} + \vartheta,$$

$$\psi_1(a\cos\alpha) = \frac{1}{4}\beta_1 a^3(\cos 3\alpha + 3\cos\alpha),$$

$$C_1(a) = \frac{3}{4}\beta_1 a^3, \quad C_3(a) = \frac{1}{4}\beta_1 a^3,$$

and all other coefficients $C_k(a)$, $D_k(a)$ vanish. Next, from (19.26)—(19.28) it follows that

$$\left.\begin{aligned} &A_1(a, \vartheta) = 0, \qquad B_1(a, \vartheta) = \frac{3\Delta}{2\nu} + \frac{9\beta_1 a^2}{8\nu}, \\ &\bar{u}_1(a, \alpha) = \frac{9\beta_1 a^3}{32\nu^2}\cos 3\alpha, \quad \tilde{u}_1(\tau) = -\frac{9P_1}{8\nu^2}\cos\tau, \end{aligned}\right\} \tag{20.60}$$

$$u_1(a, \alpha, \tau) = \frac{9\beta_1 a^3}{32\nu^2}\cos 3\alpha - \frac{9P_1}{8\nu^2}\cos\tau, \tag{20.61}$$

$$\left.\begin{aligned} U(a, \alpha) &= \frac{9\beta_1 a^3}{32\nu^2} \cos 3\alpha - \frac{9P_1}{8\nu^2} \cos 3(\alpha - \vartheta), \\ V(a, \alpha) &= -\frac{9\beta_1 \omega a^3}{32\nu^2} \sin 3\alpha + \frac{9P_1}{8\nu^2} \sin 3(\alpha - \vartheta). \end{aligned}\right\} \tag{20.62}$$

Substituting (20.61) into (17.45), using (20.57), and taking into account that $\nu^2 = 9\omega^2 + 0(\varepsilon)$, we find the second approximation for the subharmonic solution

$$x(t) = a \cos\left(\frac{\nu t}{3} + \vartheta\right) + \frac{1}{32}\beta a^3 \cos(\nu t + 3\vartheta) - \frac{P}{8\omega^2} \cos \nu t + 0(\varepsilon^2). \tag{20.63}$$

Introducing now (20.60) and (20.62) into (19.31), taking into consideration that $\nu/\omega = 3 + 0(\varepsilon)$, and calculating the integrals involved, we obtain the equations to be satisfied by the amplitude and phase of the subharmonic vibration, namely

$$M(a, \vartheta, \nu) \equiv -\frac{3\beta a^2 P}{64\omega} \sin 3\vartheta = 0, \tag{20.64}$$

$$N(a, \vartheta, \nu) \equiv \frac{3}{2\nu}\left(\omega^2 - \frac{\nu^2}{9} + \frac{3}{4}\beta\omega^2 a^2 + \frac{3}{128}\beta^2\omega^2 a^4 - \right.$$
$$\left. - \frac{3}{32}\beta a P \cos 3\vartheta\right) = 0. \tag{20.65}$$

From (20.64) it results that $\vartheta = k\pi/3$, $k = 0, 1, 2, 3, 4, 5$. Since

$$\cos\left(\frac{\nu t}{3} + \pi\right) = -\cos\frac{\nu t}{3}, \quad \cos\left(\frac{\nu t}{3} + \frac{5\pi}{3}\right) = -\cos\left(\frac{\nu t}{3} + \frac{2\pi}{3}\right),$$
$$\cos\left(\frac{\nu t}{3} + \frac{\pi}{3}\right) = -\cos\left(\frac{\nu t}{3} + \frac{4\pi}{3}\right),$$

the approximate solution (20.63) and the equation of the resonance curve (20.65) may be rewritten in the form

$$x(t) = \pm a \cos\left(\frac{\nu t}{3} + \frac{2n\pi}{3}\right) + \left(\pm\frac{\beta a^3}{32} - \frac{P}{8\omega^2}\right) \cos \nu t, \tag{20.66}$$

$$\omega^2 - \frac{\nu^2}{9} + \frac{3}{4}\beta\omega^2 a^2 + \frac{3}{128}\beta^2\omega^2 a^4 \mp \beta a P = 0, \tag{20.67}$$

where $n = 0, 1, 2$, and where the upper (as well as the lower) signs in both equations correspond to each other.

In order for Eq. (20.56) to admit a *simple* subharmonic solution of the form

$$x(t) = a \cos \frac{\nu t}{3} \tag{20.68}$$

it is necessary, by (20.66), that

$$n = 0, \qquad \frac{\beta a^3}{32} - \frac{P}{8\omega^2} = 0.$$

The last relation gives

$$a = \sqrt[3]{\frac{4P}{\beta\omega^2}} \tag{20.69}$$

and, introducing into (20.65), it follows that

$$\nu^2 = 9\omega^2\left(1 + 3\sqrt[3]{\frac{\beta P^2}{4\omega^4}}\right). \tag{20.70}$$

Equations (20.69) and (20.70) are identical with the exact conditions (19.12) obtained in § 19 a, by using CHEBYSHEV's polynomials.

Let us now come back to the equation (20.67) of the resonance curve, which may be also written as

$$F\left(\frac{\nu^2}{\omega^2}, a\right) \equiv \frac{\nu^2}{\omega^2} - 9\left(1 + \frac{3}{4}\beta a^2 + \frac{3}{128}\beta^2 a^4 \mp \frac{3\beta a P}{32\,\omega^2}\right) = 0. \tag{20.71}$$

Since $3\beta^2 a^4/128 \ll 1$, the graph of the resonance curve in the plane $(\nu^2/\omega^2, a)$ is very accurately approximated by the semiparabolas of equations

$$\frac{\nu^2}{\omega^2} = 9\left(1 + \frac{3}{4}\beta a^2 \mp \frac{3\beta a P}{32\omega^2}\right), \qquad a > 0.$$

The two branches of the resonance curve are very close to each other, because the term $3\beta a P/(32\,\omega^2)$ is of order $0(\varepsilon^2)$, i.e., much less than unity. Consequently, the values of a derived from (20.71) for $\nu^2/\omega^2 > 9$ will be also very close to each other. By introducing these values into (20.66), one finds six subharmonic solutions of order 1/3, which are pairwise in approximately opposition of phase, as shown by the principal term of frequency $\nu/3$.

The resonance curve of Eq. (20.71) may be easily plotted by taking a as the independent variable. For each value of a, one obtains two points of the resonance curve situated at equal distance from the skeleton curve, whose equation is

$$\frac{\nu^2}{\omega^2} = 9\left(1 + \frac{3}{4}\beta a^2 + \frac{3}{128}\beta^2 a^4\right). \tag{20.72}$$

Figure 51 shows the resonance curves corresponding to $P_1/\omega^2 = 1$ cm, $P_2/\omega^2 = 2$cm, $P_3/\omega^2 = 3$ cm, for $\beta = 0.04\,\text{cm}^{-2}$. The skeleton curve is represented by the broken line C_s.

We now consider the stability problem. As already shown at the end of § 19, the stable conditions are separated from the unstable ones by the points at which the resonance curve has vertical tangent, i.e., where $d(\nu^2/\omega^2)/da = 0$. By considering (20.71), this condition becomes

$$a + \frac{1}{16}\beta a^3 \mp \frac{P}{16\omega^2} = 0. \tag{20.73}$$

Eliminating P between (20.71) and (20.73) yields the equation of the geometric locus of the points at which the resonance curves have vertical tangent:

$$\frac{\nu^2}{\omega^2} = 9\left(1 - \frac{3}{4}\beta a^2 - \frac{9}{128}\beta^2 a^4\right). \tag{20.74}$$

In Fig. 51 this geometric locus is represented by the broken line L_v, which cuts each resonance curve in exactly one point. To better illustrate this fact, we have used a larger scale in Fig. 52 for the graphical representation of the curves L_v, C_s, and P_2 in the neighborhood of the point B at which L_v cuts P_2, and hence P_2 has a vertical tangent.

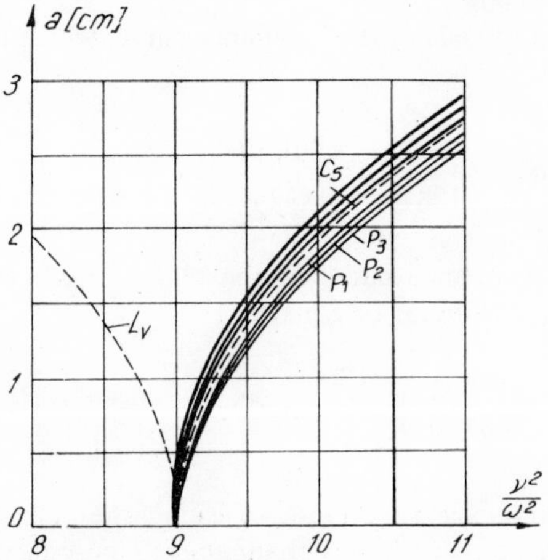

Fig. 51

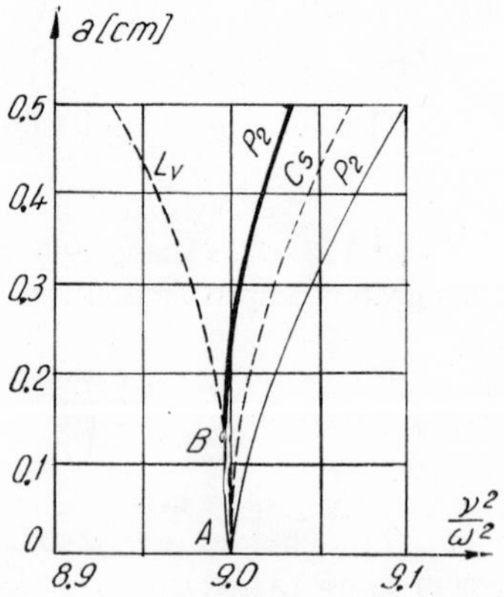

Fig. 52

Let us now see what become the conditions (19.37) in the present case. We first derive from (20.64) and (20.65):

$$\frac{\partial M}{\partial a} = -\frac{3\beta a P}{32\,\omega}\sin 3\vartheta, \qquad \frac{\partial M}{\partial \vartheta} = -\frac{9\beta a^2 P}{64\,\omega}\cos 3\vartheta,$$

$$\frac{\partial N}{\partial \vartheta} = \frac{9\beta a P}{54\nu}\sin 3\vartheta, \qquad \frac{\partial N}{\partial a} = \frac{3}{2\nu}\left(\frac{3}{2}\beta\omega^2 a + \frac{3}{32}\beta^2\omega^2 a^3 - \frac{3\beta P}{32}\cos 3\vartheta\right).$$

Next, by replacing sin 3ϑ and cos 3ϑ in these relations with their values given by Eqs. (20.44) and (20.65), we find

$$\frac{\partial M}{\partial a} + \frac{\partial N}{\partial \vartheta} = 0,$$

$$\frac{\partial M}{\partial a}\frac{\partial N}{\partial \vartheta} - \frac{\partial M}{\partial \vartheta}\frac{\partial N}{\partial a} = \pm\left(\frac{\nu^2}{9} - \omega^2 + \frac{3}{4}\beta\omega^2 a^2 + \frac{9}{128}\beta^2\omega^2 a^4\right).$$

The first of these equations shows that the sign of the expression $\partial M/\partial a + \partial N/\partial\vartheta$ is not determined up to the second stage of approximation. However, as will be seen in § 24, if the equation of motion contains a dissipative term (which is always the case for real systems), then the expression $\partial M/\partial a + \partial N/\partial\vartheta$ is positive, and hence condition $(19.37)_1$ is satisfied. Next, we see that the stability condition $(19.37)_2$ is fulfilled only if

$$\pm\left(\frac{\nu^2}{9} - \omega^2 + \frac{3}{4}\beta\omega^2 a^2 + \frac{9}{128}\beta^2\omega^2 a^4\right) > 0, \tag{20.75}$$

i.e., only on the left branch of the resonance curve (corresponding to the upper sign in the inequality above), and only if

$$\frac{\nu^2}{\omega^2} > 9\left(1 - \frac{3}{4}\beta a^2 - \frac{9}{128}\beta^2 a^4\right),$$

i.e., on the right of L_v, or, equivalently, above the point B. The arc of the resonance curve corresponding to stable periodic vibrations is represented in Fig. 52 by a thickened solid line. There are three periodic vibrations, phase-shifted by $2\pi/3$, corresponding to each branch of the resonance curve; hence, there exist three stable and three unstable subharmonic vibrations of order 1/3 for any value of ν^2/ω^2 greater than the abscissa of B.

Assume now that the exciting frequency ν gradually increases from 0. From the reasoning above it follows that a stationary subharmonic vibration of order 1/3 cannot set in until ν^2/ω^2 equals the abscissa of B. To determine this value, we first note that the ordinate of B has to satisfy Eq. (20.73) with the upper sign. It is apparent from (20.73) that $a = 0(\varepsilon)$, and hence the ordinate of B is

$$a = \frac{P}{14\omega^2} + 0(\varepsilon^4).$$

Introducing this into Eq. (20.74) yields for the abscissa of B

$$\frac{\nu^2}{\omega^2} = 9 + 0(\varepsilon^4).$$

By summarizing the above, we conclude that the point B separates the stable from the unstable periodic vibrations and marks the occurrence of the stationary subharmonic vibrations as ν continuously increases from 0.

— *Use of the method of finite sums of trigonometric functions*

Consider again the equation of motion

$$\ddot{x} + \omega^2 x(1 + \beta x^2) = P \cos \nu t \tag{20.76}$$

and seek an approximate subharmonic solution of order 1/3 of this equation in the form (18.1) with $q = 3$.

Since Eq. (20.76) does not change when replacing t by $-t$, we may seek the solution as a sum of cosines. On the other hand, we see that the equation does not change when simultaneously replacing t by $t + 3\pi/\nu$ and x by $-x$; hence, we may retain in (18.1) only the cosines whose arguments are odd multiples of $\nu t/3$. Consequently, if we content ourselves with the first two nonzero terms of the sum (18.1), we may try an approximate subharmonic solution of the form

$$\bar{x}(t) = x_{1/3} \cos \frac{\nu t}{3} + x_1 \cos \nu t. \tag{20.77}$$

It is worth noting that when $x(t)$ is a solution of Eq. (20.76), $x(t + 2n\pi/\nu)$ with n an integer is also a solution. Consequently, if $\bar{x}(t)$ is an approximate solution of Eq. (20.76), then

$$\left.\begin{aligned} \bar{x}\left(t + \frac{2\pi}{\nu}\right) &= x_{1/3} \cos \left(\frac{\nu t}{3} + \frac{2\pi}{3}\right) + x_1 \cos \nu t, \\ \bar{x}\left(t + \frac{4\pi}{\nu}\right) &= x_{1/3} \cos \left(\frac{\nu t}{2} + \frac{4\pi}{3}\right) + x_1 \cos \nu t \end{aligned}\right\} \tag{20.78}$$

are also approximate subharmonic solution of Eq. (20.76), different from $\bar{x}(t)$. Therefore, subharmonic solutions of order 1/3 of Eq. (20.76) may be divided into groups, each of which contains three subharmonic vibrations with the same amplitude but phase-shifted by $2\pi/3$ (see also Mc Lachlan [87]).

Introducing now (20.77) into (20.76) yields

$$\begin{aligned} & x_{1/3}\left[\omega^2 - \frac{\nu^2}{9} + \frac{3}{4}\beta\omega^2(x_{1/3}^2 + x_{1/3}x_1 + 2x_1^2)\right] \cos \frac{\nu t}{3} + \\ & + \left[(\omega^2 - \nu^2)x_1 + \frac{1}{4}\beta\omega^2(x_{1/3}^3 + 6x_{1/3}^2 x_1 + 3x_1^3) - P\right] \cos \nu t + \\ & + \frac{3}{4}\beta\omega^2 x_{1/3}x_1(x_{1/3} + x_1) \cos \frac{5\nu t}{3} + \frac{3}{4}\beta\omega^2 x_{1/3}x_1^2 \cos \frac{7\nu t}{3} + \\ & + \frac{1}{4}\beta\omega^2 x_1^3 \cos 3\nu t = 0. \end{aligned} \tag{20.79}$$

Since only two free coefficients, x_1 and x_3, are available, we may require only the vanishing of the coefficients of cos ($\nu t/3$) and cos νt in (20.79), which gives

$$\left.\begin{aligned} &\left[\omega^2 - \frac{\nu^2}{9} + \frac{3}{4}\,\beta\omega^2(x_{1/3}^2 + x_{1/3}x_1 + 2x_1^2)\right] x_{1/3} = 0, \\ &(\omega^2 - \nu^2)\,x_1 + \frac{1}{4}\,\beta\omega^2(x_{1/3}^3 + 6x_{1/3}^2x_1 + 3x_1^3) - P = 0. \end{aligned}\right\} \tag{20.80}$$

We remark that, if $x_{1/3} = 0$, the first equation (20.80) is identically satisfied, and the second one leads to

$$\omega^2 - \nu^2 + \frac{3}{4}\,\beta\omega^2x_1^2 - \frac{P}{x_1} = 0, \tag{20.81}$$

i.e., exactly the same as Eq. (20.43) obtained for the resonance curve of the harmonic vibration in the neighborhood of the principal "resonance." Hence, the subharmonic solution results by bifurcation from the harmonic one. We shall return to this observation later.

To obtain a subharmonic vibration we assume in the following that $x_{1/3} = 0$ and divide out a factor of $x_{1/3}$ from $(20.80)_1$. A complete study of system (20.80) was undertaken by STOKER [184], KAUDERER [82], and MC LACHLAN [116]. However, the consideration of all terms of this system seems not to be very profitable, for it renders the calculation much more difficult and brings in only unimportant corrections of the results. To simplify Eqs. (20.80), we first remark that the terms $\beta x_{1/3}^2x_1$, $\beta x_{1/3}x_1^2$, and βx_1^3 have been neglected in Eq. (20.79).[1] For a consistent approximation, we ought to neglect the same terms in Eqs. (20.80). However, to obtain a more accurate equation of the resonance curve we will adopt a midway procedure (cf. the end of § 20 c), by retaining in $(20.80)_1$ the term (3/4) $\beta\omega^2x_{1/3}x_1$, too. We then obtain

$$\omega^2 - \frac{\nu^2}{9} + \frac{3}{4}\,\beta\omega^2(x_{1/3}^2 + x_{1/3}x_1) = 0, \tag{20.82}$$

$$(\omega^2 - \nu^2)\,x_1 + \frac{1}{4}\,\beta\omega^2x_{1/3}^3 - P = 0. \tag{20.83}$$

Till now we have supposed that β is small, but no assumption has been made about P. If we further assume [2] that both P and β are of order of magnitude $0(\varepsilon)$,

[1] This approximation in justified only if $\beta x_{1/3}^2$ and $x_1/x_{1/3}$ are small in comparison with unity.

[2] Although it introduces only corrections of second order in ε, this hypothesis influences some of the qualitative features of the resonance curve, e.g., the form of the geometric locus of the points at which the resonance curve has a vertical tangent.

where ε is a small parameter, we deduce from (20.82) that $\omega^2 - \nu^2/9 = 0(\varepsilon)$, and then Eq. (20.83) yields

$$x_1 = \frac{1}{32}\beta x_{1/3}^3 - \frac{P}{8\omega^2} + 0(\varepsilon^2). \tag{20.84}$$

Substituting (20.84) into (20.77) and (20.78), we obtain the approximate subharmonic solutions

$$\bar{x}(t) = x_{1/3}\cos\left(\frac{\nu t}{3} + \frac{2n\pi}{3}\right) + \left(\frac{1}{32}\beta x_{1/3}^3 - \frac{P}{8\omega^2}\right)\cos \nu t, \tag{20.85}$$

where $n = 0, 1, 2$. The coefficient $x_{1/3}$ is to be determined from the equation (20.82) of the resonance curve, which, considering (20.84), becomes

$$\frac{\nu^2}{\omega^2} = 9\left(1 + \frac{3}{4}\beta x_{1/3}^2 + \frac{3}{128}\beta^2 x_{1/3}^4 - \frac{3\beta P x_{1/3}}{32\omega^2}\right) + 0(\varepsilon^3). \tag{20.86}$$

As $3\beta^2 x_{1/3}^4 \ll 1$, the resonance curve has a parabolic form in the plane $(\nu^2/\omega^2, x_{1/3})$. For each value of $\nu^2/\omega^2 \geqslant 9$ this equation yields two real values of $x_{1/3}$, which, by substitution into (20.85) give six approximate subharmonic solutions, i.e., as many as those obtained by the method of KRYLOV and BOGOLYUBOV. Moreover, if we denote $|x_{1/3}| = a$, and hence $x_{1/3} = \pm a$, then Eqs. (20.85) and (20.86) reduce to the expression (20.66) of the subharmonic solution and, respectively, to the equation (20.67) of the resonance curve, as obtained by the method of KRYLOV and BOGOLYUBOV.

Figure 53 shows the graph of the resonance curve of Eq. (20.86), corresponding to $\beta = 0.04$ cm^{-2} and $P/\omega^2 = 2$ cm, i.e., to the same values adopted for the curve by P_2 in Figs. 51 and 52. In Fig. 51, however, the quantity $a = |x_{1/3}|$ has been taken as ordinate and, consequently, a second branch of the resonance curve has been obtained in the upper half-plane, which is symmetric to the part of the resonance curve situated in the lower half-plane in Fig. 53 with respect to the (ν^2/ω^2)-axis.

From (20.86) we deduce that

$$\frac{d(\nu^2/\omega^2)}{dx_{1/3}} = \frac{27}{2}\beta\left(x_{1/3} + \frac{1}{16}\beta x_{1/3}^3 - \frac{P}{16\,\omega^2}\right), \tag{20.87}$$

and hence the ordinate of the point B at which the resonance curve has vertical tangent satisfies the equation

$$x_{1/3} + \frac{1}{16}\beta x_{1/3}^3 - \frac{P}{16\omega^2} = 0.$$

It follows from this relation and Eq. (20.86) that the approximate coordinates of B are

$$\frac{\nu^2}{\omega^2} = 9 + 0(\varepsilon^4), \quad x_{1/3} = \frac{P}{16\omega^2} + 0(\varepsilon^4), \tag{20.88}$$

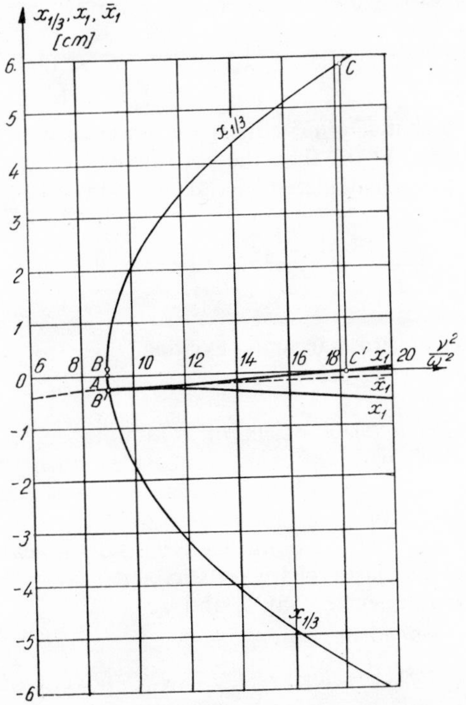

Fig. 53

a result already obtained by the method of KRYLOV and BOGOLYUBOV. If ν continuously increases from 0, the subharmonic vibration sets in when ν^2/ω^2 equals the abscissa of B. Let us now consider more attentively the bifurcation of the subharmonic solution from the harmonic one. The coefficient (20.84) of the harmonic part of the solution corresponding to the values (20.88) of ν^2/ω^2 and $x_{1/3}$ is

$$x_1 = -\frac{P}{8\omega^2} + 0(\varepsilon^4) = -2x_{1/3} + 0(\varepsilon^4).$$

For the same value $(20.88)_1$ of ν^2/ω^2 there possibly exists a simple harmonic vibration of the oscillator. Its coefficient, say $\bar{x}_1$, has to fulfill Eq. (20.81), from which, by $(20.88)_1$, we find

$$\bar{x}_1 = -\frac{P}{8\omega^2} + 0(\varepsilon^4) = x_1 + 0(\varepsilon^4).$$

This last relation again shows that the subharmonic vibration of order 1/3 appears by bifurcation from the harmonic one for $\nu^2/\omega^2 \approx 9$. At the bifurcation point we have $x_{1/3} = -x_1/2$. Figure 53 illustrates the graphs of x_1 and $\bar{x}_1$ versus ν^2/ω^2, as given by Eqs. (20.81), (20.84), and (20.86), $x_{1/3}$ being used as parameter in the last two of these equations. In the numerical case considered in Fig. 53 the coordinates of the bifurcation point B' are:

$$\frac{\nu^2}{\omega^2} = 9 + 0(\varepsilon^4), \qquad x_1 = \bar{x}_1 = -\frac{P}{8\omega^2} + 0(\varepsilon^4) \approx -0.25.$$

We finally remark that if ν^2/ω^2 equals the abscissa of the point at which the x_1 — curve cuts the (ν^2/ω^2)-axis, then the coordinates of the corresponding point C on the resonance curve, calculated by (20.69) and (20.70), are

$$\frac{\nu^2}{\omega^2} = 18.243, \quad x_{1/3} = \sqrt[3]{200} \approx 5.848.$$

f) Simultaneous action of two harmonic exciting forces with different frequencies

As is well known, if a *linear* oscillator is simultaneously acted on by several harmonic external forces with different frequencies, then the resulting vibration equals the algebraic sum of the vibrations that would be excited by the separate action of these external forces. For a *nonlinear* oscillator this simple superposition is no longer valid. To illustrate this point, let us consider a weakly nonlinear conservative oscillator with cubic elastic characteristic, acted on by two harmonic external forces with different frequencies, say ν_1 and ν_2.

The governing equation is

$$\ddot{x} + \omega^2 x(1 + \beta x^2) = P_1 \cos \nu_1 t + P_2 \cos \nu_2 t, \tag{20.89}$$

where β is supposed to be small. Since for $\beta = 0$ the linear superposition holds, we seek a solution of Eq. (20.89) of the form

$$\bar{x}(t) = x_1 \cos \nu_1 t + x_2 \cos \nu_2 t + \beta u(t), \tag{20.90}$$

where $\beta u(t)$ denotes a correction introduced to account for the elastic nonlinearity

Substituting (20.90) into (20.89) yields

$$\left\{\left[\omega^2 - \nu_1^2 + \frac{3}{4}\beta\omega^2(x_1^2 + 2x_2^2)\right]x_1 - P_1\right\}\cos \nu_1 t + \left\{\left[\omega^2 - \nu_2^2 + \frac{3}{4}\beta\omega^2(2x_1^2 + x_2^2)\right]x_2 - P_2\right\}\cos \nu_2 t + \beta\left\{\ddot{u} + \omega^2 u + \frac{\omega^2}{4}[x_1^3 \cos 3\nu_1 t + 3x_1^2 x_2 \cos(2\nu_1 + \nu_2)t + 3x_1^2 x_2 \cos(2\nu_1 - \nu_2)t + 3x_1 x_2^2 \cos(\nu_1 + 2\nu_2)t + 3x_1 x_2^2 \cos(\nu_1 - 2\nu_2)t + x_2^3 \cos 3\nu_2 t]\right\} + 0(\beta^2) = 0.$$

Requiring the coefficients of cos $\nu_1 t$, cos $\nu_2 t$, and β to vanish, we obtain

$$\left.\begin{aligned} \omega^2 - \nu_1^2 + \frac{3}{4}\,\beta\omega^2(x_1^2 + 2x_2^2) - \frac{P_1}{x_1} = 0, \\ \omega^2 - \nu_2^2 + \frac{3}{4}\,\beta\omega^2(2x_1^2 + x_2^2) - \frac{P_2}{x_2} = 0, \end{aligned}\right\} \tag{20.91}$$

$$\ddot{u} + \omega^2 u = -\frac{\omega^2}{4}\,[x_1^3 \cos 3\nu_1 t + 3x_1^2 x_2 \cos(2\nu_1 + \nu_2)\,t + 3x_1^2 x_2 \cos(2\nu_1 - \nu_2)\,t + \\ + 3x_1 x_2^2 \cos(\nu_1 + 2\nu_2)\,t + 3x_1 x_2^2 \cos(\nu_1 - 2\nu_2)\,t + x_2^3 \cos 3\nu_2 t]. \tag{20.92}$$

By means of Eq. (20.91), it may be shown that x_1 and x_2 remain finite for all finite ν_1, ν_2, P_1, and P_2, a specific property of nonlinear oscillators, which has been already noticed in the case of a single harmonic exciting force. For given ν_1, ν_2, P_1, and P_2, the amplitudes x_1 and x_2 can be calculated from (20.91). Equation (20.92) has the particular solution

$$u(t) = -\frac{\omega^2}{4}\left[\frac{x_1^3}{\omega^2 - 9\nu_1^2}\cos 3\nu_1 t + \frac{3x_1^2 x_2}{\omega^2 - (2\nu_1 + \nu_2)^2}\cos(2\nu_1 + \nu_2)\,t + \right. \\ + \frac{3x_1^2 x_2}{\omega^2 - (2\nu_1 - \nu_2)^2}\cos(2\nu_1 - \nu_2)\,t + \frac{3x_1 x_2^2}{\omega^2 - (\nu_1 + 2\nu_2)^2}\cos(\nu_1 + 2\nu_2)\,t + \\ \left. + \frac{3x_1 x_2^2}{\omega^2 - (\nu_1 - 2\nu_2)^2}\cos(\nu_1 - 2\nu_2)\,t + \frac{x_2^3}{\omega^2 - 9\nu_2^2}\cos 3\nu_2 t\right]. \tag{20.93}$$

Finally, by introducing x_1, x_2, and $u(t)$ into (20.90), one obtains the expression of the approximate solution.

Inspection of (20.93) reveals a new specific feature of forced nonlinear vibrations: both harmonic external forces excite not only forced vibrations having the same frequencies as them, but also vibrations with frequencies $3\nu_1$, $2\nu_1 + \nu_2$, $2\nu_1 - \nu_2$, $\nu_1 + 2\nu_2$, $\nu_1 - 2\nu_2$, and $3\nu_2$. This phenomenon may lead to undesired effects in acoustical devices. Since the characteristic of a loud speaker is usually nonlinear, two periodic electromagnetic forces with frequencies ν_1 and ν_2 acting on a loud speaker will excite forced vibrations having not only the frequencies ν_1 and ν_2 but also combined frequencies. If the intensity of the primary tones is high enough, the intensity of the tones with combined frequencies will be above the threshold of audibility, thus introducing unpleasant distortions.

§ 21. THE OSCILLATOR WITH PIECEWISE LINEAR ELASTIC CHARACTERISTIC

Piecewise linear elastic characteristics are frequently used, for they may be easily obtained by assembling in parallel several linear springs that begin to work at different values of the displacement. For example, truck suspensions often comprise two leaf type springs, the weaker of which operates beginning at a displacement, say x_1, which is usually a little larger than the static deflection. A further slope increment of the elastic characteristic is achieved by means of rubber buffers which begin to work at some value $x_2 > x_1$ of the displacement. Figure 54 shows the elastic characteristic of such a suspension, which is, obviously, asymmetric and piecewise linear.

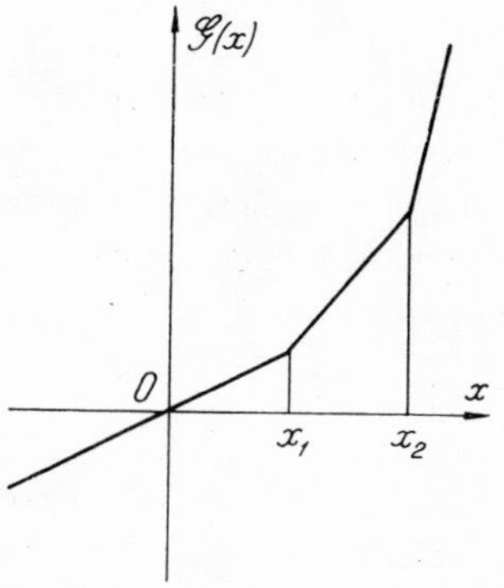

Fig. 54

Forced vibrations of oscillators with piecewise linear elastic characteristics excited by harmonic external forces have been studied by DEN HARTOG and HEILES [69] by using an exact but rather laborious method, by LURE and CHEKMAREV [110] by means of GALERKIN's method, and by BOGOLYUBOV and MITROPOLSKY [16] with the aid of the method of KRYLOV and BOGOLYUBOV.

Let us now consider as an illustration the oscillator whose piecewise linear and symmetric elastic characteristic is represented in Fig. 55. The governing equation of motion is

$$\ddot{x} + g(x) = P \cos \nu t, \tag{21.1}$$

where

$$g(x) \equiv \begin{cases} \omega^2[x + c(x + x_1)] & \text{for} \quad x \in (-\infty, -x_1], \\ \omega^2 x & \text{for} \quad x \in (-x_1, x_1), \\ \omega^2[x + c(x - x_1)] & \text{for} \quad x \in [x_1, \infty), \end{cases} \tag{21.2}$$

$$\omega^2 = \frac{2k_1}{m}, \qquad c = \frac{k_2}{k_1}, \tag{21.3}$$

m is the mass of the oscillator, k_1 is the constant of the basic spring, and k_2 is the constant of the supplementary spring. P is taken as positive. For the hardening characteristic shown in Fig. 55 a we have $c > 1$, and for the softening characteristic

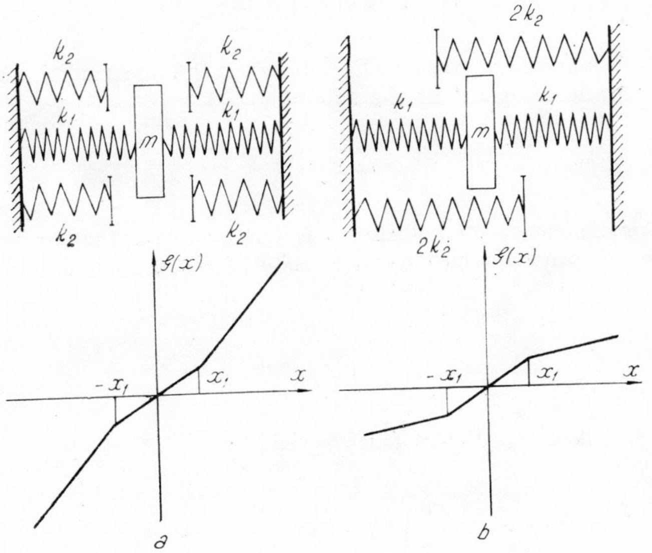

Fig. 55

shown in Fig. 55 b, $c < 1$. We make use of the method of KRYLOV and BOGOLYUBOV. Denoting

$$c = \varepsilon c_1, \quad P = \varepsilon P_1 \tag{21.4}$$

and assuming that ε is a small parameter, Eq. (21.1) may be rewritten as

$$\ddot{x} + \omega^2 x = -\varepsilon[\psi_1(x) - P_1 \cos \nu t], \tag{21.5}$$

where

$$\psi_1(x) \equiv \begin{cases} \omega^2 c_1(x + x_1) & \text{for} \quad x \in (-\infty, -x_1], \\ 0 & \text{for} \quad x \in (-x_1, x_1), \\ \omega^2 c_1 (x - x_1) & \text{for} \quad x \in [x_1, \infty). \end{cases} \tag{21.6}$$

Consider now the periodic vibration of the oscillator in the neighborhood of the principal "resonance", i.e., for $p = q = 1$ and

$$\omega^2 - \nu^2 = \varepsilon\Delta. \tag{21.7}$$

Note first that, by replacing ω with ν into $(17.10)_3$, one obtains $D_k(a) = 0$ for any k, since $f_1(\dot{x}) \equiv 0$ and $\psi_1(a \cos \alpha)$ is an even function of α. If now $0 < a \leqslant x_1$, then from (21.6) it follows that

$$\psi_1(a \cos \alpha) \equiv 0,$$

whence $C_k(a) = 0$, and hence the oscillator behaves as a linear undamped oscillator. Its periodic vibration is given by the equation

$$x_{\mathrm{I}} = a \cos (\nu t + \vartheta), \tag{21.8}$$

where $\vartheta = 0$ or π, depending on whether ν is less or greater than ω, and the amplitude a, which is always assumed positive, results from the relation

$$a = \frac{P}{|\omega^2 - \nu^2|}. \tag{21.9}$$

Alternatively, if $a > x_1$, then (17.10) yields

$$C_k(a) = \frac{1}{\pi} \int_0^{2\pi} \psi_1 \, (a \cos \alpha) \cos k\alpha \, d\alpha. \tag{21.10}$$

As $\psi_1(-x) = -\psi_1(x)$, the function $\psi_1(a \cos \alpha)$ is an odd function of $\cos \alpha$, and hence $C_k(a) = 0$ or

$$C_k(a) = \frac{4}{\pi} \int_0^{\frac{\pi}{2}} \psi_1(a \cos \alpha) \cos k\alpha \, d\alpha, \tag{21.11}$$

according as k is even or odd. Next, by (21.6), we deduce for $a > x_1$ that

$$\psi_1(a \cos \alpha) = \begin{cases} \omega^2 c_1 a \left(\cos \alpha - \dfrac{x_1}{a} \right) & \text{for} \quad \alpha \in \left[0, \ \arccos \dfrac{x_1}{a} \right], \\ 0 & \text{for} \quad \alpha \in \left(\arccos \dfrac{x_1}{a}, \dfrac{\pi}{2} \right] \end{cases} \tag{21.12}$$

and substituting this into (21.11), we obtain

$$C_1(a) = \frac{2\omega^2 a c_1}{\pi} \left[\arccos \left[\frac{x_1}{a} - \frac{x_1}{a} \sqrt{1 - \left(\frac{x_1}{a} \right)^2} \right] \right], \tag{21.13}$$

and

$$C_k(a) = \frac{2\omega^2 a c_1}{\pi} \left\{ \frac{1}{k+1} \sin\left[(k+1) \arc\cos \frac{x_1}{a} \right] + \right.$$

$$\left. + \frac{1}{k-1} \sin\left[(k-1) \arc\cos \frac{x_1}{a} \right] - \frac{2x_1}{ka} \sin\left(k \arc\cos \frac{x_1}{a} \right) \right\}, \qquad (21.14)$$

for $k = 3, 5, 7, \ldots$.

Introducing (21.13) into (17.37) gives

$$\left.\begin{aligned} A_1(a, \vartheta) &= -\frac{P_1 \sin \vartheta}{2\nu}, \\ B_1(a, \vartheta) &= \frac{\Delta}{2\nu} + \frac{\omega^2 c_1}{\pi\nu} \left[\arc\cos \frac{x_1}{a} - \frac{x_1}{a} \sqrt{1 - \left(\frac{x_1}{a}\right)^2} \right] - \frac{P_1 \cos \vartheta}{2\nu a}. \end{aligned}\right\} \qquad (21.15)$$

The equations of periodic vibrations:

$$\varepsilon A_1(a, \vartheta) = 0, \qquad \varepsilon B_1(a, \vartheta) = 0$$

now yield by considering (21.4) and (21.7),

$$\sin \vartheta = 0, \qquad \vartheta = 0 \quad \text{or} \quad \pi, \qquad (21.16)$$

$$\omega^2 - \nu^2 + \frac{2\omega^2 c}{\pi} \left[\arc\cos \frac{x_1}{a} - \frac{x_1}{a} \sqrt{1 - \left(\frac{x_1}{a}\right)^2} \right] \mp \frac{P}{a} = 0. \qquad (21.17)$$

Summarizing the results obtained, we conclude that, at the first stage of approximation, the periodic vibration is given by (21.8). The relation between the amplitude $a > 0$ and the exciting frequency ν, i.e., the equation of the *resonance curve*, is given by (21.9) for $a/x_1 \leqslant 1$, and by (21.17) for $a/x_1 \geqslant 1$; it may be also written in the form

$$\frac{\nu^2}{\omega^2} = \begin{cases} 1 \mp \dfrac{P}{\omega^2 x_1} \cdot \dfrac{x_1}{a} & \text{for } \dfrac{a}{x_1} \leqslant 1, \\[2ex] 1 + \dfrac{2c}{\pi} \left[\arc\cos \dfrac{x_1}{a} - \dfrac{x_1}{a} \sqrt{1 - \left(\dfrac{x_1}{a}\right)^2} \right] \mp \dfrac{P}{\omega^2 x_1} \cdot \dfrac{x_1}{a} = 0 & \text{for } \dfrac{a}{x_1} \geqslant 1. \end{cases}$$

(21.18)

where the minus sign corresponds to $\vartheta = 0$, and the plus sign to $\vartheta = \pi$. The equation of the skeleton curve (C_s) results from (21.18) by setting $P = 0$ (nonlinear free vibrations).

Next, by (21.18), we deduce

$$\frac{d(\nu^2/\omega^2)}{d(a/x_1)} = \begin{cases} \pm \dfrac{P}{\omega^2 x_1}\left(\dfrac{x_1}{a}\right)^2 & \text{for } \dfrac{a}{x_1} \leqslant 1, \\ \left(\dfrac{x_1}{a}\right)^2\left[\dfrac{4c}{\pi}\sqrt{1-\left(\dfrac{x_1}{a}\right)^2} \pm \dfrac{P}{\omega^2 x_1}\right] & \text{for } \dfrac{a}{x_1} \geqslant 1. \end{cases} \tag{21.19}$$

We note that the tangent to each resonance curve is continuous for $a/x_1 = 1$. By equating to zero the right-hand side of (21.19) we derive the condition to be satisfied at the points of vertical tangency:

$$\frac{P}{\omega^2 x_0} = \pm\frac{4c}{\pi}\sqrt{1-\left(\frac{x_1}{a}\right)^2}, \qquad \frac{a}{x_1} \geqslant 1.$$

Substituting this relation into $(21.18)_2$ we find the equation of the geometric locus L_v of the points at which the resonance curves have vertical tangent:

$$\frac{\nu^2}{\omega^2} = 1 + \frac{2c}{\pi}\left[\arccos\frac{x_1}{a} + \frac{x_1}{a}\sqrt{1-\left(\frac{x_1}{a}\right)^2}\right], \qquad \frac{a}{x_1} \geqslant 1.$$

In Fig. 56 the resonance curves obtained for $P_1/(\omega^2 x_1) = 0.1$, $P_2/(\omega^2 x_1) = 0.2$, $P_3/(\omega^2 x_1) = 0.3$, and $c = 2/3$, have been plotted by using the dimensionless variables ν^2/ω^2 and a/x_1. The branches of the resonance curve situated on the left of C_s correspond to periodic vibrations in phase with the exciting force ($\vartheta = 0$), and those situated on the right of C_s correspond to periodic vibrations in opposition of phase with the exciting force ($\vartheta = \pi$). Figure 57 shows the resonance curves obtained for the same values of the external force as in Fig. 56, but for $c = -2/3$.

As shown in § 17 c, stable periodic vibrations correspond on the left of C_s to arcs of the resonance curve on which the amplitude a is an increasing function of ν^2, and on the right of C_s to arcs on which the amplitude a is a decreasing function of ν^2. The curve L_v separates stable from unstable conditions. If one of the quantities P or ν varies continuously, the other one remaining constant, jump phenomena in the variation of a and ϑ occur; these are quite similar to those encountered in § 20d for the oscillator with cubic elastic characteristic.

Let us now determine the second approximation. For $a/x_1 \leqslant 1$ the second approximation brings nothing new. For $a/x_1 > 1$ we obtain from (17.39), (17.42), and (21.19)

$$\tilde{u}_1(\tau) = 0, \qquad u_1(a, \alpha, \tau) = \bar{u}_1(a, \alpha) = U(a, \alpha) = \sum_{k=3,5,\ldots} \frac{C_k(a)\cos k\alpha}{\nu^2(k^2-1)}. \tag{21.20}$$

Since $\omega^2/\nu^2 = 1 + 0(\varepsilon)$, Eqs. (17.45) and (21.20) yield for the second approximation

$$x_{II}(t) = a \cos(\nu t + \vartheta) + \sum_{k=3,5,\ldots} \frac{C_k(a)}{\omega^2(k^2-1)} \cos k(\nu t + \vartheta),$$

where $C_k(a)$ are given by (21.14).

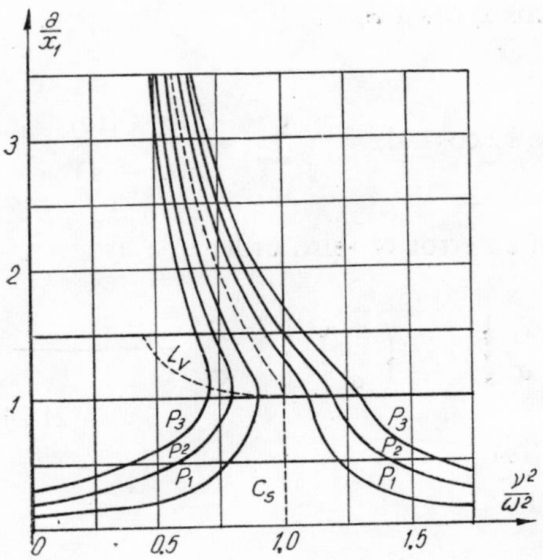

Fig. 56

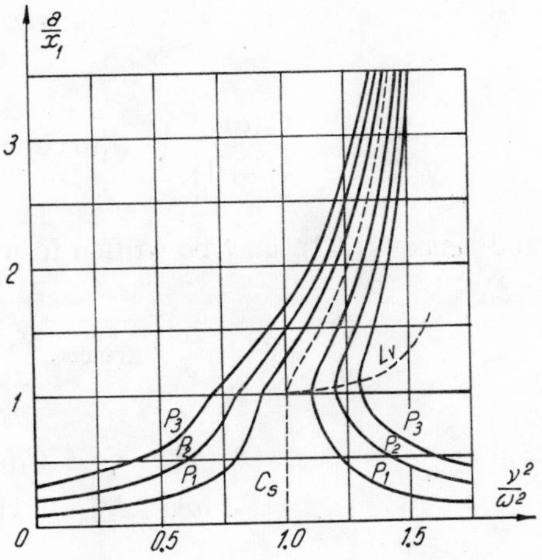

Fig. 57

From (17.46) it follows that for periodic vibrations

$$\varepsilon A_1(a, \vartheta) + \varepsilon^2 A_2(a, \vartheta) = 0, \qquad \varepsilon B_1(a, \vartheta) + \varepsilon^2 B_2(a, \vartheta) = 0. \tag{21.21}$$

Therefore, $A_1(a, \vartheta) = 0(\varepsilon)$, $B_1(a, \vartheta) = 0(\varepsilon)$, and hence we can neglect terms containing $A_1(a, \vartheta)$ and $B_1(a, \vartheta)$ in the expressions (17.40) and (17.41) of $A_2(a, \vartheta)$ and $B_2(a, \vartheta)$. Consequently, Eqs. (21.21) may be written as

$$\left.\begin{aligned} &P \sin \vartheta + \frac{1}{\pi} \int_0^{2\pi} U(a, \alpha)\, \psi_1'(a \cos \alpha) \sin \alpha \, d\alpha = 0, \\ &\omega^2 - \nu^2 + \frac{2\omega^2 c}{\pi} \left[\arccos \frac{x_1}{a} - \frac{x_1}{a} \sqrt{1 - \left(\frac{x_1}{a}\right)^2} \right] - \frac{P \cos \vartheta}{a} + \\ &+ \frac{\varepsilon^2}{\pi a} \int_0^{2\pi} U(a, \alpha)\, \psi_1'(a \cos \alpha) \cos \alpha \, d\alpha = 0. \end{aligned}\right\} \tag{21.22}$$

Since $U(a, \alpha)$ and $\psi_1'(a \cos \alpha)$ are even functions of α, the integral in $(21.22)_1$ vanishes, and we again obtain $\sin \vartheta = 0$, whence

$$\vartheta = 0 \quad \text{or} \quad \pi. \tag{21.23}$$

On the other hand, we have by (21.20) and (21.10):

$$\frac{1}{\pi}\int_0^{2\pi} U(a, \alpha)\, \psi_1'(a \cos \alpha) \cos \alpha \, d\alpha =$$

$$= \sum_{k=3,5,\ldots} \frac{C_k(a)}{\nu^2(k^2 - 1)} \int_0^{2\pi} \psi_1'(a \cos \alpha) \cos k\alpha \cos \alpha \, d\alpha = \sum_{k=3,5,\ldots} \frac{C_k(a)\, C_k'(a)}{\nu^2(k^2 - 1)},$$

and hence $(21.22)_2$ may be written to within an error of third order in ε as

$$\frac{\nu^2}{\omega^2} = 1 + \frac{2c}{\pi}\left[\arccos \frac{x_1}{a} - \frac{x_1}{a}\sqrt{1 - \left(\frac{x_1}{a}\right)^2}\,\right] \mp \frac{P}{a} + \frac{\varepsilon^2}{\omega^4 a} \sum_{k=3,5,\ldots} \frac{C_k(a)\, C_k'(a)}{k^2 - 1}, \tag{21.24}$$

where the minus sign corresponds to $\vartheta = 0$, and the plus sign to $\vartheta = \pi$. From (21.14) it follows that

$$C_k'(a) = \frac{2\omega^2 c_1}{\pi}\left\{\frac{1}{k+1} \sin\left[(k+1) \arccos \frac{x_1}{a}\right] + \frac{1}{k-1} \sin\left[(k-1) \arccos \frac{x_1}{a}\right]\right\}. \tag{21.25}$$

By means of Eqs. (21.14) and (21.15) one could easily calculate the sum in (21.24) Since the denominators very rapidly increase with k, it is sufficient to retain only the first two terms of the sum. However, the correction obtained is so small that it cannot be graphically illustrated on the scale used in Fig. 56. We thus convince ourselves once more that, when studying the principal "resonance" by the method of KRYLOV and BOGOLYUBOV, it is generally sufficient to consider only the first approximation.

Chapter VI

Forced vibrations of dissipative systems

§ 22. DISSIPATIVE SYSTEMS WITH DRY FRICTION

a) General properties of the oscillators with dry friction

The equation of motion of a nonlinear oscillator with dry friction has the form

$$\ddot{x} + F(\dot{x}) + R \operatorname{sgn} \dot{x} + g(x) = p(t), \tag{22.1}$$

where $-F(\dot{x})$ is the viscous damping force, $-g(x)$ is the elastic restoring force, $d(t)$ is the external force, and $R > 0$ is the intensity of the force of dry friction, all these forces being measured per unit mass of the oscillator. Since the function $f(\dot{x}) \equiv F(\dot{x}) + R \operatorname{sgn} \dot{x}$ is discontinuous for $\dot{x} = 0$, we cannot directly apply to Eq. (22.1) the results set forth in § 15. However, as we shall see below, a good deal of these results may be generalized in an indirect way to include the case of dry friction as well.

We begin by some comments concerning the structure of Eq. (22.1) and the particularities of the motion with dry friction.

First, it is necessary to reconsider the meaning of the displacement for the oscillator with dry friction. We have agreed in the introduction to measure the displacement from the equilibrium position of the oscillator. In the presence of dry friction, however, the unique equilibrium position is replaced, as is well-known, by an interval of equilibrium. It then proves advantageous to measure further the displacement from the equilibrium position corresponding to the lack of dry friction, so that the condition $g(0) = 0$ is still satisfied. At the same time, it should be noted that the oscillator with dry friction remains at rest for $p(t) \equiv 0$, $\dot{x} = 0$, if x satisfies the inequality

$$-R \leqslant g(x) \leqslant R, \tag{22.2}$$

since, for these values of x, the magnitude of the elastic restoring force is less than the magnitude of the maximal friction force. Moreover, the motion of the oscillator may have pauses (stops) even for $p(t) \not\equiv 0$, if there exist values of t and $x(t)$ such that

$$\dot{x} = 0, \qquad p(t) - R \leqslant g(x) \leqslant p(t) + R. \tag{22.3}$$

Another point to be clarified is that concerning the term $R \operatorname{sgn} \dot{x}$ in Eq. (22.1). It is well-known that the function $\operatorname{sgn} \dot{x}$ is equal to 1 or to -1, depending on whether $\dot{x} > 0$ or $\dot{x} < 0$, but it is not defined for $\dot{x} = 0$. In order to avoid this indeterminacy we shall seek a solution of Eq. (22.1) by matching the solutions of the equations

$$\ddot{x} + F(\dot{x}) + R + g(x) = p(t) \quad \text{for} \quad \dot{x} > 0, \tag{22.4}$$

$$\ddot{x} + F(\dot{x}) + R + g(x) = p(t) \quad \text{for} \quad \dot{x} < 0, \tag{22.5}$$

on requiring that the solution obtained be continuous together with its first derivative for $\dot{x} = 0$.

Let us examine this matching procedure in more detail. We assume that the initial conditions are

$$x(t_0) = x_0, \qquad \dot{x}(t_0) = v_0 \tag{22.6}$$

and that the functions $F(\dot{x})$, $g(x)$, and $p(t)$ fulfill the hypotheses H_1 and H_2, which assure the existence and uniqueness of the solutions of Eqs. (22.4) and (22.5) satisfying initial conditions of the form (22.6). We also suppose that $p(t)$ is periodic with minimum period T and denote by M the least upper bound and by m the greatest lower bound of $p(t)$ on each period.

Assume that the motion begins in the positive direction of the x-axis, i.e., $\dot{x} > 0$. Let $x(t)$ be the solution of Eq. (22.4) satisfying the initial condition (22.6). If $\dot{x}(t)$ does not vanish for $t \in [t_0, \infty)$, then $x(t)$ is the solution sought. In the opposite case, there exists a $t_1 > t_0$ such that $\dot{x}(t_1) = 0$, $\dot{x}(t) > 0$ for $t \in (t_0, t_1)$.

Denote $x(t_1) = x_1$. There are now three alternatives. If t_1 and x_1 do not satisfy the condition $(22.3)_2$, i.e., if

$$p(t_1) < g(x_1) - R \quad \text{or} \quad p(t_1) > g(x_1) + R, \tag{22.7}$$

then the motion continues without stop. Since $\dot{x}$ changes of sign when passing through 0, we have to use now Eq. (22.5). For the sake of simplicity, we also denote by $x(t)$ the solution of Eq. (22.5) satisfying the initial conditions

$$x(t_1) = x_1, \qquad \dot{x}(t_1) = 0. \tag{22.8}$$

These conditions assure the continuity of the matched solution and of its first derivative for $t = t_1$. If $\dot{x}(t)$ does not vanish for $t > t_1$, then $x(t)$ is the sought solution for $t \in [t_1, \infty)$. In the opposite case, there exists some $t_2 > t_1$ such that $\dot{x}(t_2) = 0$, $\dot{x}(t) < 0$ for $t \in (t_1, t_2)$, and the discussion must be similarly repeated for $t = t_2$.

The second alternative is that the condition (22.3) be fulfilled for $t = t_1$, i.e.,

$$p(t_1) - R \leqslant g(x_1) \leqslant p(t_1) + R. \tag{22.9}$$

In this case the motion ceases until, for some $t_1' > t_1$, the magnitude of the exciting force is large enough to reverse the inequality (22.9). Then

$$p(t_1') < g(x_1) - R \quad \text{or} \quad p(t_1') > g(x_1) + R, \tag{22.10}$$

the motion resumes, and we have to determine the function $x(t)$ that satisfies Eq. (22.5) and initial conditions similar to (22.8):

$$x(t_1') = x_1, \qquad \dot{x}(t_1') = 0. \tag{22.11}$$

Finally, a third alternative would be that the rest continues for $t > t_1$, i.e.,

$$\dot{x}(t) \equiv 0, \qquad p(t) - R \leqslant g(x_1) \leqslant p(t) + R, \tag{22.12}$$

for any $t \geqslant t_1$. Obviously, the last condition can be fulfilled if and only if

$$M - R \leqslant g(x_1) \leqslant m + R,$$

from which it follows that

$$2R > M - m, \tag{22.13}$$

since $x_1 \neq 0$. When the inequality (22.13) is satisfied, we say, after REISSIG, that the dry friction is *large*.

From the discussion above it follows that an oscillator with dry friction acted on by a periodic exciting force may perform an oscillatory motion with or without pauses, or which ceases at a certain time. Denoting by $x(t)$ the dependence of the displacement on time, we see that $x(t)$ satisfies Eq. (22.4) for $\dot{x} > 0$, Eq. (22.5) for $\dot{x} < 0$, and the inequality $(22.3)_2$ for $\dot{x} = 0$. With this caution we shall call $x(t)$ a solution of Eq. (22.1), although, more accurately, we ought to use different notations for the dependence of the displacement on time and for solutions of Eqs. (22.4) and (22.5)[1)].

By means of the matching method explained above, DEN HARTOG [68] found the *exact* periodic solution of the equation

$$\ddot{x} + 2h\dot{x} + R \operatorname{sgn} \dot{x} + \omega^2 x = P \cos \nu t \quad (h, R, P > 0), \tag{22.14}$$

1) REISSIG [155—158] denotes the general dependence of the displacement on time by $u(t)$, by using the notation $x(t)$ either for $u(t)$ itself when the motion proceeds without stops, or for solutions of Eqs. (22.4) and (22.5) defined between two stops, in the case of a motion with pauses.

and SZABLEWSKI [187] that of the equation

$$\ddot{x} + 2h\dot{x} + R \operatorname{sgn} \dot{x} + \omega^2 x = p(t) \quad (h, R > 0), \tag{22.15}$$

where $p(t)$ is a periodic function of t with two or four extrema in each period. SZABLEWSKI used the results obtained to substantiate a method for eliminating the errors introduced by dry friction in vibration measuring instruments.

By studying periodic solutions of Eq. (22.14), DEN HARTOG [68] has shown that, for R sufficiently small, there exists an uninterrupted periodic vibration, which is asymptotically stable in the large. As R increases, periodic pauses occur in the periodic vibration, which extend more and more over each period, and, for $R > P$, there exists no other periodic solution than the perpetual rest. For comparatively large values of the dry friction and small values of the exciting frequency, two or even more pauses may occur on each half-oscillation.

For the complete expression of the solution and for other supplementary results concerning the oscillator governed by Eq. (22.14), we refer to the paper by DEN HARTOG just cited, and we confine ourselves to indicating in the next section how one can calculate the approximate periodic solution by the method of KRYLOV and BOGOLYUBOV in the case of a weak dry friction. In exchange, we expound here, in a somewhat simplified form, some of the qualitative results obtained by REISSIG [155—158] by a systematic study of the nonlinear equation (22.1). These results extend significantly over Eq. (22.1) the conclusions deduced by DEN HARTOG about the linear oscillator with dry friction. For instance, REISSIG (156], pp. 119—128) has shown that, when the dry friction is large, i.e., when condition (22.13) is fulfilled, then, irrespective of the initial conditions, the oscillator approaches a position of equilibrium as $t \to \infty$. More precisely, he has proved the following theorem:

Theorem 1. *If $g(x)$ and $F(\dot{x})$ are continuous and their derivatives are continuous and strictly positive for $x \in (-\infty, \infty)$ and $\dot{x} \in (-\infty, \infty)$, respectively, if $p(t)$ is continuous and periodic for $t \in [t_0, \infty)$, if $xg(x) > 0$ for $x \neq 0$ and $\dot{x}F(\dot{x}) > 0$ for $\dot{x} \neq 0$, if there exists a value of x such that $g(x) = (M + m)/2$, where M is the least upper bound and m is the greatest lower bound of $p(t)$ over each period, and finally, if*

$$2R > M - m,$$

then the oscillator asymptotically approaches an equilibrium position as $t \to \infty$, i.e.

$$\lim_{t\to\infty} \dot{x}(t) = 0, \qquad \lim_{t\to\infty} x(t) = \bar{x}, \quad \text{where} \quad M - R \leqslant g(\bar{x}) \leqslant m + R, \tag{22.16}$$

irrespective of the initial conditions (22.6) *of the motion.*

This theorem still holds if $p(t)$ is not periodic, in which case M means the least upper bound, and m the greatest lower bound of $p(t)$ for $t \in [t_0, \infty)$. It is easily seen that the hypothesis of Theorem 1 is covered by the assumptions $H_1 - H_5$

made in the introduction about the functions $F(\dot{x})$, $g(x)$, and $p(t)$, if, in addition, the derivatives of $F(\dot{x})$ and $g(x)$ are continuous, and $F'(\dot{x})$ does not vanish.

Next, consider the case of the moderate dry friction, i.e.,

$$2R < M - m. \tag{22.17}$$

The oscillator may now perform a periodic vibration, with or without pauses. The question arises whether this periodic vibration is asymptotically stable in the large, like the equilibrium solution in the case of the large dry friction. The case when $g(x)$ is a nonlinear function has not been studied so far. However, considering the results expounded in § 15 for oscillators with viscous damping, it is to be expected that periodic vibrations are asymptotically stable in the large only if the elastic nonlinearity is small enough. For the particular case $g(x) = \omega^2 x$, REISSIG ([158], pp. 231–245) has shown that most of the results reviewed in § 15 still hold in the presence of dry friction. We will group these results into two theorems, the first of which assures the boundedness of the solution together with its derivative for $t \in [t_0, \infty)$, and the second one the asymptotic stability in the large of the periodic solution. We deliberately adopt more restrictive hypotheses than those of REISSIG, in order to simplify the exposition, as well as to facilitate the comparison with the discussion in § 15.

Theorem 2. *If $F(\dot{x})$ is continuous together with its first derivative for $\dot{x}(-\infty, \infty)$ and satisfies the conditions $F(0) = 0$ and $F'(\dot{x}) \geqslant 2h > 0$, if $p(t)$ is continuous and periodic with minimum period T for $t \in [t_0, \infty)$, and if*

$$2R < M - m,$$

where M is the least upper bound and m is the greatest lower bound of $p(t)$ on each period, then, corresponding to any solution $x(t)$ of the equation

$$\ddot{x} + F(\dot{x}) + R \operatorname{sgn} \dot{x} + \omega^2 x = p(t), \tag{22.18}$$

there exists a T_0 such that

$$A' \leqslant x(t) \leqslant A'', \qquad |\dot{x}(t)| \leqslant B, \tag{22.19}$$

for $t \geqslant T_0$, where, by denoting $\zeta = h/\omega$,

$$A' = \frac{1}{\omega^2}\left[m + R - \frac{M - m - 2R}{\exp(\pi\zeta/\sqrt{1-\zeta^2}) - 1}\right],$$

$$A'' = \frac{1}{\omega^2}\left[M - R + \frac{M - m - 2R}{\exp(\pi\zeta/\sqrt{1-\zeta^2}) - 1}\right],$$

$$B = \frac{M - m - 2R}{\omega} \cdot \frac{\exp\left(-\dfrac{\zeta}{\sqrt{1-\zeta^2}}\arcsin\sqrt{1-\zeta^2}\right)}{1 - \exp(-\pi\zeta/\sqrt{1-\zeta^2})}$$

for $0 < \zeta \leqslant 1$, *and*

$$A' = \frac{m + R}{\omega^2}, \qquad A'' = \frac{M - R}{\omega^2},$$

$$B = \frac{M - m - 2R}{\omega} \cdot \exp\left(- \frac{\zeta}{\zeta^2 - 1} \operatorname{arg\,sh} \sqrt{\zeta^2 - 1}\right)$$

for $\zeta \geqslant 1$.

We note that the above statement, unlike Theorem 3, § 15, not only assures the uniform boundedness in the future of the solutions and of their first derivatives, but also provides comparatively simple evaluations of the bounds. As functions of $\zeta : A''$ monotonically decreases from $+\infty$ to $(M - R)/\omega^2$ when ζ varies from 0 to 1, and then remains constant for $\zeta \in [1, \infty)$; A' monotonically increases from $-\infty$ to $(m - R)/\omega^2$ when ζ varies from 0 to 1, and then remains constant for $\zeta \in [1, \infty)$; and finally, B monotonically decreases from $+\infty$ to 0 as ζ increases from 0 to $+\infty$. For $\zeta = 1$, $B = (M - m - 2R)/\omega$. The condition $F'(\dot{x}) \geqslant 2h > 0$ may be further relaxed, but then the use of the evaluations above becomes rather cumbersome.

Theorem 3. *If $F(\dot{x})$ and $p(t)$ satisfy the hypotheses of theorem 2, then there exists exactly one periodic vibration of the oscillator, which is asymptotically stable in the large and whose frequency equals the exciting frequency ($\nu = 2\pi/T$), or a multiple of it. If this stationary vibration proceeds without stops, then its frequency is equal to the exciting frequency.*

The possible existence of a superharmonic vibration with pauses also results from the following reasoning. Assume that a stationary vibration with pauses has the same minimum period T as the exciting force. Arbitrarily change the exciting force during the pauses, but perform this change every p^{th} pause. The exciting force will then have the minimum period pT, whereas the stationary vibration will remain unchanged, and hence its frequency will be p times greater than the exciting frequency.

b) The oscillator with dry friction and linear viscous damping

Theorem 3 assures the existence of a stationary vibration in the case of a moderate dry friction ($2R < M - m$). We will now determine this vibration in the neighborhood of the principal "resonance" ($\omega \approx \nu$), by means of the method of Krylov and Bogolyubov, under the simplifying hypothesis that R is small enough for the motion to proceed *without pauses*.

The equation of motion of an oscillator with dry friction and linear viscous damping has the form

$$\ddot{x} + 2h\dot{x} + R \operatorname{sgn} \dot{x} + \omega^2 x = P \cos \nu t. \tag{22.20}$$

Denoting

$$h = \varepsilon h_1, \qquad R = \varepsilon R_1, \qquad P = \varepsilon P_1, \qquad \omega^2 - \nu^2 = \varepsilon\Delta, \tag{22.21}$$

where ε is a small parameter, this equation becomes

$$\ddot{x} + \nu^2 x = -\varepsilon(x\Delta + 2h_1\dot{x} + R_1 \operatorname{sgn} \dot{x} - P_1 \cos \nu t). \tag{22.22}$$

Since, by assumption, the motion has no stops, it may be represented in the first approximation as

$$x(t) = a \cos (\nu t + \vartheta). \tag{22.23}$$

We suppose as before that $P > 0$ and $a > 0$, by including the sign of the displacement in the difference of phase ϑ between the stationary vibration and the external force.

Putting $\psi_1(x) \equiv 0$, $f_1(\dot{x}) = 2h_1\dot{x} + R_1 \operatorname{sgn} \dot{x}$, and $p = 1$ into (17.37) and (17.38) yields

$$A_1(a, \vartheta) = -ah_1 - \frac{2R_1}{\pi\nu} - \frac{P_1 \sin \vartheta}{2\nu}, \qquad B_1(a, \vartheta) = \frac{\Delta}{2\nu} - \frac{P_1 \cos \vartheta}{2\nu a}. \tag{22.24}$$

For the periodic vibration ($\dot{a} = \dot{\vartheta} = 0$) we have from (17.44)

$$\varepsilon A_1(a, \vartheta) = 0, \qquad \varepsilon B_1(a, \vartheta) = 0$$

and, by taking into account (22.21) and (22.24), there results:

$$\sin \vartheta = -\frac{2\nu ah}{P} - \frac{4R}{\pi P}, \qquad \cos \vartheta = \frac{a(\omega^2 - \nu^2)}{P}. \tag{22.25}$$

Eliminating ϑ between the last two relations we derive the equation of the *resonance curve*

$$a^2(\omega^2 - \nu^2)^2 + \left(2\nu ah + \frac{4R}{\pi}\right)^2 - P^2 = 0. \tag{22.26}$$

By solving this equation for a, we obtain the dependence of the response amplitude on the frequency and amplitude of the exciting force:

$$a = \frac{P}{\omega^2} \cdot \frac{-2c\zeta\dfrac{\nu}{\omega} + \sqrt{4c^2\zeta^2\dfrac{\nu^2}{\omega^2} + (1 - c^2)\left[\left(\dfrac{\nu^2}{\omega^2} - 1\right)^2 + 4\zeta^2\dfrac{\nu^2}{\omega^2}\right]}}{\left(\dfrac{\nu^2}{\omega^2} - 1\right)^2 + 4\zeta^2\dfrac{\nu^2}{\omega^2}}, \tag{22.27}$$

where

$$c = \frac{4R}{\pi P} \geqslant 0, \qquad \zeta = \frac{h}{\omega}; \tag{22.28}$$

in (22.27) only the plus sign has been taken before the radical, because $a > 0$.

Since a is positive, the stationary motion exists only if $c < 1$. This condition must be further strengthened, by requiring the motion to have no pauses. To this end we first note that, by (22.23),

$$\dot{x}(t) = -\nu a \sin(\nu t + \vartheta), \tag{22.29}$$

and hence, for $\dot{x} = 0$, we have

$$\nu t_1 + \vartheta = 0 \quad \text{or} \quad \pi, \qquad \text{and} \quad x(t_1) = \pm a. \tag{22.30}$$

The condition (22.7), which assures the continuation of the motion for $t = t_1$, gives

$$P \cos \nu t - R > \omega^2 x \quad \text{or} \quad P \cos \nu t + R < \omega^2 x \quad \text{for} \quad t = t_1. \tag{22.31}$$

By $(22.25)_2$, Eqs. (22.31) reduce in both cases expressed by (22.30) to the unique condition

$$R < a\nu^2. \tag{22.32}$$

Figure 58 illustrates the dependence of a on the exciting frequency for $\zeta = 0.2$, and $c = 0$, 0.2, 0.4, and 0.6, as given by (22.27) when using the dimensionless variables $\omega^2 a/P$ and ν/ω. The broken line is the lower limit of the region within which the lack of stops is assured by (22.32). To find its equation we put $R = a\nu^2$ into (22.26), obtaining

$$a\left(\frac{\nu^2}{\omega^2} - 1\right)^2 + \left(2a\zeta\frac{\nu}{\omega} + \frac{4a\nu^2}{\pi\omega^2}\right)^2 = \frac{P^2}{\omega^4},$$

whence

$$\frac{\omega^2 a}{P} = \frac{1}{\sqrt{\left(\frac{\nu^2}{\omega^2} - 1\right)^2 + \frac{4\nu^2}{\omega^2}\left(\zeta + \frac{2}{\pi}\frac{\nu}{\omega}\right)^2}}. \tag{22.33}$$

Figure 59 shows the dependence of the phase difference ϑ on the exciting frequency, obtained for the same values of ζ and c as those in Fig. 58. We see from (22.25) that $\sin \vartheta \leqslant 0$, and that $\cos \vartheta$ is positive or negative according as $\nu < \omega$

or $\nu > \omega$. Consequently, $\vartheta \in (-\pi/2, 0)$ for $\nu < \omega$, and $\vartheta \in (-\pi, -\pi/2)$ for $\nu > \omega$. The broken line limits the region where condition (22.32) is fulfilled. Its equation can be obtained by substituting $R = a\nu^2$ into $(22.25)_1$ and eliminating a between the two relations (22.25). There results

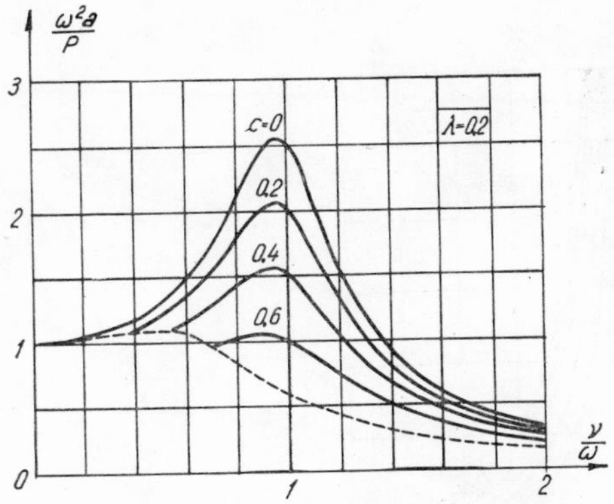

Fig. 58

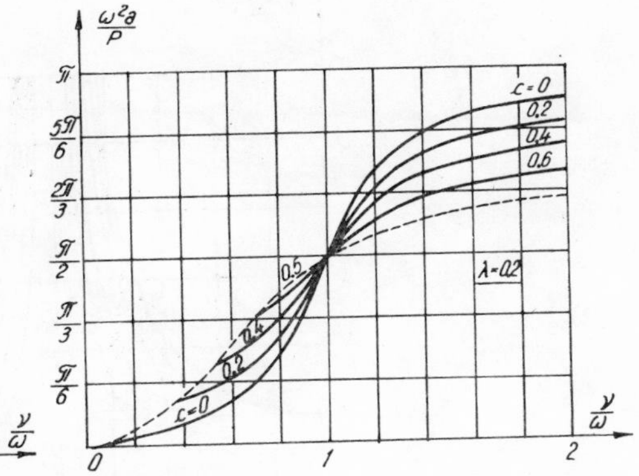

Fig. 59

$$\tan \vartheta = \frac{(2\nu/\omega)(\zeta + 2\nu/\pi\omega)}{\nu^2/\omega^2 - 1}, \tag{22.34}$$

and hence the equation of the limit curve is

$$\left.\begin{aligned} \vartheta &= -\arctan \frac{(2\nu/\omega)(\zeta + 2\nu/\pi\omega)}{1 - \nu^2/\omega^2} \quad \text{for} \quad \nu < \omega, \\ \vartheta &= -\pi + \arctan \frac{(2\nu/\omega)(\zeta + 2\nu/\pi\omega)}{\nu^2/\omega^2 - 1} \quad \text{for} \quad \nu > \omega. \end{aligned}\right\} \tag{22.35}$$

For the oscillator with dry friction but without viscous damping ($\zeta = 0$), we find from (22.25) and (22.26) that

$$\sin \vartheta = -c, \qquad \cos \vartheta = \frac{a(\omega^2 - \nu^2)}{P}, \tag{22.36}$$

$$a = \frac{P}{\omega^2} \cdot \frac{\sqrt{1 - c^2}}{|\nu^2/\omega^2 - 1|}. \tag{22.37}$$

The dependence of the amplitude a and of the phase angle ϑ on the exciting frequency is now illustrated by Figs 60 and 61. The equation of the curves limiting the motions without pauses may be easily derived by setting $\zeta = 0$ in (22.33)

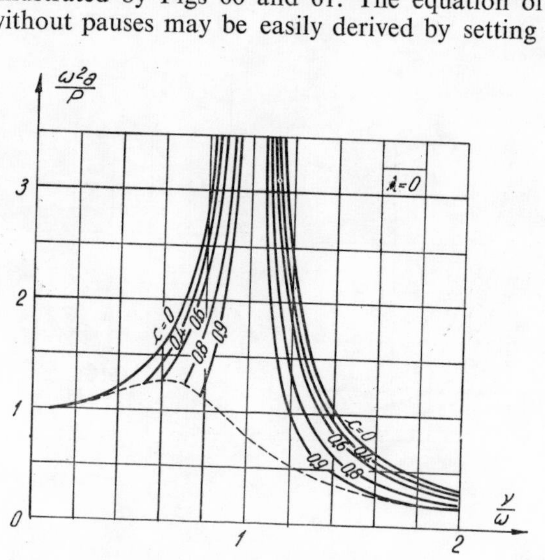

Fig. 60

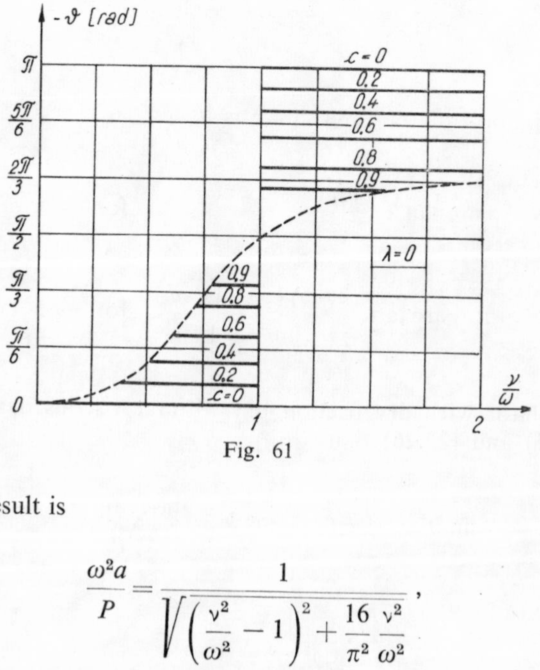

Fig. 61

and (22.35). The result is

$$\frac{\omega^2 a}{P} = \frac{1}{\sqrt{\left(\frac{\nu^2}{\omega^2} - 1\right)^2 + \frac{16}{\pi^2}\frac{\nu^2}{\omega^2}}}, \tag{22.38}$$

$$\vartheta = -\arctan\frac{4\nu^2/\pi\omega^2}{1-\nu^2/\omega^2} \quad \text{for} \quad \nu < \omega,$$

$$\vartheta = \pi + \arctan\frac{4\nu^2/\pi\omega^2}{1-\nu^2/\omega^2} \quad \text{for} \quad \nu > \omega. \tag{22.39}$$

It is worth noting that in the absence of viscous damping the amplitude $a \to \infty$ as $\nu \to \omega$, a resonance behavior quite similar to that of a linear conservative oscillator. Accordingly, there also appears a discontinuity in the variation of ϑ with ν for $\nu/\omega = 1$.

Comparing the curves represented in Figs 58—61 with those given by DEN HARTOG for the exact solution obtained by the matching method, it may be shown that the approximate solution is sufficiently accurate only in the neighborhood of the resonance. However, as this region is practically the most interesting, the restricted range of applicability of the approximate solution should not be considered as very unsatisfactory.

§ 23. THE OSCILLATOR WITH LINEAR VISCOUS DAMPING AND CUBIC ELASTIC RESTORING FORCE

The forced vibrations of an oscillator with linear viscous damping and cubic elastic restoring force are governed by the equation

$$\ddot{x} + 2h\dot{x} + \omega^2 x(1 + \beta x^2) = p(\nu t), \tag{23.1}$$

where $p(\nu t)$ is a periodic function of νt with minimum period 2π. In the following we shall content ourselves to considering as in § 20 only the equation

$$\ddot{x} + 2h\dot{x} + \omega^2 x(1 + \beta x^2) = P\cos\nu t, \qquad h > 0, \tag{23.2}$$

which is also called *generalized Duffing's equation with damping*. We assume as before that $g'(x) = \omega^2(1 + 3\beta x^2) > 0$ for any x, a condition which is fulfilled either for $\beta > 0$ and any x, or for $\beta < 0$ and $|x| < 1/\sqrt{-3\beta}$. As in the case of the conservative oscillator (cf. § 20), we will comparatively apply the perturbation method, the method of KRYLOV and BOGOLYUBOV, and the method of finite sums of trigonometric functions, in order to obtain approximate periodic solutions of Eq. (23.2). In the last part of this section we shall examine the stability of periodic solutions, the jump phenomena and the subharmonic "resonance."

a) Use of the perturbation method

— *"Nonresonance" case* ($\omega/\nu \neq p$ for any integer p)

Applying the transformation

$$\tau = \nu t \tag{23.3}$$

and denoting

$$x' = \frac{dx}{d\tau}, \qquad x'' = \frac{d^2x}{d\tau^2}, \qquad \eta = \frac{\omega}{\nu}, \tag{23.4}$$

Eq. (23.2) becomes

$$x'' + \frac{h}{\nu}x' + \eta^2 x(1 + \beta x^2) = \frac{P}{\nu^2}\cos\tau. \tag{23.5}$$

Let us seek a harmonic solution of this equation. Assume that the damping and the departure of the elastic restoring force from linearity are small, and set

$$\frac{h}{\nu} = \varepsilon\bar{h}, \qquad \eta^2\beta = \varepsilon\bar{\beta}, \tag{23.6}$$

where ε is a small parameter. The equation of motion takes the form

$$x'' + \eta^2 x = \frac{P}{\nu^2}\cos\tau - \varepsilon(\bar{\beta}x^3 + 2\bar{h}x'). \tag{23.7}$$

Since, by assumption, η is not an integer, Eq. (23.7) has for $\varepsilon = 0$ a unique periodic solution with minimum period 2π, namely

$$\bar{x}(\tau) = \frac{P}{\nu^2(\eta^2 - 1)}\cos\tau = \frac{P}{\omega^2 - \nu^2}\cos\tau. \tag{23.8}$$

For $\varepsilon \neq 0$ the solution of Eq. (23.7) may be sought as an expansion in powers of ε:

$$x(\tau; \varepsilon) = \bar{x}(\tau) + x_1(\tau)\varepsilon + x_2(\tau)\varepsilon^2 + \ldots \tag{23.9}$$

whose coefficients are periodic functions of τ with minimum period 2π, and which is convergent for ε small enough, i.e., for sufficiently small β and h (cf. § 16 a). Substituting (23.9) into (23.7) and equating coefficients of like powers of ε, we obtain

$$x_1'' + \eta^2 x_1 = -\bar{\beta}\bar{x}^3 - 2\bar{h}\bar{x}', \tag{23.10}$$

$$x_2'' + \eta^2 x_2 = -3\bar{\beta}x_1 x^2 - 2\bar{h}x', \text{ etc.} \tag{23.11}$$

Introducing now (23.8) into (23.10) yields

$$x_1'' + \eta^2 x_1 = -\frac{\bar{\beta}P^3}{4(\omega^2 - \nu^2)^3}(\cos 3\tau + 3\cos\tau) + \frac{2P\bar{h}}{\omega^2 - \nu^2}\sin\tau.$$

Since η is not an integer, this equation admits of exactly one periodic solution with minimum period 2π:

$$x_1(\tau) = \frac{P}{\omega^2 - \nu^2}\left[-\frac{\bar{\beta}P^2}{4(\omega^2 - \nu^2)^2}\left(\frac{1}{\eta^2 - 9}\cos 3\tau + \frac{3}{\eta^2 - 1}\cos\tau\right) + \frac{2\bar{h}}{\eta^2 - 1}\sin\tau\right]. \tag{23.12}$$

Then, by substituting (23.8) and (23.12) into (23.11), we can derive $x_2(\tau)$, and so on. From (23.8), (23.9), (23.11), and considering (23.3), $(23.4)_3$, and (23.6), we deduce that

$$x(t;\varepsilon) = \frac{P}{\omega^2 - \nu^2}\left\{\left[1 - \frac{3\beta\omega^2P^2}{4(\omega^2 - \nu^2)^3}\right]\cos\nu t + \frac{2h\nu}{\omega^2 - \nu^2}\sin\nu t - \right.$$
$$\left. - \frac{\beta\omega^2P^2}{4(\omega^2 - \nu^2)^2(\omega^2 - 9\omega^2)}\cos 3\nu t\right\} + \ldots \tag{23.13}$$

– *Principal "resonance"* ($\omega \approx \nu$)

We will study here only the principal "resonance" ($\eta \approx 1$), putting off the discussion of subharmonic "resonance" till the end of this section. We accordingly assume

$$\frac{\omega^2}{\nu^2} = \eta^2 = 1 + \varepsilon\bar{\Delta} \tag{23.14}$$

and suppose that the magnitude of the exciting force is also of order $0(\varepsilon)$, i.e.,

$$\frac{P}{\nu^2} = \varepsilon\bar{P}, \tag{23.15}$$

this last condition being necessary for the existence of a periodic solution in the "resonance" case. With these notations, Eq. (23.7) becomes

$$x'' + x = -\varepsilon(x\bar{\Delta} + \bar{\beta}x^3 + 2\bar{h}x' - \bar{P}\cos\tau) \tag{23.16}$$

and has for $\varepsilon = 0$ the general solution

$$\bar{x}(\tau) = M\cos\tau + N\sin\tau, \tag{23.17}$$

which depends on the arbitrary constants M and N, and will be taken as generating solution. For $\varepsilon \neq 0$, we seek a solution of the form

$$x(\tau; \varepsilon) = \bar{x}(\tau) + x_1(\tau)\varepsilon + x_2(\tau)\varepsilon^2 + \ldots \tag{23.18}$$

Substituting this expansion into (23.16) and equating coefficients of like powers of ε, we obtain a recursive set of linear differential equations

$$\begin{aligned} x_1'' + x_1 &= -\bar{x}\bar{\Delta} - \bar{\beta}\bar{x}^3 - 2\bar{h}\bar{x}' + \bar{P}\cos\tau, \\ x_2'' + x_2 &= -x_1\bar{\Delta} - 3\bar{\beta}x_1\bar{x}^2 - 2\bar{h}x_1', \text{ etc.} \end{aligned} \tag{23.19}$$

Introducing $\bar{x}(\tau)$ into $(23.19)_1$ gives

$$\begin{aligned} x_1'' + x_1 = &-\left(M\bar{\Delta} + \frac{3}{4}\bar{\beta}M^3 + \frac{3}{4}\bar{\beta}MN^2 + 2\bar{h}N - \bar{P}\right)\cos\tau \\ &- \frac{1}{4}\bar{\beta}M(M^2 - 3N^2)\cos 3\tau - \left(N\bar{\Delta} + \frac{3}{4}\bar{\beta}M^2N + \frac{3}{4}\bar{\beta}N^3 - 2\bar{h}M\right)\sin\tau \\ &- \frac{1}{4}\bar{\beta}N(3M^2 - N^2)\sin 3\tau. \end{aligned}$$

The periodicity of $x_1(\tau)$ requires the vanishing of the terms in $\cos\tau$ and $\sin\tau$ in the right-hand side, because, otherwise, $x_1(\tau)$ would contain secular terms. This condition yields

$$\begin{aligned} M\bar{\Delta} + \frac{3}{4}\bar{\beta}M(M^2 + N^2) - 2\bar{h}N - \bar{P} &= 0, \\ N\bar{\Delta} + \frac{3}{4}\bar{\beta}N(M^2 + N^2) + 2\bar{h}M \qquad &= 0. \end{aligned} \tag{23.20}$$

The remaining part of the differential equation is

$$x_1'' + x_1 = -\frac{1}{4}\bar{\beta}M(M^2 - 3N^2)\cos 3\tau - \frac{1}{4}\bar{\beta}N(3M^2 - N^2)\sin 3\tau$$

and has the particular solution

$$x_1(\tau) = M_1\cos\tau + N_1\sin\tau + \frac{1}{32}\bar{\beta}M(M^2 - 3N^2)\cos 3\tau + \frac{1}{32}\bar{\beta}N(3M^2 - N^2)\sin 3\tau,$$

where M_1 and N_1 are arbitrary constants to be determined by requiring the periodicity of $x_2(\tau)$, and so on. However, it is readily seen that the terms grow more and more complicated at the successive stages of approximation, and, therefore, we content ourselves with the first approximation. To within terms of second order in ε, we have from (23.17) and (23.18)

$$x(t;\varepsilon) = M\cos\tau + N\sin\tau.$$

Considering (23.3) and setting

$$M = a\cos\vartheta, \qquad N = -a\sin\vartheta, \qquad a > 0,$$

the approximate solution assumes the form

$$x(t;\varepsilon) = a\cos(\nu t + \vartheta) \tag{23.21}$$

and (23.20) yields

$$\cos\vartheta = \frac{a}{\bar{P}}\left(\bar{\Delta} + \frac{3}{4}\bar{\beta}a^2\right), \qquad \sin\vartheta = -\frac{2\bar{h}a}{\bar{P}},$$

$$\left(\bar{\Delta} + \frac{3}{4}\bar{\beta}a^2\right)^2 + 4\bar{h}^2 = \frac{\bar{P}^2}{a^2}.$$

Finally, by taking into account (23.6), (23.14), and (23.15), the last two relations become

$$\cos\vartheta = \frac{a}{P}\left(\omega^2 - \nu^2 + \frac{3}{4}\beta\omega^2a^2\right), \qquad \sin\vartheta = -\frac{2\nu ha}{P}, \tag{23.22}$$

$$\left(\omega^2 - \nu^2 + \frac{3}{4}\beta\omega^2a^2\right)^2 + 4\nu^2h^2 = \frac{P^2}{a^2}. \tag{23.23}$$

b) Use of the method of KRYLOV and BOGOLYUBOV

— *"Nonresonance" case* ($\omega \neq \nu$)

We again consider Eq. (23.2) and, assuming that the damping, the elastic nonlinearity, and the magnitude of the exciting force are small, we denote

$$h = \varepsilon h_1, \qquad \omega^2\beta = \varepsilon\beta_1, \qquad P = \varepsilon P_1, \tag{23.24}$$

where ε is a small parameter. The equation of motion becomes

$$\ddot{x} + \omega^2 x = -\varepsilon(\beta_1 x^3 + 2h_1\dot{x} - P_1 \cos \nu t). \tag{23.25}$$

Next, supposing $\omega \neq \nu$ and putting $f_1(\dot{x}) = 2h_1\dot{x}$, $\psi_1(x) = \beta_1 x^3$ into Eqs. (17.10), gives

$$C_1(a) = \frac{3}{4}\beta_1 a^3, \qquad C_3(a) = \frac{1}{4}\beta_1 a^3, \qquad D_1(a) = -2h_1\omega a,$$

and all other coefficients $C_k(a)$, $D_k(a)$ vanish. Introducing these values into Eqs. (17.15), (17.16), (17.21), and (17.22), it follows that

$$A_1(a) = -h_1 a, \qquad B_1(a) = \frac{3\beta_1 a^2}{8\omega}, \qquad x_{\mathrm{I}}(t) = a\cos\alpha,$$

$$\dot{a} = -ha, \qquad \dot{\alpha} = \omega\left(1 + \frac{3}{8}\beta a^2\right).$$

Integrating the last two equations, we obtain as first approximation for the solution:

$$x_{\mathrm{I}}(t) = a_0 e^{-ht}\cos\left[\omega\left(t - \frac{3\beta a_0^2}{16h}e^{-2ht}\right) + \gamma\right]. \tag{23.26}$$

It is worth noting that no forcing terms occur in the first approximation.

To derive the second approximation, we make use of Eqs. (17.17), (17.20), (17.23), and (17.24). There then successively results:

$$\bar{u}_1(a, \alpha) = \frac{\beta_1 a^3}{32\omega^2}\cos 3\alpha, \qquad \tilde{u}_1(\nu t) = \frac{P_1}{\omega^2 - \nu^2}\cos\nu t,$$

$$A_2(a) = \frac{3\beta_1 h_1 a^3}{8\omega^2}, \qquad B_2(a) = -\frac{15\beta_1^2 a^4}{256\omega^3} - \frac{h_1^2}{2\omega},$$

$$\dot{a} = -ha + \frac{3}{8}\beta h a^3, \qquad \dot{\alpha} = \omega\left(1 + \frac{3}{8}\beta a^2 - \frac{15}{256}\beta^2 a^4 - \frac{h^2}{2\omega^2}\right), \tag{23.27}$$

$$x_{\mathrm{II}}(t) = a\cos\alpha + \frac{1}{32}\beta a^3\cos 3\alpha + \frac{P}{\omega^2 - \nu^2}\cos\nu t; \tag{23.28}$$

$a(t)$ and $\alpha(t)$ are to be calculated by solving Eqs. (23.27). The first of these equations may be integrated in an elementary way, but the resulting transcendental algebraic equation can hardly be solved for a. To avoid this difficulty, we note that, by (23.26), $a(t) = a_0 + 0(\varepsilon)$, and hence Eq. $(23.27)_1$ may be written to within terms of third order in ε in the form

$$\dot{a} = -ha\left(1 - \frac{3}{8}\beta a_0^2\right),$$

from which, by integration, it follows that

$$a(t) = a_0\,\mathrm{e}^{-ht\left(1 - \frac{3}{8}\beta a_0^2\right)}.$$

Finally, by substituting this relation into $(23.27)_2$ and integrating, we obtain

$$\alpha(t) = \omega\left[\left(1 - \frac{h^2}{2\omega^2}\right)t - \frac{3\beta a_0^2}{16h\left(1 - \frac{3}{8}\beta a_0^2\right)}\mathrm{e}^{-2ht\left(1 - \frac{3}{8}\beta a_0^2\right)} + \frac{15\beta^2 a_0^4}{1024h\left(1 - \frac{3}{8}\beta a_0^2\right)}\cdot\mathrm{e}^{-4ht\left(1 - \frac{3}{8}\beta a_0^2\right)}\right] + \gamma.$$

The last two equations and (23.28) give the second approximation for the solution. We cannot directly compare this result with that obtained by the perturbation method, since this last method has been used to study only periodic vibrations and under the assumption that P is independent of ε. However, by noting that $a(t) \to 0$ as $t \to \infty$, we derive from (23.28) for the periodic vibration

$$x(t) = \frac{P}{\omega^2 - \nu^2}\cos\nu t,$$

i.e., just the generating solution (23.8) in the perturbation method. The difference in the number of steps used to obtain the approximate solution may be understood when remembering that the method of Krylov and Bogolyubov assumes from the very beginning that $P = 0(\varepsilon)$. This difference disappears in the "resonance" case (cf. § 20).

— *"Resonance" case* ($\omega \approx \nu$)

To study the principal "resonance," assume that ν is close to ω and denote $\omega^2 - \nu^2 = \varepsilon\Delta$. The governing equation becomes

$$\ddot{x} + \nu^2 x = -\varepsilon(x\Delta + \beta_1 x^3 + 2h_1\dot{x} - P_1\cos\nu t). \tag{23.29}$$

By putting $p = q = 1$ into (17.37), (17.38), and (17.44), we obtain as the first approximation

$$A_1(a, \vartheta) = -h_1 a - \frac{P_1 \sin \vartheta}{2\nu}, \qquad B_1(a, \vartheta) = \frac{\Delta}{2\nu} + \frac{3\beta_1 a^2}{8\nu} - \frac{P_1 \cos \vartheta}{2\nu a}.$$

$$\dot{a} = \varepsilon A_1(a, \vartheta), \qquad \dot{\vartheta} = \varepsilon B_1(a, \vartheta).$$

For the periodic vibration ($\dot{a} = \dot{\vartheta} = 0$) and using the notations (23.24), we deduce that

$$\cos \vartheta = \frac{a}{P}\left(\omega^2 - \nu^2 + \frac{3}{4}\beta\omega^2 a^2\right), \qquad \sin\vartheta = -\frac{2\nu h a}{P}, \tag{23.30}$$

$$\left(\omega^2 - \nu^2 + \frac{3}{4}\beta\omega^2 a^2\right)^2 + 4\nu^2 h^2 = \frac{P^2}{a^2}. \tag{23.31}$$

Next, from (17.43) there results:

$$x_1(t) = a \cos(\nu t + \vartheta), \tag{23.32}$$

where a and ϑ are to be determined by (23.31) and (23.30). We consider, as usually, that a and P are positive, by including the eventual difference of sign between the displacement and the external force in the phase difference ϑ. We note that Eqs. (23.30)—(23.32) coincide with Eqs. (23.21)—(23.23) obtained for the generating solution in the perturbation method. It is also worth noting that Eqs. (23.22) and (23.23), which determine the periodic vibration, could be directly obtained from Eqs. (17.70) and (17.71) by the rule given in § 17 c, and taking into account that in our case the parameters of the linear equivalent system are (see § 13 d)

$$\bar{h}(a) = h, \qquad \bar{\omega}^2(a) = \omega^2 + \frac{3}{4}\beta\omega^2 a^2. \tag{23.33}$$

For the second approximation Eqs. (17.39)—(17.41) yield

$$\bar{u}_1(a, \alpha) = \frac{\beta_1 a^3}{32\nu^2}\cos 3\alpha, \qquad \tilde{u}_1(\nu t) = 0, \tag{23.34}$$

$$A_2(a, \vartheta) = \frac{1}{4\nu^2}\left[\left(h_1 a + \frac{P_1 \sin\vartheta}{2\nu}\right)\left(\frac{3}{2}\beta_1 a^2 + \frac{P_1 \cos\vartheta}{a}\right) + \right.$$

$$\left. + \left(\Delta + \frac{3}{4}\beta_1 a^2 - \frac{P_1\cos\vartheta}{a}\right)\frac{P_1 \sin\vartheta}{2\nu}\right],$$

$$B_2(a, \vartheta) = -\frac{1}{8\nu^3}\left(\Delta + \frac{3}{4}\beta_1 a^2 - \frac{P_1 \cos\vartheta}{a}\right)\left(\Delta + \frac{3}{4}\beta_1 a^2\right) -$$

$$-\frac{h_1}{2\nu a}\left(h_1 a + \frac{P_1 \sin\vartheta}{2\nu}\right) + \frac{3\beta_1^2 a^4}{256\nu^3}.$$

Substituting the last two relations into (17.46) gives in the case of periodic vibrations ($\dot{a} = \dot{\vartheta} = 0$)

$$-\varepsilon\left(h_1 a + \frac{P_1 \sin\vartheta}{2\nu}\right) + \frac{\varepsilon^2}{4\nu^2}\left[\left(h_1 a + \frac{P_1 \sin\vartheta}{2\nu}\right)\left(\frac{3}{2}\beta_1 a^2 + \frac{P_1 \cos\vartheta}{a}\right) +\right.$$

$$\left. + \left(\Delta + \frac{3}{4}\beta_1 a^2 - \frac{P_1 \cos\vartheta}{a}\right)\frac{P_1 \sin\vartheta}{2\nu}\right] = 0,$$

$$\varepsilon\left(\Delta + \frac{3}{4}\beta_1 a^2 - \frac{P_1 \cos\vartheta}{a}\right) + \varepsilon^2\left[-\frac{1}{4\nu^2}\left(\Delta + \frac{3}{4}\beta_1 a^2 - \frac{P_1 \cos\vartheta}{a}\right) \times\right.$$

$$\left. \times \left(\Delta + \frac{3}{4}\beta_1 a^2\right) - \frac{h_1}{a}\left(h_1 a + \frac{P_1 \sin\vartheta}{2\nu}\right) + \frac{3\beta^2 a^4}{128\nu^2}\right] = 0.$$

From these equations it first follows that

$$h_1 a + \frac{P_1 \sin\vartheta}{2\nu} = 0(\varepsilon), \qquad \Delta + \frac{3}{4}\beta_1 a^2 - \frac{P_1 \cos\vartheta}{a} = 0(\varepsilon),$$

and hence, neglecting terms of third order in ε, we obtain

$$\varepsilon\left(h_1 a + \frac{P_1 \sin\vartheta}{2\nu}\right) = 0, \qquad \varepsilon\left(\Delta + \frac{3}{4}\beta_1 a^2 - \frac{P_1 \cos\vartheta}{a}\right) + \frac{3\varepsilon^2\beta^2 a^4}{128\nu^2} = 0.$$

With the notations (23.24), and taking into consideration that $\omega^2/\nu^2 = 1 + \varepsilon\Delta$, we now deduce

$$\sin\vartheta = -\frac{2\nu h a}{P}, \qquad \cos\vartheta = \frac{a}{P}\left(\omega^2 - \nu^2 + \frac{3}{4}\beta\omega^2 a^2 + \frac{3}{128}\beta^2\omega^2 a^4\right), \tag{23.35}$$

$$\left(\omega^2 - \nu^2 + \frac{3}{4}\beta\omega^2 a^2 + \frac{3}{128}\beta^2\omega^2 a^4\right)^2 + 4\nu^2 h^2 = \frac{P^2}{a^2}. \tag{23.36}$$

Finally, by (17.43) and (23.34), and neglecting terms of third order in ε, we derive the second approximation

$$x_{\text{II}}(t) = a\cos(\nu t + \vartheta) + \frac{1}{32}\beta a^3 \cos(\nu t + \vartheta). \tag{23.37}$$

It is easily seen that the second approximation differs from the first one only through the term $(3/128)\,\beta^2\omega^2 a^4$, whose influence is generally negligible.

c) Use of the method of finite sums of trigonometric functions

We will now seek an approximate solution of Eq. (23.2) by means of the method of finite sums of trigonometric functions, under the assumption that h and β are small. In the presence of viscous damping the periodic response is no longer in phase or in opposition of phase with the excitation. When using the method of finite sums of trigonometric functions it is more convenient to take the phase of the periodic solution as zero and to introduce the phase difference into the argument of the external force. Accordingly, the equation of motion will be written in the form:

$$\ddot{x} + 2h\dot{x} + \omega^2 x(1 + \beta x^2) = P\cos(\nu t - \vartheta), \quad P > 0. \tag{23.38}$$

Let us look for an approximate harmonic solution of this equation. Setting $q = 1$ and retaining only the first term of the sum (18.1), we may try an approximate solution of the form

$$\bar{x}(t) = x_1 \cos \nu t. \tag{23.39}$$

Substituting (23.39) into (23.38) yields

$$\left[(\omega^2 - \nu^2)\,x_1 + \frac{3}{4}\beta\omega^2 x_1^3 - P\cos\vartheta\right]\cos\nu t -$$

$$-\,(2\nu h\,x_1 + P\sin\vartheta)\sin\nu t + \frac{1}{4}\beta\omega^2 x^3 = 0. \tag{23.40}$$

Since only two free constants, x_1 and ϑ, are available, we can require only the vanishing of the coefficients of $\cos\nu t$ and $\sin\nu t$ in (23.40), which gives

$$(\omega^2 - \nu^2)\,x_1 + \frac{3}{4}\beta\omega^2 x_1^3 - P\cos\vartheta = 0, \tag{23.41}$$

$$2\nu h x_1 + P\sin\vartheta = 0. \tag{23.42}$$

Thus, the term βx_1^3 has been neglected in (23.40) in comparison with $(\omega^2 - \nu^2)\, x_1$. However, to obtain a more accurate equation of the resonance curve, we will still retain this term in Eq. (23.41) (cf. § 20 c).

From (23.41) and (23.42) it follows that

$$\sin \vartheta = -\frac{2\nu h x_1}{P}, \qquad \cos \vartheta = \frac{x_1}{P}\left(\omega^2 - \nu^2 + \frac{3}{4}\beta\omega^2 x_1^2\right), \tag{23.43}$$

$$\left(\omega^2 - \nu^2 + \frac{3}{4}\beta\omega^2 x_1^2\right)^2 + 4\nu^2 h^2 = \frac{P^2}{x_1^2}. \tag{23.44}$$

Finally, we remark that x_1 and P may be assumed positive, by using the phase angle ϑ to take into account the eventual difference of sign between the displacement and the external force. Then, setting $x_1 = a$, the last two relations coincide with Eqs. (23.30) and (23.31), obtained in the first approximation by the method of KRYLOV and BOGOLYUBOV.

d) Resonance curve. Stability of periodic vibrations. Jump phenomena

Let us study now in more detail the relation between the amplitude a of the periodic response and the exciting frequency ν, which may be written in the first approximation as

$$F\left(\frac{\nu^2}{\omega^2}, a\right) \equiv \left(\frac{\nu^2}{\omega^2} - 1 - \frac{3}{4}\beta a^2\right)^2 + 4\zeta^2\frac{\nu^2}{\omega^2} - \frac{P^2}{\omega^4 a^2} = 0, \tag{23.45}$$

where $\zeta = h/\omega$. Solving this equation for ν^2/ω^2 gives

$$\frac{\nu^2}{\omega^2} = 1 + \frac{3}{4}\beta a^2 - 2\zeta^2 \pm \sqrt{\frac{P^4}{\omega^4 a^2} - 4\zeta^2\left(1 + \frac{3}{4}\beta a^2 - \zeta^2\right)}.$$

The graph of this function in the $(\nu^2/\omega^2, a)$ – plane for $P = \text{const}$ is said to be a *resonance curve*. Its points are situated at equal horizontal distances from the *skeleton curve*, whose equation is

$$\frac{\nu^2}{\omega^2} = 1 + \frac{3}{4}\beta a^2 - 2\zeta^2. \tag{23.46}$$

It is interesting to find the geometric locus of the points of vertical and horizontal tangency of the resonance curves. From (23.45) we deduce that

$$\frac{da}{d(\nu^2/\omega^2)} = -\frac{\dfrac{\partial F}{\partial(\nu^2/\omega^2)}}{\dfrac{\partial F}{\partial a}} = \frac{\dfrac{\nu^2}{\omega^2} - 1 - \dfrac{3}{4}\beta a^2 + 2\zeta^2}{\dfrac{1}{a}\left[\dfrac{3}{2}\beta a^2\left(\dfrac{\nu^2}{\omega^2} - 1 - \dfrac{3}{4}\beta a^2\right) - \dfrac{P^2}{\omega^4 a^2}\right]} \cdot \qquad (23.47)$$

The geometric locus of the points at which the resonance curves have horizontal tangent may be obtained by equating to zero the numerator of the fraction in (23.47), and hence it coincides with the skeleton curve. By eliminating ν^2/ω^2 between the equation of the skeleton curve and that of the resonance curve, we obtain the equation determining the extreme values of the response amplitude, say A, for $P =$ const. and ν variable, namely

$$3\beta\zeta^2 A^4 + 4\zeta^2(1 - \zeta^2)\, A^2 - \frac{P^2}{\omega^4} = 0, \qquad (23.48)$$

from which there results

$$A^2 = -\frac{2(1-\zeta^2)}{3\beta}\left[1 \pm \sqrt{1 + \frac{3\beta P^2}{4\omega^4\zeta^2(1-\zeta^2)^2}}\right]. \qquad (23.49)$$

If $\beta < 0$, then for

$$\beta < -\frac{4\omega^4\zeta^2(1-\zeta^2)^2}{3P^2} \qquad (23.50)$$

the values (23.49) of A are imaginary, i.e., there exist no points of horizontal tangency on the resonance curve, and hence the skeleton curve does not intersect the resonance curves. For example, taking $P/\omega^2 = 1$ cm, $\zeta = 0.2$, and $\beta = -0.06$ cm^{-2}, the condition (23.50) is fulfilled; the corresponding resonance curve is shown in Fig. 62 *a*. For

$$-\frac{4\omega^4\zeta^2(1-\zeta^2)^2}{3P^2} < \beta < 0 \qquad (23.51)$$

Eq. (23.49) yields two extreme values of a, a minimum and a maximum one. Taking, e.g., $P/\omega^2 = 1$ cm, $\zeta = 0.2$, and $\beta = -0.04$ cm^{-2}, the inequality (23.51) is satisfied, and we obtain from (23.49): $A_1 = 3.02$ cm, $A_2 = 4.79$ cm; the corresponding resonance curve is illustrated in Fig. 62*b*. The restriction $g'(x) > 0$ implies $a < 1/\sqrt{-3\beta}$ for $\beta < 0$. If $\zeta < 0.7$ this restriction comes to taking only the minus sign before the radical in (23.49), and to considering only the lower branch of the resonance curve.

Finally, for $\beta > 0$, we have to choose only the minus sign before the radical in (23.49), since $A^2 > 0$. We thus obtain an evaluation of the maximum amplitude, when $P = \text{const}$ and ν varies, namely

$$a^2 < A^2 = \frac{2(1-\zeta^2)}{3\beta}\left[\sqrt{1+\frac{3\beta P^2}{4\omega^4\zeta^2(1-\zeta^2)^2}}-1\right]. \tag{23.52}$$

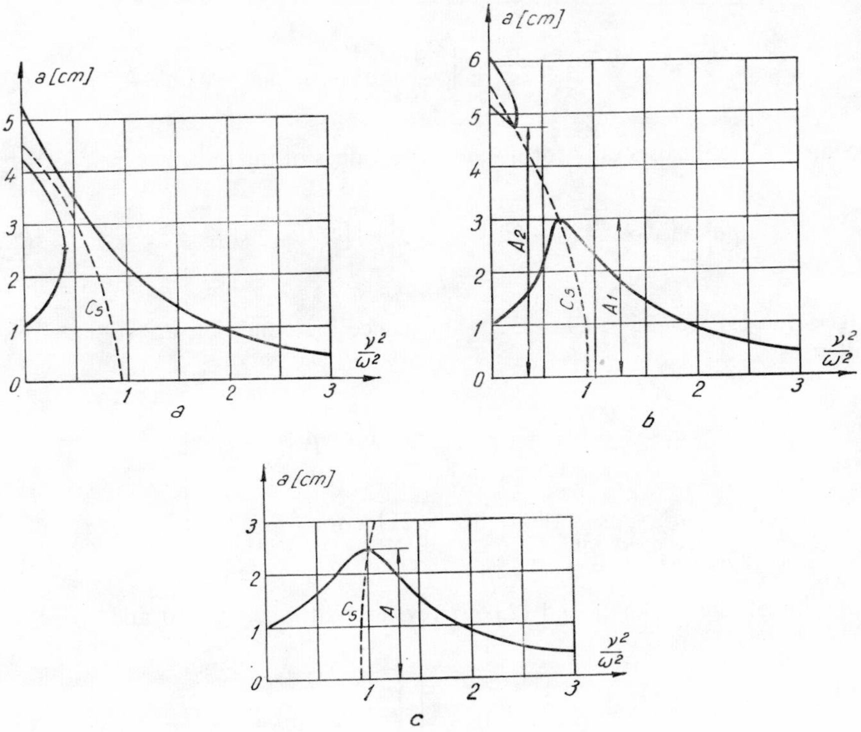

Fig. 62

For instance, by taking $P/\omega^2 = 1$ cm, $\zeta = 0.2$, and $\beta = 0.02$ cm^{-2}, we obtain $a < A = 9.2$ cm and the resonance curve represented in Fig. 62 *c*. We recall that, for the same values of P, ω, and h, we found in § 15 b the weaker estimation $A = 11$ cm, by making use of LOUD's result, which is independent of β in the case of DUFFING's equation.

Let us now return to Eq. (23.47). At the points where the resonance curve has a vertical tangent, the denominator of the fraction in the right-hand side must vanish, and hence

$$\frac{3}{2}\beta a^2\left(\frac{\nu^2}{\omega^2}-1-\frac{3}{4}\beta a^2\right)-\frac{P^2}{\omega^4 a^2}=0. \tag{23.53}$$

Eliminating P between this condition and Eq. (23.45), we find the equation of the geometric locus of the points at which the resonance curve has a vertical tangent

$$\left(\frac{\nu^2}{\omega^2} - 1 - \frac{3}{4}\beta a^2\right)\left(\frac{\nu^2}{\omega^2} - 1 - \frac{9}{4}\beta a^2\right) + 4\zeta^2\frac{\nu^2}{\omega^2} = 0. \tag{23.54}$$

Solving this equation for ν^2/ω^2 gives

$$\frac{\nu^2}{\omega^2} = 1 + \frac{3}{2}\beta a^2 - 2\zeta^2 \pm \sqrt{\frac{9}{16}\beta^2 a^4 - 6\zeta^2\beta a^2 - 4\zeta^2(1-\zeta^2)}. \tag{23.55}$$

The curve (23.55) exists only for nonnegative values of the radicand, i.e., for

$$\frac{3}{4}\beta a^2 \in (-\infty,\ 4\zeta^2 - 2\zeta\sqrt{\zeta^2+3}\,] \cup [\,4\zeta^2 + 2\zeta\sqrt{\zeta^2+3},\ \infty). \tag{23.56}$$

As $\zeta < 1$, we have $2\zeta < \sqrt{1+3\zeta^2}$, and hence the condition above may be re-written as

$$\left.\begin{aligned} a^2 &> \frac{8\zeta}{3\beta}(2\zeta + \sqrt{1+3\zeta^2}) \quad \text{for} \quad \beta > 0, \\ a^2 &> -\frac{8\zeta}{3\beta}(\sqrt{1+3\zeta^2} - 2\zeta) \quad \text{for} \quad \beta < 0. \end{aligned}\right\} \tag{23.57}$$

Inspection of Eqs. $(23.57)_1$ and (23.52) reveals that when $\beta > 0$ and

$$\frac{8\zeta}{3\beta}(2\zeta + \sqrt{1+3\zeta^2}) > \frac{2(1-\zeta^2)}{3\beta}\left[\sqrt{1 + \frac{3\beta P^2}{\omega^4\zeta^2(1-\zeta^2)^2}} - 1\right],$$

or, equivalently,

$$\frac{\beta P^2}{\omega^4} < \frac{8\omega^3}{3}\left[4 + 12\zeta^2 + (1+7\zeta^2)\sqrt{1+3\zeta^2}\right], \tag{23.58}$$

then there are no points of vertical tangency on the resonance curve. This result shows that, if β is small enough, or if ζ is sufficiently large, then the periodic vibration is asymptotically stable in the large, a conclusion in complete agreement with Theorem 9, § 15. Choosing, e.g., $P/\omega^2 = 1$ cm and $\zeta = 0.2$, we obtain from (23.58) the stability condition $\beta < 0.125\ \text{cm}^{-2}$. Since in Fig. 62 c we have taken $\beta = 0.02\ \text{cm}^{-2}$, the resonance curve illustrated in that figure had no points of vertical tangency. It is worth noting that in § 15 b we obtained for the same values of P, ω, and ζ, the more restrictive condition $\beta < 0.00214$ cm^{-2}, by using OPIAL's theorem and the bounds given by LOUD for the displacement and velocity magnitudes.

For most of the nonlinear mechanical oscillators in practical use, the damping is comparatively strong, and hence the stability condition (23.58) is generally

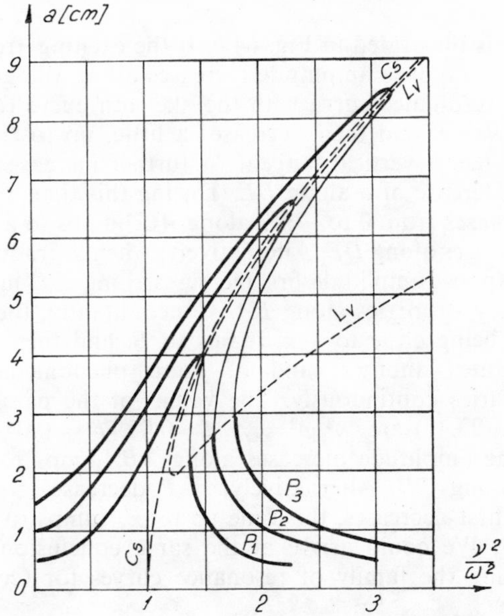

Fig. 63

fulfilled. However, for the sake of completeness, and considering the wide range of applicability of DUFFING's equation, we will also study oscillators with cubic elastic restoring force that are liable to pass through unstable states. To this end we have represented in Fig. 63 the resonance curves corresponding to the values $\zeta = 0.05$, $\beta = 0.04$ cm^{-2}, $P_1/\omega^2 = 0.5$ cm, $P_2/\omega^2 = 1$ cm, $P_3/\omega^2 = 1.5$ cm of the parameters. For all these three choices of the excitation magnitude, condition (23.58) is no longer fulfilled, and hence the periodic vibrations are no longer stable for any values of a and ν satisfying Eq. (23.45). The geometric locus of the points at which the resonance curves have vertical tangent is represented in Fig. 63 by the broken line L_v, which separates stable from unstable periodic vibrations. The arcs of the resonance curves corresponding to stable periodic vibrations are represented by a thickened solid line (cf. § 17 c).

The existence of unstable conditions and of points of vertical tangency of a resonance curve originates discontinuities in the variation of the amplitude and phase of the periodic vibration when the frequency or the amplitude of the exciting force varies continuously.

Assume first that $P =$ const. and ν varies continuously. Figure 64 *a* shows the resonance curve obtained by (23.45) for $P/\omega^2 = 1$ cm, $\zeta = 0.05$, $\beta = 0.04$ cm^{-2}. The corresponding variation of the phase angle, as given by the equations

$$\sin \vartheta = -\frac{2\nu h a}{P}, \qquad \cos \vartheta = \frac{a}{P}\left(\omega^2 - \nu^2 + \frac{3}{4}\,\beta\omega^2 a^2\right),$$

where $a > 0$, $P > 0$ is illustrated in Fig. 64 *b*. If the exciting frequency ν gradually increases from 0, the response amplitude increases along the arc AB. At the intersecting point of the resonance curve with the skeleton curve (C_s), the amplitude a takes its maximum value, and then decreases a little, up to the point C at which the resonance curve has a vertical tangent. A further increase of ν causes a jump to D and then the decrease of a along DE. During this time, as seen in Fig. 64 *b*, the phase angle decreases from 0 to $-\pi/2$ along AC, jumps to a value close to $-\pi$, and then slowly decreases along DE. Alternatively, when ν gradually decreases from a larger value, the response amplitude first increases along EF, jumps from E to B, and then continuously decreases along BA. Concomitantly, the phase angle slowly increases along EF being close to $-\pi$, jumps to B, and then increases along BA.

Figure 65 illustrates another kind of jump phenomena that occur when $\nu =$ const. and P varies continuously. The values of the parameters used now in connection with Eq. (23.45) are: $\nu^2/\omega^2 = 1.5$, $\zeta = 0.05$, $\beta = 0.04$ cm^{-2}. If P gradually increases from 0, the amplitude increases along AB, jumps to C, and then continuously increases along CD. Alternatively, if P decreases from a comparatively large value, then a first decreases, this time up to E, jumps to F, and then further decreases along FA. We could arrive at the same conclusions about these jump phenomena by tracing the family of resonance curves for (constant) values of P

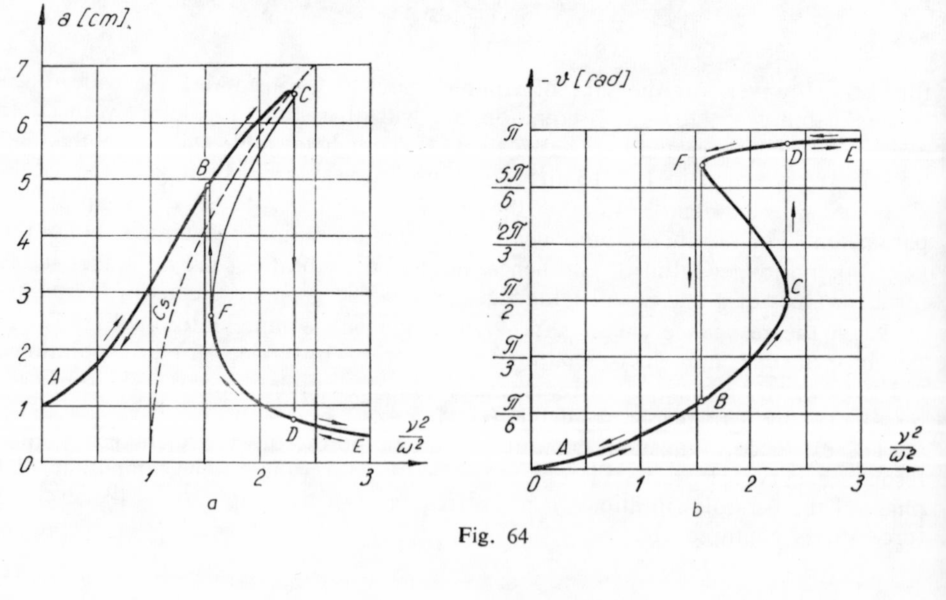

Fig. 64

sufficiently close to each other, and then considering their points of intersection with the vertical line of equation $\nu^2/\omega^2 = 1.5$. The points E and B in Fig. 65 correspond to the points in Fig. 63 where the vertical line of equation $\nu^2/\omega^2 = 1.5$ cuts the curve L_v (the geometric locus of the points at which the resonance curves have

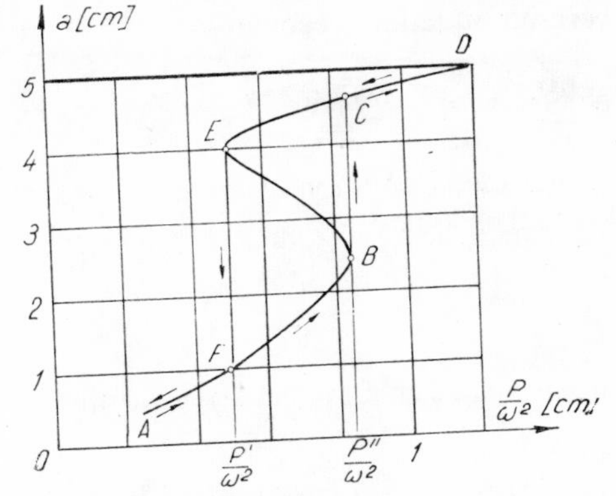

Fig. 65

vertical tangent). Putting $\nu^2/\omega^2 = 1.5$, $\zeta = 0.05$, $\beta = 0.04\ \text{cm}^{-2}$ into Eq. (23.54) yields the coordinates of the points E and B, namely $a' = 4.02$ cm, $P'/\omega^2 = 0.496$ cm and, respectively, $a'' = 2.46$ cm, $P''/\omega^2 = 0.840$ cm.

When considering jump phenomena we must bear in mind that the discontinuities refer to the stationary values of the response amplitude and phase. Actually, when ν (or P) approaches a critical value corresponding to the transition from a stable to an unstable condition, the response amplitude and phase vary at a *finite* rate, and transitory vibrations take place, which depend on how rapidly ν (or P) passes through the critical value.

The existence of jump phenomena as those described above has been experimentally proved by many authors (see, e.g., MARTIENSSEN [115], DUFFING [55], APPLETON [5], and LUDEKE [108]).

e) Subharmonic vibrations

We will now consider the subharmonic vibrations of order 1/3 of the equation

$$\ddot{x} + 2h\dot{x} + \omega^2 x(1 + \beta x^2) = P \cos \nu t, \tag{23.59}$$

under the assumption that h, β, and P are of order $0(\varepsilon)$, where ε is a small parameter, and using both the method of KRYLOV and BOGOLYUBOV and the method of finite sums of trigonometric functions. The calculation will show that a subharmonic vibration of order 1/3 cannot appear unless $h = 0(\varepsilon^2)$.

— *Use of the method of* KRYLOV *and* BOGOLYUBOV

By putting

$$h = \varepsilon h_1, \quad \omega^2 \beta = \varepsilon \beta_1, \qquad P = \varepsilon P_1, \qquad \nu t = \tau, \tag{23.60}$$

into (23.59), the governing equation becomes

$$\ddot{x} + \omega^2 x = -\varepsilon(2h_1 \dot{x} + \beta_1 x^3 - P_1 \cos \tau). \tag{23.61}$$

Let us seek a strong subharmonic solution of order 1/3 of this equation, by making use of the results obtained in § 19 c. By setting $p = 1$, $q = 3$, and

$$\omega^2 - \frac{\nu^2}{9} = \varepsilon \Delta, \tag{23.62}$$

and using the same notations as in §§ 17 and 19, we find

$$f_1(\dot{x}) = 2h_1 \dot{x}, \qquad \psi_1(x) = \beta_1 x^3, \qquad \alpha = \frac{\nu t}{3} + \vartheta,$$

$$\psi_1(a \cos \alpha) + f_1\left(-\frac{\nu}{3} a \sin \alpha\right) = \frac{1}{4} \beta_1 a^3 (\cos 3\alpha + 3 \cos \alpha) - \frac{2}{3} h_1 \nu\, a \sin \alpha,$$

$$C_1(a) = \frac{3}{4} \beta_1 a^3, \qquad C_3(a) = \frac{1}{4} \beta_1 a^3, \qquad D_1(a) = -\frac{2}{3} h_1 \nu a,$$

and all other coefficients $C_k(a)$, $D_k(a)$ vanish. From (19.26)–(19.28) it then follows that

$$A_1(a, \vartheta) = -h_1 a, \qquad B_1(a, \vartheta) = \frac{3\Delta}{2\nu} + \frac{9\beta_1 a^2}{8\nu}, \tag{23.63}$$

$$\bar{u}_1(a, \alpha) = \frac{9\beta_1 a^3}{32\nu^2} \cos 3\alpha, \qquad \tilde{u}_1(\tau) = -\frac{9P_1}{8\nu^2} \cos \tau, \tag{23.64}$$

$$u_1(a, \alpha, \tau) = \frac{9\beta_1 a^3}{32\nu^2} \cos 3\alpha - \frac{9P_1}{8\nu^2} \cos \tau, \tag{23.65}$$

$$\left.\begin{aligned} U(a, \alpha) &= \frac{9\beta_1 a^3}{32\nu^2} \cos 3\alpha - \frac{9P_1}{8\nu^2} \cos 3(\alpha - \vartheta), \\ V(a, \alpha) &= -\frac{27\beta_1 \omega a^3}{32\nu^2} \sin 3\alpha + \frac{27P_1}{8\nu} \sin 3(\alpha - \vartheta). \end{aligned}\right\} \tag{23.66}$$

Substituting now (23.65) into (17.45), taking into account that $\nu^2 = 9\omega^2 + 0(\varepsilon)$, and considering (23.60), we directly obtain the second approximation for the subharmonic solution

$$x(t) = a \cos\left(\frac{\nu t}{3} + \vartheta\right) + \frac{1}{32}\,\beta\, a^3 \cos(\nu t + 3\vartheta) - \frac{P}{8\omega^2}\cos \nu t + 0(\varepsilon^2). \qquad (23.67)$$

Next, by introducing (23.63) and (23.66) into (19.31), calculating the integrals, and taking into consideration that $\nu/\omega = 3 + 0(\varepsilon)$, we find the equations which determine the amplitude and phase of the periodic vibration

$$\left.\begin{aligned} M(a,\ \vartheta,\ \nu) &\equiv -a\left(h + \frac{3\beta a P}{64\omega}\sin 3\vartheta\right) = 0,\\ N(a,\ \vartheta,\ \nu) &\equiv \frac{3}{2\nu}\left(\omega^2 - \frac{\nu^2}{9} + \frac{3}{4}\beta\omega^2 a^2 + \right.\\ &\left. + \frac{3}{128}\beta^2\omega^2 a^4 - \frac{3}{32}\beta a P \cos 3\vartheta\right) = 0. \end{aligned}\right\} \qquad (23.68)$$

Since, by assumption, $\beta = 0(\varepsilon)$, it follows from $(23.68)_1$ that a subharmonic vibration cannot appear unless h is of second order in ε, i.e., unless the damping is sufficiently small with respect to the elastic nonlinearity. This result is, obviously, in agreement with OPIAL's theorem (cf. § 15 b), which implies that, when the damping is strong enough, the only possible periodic motion of the oscillator is harmonic with respect to the exciting frequency and is asymptotically stable in the large. From (23.68) we further deduce that

$$\left.\begin{aligned} \sin 3\vartheta &= -\frac{64 h\omega}{3\beta a P},\\ \cos \vartheta &= \frac{32}{3\beta a P}\left(\omega^2 - \frac{\nu^2}{9} + \frac{3}{4}\beta\omega^2 a^2 + \frac{3}{128}\beta^2\omega^2 a^2\right) \end{aligned}\right\} \qquad (23.69)$$

and, by eliminating ϑ between these relations, we find the equation of the *resonance curve*

$$F\left(\frac{\nu^2}{\omega^2},\ a\right) \equiv \left(\frac{\nu^2}{9\omega^2} - 1 - \frac{3}{4}\beta a^2 - \frac{3}{128}\beta^2 a^4\right)^2 + 4\zeta^2 - \frac{9\beta^2 a^2 P^2}{1024\,\omega^4} = 0, \qquad (23.70)$$

where $\zeta = h/\omega$.

The last equation allows the calculation of the response amplitude a for given ν and P. The best way of doing this is to solve the equation for ν^2/ω^2

$$\frac{\nu^2}{\omega^2} = 9\left(1 + \frac{3}{4}\beta a^2 + \frac{3}{128}\beta^2 a^4 \pm \sqrt{\frac{9\beta^2 P^2 a^2}{10\,24\omega^4} - 4\zeta^2}\right) \tag{23.71}$$

and then to graphically represent ν^2/ω^2 as a function of a. We see that ν^2/ω^2 takes real values only if the radicand is nonnegative, i.e., only if

$$a \geqslant a_{\min} = \frac{64\zeta\omega^2}{3\beta P}. \tag{23.72}$$

To any value of $a > a_{min}$ correspond two points of the resonance curve situated at equal horizontal distances from the *skeleton curve*, whose equation is

$$\frac{\nu^2}{\omega^2} = 9\left(1 + \frac{3}{4}\beta a^2 + \frac{3}{128}\beta^2 a^4\right), \tag{23.73}$$

For any value of a one obtains, by Eqs. (23.69), three values of ϑ, which are phase-shifted by $2\pi/3$. Hence, to any value of ν, P being constant, correspond six subharmonic solutions of order 1/3 of Eq. (23.59).

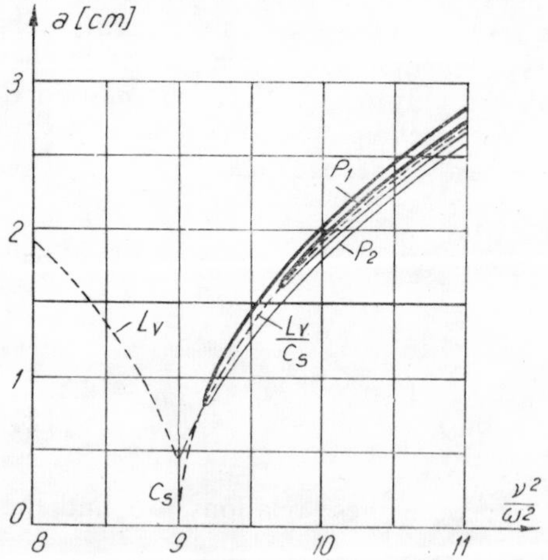

Fig. 66

Figure 66 shows the resonance curves obtained for $P_1/\omega^2 = 1$ cm, $P_2/\omega^2 = 2$ cm, $\beta = 0.04$ cm^{-2}, and $\zeta = 0.003$. The skeleton curve is represented by the broken line C_s.

Next, by differentiating (23.70), we derive

$$\frac{da}{d(\nu^2/\omega^2)} = - \frac{\dfrac{\partial F}{\partial(\nu^2/\omega^2)}}{\dfrac{\partial F}{\partial a}} =$$

(23.74)

$$= \frac{\dfrac{2}{9}\left(\dfrac{\nu^2}{9\omega^2} - 1 - \dfrac{3}{4}\beta a^2 - \dfrac{3}{128}\beta^2 a^4\right)}{2\left(\dfrac{\nu^2}{9\omega^2} - 1 - \dfrac{3}{4}\beta a^2 - \dfrac{3}{128}\beta^2 a^4\right)\left(\dfrac{3}{2}\beta a + \dfrac{3}{32}\beta^2 a^3\right) + \dfrac{9\beta^2 P^2 a}{512\,\omega^4.}} .$$

The geometric locus of the points at which the resonance curves have horizontal tangent results by equating to zero the numerator of the fraction in the right-hand side, and hence it coincides with the skeleton curve.

At the points where the resonance curve has vertical tangent, the denominator of the fraction in (23.74) has to vanish, i.e.,

$$\left(\frac{\nu^2}{9\omega^2} - 1 - \frac{3}{4}\beta a^2 - \frac{3}{128}\beta^2 a^4\right)\left(a + \frac{1}{16}\beta a^3\right) + \frac{3\beta P^2 a^2}{512\,\omega^2} = 0 \qquad (23.75)$$

By eliminating P between (23.75) and (23.70), we find the equation of the geometric locus of the points at which the resonance curves have vertical tangent

$$\left(\frac{\nu^2}{9\omega^2} - 1 - \frac{3}{4}\beta a^2 - \frac{3}{128}\beta^2 a^4\right)\left(\frac{\nu^2}{9\omega^2} - 1 + \frac{3}{4}\beta a^2 + \frac{9}{128}\beta^2 a^4\right) + 4\zeta^2 = 0.$$

(23.76)

Solving this equation for ν^2/ω^2 gives [1)]

$$\frac{\nu^2}{\omega^2} = 9\left[1 - \frac{3}{128}\beta^2 a^4 \pm \frac{3}{4}\beta a^2 \sqrt{\left(1 + \frac{1}{16}\beta a^2\right)^2 - \left(\frac{8\zeta}{3\beta a^2}\right)^2}\,\right],$$

In Fig. 66 this geometric locus is represented by the broken line L_v. The right-hand branch of L_v is very close to C_s, and hence the tangent to the resonance curve varies very rapidly from a horizontal to a vertical direction in the neighborhood of the point corresponding to the minimum value of a.

1) If instead of assuming $P = 0(\varepsilon)$ we had supposed that P is independent of ε, we should have obtained a different form of L_v, although the resonance curve itself and their points of intersection with L_v would have been only negligibly changed. This may be easily seen, by comparing the equations above and Fig. 66 to the corresponding equations deduced by KAUDERER and to Fig. 183 in his book [82]. § 69.

As already seen at the end of § 19, the curve L_v separates stable from unstable periodic vibrations. To see which of the two branches of the resonance curve determined by L_v corresponds to stable conditions, we first calculate from (23.68)

$$\frac{\partial M}{\partial a} = -\left(h + \frac{3\,\beta a P}{32\,\omega}\sin 3\,\vartheta\right), \qquad \frac{\partial M}{\partial \vartheta} = -\frac{9\,\beta a^2 P}{64\,\omega}\cos 3\vartheta,$$

$$\frac{\partial N}{\partial \vartheta} = \frac{9\,\beta a P}{64\omega}\sin 3\vartheta,$$

$$\frac{\partial N}{\partial a} = \frac{3}{2\nu}\left(\omega^2 - \frac{\nu^2}{9} + \frac{3}{4}\beta\omega^2 a + \frac{3}{128}\beta^2\omega^2 a^4 - \frac{3}{32}\beta a\, P\cos 3\vartheta\right).$$

Next, by substituting Eqs. (23.69) into these equations and introducing the result obtained into (19.37), we derive the stability conditions

$$\frac{\partial M}{\partial a} + \frac{\partial N}{\partial \vartheta} = -\,2h < 0, \tag{23.77}$$

$$\frac{\partial M}{\partial a}\frac{\partial N}{\partial \vartheta} - \frac{\partial M}{\partial \vartheta}\frac{\partial N}{\partial a} = -\,3h^2 + \frac{27}{4\nu^2}\left(\omega^2 - \frac{\nu^2}{9} + \frac{3}{4}\beta\omega^2 a^2 + \right.$$
$$\left. + \frac{3}{128}\beta^2\omega^2 a^4\right)\left(\frac{\nu^2}{9} - \omega^2 + \frac{3}{4}\beta\omega^2 a^2 + \frac{9}{128}\beta^2\omega^2 a^4\right) > 0. \tag{23.78}$$

The condition (23.77) is automatically fulfilled since $h > 0$. Next, as $\nu^2 = 9\omega^2 + 0(\varepsilon)$, we may write (23.78) in the form

$$\left(\frac{\nu^2}{9\omega^2} - 1 - \frac{3}{4}\beta a^2 - \frac{3}{128}\beta^2 a^4\right)\left(\frac{\nu^2}{9\omega^2} - 1 + \frac{3}{4}\beta a^2 + \frac{9}{128}\beta^2 a^4\right) + 4\zeta^2 < 0. \tag{23.79}$$

The left-hand side of this inequality obviously coincides with the left-hand side of Eq. (23.76) of L_v. It may be easily shown, by using only the terms of order $0(\varepsilon^2)$ in (23.79) that the stability region is situated above L_v. Consequently, the branch of the resonance curve situated on the left of C_s corresponds to stable periodic vibrations, and that situated on the right of C_s to unstable ones. On the other hand, since either branch corresponds to three subharmonic vibrations phase-shifted by $2\pi/3$, we conclude that for each value of ν, P being constant, there exist three stable subharmonic solutions of order 1/3 of Eq. (23.59) and another three unstable ones.

– Use of the method of finite sums of trigonometric functions

Consider again the equation of motion

$$\ddot{x} + 2h\dot{x} + \omega^2 x(1 + \beta x^2) = P \cos \nu t, \tag{23.80}$$

and try to find an approximate subharmonic solution of order 1/3 of this equation, this time by using a finite sum of trigonometric functions of the type (18.1) with $q = 3$.

Unlike the case of the conservative oscillator considered in § 20 e, Eq. (23.80) no longer remains invariant when replacing t by $-t$, and hence we have to retain also the sines in (18.1). In exchange, the equation still does not change when simultaneously replacing t by $t + 3\pi/\nu$ and x by $-x$. Consequently, we may retain in the approximate solution (18.1) only the trigonometric functions whose arguments are odd multiples of $\nu t/3$. Then, by limiting ourselves to consider only the first two nonzero harmonics of (18.1), we seek an approximate solution of the form

$$\bar{x}(t) = x_{1/3} \cos \left(\frac{\nu t}{3} + \vartheta\right) + x_1 \cos (\nu t + \vartheta'). \tag{23.81}$$

The calculation may be further simplified by choosing the time origin such that the phase of the principal term of the subharmonic solution (with frequency $\nu/3$) be zero. This can be achieved by a suitable translation on the time axis, which comes to replacing t by $t - 3\vartheta/\nu$ in (23.81). Then, the approximate solution assumes the form

$$\bar{x}(t) = x_{1/3} \cos \frac{\nu t}{3} + x_1 \cos(\nu t + \vartheta_1), \tag{23.82}$$

where $\vartheta_1 = \vartheta' - 3\vartheta$, and the governing equation becomes

$$\ddot{x} + 2h\dot{x} + \omega^2 x(1 + \beta x^2) = P \cos(\nu t - 3\vartheta). \tag{23.83}$$

Introducing now (23.82) into (23.83) gives

$$x_{1/3} \left\{\left[\omega^2 - \frac{\nu^2}{9} + \frac{3}{4} \beta\omega^2 (x_{1/3}^2 + x_{1/3} x_1 \cos \vartheta_1 + 2x_1^2)\right] \cos \frac{\nu t}{3} - \left(\frac{2}{3} h\nu + \right.\right.$$

$$\left.\left. + \frac{3}{4} \beta\omega^2 x_{1/3} x_1 \sin \vartheta_1 \right)\sin \frac{\nu t}{3}\right\} + \left[x_1(\omega^2 - \nu^2) \cos \vartheta_1 - 2\nu h x_1 \sin \vartheta_1 - \right.$$

$$\left. - P \cos 3\vartheta + \frac{1}{4} \beta\omega^2 (x_{1/3}^3 + 6 x_{1/3}^2 x_1 \cos \vartheta_1 + 3 x_1^3 \cos \vartheta_1)\right] \cos \nu t -$$

$$- \left[x_1(\omega^2 - \nu^2)\sin\vartheta_1 + 2\nu h x_1 \cos\vartheta_1 + P\sin 3\vartheta + \frac{3}{4}\beta\omega^2 x_1(2x_{1/3}^2 + x_1^2)\sin\vartheta_1\right]\sin\nu t +$$

$$+ \frac{1}{4}\beta\omega^2 x_1\left[3x_{1/3}^2\cos\left(\frac{5\nu t}{3} + \vartheta_1\right) + 3x_{1/3}x_1\cos\left(\frac{5\nu t}{3} + 2\vartheta_1\right) +\right.$$

$$\left. + 3x_{1/3}x_1\cos\left(\frac{7\nu t}{3} + 2\vartheta_1\right) + x_1^2\cos 3(\nu t + \vartheta_1)\right] = 0. \tag{23.84}$$

Since only four free constants are available, namely $x_{1/3}$, x_1, ϑ, and ϑ_1, we can require only the vanishing of the coefficients of $\cos(\nu t/3)$, $\sin(\nu t/3)$, $\cos \nu t$, and $\sin \nu t$ in (23.84), thus obtaining

$$x_{1/3}\left[\omega^2 - \frac{\nu^2}{9} + \frac{3}{4}\beta\omega^2(x_{1/3}^2 + 2x_1^2)\right] = -\frac{3}{4}\beta\omega^2 x_{1/3}^2 x_1\cos\vartheta_1, \tag{23.85}$$

$$\frac{2}{3}h\nu x_{1/3} = -\frac{3}{4}\beta\omega^2 x_{1/3}^2 x_1 \sin\vartheta_1, \tag{23.86}$$

$$x_1\left\{\left[\omega^2 - \nu^2 + \frac{3}{4}\beta\omega^2(2x_{1/3}^2 + x_1^2)\right]\cos\vartheta_1 - 2\nu h\sin\vartheta_1\right\} +$$

$$+ \frac{1}{4}\beta\omega^2 x_{1/3}^3 = P\cos 3\vartheta, \tag{23.87}$$

$$x_1\left\{\left[\omega^2 - \nu^2 + \frac{3}{4}\beta\omega^2(2x_{1/3}^2 + x_1^2)\right]\sin\vartheta_1 + 2\nu h\cos\vartheta_1\right\} = -P\sin 3\vartheta. \tag{23.88}$$

Consider first the cases $x_1 = 0$ and $x_{1/3} = 0$, i.e., the possible occurrence of pure vibrations of frequencies $\nu/3$ and ν, respectively. For $x_1 = 0$, Eq. (23.86) yields $h = 0$. This means that pure subharmonic vibrations of order 1/3 of the oscillator with cubic elastic restoring force can appear only in the absence of damping. If we now set $x_{1/3} = 0$, then Eqs. (23.85) and (23.86) are identically satisfied, and we may take, for the sake of simplicity, $\vartheta_1 = 0$. Next, we note that the equation of motion (23.83), the approximate solution (23.82), and Eqs. (23.87), (23.88) reduce, respectively, to Eqs. (23.38), (23.39), (23.41), and (23.42) obtained for the harmonic response in the neighborhood of the principal "resonance", provided 3ϑ be replaced by ϑ.

From this point onward in this section we assume that h, β, and P are of order $0(\varepsilon)$, where ε is a small parameter, and that $x_{1/3} \neq 0$, which allows to divide out a factor of $x_{1/3}$ from the first two equations above. It then follows by (23.85) that

$$\omega^2 - \frac{\nu^2}{9} = 0(\varepsilon). \tag{23.89}$$

and, introducing this into (23.87) and (23.88) gives

$$\left.\begin{aligned} x_1 \cos \vartheta_1 &= \frac{1}{32} \beta x_{1/3}^3 - \frac{P}{8\,\omega^2} \cos 3\vartheta + 0(\varepsilon^2), \\ x_1 \sin \vartheta_1 &= \frac{P}{8\omega^2} \sin 3\vartheta + 0(\varepsilon^2). \end{aligned}\right\} \tag{23.90}$$

By using these relations, the approximate solution (23.82) may be rewritten as

$$\bar{x}(t) = x_{1/3} \cos \frac{\nu t}{3} + \frac{1}{32} \beta x_{1/3}^3 \cos \nu t - \frac{P}{8\omega^2} \cos (\nu t - 3\vartheta) + 0(\varepsilon^2), \tag{23.91}$$

and Eqs. (23.87), (23.88) yield, by neglecting terms of third and higher order in ε,

$$\left.\begin{aligned} \omega^2 - \frac{\nu^2}{9} + \frac{3}{4} \beta\omega^2 x_{1/3}^2 + \frac{3}{128} \beta^2\omega^2 x_{1/3}^4 - \frac{3}{32} \beta P x_{1/3} \cos 3\vartheta = 0, \\ h + \frac{3\,\beta P}{64\,\omega} x_{1/3} \sin 3\vartheta = 0. \end{aligned}\right\} \tag{23.92}$$

Finally, by eliminating ϑ between the last two relations, we deduce the equations of the *resonance curve*

$$\left(\frac{\nu^2}{9\omega^2} - 1 - \frac{3}{4} \beta x_{1/3}^2 - \frac{3}{128} \beta^2 x_{1/3}^4\right) + 4\zeta^2 - \frac{9\,\beta^2 P^2}{1024\,\omega^4} x_{1/3}^2 = 0, \tag{23.93}$$

where $\zeta = h/\omega$. One may always assume that $x_{1/3} > 0$, $P > 0$, by introducing the eventual difference of sign between the first term of the approximate solution and the exciting force into the phase difference ϑ. Then, by denoting $x_{1/3} = a > 0$, we see that Eqs. (23.92) and (23.93) which determine the response amplitude and phase are identical with Eqs. (23.68) and (23.70) already obtained by the method of Krylov and Bogolyubov. By taking into account that t has been replaced by $t - 3\vartheta/\nu$, it is apparent that the approximate solution (23.91) also coincides with the solution (23.67) found by the method of Krylov and Bogolyubov.

§ 24. SYSTEMS WITH QUADRATIC VISCOUS DAMPING

This section will be entirely devoted to the study of forced vibrations of dissipative oscillators with quadratic viscous damping.

To this end we shall use in all examples considered the first approximation given by the method of KRYLOV and BOGOLYUBOV, which provides, as already seen, sufficient information about the periodic response in the neighborhood of the principal "resonance." Throughout this section we assume that the damping, the elastic nonlinearity, and the magnitude of the exciting force are small.

To facilitate further reference, we recollect here the principal results concerning the first approximation, which have been obtained in § 17 c. In the neighborhood of the principal "resonance," i.e., for $\omega^2 - \nu^2 = 0(\varepsilon)$, the periodic response of a weakly nonlinear oscillator governed by the equation

$$\ddot{x} + \omega^2 x = - \varepsilon [\psi_1(x) + f_1(\dot{x})] + \varepsilon P_1 \cos \nu t \tag{24.1}$$

is given in the first approximation by

$$x_1(t) = a \cos(\nu t + \vartheta). \tag{24.2}$$

The phase angle ϑ results from the relations

$$\sin \vartheta = - \frac{2 \nu a \bar{h}(a)}{P}, \qquad \cos \vartheta = \frac{a}{P} [\bar{\omega}^2(a) - \nu^2], \tag{24.3}$$

and the response amplitude from the equation

$$[\nu^2 - \bar{\omega}^2 (a)]^2 + 4\nu^2 \bar{h}^2(a) = \frac{P^2}{a^2}, \tag{24.4}$$

where $P = \varepsilon P_1$, and $\bar{h}(a)$, $\bar{\omega}(a)$ are the parameters of the equivalent linear system, which are given by

$$\bar{h}(a) = - \frac{1}{2 \pi \omega a} \int_0^{2\pi} \varepsilon f_1(- \omega a \sin \alpha) \sin \alpha \, d\alpha, \tag{24.5}$$

$$\bar{\omega}^2(a) = \omega^2 + \frac{1}{\pi a} \int_0^{2\pi} \varepsilon \psi_1(a \cos \alpha) \cos \alpha \, d\alpha. \tag{24.6}$$

We assume, as usually, that $a > 0$, the sign of x being taken into account by ϑ.

a) The oscillator with linear elastic restoring force and asymmetric quadratic damping

The governing equation of this oscillator under the action of a simple harmonic excitation is (cf. § 13 b)

$$\left.\begin{aligned} \ddot{x} + a_e\dot{x}^2 + \omega^2 x &= P\cos\nu t \quad \text{for} \quad \dot{x} \geqslant 0, \\ \ddot{x} - a_c\dot{x}^2 + \omega^2 x &= P\cos\nu t \quad \text{for} \quad \dot{x} < 0. \end{aligned}\right\} \tag{24.7}$$

Since a_e and a_c are positive, by applying the first corollary of theorem 9, § 15, we deduce that, for any finite values of P and ν, the equation of motion admits exactly one periodic solution with minimum period $2\pi/\nu$, and this solution is asymptotically stable in the large, hence stationary.

The parameters of the equivalent linear system have been calculated in § 17 c; they are

$$\bar{h}(a) = h + \frac{2\omega}{3\pi}(a_e + a_c), \qquad \bar{\omega}(a) = \omega. \tag{24.8}$$

Substituting (24.8) into (24.4) yields as the first approximation

$$\left.\begin{aligned} &x_{\mathrm{I}}(t) = a\cos(\nu t + \vartheta), \\ &\sin\vartheta = -\frac{4}{3\pi P}\omega\nu a^2(a_e + a_c)^2, \qquad \cos\vartheta = \frac{a}{P}(\omega^2 - \nu^2), \\ &(\omega^2 - \nu^2)^2 + \frac{16}{9\pi^2}\omega^2\nu^2 a^2(a_e + a_c)^2 = \frac{P^2}{a^2}. \end{aligned}\right\} \tag{24.9}$$

Denoting

$$k = \frac{8}{9\pi^2}(a_e + a_c)^2 > 0, \tag{24.10}$$

the last equation may be rewritten as

$$F\left(\frac{\nu^2}{\omega^2}, a\right) \equiv \left(\frac{\nu^2}{\omega^2} - 1\right)^2 + 2ka^2\frac{\nu^2}{\omega^2} - \frac{P^2}{\omega^4 a^2} = 0. \tag{24.11}$$

Next, by solving Eq. (24.11) for ν^2/ω^2, we find

$$\frac{\nu^2}{\omega^2} = 1 - ka^2 \pm \sqrt{\frac{P^2}{\omega^4 a^2} - 2ka^2 + k^2a^4}. \tag{24.12}$$

The graph of this functions for $P = \text{const}$ is a *resonance curve*. Its points are situated pairwise at equal horizontal distances from the *skeleton curve*, whose equation is

$$\frac{\nu^2}{\omega^2} = 1 - ka^2. \tag{24.13}$$

Since $k > 0$, the skeleton curve is bent to the left from the vertical.

From (24.11) we now obtain

$$\frac{\mathrm{d}a}{\mathrm{d}(\nu^2/\omega^2)} = -\frac{\dfrac{\nu^2}{\omega^2} - 1 + ka^2}{\dfrac{1}{a}\left(2ka^2\dfrac{\nu^2}{\omega^2} + \dfrac{P^2}{\omega^4 a^2}\right)}. \tag{24.14}$$

Since the denominator of the fraction in the right-hand side is nonzero for any values of the variables, we conclude that there are no points of vertical tangency on the resonance curves. This result is in agreement with the fact that the periodic solution is stable for any finite P and ν.

From (24.14) we also deduce that the geometric locus of the points at which the resonance curves have horizontal tangent, and hence at which the numerator of the fraction vanishes, coincides with the skeleton curve. The common roots of Eqs. (24.12) and (24.13), i.e., the values of a for which the radicand in (24.12) vanishes, are the extreme values of the response amplitude, say A. These values can be obtained by solving the equation

$$k^2A^4 - 2kA^2 + \frac{P^2}{\omega^4 A^2} = 0,$$

which may be also written in the form

$$(kA^2)^3 - 2(kA^2)^2 + \frac{kP^2}{\omega^4} = 0. \tag{24.15}$$

Since the value of ν^2/ω^2 corresponding to kA^2 results from (24.13), and since both ν^2/ω^2 and kA^2 must be positive, we are interested only in the roots of Eq. (24.15) belonging to the interval [0, 1]. By using ROLLE's theorem, it may be shown that, if $kP^2/\omega^4 < 0$, Eq. (24.15) has exactly *one* solution in the interval [0, 1]. For $kP^2/\omega^4 > 1$, Eq. (24.15) has no solutions in the interval [0, 1], and hence the resonance curve does not cut the skeleton curve. The roots of Eq. (24.15) may be easily found by a graphical procedure. By solving this equation for kP^2/ω^4, we may graphically represent the variation of kP^2/ω^4 with kA^2 for $0 \leqslant kA^2 \leqslant 1$.

Figure 67 shows the diagram obtained in this way. Then, for any given value of kP^2/ω^4 in the interval [0, 1] we can immediately find the corresponding value of kA^2.

For instance, in the case of road vehicle suspensions equipped by hydraulic shock absorbers one has, usually, $a_e + a_c = (0.2 \div 0.5)\ \text{cm}^{-1}$, $k = (0.004 \div 0.020)\text{cm}^{-2}$, and, $P/\omega^2 = (0 \div 5)$ cm. Hence, in these cases, it follows that $kP^2/\omega^4 = 0 \div 0.5$.

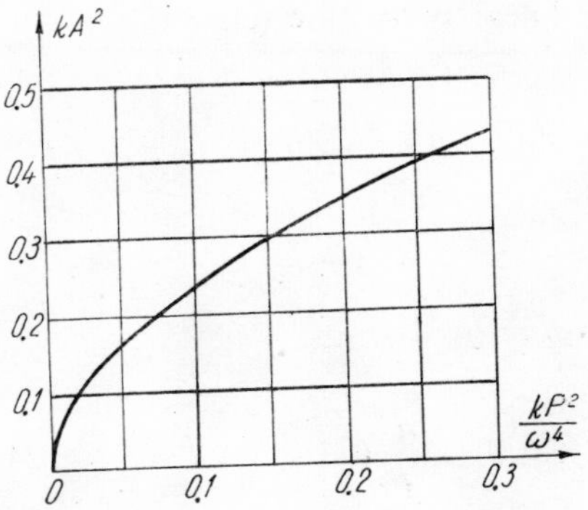

Fig. 67

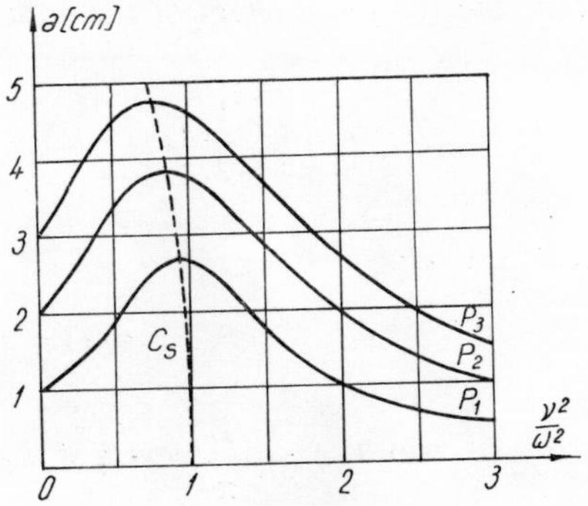

Fig. 68

Figure 68 illustrates the resonance curves obtained for $P_1/\omega^2 = 1$ cm, $P_2/\omega^2 = 2$ cm, $P_3/\omega^2 = 3$ cm, and $k = 0.01\ \text{cm}^{-2}$. The skeleton curve is represented by the broken line C_s. After determining the amplitude by using the resonance curve, the phase ϑ results from (24.9) and the first approximation of the solution from (24.2).

b) The oscillator with linear elastic restoring force and asymmetric parabolic damping (with linear and quadratic terms)

The governing equation of this oscillator under the action of a simple harmonic force may be written in the form

$$\left.\begin{aligned} \ddot{x} + 2h_e\dot{x} + a_e\dot{x}^2 + \omega^2 x &= P\cos\nu t \quad \text{for} \quad \dot{x} \geqslant 0, \\ \ddot{x} + 2h_c\dot{x} - a_c\dot{x}^2 + \omega^2 x &= P\cos\nu t \quad \text{for} \quad \dot{x} < 0, \end{aligned}\right\} \tag{24.16}$$

where the coefficients h_e, h_c, a_e ,and a_c are supposed to be positive. Equations (24.16) provide a good model for the vibrations of vehicle suspensions equipped by hydraulic shock absorbers. Indeed, for small values of $\dot{x}$, the flow of the oil in the shock absorber is laminar, and the damping force is approximately proportional to the velocity; as $\dot{x}$ increases, the flow of the oil becomes turbulent, and the magnitude of the damping force becomes proportional to the square of $\dot{x}$ (cf.§ 14). In fact, there are also secondary uncontrollable sources of dissipation, so that Eqs. (24.16) actually provide a four-parameter model for the approximation of the real dissipative properties of the suspension.

As in the preceding example, since the damping coefficients are positive, the Corollary 1, § 15 assures the existence of a unique periodic solution of the equation of motion, which has the same minimum period as the exciting force and is asymptotically stable in the large, hence stationary. If the magnitude of the exciting force and the damping coefficients are small enough, we can obtain a satisfactory first approximation for the stationary solution by means of Eqs. (24.2)—(24.4). Since in our case the parameters of the equivalent linear system are (cf. § 17 c):

$$\bar{h}(a) = \frac{1}{2}(h_e + h_c) + \frac{2}{3\pi}\omega a(a_e + a_c), \quad \bar{\omega}(a) = \omega, \tag{24.17}$$

it results that

$$x_I(t) = a\cos(\nu t + \vartheta),$$

$$\sin\vartheta = -\frac{2}{\pi}\nu\omega a(\zeta + k_1 a), \quad \cos\vartheta = \frac{a}{P}(\omega^2 - \nu^2), \tag{24.18}$$

$$F\left(\frac{\nu^2}{\omega^2}, a\right) \equiv \left(\frac{\nu^2}{\omega^2} - 1\right)^2 + 4(\zeta + k_1 a)^2\frac{\nu^2}{\omega^2} - \frac{P^2}{\omega^4 a^2} = 0, \tag{24.19}$$

with the notations

$$\zeta = \frac{1}{2\omega}(h_e + h_c), \quad k_1 = \frac{2}{3\pi}(a_e + a_c). \tag{24.20}$$

For suspensions of road vehicles, the parameters (24.20) usually have the values $\zeta = 0.1 \div 0.4$, $k_1 = (0.04 \div 0.10)\ \text{cm}^{-1}$.

By solving the equation (24.19) of the *resonance curve* for ν^2/ω^2 we obtain

$$\frac{\nu^2}{\omega^2} = 1 - 2(\zeta + k_1 a)^2 \pm \sqrt{\frac{P^2}{\omega^4 a^2} - 4(\zeta + k_1 a)^2[1 - (\zeta + k_1 a)^2]} \tag{24.21}$$

From this relation it follows that the points of the resonance curve are situated at equal horizontal distances from the *skeleton curve*, which is given by the equation

$$\frac{\nu^2}{\omega^2} = 1 - 2(\zeta + k_1 a)^2. \tag{24.22}$$

We see that the skeleton curve is bent to the left from the vertical as in the preceding example.

By using the differentiation rule for implicit functions, we deduce from (24.19) that

$$\frac{da}{d(\nu^2/\omega^2)} = -\frac{\dfrac{\nu^2}{\omega^2} - 1 - 2(\zeta + k_1 a)^2}{4k_1(\zeta + k_1 a)\dfrac{\nu^2}{\omega^2} + \dfrac{P^2}{\omega^4 a^3}}. \tag{24.23}$$

Since $a > 0$, the denominator of the fraction in right-hand side is strictly positive for any values of the variables, and hence the resonance curves have no vertical tangents, in agreement with the fact that the periodic response corresponding to any finite values of P and ν is stable. From (24.23) it also follows that the geometric locus of the points at which the resonance curves have horizontal tangent coincides with the skeleton curve. Therefore, the maximum value of the amplitude on each resonance curve corresponds to the point of intersection with the skeleton curve. This maximum value, say A, has to satisfy Eqs. (24.21) and (24.22), and hence is given by the equation

$$\frac{P^2}{\omega^4 A^2} - 4(\zeta + k_1 A)^2 [1 - (\zeta + k_1 A)^2] = 0. \tag{24.24}$$

To solve this transcendental equation we first rewrite it in the form

$$\frac{k_1 P}{\omega^2} = 2k_1 A(\zeta + k_1 A)\sqrt{1 - (\zeta + k_1 A)^2} \tag{24.25}$$

and, for constant values of ζ, we plot k_1P/ω^2 versus k_1A, as shown in Fig. 69. Then, for given k_1P/ω^2, one may easily determine k_1A, and hence also the maximum value A of the amplitude. Since A must satisfy Eq. (24.22) we are interested only in the solutions of (24.25) for which

$$\zeta + k_1 A \leqslant \frac{1}{2}. \tag{24.26}$$

Therefore, the curves in Fig. 69 have been interrupted at the points where condition (24.26) is satisfied with equality signs. The slopes of the tangents to these curves at the origin equal $(1/2)\ \zeta/\sqrt{1 - \zeta^2}$.

Figure 70 illustrates the resonance curves corresponding to $P_1/\omega^2 = 1$ cm, $P_2/\omega^2 = 2$ cm, $P_3/\omega^2 = 3$ cm, and $\zeta = 0.1$, $k_1 = 0.1\ \text{cm}^{-1}$. The skeleton curve is represented by the broken line C_s. From (24.22) it follows that the skeleton curve cuts the (ν^2/ω^2)-axis at the point of abscissa $1 - 2\zeta^2$, and that the slope of the tangent to the skeleton curve at this point equals $-(1/4)\zeta k_1$. After determining the response amplitude by using the resonance curve corresponding to given values of ν and P, one may easily calculate the phase angle from (24.18) and the approximate stationary solution $x_{\text{I}}(t)$ from (24.2).

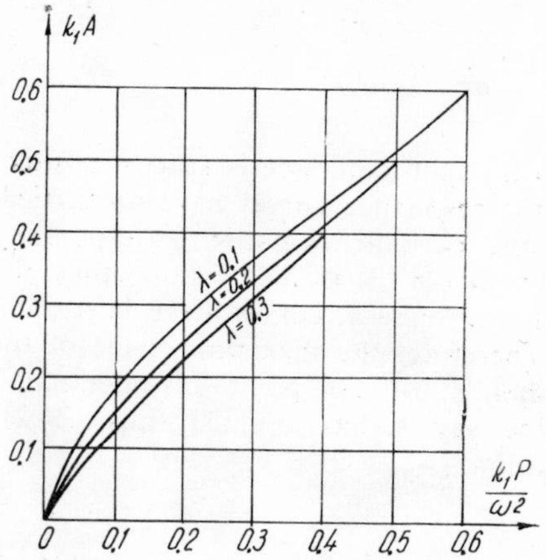

Fig. 69

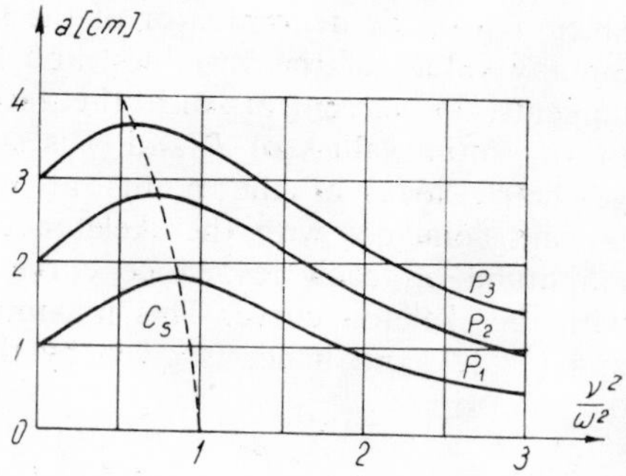

Fig. 70

c) The oscillator with cubic elastic restoring force and asymmetric quadratic damping

The equation governing the forced vibrations of the oscillator with cubic elastic restoring force and asymmetric quadratic damping acted on by a harmonic excitation is

$$\left.\begin{aligned} \ddot{x} + a_e\dot{x}^2 + \omega^2 x(1 + \beta x^2) &= P\cos\nu t \quad \text{for} \quad \dot{x} \geqslant 0, \\ \ddot{x} - a_c\dot{x}^2 + \omega^2 x(1 + \beta x^2) &= P\cos\nu t \quad \text{for} \quad \dot{x} < 0. \end{aligned}\right\} \tag{24.27}$$

By Theorem 8, § 15, we deduce that there exists at least one periodic solution of Eqs. (24.27) with the same minimum period as the exciting force. However, Theorem 9, § 15, no longer assures that this solution is unique and asymptotically stable in the large unless the elastic nonlinearity is sufficiently small and/or the damping is strong enough. The relation (15.17) provides a sufficient condition for stability, which becomes in our case

$$a_c > \frac{3}{2}\max\left(\frac{1}{2}\sqrt{\frac{\beta}{3}},\quad \frac{\beta A}{1 + 3\beta A^2}\right), \tag{24.28}$$

where A is the maximum value of the response amplitude. The study of the periodic vibration in the first approximation will allow us to obtain still other sufficient conditions for stability, which hold if the damping, the elastic nonlinearity, and the magnitude of the exciting force are small.

The parameters of the equivalent linear system, as found in § 17 c, are

$$\bar{h}(a) = \frac{2}{3\pi}\,\omega a\,(a_e + a_c), \qquad \bar{\omega}^2(a) = \omega^2\left(1 + \frac{3}{4}\,\beta a^2\right). \tag{24.29}$$

Substituting them into (24.2)—(24.4), we obtain as first approximation

$$x_{\text{I}}(t) = a\cos(\nu t + \vartheta), \tag{24.30}$$

$$\sin\vartheta = -\frac{4}{3\pi P}\,\omega\nu a^2(a_e + a_c), \qquad \cos\vartheta = \frac{a}{P}\left(\omega^2 + \frac{3}{4}\,\beta\omega^2 a^2 - \nu^2\right), \tag{24.31}$$

$$\left[\nu^2 - \omega^2\left(1 + \frac{3}{4}\,\beta a^2\right)\right]^2 + \frac{16}{9\pi^2}\,\omega^2\nu^2 a^2(a_e + a_c)^2 = \frac{P^2}{a^2}.$$

By denoting

$$k = \frac{8}{9\pi^2}\,(a_e + a_c)^2, \qquad \beta_1 = \frac{3\beta}{4k}, \tag{24.32}$$

the last equation may be rewritten as

$$F\left(\frac{\nu^2}{\omega^2},\, a\right) \equiv \left(\frac{\nu^2}{\omega^2} - 1 - \beta_1 k a^2\right)^2 + 2ka^2\,\frac{\nu^2}{\omega^2} - \frac{P^2}{\omega^4 a^2} = 0. \tag{24.33}$$

The graph of this function for $P = \text{const}$ is said to be a *resonance curve*. By solving (24.33) for ν^2/ω^2, we obtain

$$\frac{\nu^2}{\omega^2} = 1 + (\beta_1 - 1)\,ka^2 \pm \sqrt{\frac{P^2}{\omega^4 a^2} - 2ka^2\left[1 + \left(\beta_1 - \frac{1}{2}\right)ka^2\right]}. \tag{24.34}$$

This equation shows that the two points of a resonance curve corresponding to any given value of a for which the radicand in (24.34) is nonnegative, are situated at equal horizontal distances from the *skeleton curve*, whose equation is

$$\frac{\nu^2}{\omega^2} = 1 + (\beta_1 - 1)\,ka^2. \tag{24.35}$$

The skeleton curve is bent to the left or to the right from the vertical line of equation $\nu^2/\omega^2 = 1$, according as $\beta_1 < 1$ or $\beta_1 > 1$. For $\beta_1 = 1$ the skeleton curve coincides with this vertical line. Finally, for $\beta_1=0$, one recovers the case considered in § 24 a.

By using the differentiation rule for implicit functions, we obtain from (24.33)

$$\frac{\mathrm{d}a}{\mathrm{d}(\nu^2/\omega^2)} = \frac{\dfrac{\nu^2}{\omega^2} - 1 - (\beta_1 - 1)\, ka^2}{\dfrac{1}{a}\left[2k\beta_1 a^2\left(\dfrac{\nu^2}{\omega^2} - 1 - k\beta_1 a^2\right) - 2ka^2\dfrac{\nu^2}{\omega^2} - \dfrac{P^2}{\omega^4 a^2}\right]} \cdot \tag{24.36}$$

This relation shows that the geometric locus of the points at which the resonance curves have horizontal tangent coincides with the skeleton curve. Denoting by A the extreme value (or values) taken by a at the point (or points) of intersection with the skeleton curve, we obtain from (24.34) and (24.35)

$$\frac{P^2}{\omega^4 A^2} - 2kA^2\left[1 + \left(\beta_1 - \frac{1}{2}\right) kA^2\right] = 0.$$

To solve this transcendental equation we write it in the form

$$\frac{kP^2}{\omega^4} = \left(\beta_1 - \frac{1}{2}\right)(kA^2)^3 + 2(kA^2)^2 \tag{24.37}$$

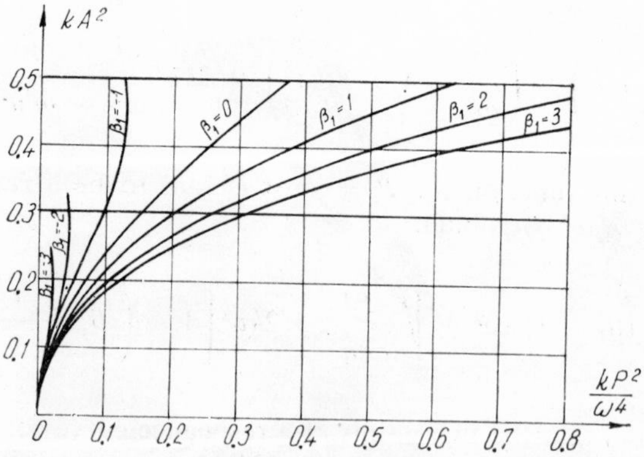

Fig. 71

and plot kP^2/ω^4 versus kA^2 for various constant values of β_1, as shown in Fig. 71. Then, for any given kP^2/ω^4 and β_1, we may easily determine the corresponding value (or values) of kA^2. The curve obtained for $\beta_1 = 0$ coincides, regardless of scale, with that illustrated in Fig. 67. Since kA^2 must satisfy Eq. (24.35), too, and since $\nu^2/\omega^2 \geqslant 0$, we conclude that for $\beta_1 < 1$ only roots of Eq. (24.37) for which

$$kA^2 \leqslant \frac{1}{1-\beta_1} \tag{24.38}$$

are to be considered. This restriction has been taken into account when tracing the curves in Fig. 71, by interrupting those curves which correspond to $\beta_1 < 1$ at the points where (24.38) becomes an equality.

The roots of Eq. (24.37) may be separated by making use of ROLLE's theorem and taking β_1 and kP^2/ω^4 as parameters. Since the calculation is rather lengthy we mention only the results.

For $\beta_1 \leqslant -(1/2)$ we distinguish between three cases. First, if

$$\frac{kP^2}{\omega^4} < \frac{1}{(1-\beta_1)^3}, \tag{24.39}$$

Eq. (24.37) has exactly one root belonging to the interval $[0, (1-\beta_1)^{-1}]$, and hence satisfying the condition (24.38). Consequently, the response amplitude takes a maximum value when ν varies and P is constant and fulfills (24.39). Second, if

$$\frac{1}{(1-\beta_1)^3} \leqslant \frac{kP^2}{\omega^4} \leqslant \frac{32}{27(1-2\beta_1)^2}, \tag{24.40}$$

Eq. (24.37) has one root in the interval $[0, 4(1-2\beta_1)^{-1/3}]$ and another one in the interval $[4(1-2\beta_1)^{-1/3}, (1-\beta_1)^{-1}]$. The resonance curve has two branches, and the response amplitude takes a maximum and a minimum value when ν varies and P is constant and satisfies (24.40). Finally, if

$$\frac{kP^2}{\omega^4} > \frac{32}{27(1-2\beta_1)^2}, \tag{24.41}$$

Eq. (24.37) has no positive roots. In this case the resonance curve does not cut the skeleton curve. To illustrate these three cases, we traced in Fig. 72 the resonance curves obtained for $\beta_1 = -3$, $k = 0.01\ \text{cm}^{-2}$, and for $P_1/\omega^2 = 1$ cm, $P_2/\omega^2 = 1.5$ cm, $P_3/\omega^2 = 2$ cm. It is easily seen that each of the three values P_1, P_2, P_3 satisfies exactly one of the relations (24.39)–(24.41), and one may check in each particular case the agreement with the results expounded above.

For $-1/2 < \beta_1 < 1$ we distinguish between two cases. First, if

$$\frac{kP^2}{\omega^4} \leqslant \frac{1}{(1-\beta_1)^3}, \tag{24.42}$$

Eq. (24.37) has exactly one root belonging to the interval $[0, (1 - \beta_1)^{-1}]$, and hence satisfying condition (24.38). The response amplitude takes a maximum value when ν varies and P is constant and satisfies (24.42). Resonance curves of this

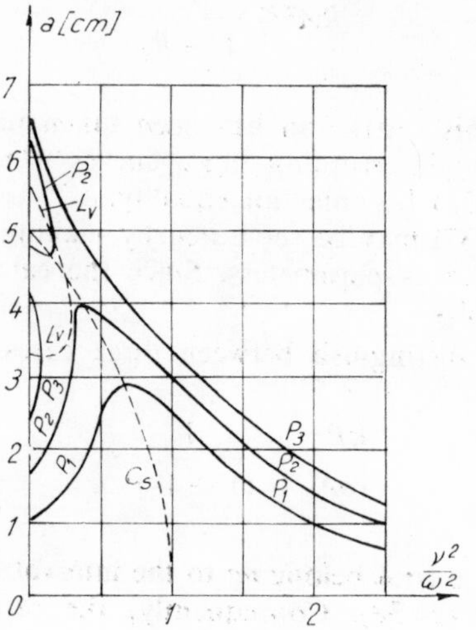

Fig. 72

type are shown in Fig. 68, as obtained for $\beta_1 = 0$, $k = 0.01$ cm^{-2}, $P_1/\omega^2 = 1$ cm, $P_2/\omega^2 = 2$ cm, $P_3/\omega^2 = 3$ cm. Second, if

$$\frac{kP^2}{\omega^4} > \frac{1}{(1 - \beta_1)^3}, \tag{24.43}$$

Eq. (24.37) has no solution in the interval $[0, (1 - \beta_1)^{-1}]$, and hence the skeleton curve no longer intersects the resonance curve.

Finally, for $\beta_1 \geqslant 1$, Eq. (24.37) has a positive solution for any finite P, and hence the response amplitude assumes a maximum value for some value of ν. Figure 73 illustrates this situation for $\beta_1 = 1$, $k = 0.01$ cm^{-2}, $P_1/\omega^2 = 1$ cm, $P_2/\omega^2 = 2$ cm, $P_3/\omega^2 = 3$ cm. In this case the skeleton curve coincides with the vertical line of equation $\nu^2/\omega^2 = 1$. Of the same type as regards the existence of extreme amplitudes are the resonance curves represented in Fig. 74, corresponding to $\beta_1 = 12$, $k = 0.01$ cm^{-2}, $P_1/\omega^2 = 1$ cm, $P_2/\omega^2 = 3$ cm, $P_3/\omega^2 = 5$ cm; we shall come back again to this case below.

Let us now return to Eq. (24.36). At the points where the resonance curves have vertical tangent the denominator of the fraction in the right-hand side must vanish, i.e.,

$$2k\beta_1 a^2 \left(\frac{\nu^2}{\omega^2} - 1 - k\beta_1 a^2 \right) - 2ka^2 \frac{\nu^2}{\omega^2} - \frac{P^2}{\omega^4 a^2} = 0.$$

By eliminating P between this relation and (24.33), we obtain the equation of the geometric locus of the points at which the resonance curves have vertical tangent, namely

$$\left(\frac{\nu^2}{\omega^2} - 1 - k\beta_1 a^2\right)\left(\frac{\nu^2}{\omega^2} - 1 - 3k\beta_1 a^2\right) + 4ka^2 \frac{\nu^2}{\omega^2} = 0. \tag{24.44}$$

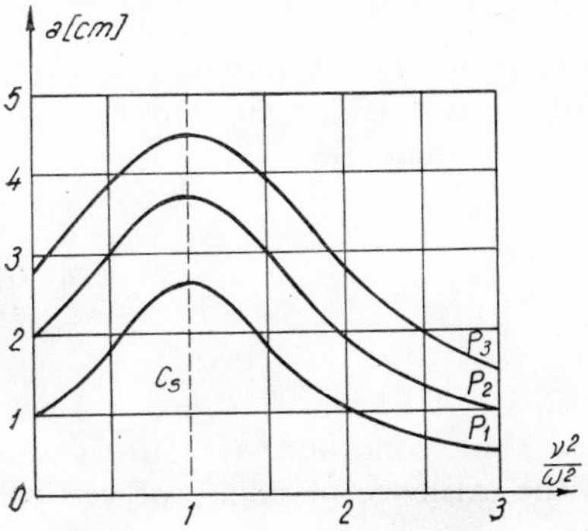

Fig. 73

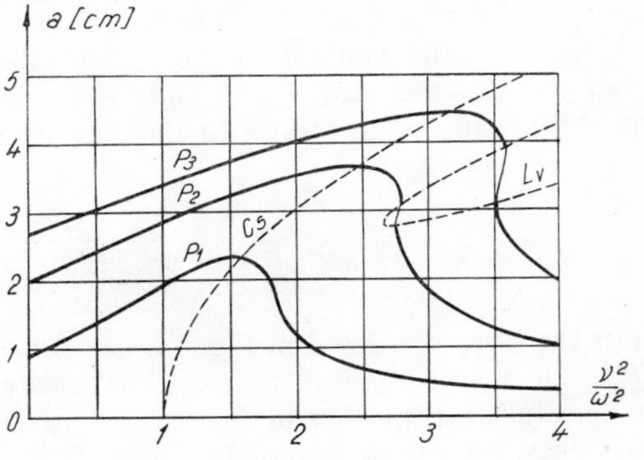

Fig. 74

Solving this equation for ν^2/ω^2 yields

$$\frac{\nu^2}{\omega^2} = 1 + 2(\beta_1 - 1)ka^2 \pm \sqrt{ka^2[(\beta_1^2 - 8\beta_1 + 4)ka^2 - 4]}. \tag{24.45}$$

By considering the radicand, we see that this curve exists only if

$$ka^2 > \frac{4}{\beta_1^2 - 8\beta_1 + 4} \tag{24.46}$$

and if $\beta_1^2 - 8\beta_1 + 4 > 0$, whence

$$\beta_1 \in (-\infty, 4 - 2\sqrt{3}) \cup (4 + 2\sqrt{3}, \infty). \tag{24.47}$$

In Figs 72 and 74 the geometric locus of the points at which the resonance curves have vertical tangent is represented by a broken line denoted by L_v, and the arcs of the resonance curves corresponding to stable periodic vibrations are represented by thickened solid lines. By combining now conditions (24.46) and (24.47) with the results about the extreme values of the amplitude we shall obtain sufficient stability conditions for the first approximation. Obviously, the ratio ν/ω must be still close to unity, in order to remain within the limits of validity of the first approximation.

From (24.45) and (24.47) it follows that the periodic vibration is stable for any P and ν if

$$\beta_1 \in [4 - 2\sqrt{3}, 4 + 2\sqrt{3}]. \tag{24.48}$$

For

$$\beta_1 > 4 + 2\sqrt{3}, \tag{24.49}$$

by requiring the root kA^2 of Eq. (24.37) that corresponds to the maximum amplitude to be smaller than $4/(\beta_1^2 - 8\beta_1 + 4)$, we obtain from (24.37) a sufficient condition for stability according to the first approximation, namely

$$\frac{kP^2}{\omega^4} < \frac{32(\beta_1^2 - 4\beta_1 + 2)}{(\beta_1^2 - 8\beta_1 + 4)^3}. \tag{24.50}$$

For the case considered in Fig. 74, i.e., for $\beta_1 = 12$, $k = 0.01 \text{ cm}^{-2}$, $P_1/\omega^2 = 1$ cm, $P_2/\omega^2 = 3$ cm, $P_3/\omega^2 = 5$ cm, it is easily seen that condition (24.50) is no longer fulfilled for P_2 and P_3, in agreement with the existence of points of vertical tangency on the corresponding resonance curves.

Finally, for

$$\beta_1 < 4 - 2\sqrt{3} \approx 0.54, \tag{24.51}$$

we have already seen that Eq. (24.37) always has a unique solution kA^2 in the interval $[0, (1-\beta_1)^{-1}]$ if

$$\frac{kP^2}{\omega^4} < \frac{1}{(1-\beta_1)^3}. \tag{24.52}$$

Requiring now that

$$\frac{4}{\beta_1^2 - 8\beta_1 + 4} \geqslant \frac{1}{1-\beta_1}.$$

and considering (24.51), gives

$$0 \leqslant \beta_1 < 4 - 2\sqrt{3}. \tag{24.53}$$

The conditions (24.52) and (24.53) assure the stability of the periodic response according to the first approximation. The discussion above does not allow to obtain sufficient conditions for stability if $\beta_1 < 0$. However, the example illustrated in Fig. 72 for $\beta_1 = -3$ shows that even if $\beta_1 < 0$ the periodic response may be stable according to the first approximation, provided P be small enough.

After the determination of the response amplitude for given P and ν, the phase angle ϑ can be derived from (24.31), and then the final expression of the periodic solution at the first stage of approximation results from (24.30).

PART III

Forced vibrations of linear and nonlinear systems acted on by random excitations

In this last part of the book we consider forced vibrations governed by the same equation of motion as in Part II, i.e.,

$$\ddot{x} + f(\dot{x}) + g(x) = p(t),$$

but where now $p(t)$ is a random function.
The main problem to be solved is: given some stochastic characteristics of the function $p(t)$, and assuming that the functions f and g satisfy the hypotheses $H_1 - H_5$ formulated in the introduction, find the corresponding stochastic characteristics of the response.

Chapter VII Introduction to the theory of random processes

§ 25. RANDOM VARIABLES

a) The notion of random variable

The theory of random processes is based upon the concept of a random variable, which is defined on a probability field; this is why we will begin by introducing the concept of probability field. The reader is supposed to be familiar with the basic notions and results concerning the algebras of events and their representation by algebras of sets, which can be found in standard books on probability theory (see, e.g., Rényi [162]). On the other hand, even a brief introduction of these notions would hardly fit in with the limited mathematical aim of this book which is mainly dedicated to vibration problems.

In the following, we will present in a concise form, and omitting extended proofs, only those elements of the theory of random processes that may assure a certain independence in considering vibration problems.

The notion of a probability field is in general axiomatically defined, by summarizing the common properties of the majority of random events occurring in practice.

The realization of a certain set of conditions is called an *experiment*, and any possible outcome of an experiment is said to be an *event*. An event that must always occur as the result of a given experiment is called *certain*. An event which certainly does not occur is said to be *impossible*. Any other event, i.e., any event which may or may not occur, is said to be a *random event*.

The nonoccurrence of an event A is itself an event, denoted by CA and called the event *complementary* to A. The occurrence of at least one of two events A and B is an event called the *sum of A and B*; the simultaneous occurrence of two events A and B is an event called the *product of A and B*. Once these operations are defined, the set of all events associated with a certain experiment may be given a structure of a Boolean algebra.

The following basic theorem, due to Stone [185], establishes a correspondence between algebras of physical events and abstract algebras of sets, enabling one to introduce axiomatically the concept of probability field.

Theorem 1. *There can be associated with every algebra of events an algebra of sets isomorphic to it.*

For a proof of this statement see, e.g., RÉNYI [162]. We confine ourselves here to noting that, in the framework of the isomophism mentioned above, to each random event corresponds a well determined subset of an abstract total set. This total set is itself associated with the certain event, while the empty set corresponds to the impossible event. The sum and the product of two events have as correspondents the union and the intersection of the two sets associated with the considered events. To the event complementary to an event A corresponds the set complementary to the set associated with A.

Denoting by the same letters the events and the sets corresponding to them, the probability field associated with an experiment generating random events can be defined by the following axioms:

I. There exist a set Ω called the *sample space* which corresponds to the given experiment and whose elements are called *elementary events*, and a set K of parts of Ω, whose elements are called *events* and which has the following properties:

I_1. If $A \in K$, then $CA \in K$.

I_2. Any finite or enumerable union of elements of K belongs to K, too.

I_3. The sure event Ω and the impossible event $\varnothing$ belong to K.

II. To each event $A \in K$ there is assigned a real number $P(A)$, called the *probability of that event*, such that:

II_1. $P(A) \geqslant 0$ for any $A \in K$.

II_2. $P(\Omega) = 1$.

II_3. For any finite or numerable union $\bigcup_n A_n$ of pairwise exclusive events $A_n (A_i \cap A_j = \varnothing$ for any $i, j, i \neq j)$ we have

$$P(\bigcup_n A_n) = \sum_n P(A_n).$$

If Ω, K, and P enjoy the above properties, then the ensemble (Ω, K) is called a *field of events* and the ensemble (Ω, K, P) is said to be a *probability field.*

Let us consider now an experiment which leads to the field of events (Ω, K), and a function $X : \Omega \to R$, where R denotes the real axis.

Definition 1. *If* $\{\omega | \omega \in \Omega,\ X(\omega) < x\} \in K$ *for any* $x \in R$, *then* X *is called a random variable.*[1)]

Before introducing other stochastic concepts, we will illustrate the notions defined above by an example. Assume that the acceleration a at a certain point of a road vehicle is measured N times at the same point of a road segment while the vehicle rides with a constant speed V. The outcomes of this repeated experiment will vary randomly because of the small variations of the real speed around V, of the departures of the traveling line from the prescribed one, and so on. Suppose that all possible values of the acceleration lie within a certain interval (a_1, a_2). We associate with the physical outcome "the acceleration takes on the value $a_0 \in (a_1, a_2)$"

1) Equivalent definitions of a random variable may be obtained by replacing the condition $X(\omega) < x$ in definition 1 with $X(\omega) > x$, $X(\omega) \in (x, y)$, $X(\omega) \in (x, y]$, or $X(\omega) \in [x, y)$, where $x, y \in R$.

the number a_0/g, where g is the gravity acceleration, and we call this number an elementary event. Then, the certain event, Ω, i.e., the set $(a_1/g, a_2/g)$ corresponds to the outcome "the acceleration takes on any value in the interval (a_1, a_2)," and the impossible event $\varnothing$, i.e., the empty set, corresponds to the outcome "the acceleration takes on a value outside (a_1, a_2)." The set of events K includes besides Ω and $\varnothing$ all proper subsets of Ω, i.e., all unions of intervals contained in $(a_1/g, a_2/g)$. It can be easily verified that K satisfies Axiom I.

Finally, the simplest random variable we may define in this case over the field of events (Ω, K) results by taking $X(\omega) = \omega$, where $\omega = a/g$ is the reduced acceleration, and $\omega \in \Omega$. Indeed,

$$\{\omega | X(\omega) < x\} = \begin{cases} \varnothing \quad \text{for} \quad x \in (-\infty, a_1/g], \\ (a_1/g, x) \ \text{for} \quad x \in (a_1/g, a_2/g), \\ \Omega \quad \text{for} \quad x \in (a_2/g, \infty), \end{cases}$$

and hence $\{\omega | X(\omega) < x\} \in K$ for any $x \in R$.

Let now (Ω, K, P) be a probability field and $X(\omega)$ a random variable defined for $\omega \in \Omega$.

Definition 2. *The function*

$$F(x) = P(\{\omega | \omega \in \Omega, \quad X(\omega) < x\}), \tag{25.1}$$

defined for $x \in R$, *is called the* distribution function *of the random variable* $X(\omega)$.

It is apparent from definition 1 that the argument of P in (25.1) belongs to K, and hence definition 2 has a precise meaning for any $x \in R$. For the sake of conciseness we shall write Eq. (25.1) in the form

$$F(x) = P(X < x). \tag{25.1'}$$

Accordingly, we shall for brevity call $P(X < x)$ "the probability of the event $X < x$" instead of "the probability of the event $\Omega' \in \Omega$ enjoying the property that $X(\omega) < x$ for any $\omega \in \Omega'$". Similar shortened notations will be used throughout this book as long as no confusion is possible.

We shall always assume that $F(x)$ is continuous together with its first derivative for $x \in (-\infty, \infty)$. The main properties of the distribution function are

$$\left.\begin{aligned} \lim_{x\to-\infty} F(x) = 0, \qquad \lim_{x\to\infty} F(x) = 1, \\ P(a \leqslant X < b) = F(b) - F(a) \quad \text{for any} \quad a, b \in R, \quad a < b, \\ F(a) \leqslant F(b) \quad \text{for any} \quad a, b \in R, \quad a < b. \end{aligned}\right\} \tag{25.2}$$

The first two of these relations are immediate consequences of the fact that $X < -\infty$ is the impossible event, whereas $X < \infty$ is the certain event.

To prove Eq. $(25.2)_3$ we consider the events Ω', Ω'', Ω''', for which $X < a$, $X < b$, and $a \leqslant X < b$, respectively. Obviously, these events are related by

$$\Omega'' = \Omega' \cup \Omega''' \tag{25.3}$$

and $\Omega' \cap \Omega''' = \varnothing$. Therefore, by II_3,

$$P(\Omega'') = P(\Omega') + P(\Omega'''), \tag{25.4}$$

or

$$P(X < b) = P(X < a) + P(a \leqslant X < b) \tag{25.5}$$

which, by (25.1′), implies $(25.2)_3$.

Finally, the last relation (25.2), which shows that $F(x)$ is a nondecreasing function, immediately results from $(25.2)_3$ and II_1.

It may be proved that, given a probability field (Ω, K, P), the distribution function of any random variable defined on Ω determines the probability of any event belonging to K.

Definition 3. *The derivative of the distribution function,*

$$f(x) = F'(x), \tag{25.6}$$

is called the frequency function *or the* probability density function *of the random variable X.*

Considering $(25.2)_1$, Eq. (25.6) may be integrated to obtain

$$F(x) = \int_{-\infty}^{x} f(u)\, du. \tag{25.7}$$

The probability density function has the following main properties

$$\left.\begin{aligned} &f(x) \geqslant 0, \\ &\int_{-\infty}^{\infty} f(u)\, du = 1, \\ &P(a \leqslant X < b) = \int_a^b f(u)\, du \quad \text{for any} \quad a, b \in R, \quad a < b, \\ &P(x \leqslant X < x + \Delta x) = f(x)\, \Delta x \quad \text{for} \quad \Delta x \text{ sufficiently small.} \end{aligned}\right\} \tag{25.8}$$

The first of these equations expresses in a different way the property that $F(x)$ is a monotone nondecreasing function, and the second relation immediately results from (25.7) and $(25.2)_2$. Equation $(25.8)_3$ is a consequence of $(25.2)_3$ and (25.6). Finally, Eq. $(25.8)_4$ may be easily proved, by setting $a = x$, $b = x + \Delta x$ in $(25.8)_3$ and making use of the mean-value theorem.

We note that the last two equations (25.8) express the probability of the random variable taking values in a finite interval and, respectively, in a sufficiently small interval of the real axis, in terms of the probability density function.

Since random variables are real-valued functions, the elementary algebraic operations between random variables defined on the same set Ω can be given the same meaning as in the theory of real functions, provided that their results are random variables, too. This really happens, as shown by the following two theorems, which are stated without proof.

Theorem 2. *If $X : \Omega \to R$ is a random variable and a is any real number, then $X + a$, aX, $|X|$, $X^2 = XX$, and $1/X$ (with $X \neq 0$) are also random variables defined on Ω.*

Theorem 3. *If $X_1 : \Omega \to R$ and $X_2 : \Omega \to R$ are random variables, then $X_1 \pm X_2$, X_1X_2, and X_1/X_2 ($X_2 \neq 0$) are also random variables defined on Ω.*

Actually, the statements above are particular cases of the following much more general theorem (see, e.g., Rényi [169], p. 180).

Theorem 4. *If $X_i : \Omega \to R$, $i = 1, \ldots, n$ are random variables, and if $H : R^n \to R$ is a continuous function*[1]*, then the function $H_{X_1, \ldots, X_n} : \Omega \to R$, defined by*

$$H_{X_1, \ldots, X_n}(\omega) = H(X_1(\omega), \ldots, X_n(\omega)),$$

is also a random variable.

Definition 4. *A property concerning the random variable $X : \Omega \to R$ is said to hold* almost everywhere *or* almost certainly *if the set $\Omega' \subset \Omega$ on which this property is not valid belongs to K and its probability is zero.*

The following theorem assures that the equality and the relation of order between random variables may be defined almost everywhere.

Theorem 5. *If $X_1 : \Omega \to R$ and $X_2 : \Omega \to R$ are random variables, then the sets $\{\omega | \omega \in \Omega, X_1(\omega) = X_2(\omega)\}$ and $\{\omega | \omega \in \Omega, X_1(\omega) > X_2(\omega)\}$ belong to K.*

From definition 4 and theorem 4 it follows that two random variables, $X_1 : \Omega \to R$ and $X_2 : \Omega \to R$, are equal *almost everywhere* if the probability of the event $X_1 \neq X_2$ is zero.

The sequences of random variables are in fact sequences of real-valued functions. However, since they are defined on sets of elementary events, their convergence has some special features which will become apparent from the following definitions.

[1] Here and in the following R denotes the n-dimensional real Euclidean space.

Definition 5. *A sequence* $\{X_n\}$ *of random variables* $X_n : \Omega \to R$ *is said to* converge almost certainly (*or* with probability one) *to the random variable* $X : \Omega \to R$ *if*

$$\lim_{n\to\infty} X_n(\omega) = X(\omega) \tag{25.9}$$

for any $\omega \in \Omega$, *except, possibly, some subset* $\Omega' \in K$ *of probability zero.*

Definition 6. *A sequence* $\{X_n\}$ *of random variables* $X_n : \Omega \to R$ *is said to* converge in probability *to the random variable* $X : \Omega \to R$ *if, given any positive real numbers* η *and* ε, *there exists a positive integer* $N(\eta, \varepsilon)$ *such that* $P(|X_n - X| \geqslant \eta) < \varepsilon$ *for any* $n > N(\eta, \varepsilon)$.

Definition 7. *A sequence* $\{X_n\}$ *of random variables* $X_n : \Omega \to R$ *is said to* converge strongly *to the random variable* $X : \Omega \to R$ *if, given any positive real numbers* η *and* ε, *there exists a positive integer* $N(\eta, \varepsilon)$ *such that* $P(\bigcup_{n>N} \{\omega | \omega \in \Omega, |X_n(\omega) - X(\omega)| \geqslant \eta\}) < \varepsilon$.

Definition 8. *A sequence* $\{X_n\}$ *of random variables* $X_n : \Omega \to R$ *with distribution functions* $F_n(x)$ *is said to* converge in distribution *to the random variable* $X : \Omega \to R$ *with distribution function* $F(x)$ *if*

$$\lim_{n\to\infty} F_n(x) = F(x) \tag{25.10}$$

for any $x \in R$.

A very useful notion in the theory of random processes is that of a random vector. This is a natural generalization of the notion of random variable.

Let $\boldsymbol{X} : \Omega \to R^n$ be a vector-valued function with components $X_1, \ldots, X_n$.

Definition 9. *If* $\{\omega | \omega \in \Omega, X_1(\omega) < x_1, \ldots, X_n(\omega) < x_n\} \in K$ *for any* $(x_1, \ldots, x_n) \in R^n$, *then* $\boldsymbol{X}$ *is called a* vector random variable *or a* random vector.

Definition 10. *The function*

$$F(x_1, \ldots, x_n) = P(X_1 < x_1, \ldots, X_n < x_n) \tag{25.11}$$

defined for $(x_1, \ldots, x_n) \in R^n$ *is called the* n-dimensional distribution function *of the random vector* $\boldsymbol{X}$.

Let (X_1, X_2) be a two-dimensional random vector whose distribution function $F(x_1, x_2)$ is continuously differentiable and has continuous mixed partial derivative of second order. Such a distribution function enjoys the following main properties

$$0 \leqslant F(x_1, x_2) \leqslant 1, \tag{25.12}$$

$$F(b, x_2) \geqslant F(a, x_2), \qquad F(x_1, b) \geqslant F(x_1, a) \text{ for any } a, b \in R, \ a < b, \tag{25.13}$$

$$\lim_{x_1 \to -\infty} F(x_1, x_2) = \lim_{x_2 \to -\infty} F(x_1, x_2) = \lim_{\substack{x_1 \to -\infty \\ x_2 \to -\infty}} F(x_1, x_2) = 0, \tag{25.14}$$

$$\lim_{\substack{x_1 \to \infty \\ x_2 \to \infty}} F(x_1, x_2) = 1. \tag{25.15}$$

The (one-dimensional) distribution functions of the random variables X_1, X_2, which are also called *marginal distribution functions* of the random vector $\boldsymbol{X}$, are given by

$$F_1(x_1) = \lim_{x_2 \to \infty} F(x_1, x_2), \qquad F_2(x_2) = \lim_{x_1 \to \infty} F(x_1, x_2). \tag{25.16}$$

Obviously, the properties (25.12)—(25.15) generalize the analogous properties of the one-dimensional distribution functions and may be proved in a similar way. It may be also shown that

$$P(a \leqslant X_1 < b, X_2 < x_2) = F(b, x_2) - F(a, x_2), \tag{25.17}$$

$$P(X_1 < x_1, a \leqslant X_2 < b) = F(x_1, b) - F(x_1, a), \tag{25.18}$$

$$P(a_1 \leqslant X_1 < b_1, a_2 \leqslant X_2 < b_2) = F(b_1, b_2) - F(a_1, b_2) - \\ - [F(b_1, a_2) - F(a_1, a_2)]. \tag{25.19}$$

Definition 11. *If the function*

$$f(x_1, \ldots, x_n) = \frac{\partial^n F(x_1, \ldots, x_n)}{\partial x_1 \ldots \partial x_n} \tag{25.20}$$

exists and is continuous it is called the n-dimensional probability density function *of the random vector* $\mathbf{X}(X_1, \ldots, X_n)$.

Let (X_1, X_2) be a two-dimensional random vector. According to (25.20) its two-dimensional probability density function is

$$f(x_1, x_2) = \frac{\partial^2 F(x_1, x_2)}{\partial x_1 \partial x_2}.$$

By taking into account Eqs. (25.14), this relation may be inverted to obtain

$$F(x_1, x_2) = \int_{-\infty}^{x_1} \int_{-\infty}^{x_2} f(u, v) \, du \, dv. \tag{25.21}$$

The main properties of the function $f(x_1, x_2)$ are similar to those of the one-dimensional probability density functions. Thus

$$f(x_1, x_2) \geqslant 0, \tag{25.22}$$

$$\int_{-\infty}^{\infty}\int_{-\infty}^{\infty} f(x_1, x_2)\, dx_1\, dx_2 = 1, \tag{25.23}$$

$$P(\mathbf{X} \in D) = \iint_D f(x_1, x_2)\, dx_1\, dx_2 \quad \text{for any} \quad D \subset R^2, \tag{25.24}$$

$$P(x_1 \leqslant X_1 < x_1 + \Delta x_1,\ x_2 \leqslant X_2 < x_2 + \Delta x_2) \approx f(x_1, x_2)\, \Delta x_1 \Delta x_2$$

$$\text{for } \Delta x_1 \text{ and } \Delta x_2 \text{ sufficiently small.} \tag{25.25}$$

The probability density functions of the random variables X_1 and X_2, which are also called the *marginal probability density functions* of the random vector X, are given by

$$f_1(x_1) = \int_{-\infty}^{\infty} f(x_1, x_2)\, dx_2, \qquad f_2(x_2) = \int_{-\infty}^{\infty} f(x_1, x_2)\, dx_1. \tag{25.26}$$

Two events are said to be *independent* if the occurrence or nonoccurrence of one of them has no influence on the occurrence or nonoccurrence of the other. It may be shown[1] that two events A and B are independent if and only if the probability of their simultaneous occurrence equals the product of the probabilities of their separate occurrences, i.e.,

$$P(A \cap B) = P(A)\, P(B).$$

According to this relation we have:

Definition 12. *The components* X_1, X_2 *of a two-dimensional random vector are said to be* independent *if*

$$P(X_1 < x_1,\ X_2 < x_2) = P(X_1 < x_1)\, P(X_2 < x_2) \tag{25.27}$$

for any $(x_1, x_2) \in R^2$.

By (25.11), the definition (25.27) may be immediately rewritten in terms of distribution functions to obtain

$$F(x_1, x_2) = F_1(x_1)\, F_2(x_2). \tag{25.28}$$

The following theorem gives a simple necessary and sufficient condition for the independence of the components of two-dimensional random vectors in terms of probability density functions.

1) See, e.g., Rényi [162].

Theorem 6. *The components X_1, X_2 of a two-dimensional random vector are independent if and only if*

$$f(x_1, x_2) = f_1(x_1) f_2(x_2). \tag{25.29}$$

Proof. Assume first that X_1 and X_2 are independent, and hence Eq. (25.28) holds. Differentiating (25.28) with respect to x_1 and x_2 gives

$$\frac{\partial^2 F}{\partial x_1 \partial x_2} = F_1'(x_1)\, F_2'(x_2), \tag{25.30}$$

from which, by considering (25.6) and (25.20), we infer (25.29). Conversely, if Eq. (25.29) is satisfied, then integrating it yields

$$\int_{-\infty}^{x_1}\int_{-\infty}^{x_2} f(u, v)\, du\, dv = \left(\int_{-\infty}^{x_1} f_1(u)\, du\right)\left(\int_{-\infty}^{x_2} f_2(v)\, dv\right), \tag{25.31}$$

and, taking into account Eqs. (25.7) and (25.21), we obtain (25.28).

The notion of independence may be easily generalized to n-dimensional random vectors, as follows.

Definition 13. *The components $X_1, \ldots, X_n$ of an n-dimensional random vector are said to be independent if*

$$P(X_1 < x_1, \ldots, X_n < x_n) = P(X_1 < x_1) \ldots P(X_n < x_n) \tag{25.32}$$

or, equivalently, if

$$F(x_1, \ldots, x_n) = F_1(x_1) \ldots F_n(x_n). \tag{25.33}$$

Theorem 7. *The components $X_1, \ldots, X_n$ of an n-dimensional random vector are independent if and only if*

$$f(x_1, \ldots, x_n) = f_1(x_1) \ldots f_n(x_n). \tag{25.34}$$

b) Mean values. Moments

Let $X : \Omega \to R$ be a random variable and $f(x)$ its probability density function.

Definition 14. *If the Riemann integral*

$$m = E[X] = \int_{-\infty}^{\infty} x f(x)\, dx \tag{25.35}$$

is absolutely convergent, then $m = E[X]$ is called the mean value *or the* mathematical expectation *of the random variable X.*

More generally, given any continuous function $H : R \to R$, the mean value of the random variable H_X with respect to the probability density function of X is defined by

$$E[H_X] = \int_{-\infty}^{\infty} H(x) f(x)\, dx, \tag{25.36}$$

whenever this integral exists.

It is, of course, possible to define the mean value of the random variable H_X with respect to its own probability density function. However, it can be shown that these two definitions lead to the same mean value of H_X only if H satisfies certain supplementary conditions, e.g., if it is monotonic and differentiable and if $H'(x) \neq 0$ for any $x \in R$. In what follows, we shall use only the definition (25.36), for it may be easily generalized to include the multidimensional case.

In particular, from (25.36) it follows that

$$E[aX + b] = aE[X] + b \tag{25.37}$$

for any $a, b \in R$, whenever $E[X]$ exists.

Definition 15. *If the mean value of the random variable X^r, i.e.,*

$$m_r = E[X^r] = \int_{-\infty}^{\infty} x^r f(x)\, dx \tag{25.38}$$

exists, then m_r is called the rth moment *of the random variable X.*

The second moment, m_2, is usually termed the *mean square value* of the random variable X.

Definition 16. *If the mean value of the random variable $|X|^r$, i.e.,*

$$\overline{m}_r = E[|X|^r] = \int_{-\infty}^{\infty} |x|^r f(x)\, dx \tag{25.39}$$

exists, then $\overline{m}_r$ is said to be the absolute r^th^ moment *of the random variable X.*

Definition 17. *If the mean value of the random variable $(X - m)^r$, i.e.,*

$$\mu_r = E[(X - m)^r] = \int_{-\infty}^{\infty} (x - m)^r f(x)\, dx \tag{25.40}$$

exists, then μ_r is called the rth central moment *of the random variable X.*

The second central moment, which is often of particular interest, is commonly denoted by

$$\sigma^2 = \mu_2 = E[(X - m)^2] \tag{25.41}$$

and given the special name *variance* or *dispersion*. The square root of the variance is known as the *standard deviation* of the random variable X. If $m = 0$, the dispersion obviously reduces to the mean square value.

Equations (25.38) and (25.40) allow the calculation of any moment or central moment associated with a particular probability density function. Conversely, if it is known before-hand that the probability density function has a definite form depending on a few parameters, and only the values of the parameters are unknown, then they may be determined in the most important cases by the first moments. But if the type of probability density function is not known, then even the knowledge of all moments may fail to determine the probability density function. However, it can be proved that a given sequence of constants $m_1, m_2, \ldots$ uniquely determines a probability density function $f(x)$ such that

$$m_r = \int_{-\infty}^{\infty} x^r f(x)\,\mathrm{d}x,$$

provided that the series $\sum_{r=1}^{\infty} \frac{m_r}{r!} \alpha^r$ is absolutely convergent for some $\alpha > 0$.

We content ourselves here to state without proof two theorems which establish relations between the distribution function of a random variable X and some of its moments.

Theorem 8. *If m and σ exist and λ is any real number greater than* 1, *then*

$$P(|X - m| \geqslant \lambda\sigma) \leqslant 1/\lambda^2. \tag{25.42}$$

This relation is known as CHEBYSHEV's inequality. It shows that the values of X fall outside the interval $(m - \lambda\sigma, m + \lambda\sigma)$ with a probability less than $1/\lambda^2$.

Theorem 9. *If $\overline{m}_r$ exists, then, given any real number $\varepsilon > 0$, we have*

$$P(|X| \geqslant \varepsilon) \leqslant \overline{m}_r/\varepsilon^r. \tag{25.43}$$

In the remaining part of this subsection we consider mean values and moments of multiple random variables.

Let $X_i : \Omega \to R$, $i = 1, \ldots, n$, be the components of a random vector whose n-dimensional probability density function $f(x_1, \ldots, x_n)$ is known.

Definition 18. *Given any continuous function $H : R^n \to R$, the mean value of the random variable $H_{X_1, \ldots, X_n}$ with respect to $f(x_1, \ldots, x_n)$ is defined by*

$$E[H_{X_1, \ldots, X_n}] = \int_{-\infty}^{\infty} \cdots \int_{-\infty}^{\infty} H(x_1, \ldots, x_n) f(x_1, \ldots, x_n)\,\mathrm{d}x_1 \ldots \mathrm{d}x_n, \tag{25.44}$$

whenever this integral exists.

It may be proved that the multidimensional mean value has the following properties:

1. If $X_1, \ldots, X_n$ are the components of an n-dimensional random vector, and if $E[X_i]$ exists for any $i = 1, \ldots, n$, then

$$E\left[\sum_{i=1}^{n} X_i\right] = \sum_{i=1}^{n} E[X_i]. \tag{25.45}$$

If, in addition, the components X_i are independent, then

$$E\left[\prod_{i=1}^{n} X_i\right] = \prod_{i=1}^{n} E[X_i], \tag{25.46}$$

$$E\left[\left(\sum_{i=1}^{n} X_i\right)^2\right] = \sum_{i=1}^{n} E[X_i^2] + 2 \sum_{\substack{i,j=1 \\ i<j}}^{n} E[X_i] E[X_j], \tag{25.47}$$

the last relation being valid provided $E[X_i^2]$ exist, too.

2. If X_1, X_2 are the components of a two-dimensional random vector and if the mean values $E[|X_1X_2|]$, $E[|X_1|^2]$, and $E[|X_2|^2]$ exist, then

$$|E[X_1X_2]| \leqslant E[|X_1X_2|] \leqslant \sqrt{E[|X_1|^2]\, E[|X_2|^2]}. \tag{25.48}$$

Definition 19. *The* (r + s)th order moment *of the random vector* (X_1, X_2) *is defined as the mean value of the random variable* $X_1^r X_2^s$, *i.e.*,

$$m_{rs} = E[X_1^r X_2^s] = \int_{-\infty}^{\infty} \int_{-\infty}^{\infty} x_1^r x_2^s f(x_1, x_2)\, dx_1\, dx_2, \tag{25.49}$$

provided the integral exists.

Definition 20. *The* (r + s)th order absolute moment *of the random vector* (X_1, X_2) *is defined as the mean value of the random variable* $|X_1|^r\, |X_2|^s$, *i.e.*,

$$\overline{m}_{rs} = E[|X_1|^r\, |X_2|^s] = \int_{-\infty}^{\infty} \int_{-\infty}^{\infty} |x_1|^r\, |x_2|^s f(x_1, x_2)\, dx_1\, dx_2, \tag{25.50}$$

provided the integral exists.

Definition 21. *The* (r + s)th order central moment *of the random vector* (X_1, X_2) *is defined as the mean value of the random variable* $(X_1 - m_{x_1})^r\, (X_2 - m_{x_2})^s$, *i.e.*,

$$\mu_{rs} = E[(X_1 - m_{x_1})^r\, (X_2 - m_{x_2})^s] =$$

$$= \int_{-\infty}^{\infty} \int_{-\infty}^{\infty} (x_1 - m_{x_1})^r\, (x_2 - m_{x_2})^s f(x_1, x_2)\, dx_1\, dx_2, \tag{25.51}$$

provided the integral exists.

The $(1 + 1)^{\text{th}}$ order moment,

$$K_{x_1x_2} = \mu_{11} = E[(X_1 - m_{x_1})(X_2 - m_{x_2})], \tag{25.52}$$

is called the *covariance* of X_1 and X_2, while the normalized covariance,

$$\rho_{x_1x_2} = K_{x_1x_2}/(\sigma_{x_1}\sigma_{x_2}) = \frac{E[(X_1 - m_{x_1})(X_2 - m_{x_2})]}{\sqrt{E[(X_1 - m_{x_1})^2]\, E[(X_2 - m_{x_2})^2]}} \tag{25.53}$$

is termed the *correlation coefficient* of X_1 and X_2.

The correlation coefficient has the following properties:

1. $|\rho_{x_1x_2}| \leqslant 1.$ (25.54)

This inequality immediately results from (25.48), by replacing X_1, X_2 with $X_1 - m_{x_1}$, $X_2 - m_{x_2}$, respectively, and taking into account (25.53).

2. A necessary and sufficient condition that two random variables X_1, X_2 be linearly related, i.e.,

$$X_2 = aX_1 + b, \tag{25.55}$$

with a, b real constants, $a \neq 0$, is that $|\rho| = 1$.

Proof. We confine ourselves here to prove that the condition is necessary. Taking the mean values of both sides of Eq. (25.55) yields

$$m_{x_2} = am_{x_1} + b$$

Substituting the last two relations into (25.53), we obtain

$$\rho_{x_1x_2} = \frac{E[(X_1 - m_{x_1})(aX_1 - am_{x_1})}{\sqrt{E[(X_1 - m_{x_1})^2]\, E[(aX_1 - am_{x_1})^2]}} = \frac{a}{|a|} = \operatorname{sgn} a,$$

and hence $|\rho_{x_1x_2}| = 1$. For a proof of the sufficiency see [92].

3. If the random variables X_1, X_2 are independent, then $\rho_{x_1x_2} = 0$.

Proof. For two independent random variables, X_1 and X_2, we have by Theorem 6

$$f(x_1, x_2) = f_1(x_1)\, f_2(x_2).$$

Then, making use of (25.51) for $r = s = 1$ and of (25.35), we obtain

$$E[(X_1 - m_{x_1})(X_2 - m_{x_2})] = \int_{-\infty}^{\infty} (x_1 - m_{x_1})\, f_1(x_1)\, dx_1 \int_{-\infty}^{\infty} (x_2 - m_{x_2})\, f_2(x_2)\, dx_2 = 0,$$

and hence $\rho_{x_1x_2} = 0$.

The converse of statement 3 is generally not true. However, as we shall see in the next subsection, when the random vector (X_1, X_2) is normally distributed, then $\rho_{x_1x_2} = 0$ implies the independence of X_1 and X_2.

Using the definition (25.44) of the mean value of a function of random variables, we may define a new type of convergence for sequences of random variables.

Definition 22. *The sequence* $\{X_n\}$ *of random variables* $X_n: \Omega \to R$ *is said to* converge in r-mean *(r a positive integer) to the random variable* $X: \Omega \to R$ *if*

$$\lim_{n\to\infty} E[|X - X_n|^r] = 0.$$

In the theory of random vibrations the most used convergence of this type is the *convergence in mean square* ($r = 2$).

c) Normal distribution

A problem frequently occurring in practice is to evaluate the distribution functions of a set of random variables defined on the sample space Ω from some statistical information obtained by experiments. To this aim it is convenient to use standard distribution functions depending on a certain number of parameters, which may approximate the real distribution functions. One of these standard distributions, which play an important role in various applications, is the normal distribution.

Definition 23. *The random variable X is said to have a* normal distribution *with parameters* α, β, *if its probability density function has the form*

$$f(x) = \frac{1}{\alpha\sqrt{2\pi}} e^{-(x-\beta)^2/2\alpha^2}. \tag{25.56}$$

Let us show that the function (25.56) actually satisfies the conditions $(25.8)_1$ and $(25.8)_2$ to be fulfilled by any probability density function. Obviously, $f(x) > 0$. To prove that $(25.8)_2$ is also satisfied, we consider the integral

$$I = \frac{1}{\alpha\sqrt{2\pi}} \int_{-\infty}^{\infty} e^{-(x-\beta)^2/2\alpha^2}\, dx \tag{25.57}$$

and apply the transformation

$$t = (x - \beta)/\alpha\sqrt{2}. \tag{25.58}$$

The result reads

$$I = \frac{2}{\sqrt{\pi}} \int_0^{\infty} e^{-t^2}\, dt = \frac{1}{\sqrt{\pi}} \Gamma(1/2) = 1. \tag{25.59}$$

The significance of the parameters α and β, which characterize a normal distribution, is given by the following theorem.

Theorem 10. *If a random variable X has a normal distribution with parameters α and β, then $\alpha = \sigma$ and $\beta = m$.*

Proof. Let us first show that $\beta = m$. Introducing (25.56) into (25.35) gives

$$m = \frac{1}{\alpha\sqrt{2\pi}}\int_{-\infty}^{\infty} x\,e^{-(x-\beta)^2/2\alpha^2}\,dx.$$

Applying now the transformation (25.58) and integrating by parts, we obtain

$$m = \beta\frac{2}{\sqrt{\pi}}\int_0^{\infty} e^{-t^2}\,dt + \alpha\sqrt{\frac{2}{\pi}}\int_{-\infty}^{\infty} t\,e^{-t^2}\,dt.$$

The first integral in the right-hand side gives as above $\sqrt{\pi}/2$, and the second one vanishes, for the integrand is an odd function of t. Hence $\beta = m$.

To prove now that $\alpha = \sigma$, we use Eq. (25.40) with $r = 2$. Replacing $f(x)$ by (25.56), we have

$$\sigma^2 = \frac{1}{\alpha\sqrt{2\pi}}\int_{-\infty}^{\infty} (x-m)^2\,e^{-(x-m)^2/2\alpha^2}\,dx.$$

The same transformation (25.58) now leads to

$$\sigma^2 = \frac{4\alpha^2}{\sqrt{\pi}}\int_0^{\infty} t^2 e^{-t^2}\,dt.$$

Integrating by parts and considering (25.59), we finally obtain $\sigma^2 = \alpha^2$, which completes the proof.

The theorem above shows that the probability density function of a random variable with normal distribution is fully determined by the mean value and the dispersion of this variable, namely

$$f(x) = \frac{1}{\sigma\sqrt{2\pi}}\,e^{-(x-m)^2/2\sigma^2}. \tag{25.60}$$

Theorem 11. *The odd central moments of a normally distributed random variable are zero, and the even ones are given by*

$$\mu_{2p} = \frac{(2p)!}{p!\,2^p}\,\sigma^{2p} \tag{25.61}$$

for any positive integer p.

Proof. Substituting (25.60) into (25.40) yields

$$\mu_r = \frac{1}{\sigma\sqrt{2\pi}} \int_{-\infty}^{\infty} (x-m)^r \, e^{-(x-m)^2/2\sigma^2} \, dx$$

for any positive integer r. By applying again the transformation (25.58), where now $\beta = m$, $\alpha = \sigma$, it results that

$$\mu_r = \frac{\sigma^r \sqrt{2^r}}{\sqrt{\pi}} \int_{-\infty}^{\infty} t^r \, e^{-t^2} \, dt. \tag{25.62}$$

It is apparent that the integral in the right-hand side vanishes for odd r. Next, when $r = 2p$, integrating (25.62) by parts leads to the recursive formula

$$\mu_{2p} = (2p-1)\, \sigma^2 \mu_{2p-2}.$$

Repeated use of this relation gives

$$\mu_{2p} = 1.3.5 \ldots (2r-1)\, \sigma^{2p},$$

which is obviously equivalent to (25.61).

Finally, we consider the distribution function of a normally distributed random variable. To this aim we recall the definition and some of the properties of the Laplace integral function:

$$\Phi(u) = \frac{1}{\sqrt{2\pi}} \int_0^u e^{-t^2/2} \, dt, \tag{25.63}$$

which is defined for any real u. Obviously,

$$\Phi(-u) = -\Phi(u) \tag{25.64}$$

for any $u \in R$, and

$$\Phi(u_1) < \Phi(u_2) \tag{25.65}$$

for any $u_1, u_2 \in R$, $u_1 < u_2$.

Theorem 12. *The distribution function of a normally distributed random variable is given by*

$$F(x) = \frac{1}{2} + \Phi\left(\frac{x-m}{\sigma}\right) \tag{25.66}$$

Proof. Introducing (25.60) into (25.7) yields

$$F(x) = \frac{1}{\sigma\sqrt{2\pi}} \int_{-\infty}^{x} e^{-(t-m)^2/2\sigma^2} \, dt.$$

The transformation $(t - m)/\sigma = u$ leads to

$$F(x) = \frac{1}{\sqrt{2\pi}} \int_{-\infty}^{0} e^{-u^2/2}\, du - \frac{1}{\sqrt{2\pi}} \int_{0}^{\frac{x-m}{\sigma}} e^{-u^2/2}\, du,$$

which, in view of (25.59) and (25.63), reduces to (25.66).

Equation (25.66) enables us to express in a simple form the probabilities of various typical events related to a normally distributed random variable, in terms of the Laplace integral function. Thus

$$P(a < X < b) = \Phi\left(\frac{b-m}{\sigma}\right) - \Phi\left(\frac{a-m}{\sigma}\right), \tag{25.67}$$

as it can be readily seen from $(25.2)_3$ and (25.66). If $m = 0$, we also have for any $\varepsilon > 0$

$$P(|X| < \varepsilon) = 2\Phi(\varepsilon/\sigma). \tag{25.68}$$

This relation immediately results from (25.67), by setting $-a = b = \varepsilon$, and considering (25.64).

To illustrate the possible application of Eq. (25.68), let us calculate the probability that X falls in the interval $(-k\sigma, k\sigma)$, where k is a positive integer. Setting $\varepsilon = k\sigma$, we derive

$$P(|X| < k\sigma) = 2\Phi(k). \tag{25.69}$$

In particular,

$$P(|X| < \sigma) = 2\Phi(1) = 0.6826,$$
$$P(|X| < 2\sigma) = 2\Phi(2) = 0.9544,$$
$$P(|X| < 3\sigma) = 2\Phi(3) = 0.9972,$$

The last relation shows that X takes on values outside the interval $(-3\sigma, 3\sigma)$ with a probability less than $3^0/_{00}$.

The great practical importance of the normal distribution derives from the fact that many distributions that are encountered in the physical world are either normal or approximately normal. This result is typical when we are dealing with a large number of observations, each of which contributes only a small amount to the outcome of an experiment. The mathematical justification of this fact is given by the so-called *central-limit* theorem, which will be briefly discussed in what follows.

Let $X_1, \ldots, X_n$ be n mutually independent random variables whose individual distributions are not specified. Denote by m_i and σ_i^2 the mean value and the variance of X_i, $i = 1, \ldots, n$, which are supposed to exist.

The central-limit theorem states that, under certain conditions, the distribution function of the sum $\sum_{i=1}^{n} X_n$ approaches a normal distribution as $n \to \infty$. For example, this is always the case when the random variables X_i are identically distributed. More generally, if the third central moments μ_{3i} of X_i exist for $i = 1, \ldots, n$, another sufficient condition for the truth of the central-limit theorem is

$$\lim_{n\to\infty} \left(\sum_{i=1}^{n} \mu_{3i} \right) \Big/ \left(\sum_{i=1}^{n} \sigma_i^2 \right)^{3/2} = 0.$$

For the proof of these and other more general statements we refer to RÉNYI [169].

The notion of normal distribution may be easily extended to n-dimensional random vectors. For the sake of simplicity, we content ourselves here with considering the case $n = 2$.

Definition 24. *The random vector* (X_1, X_2) *is said to have a* normal distribution *if its two-dimensional probability density function has the form*

$$f(x_1, x_2) = \frac{1}{2\pi \sqrt{\sigma_{x_1}^2 \sigma_{x_2}^2 - K_{x_1x_2}^2}} \times$$

$$\times \exp \left[- \frac{\sigma_{x_2}^2 (x_1 - m_{x_1})^2 - 2K_{x_1x_2}(x_1 - m_{x_1})(x_2 - m_{x_2}) + \sigma_{x_1}^2 (x_2 - m_{x_2})^2}{2(\sigma_{x_1}^2 \sigma_{x_2}^2 - K_{x_1x_2}^2)} \right]. \quad (25.70)$$

By using Eqs. (25.26) and (25.70), it may be shown that the marginal probability density functions of the random vector (X_1, X_2) are

$$f_i(x_i) = \frac{1}{\sigma_{x_i} \sqrt{2\pi}} \exp \left[-(x_i - m_{x_i})^2 / 2\sigma_{x_i}^2 \right] \qquad i = 1, 2, \qquad (25.71)$$

and hence the random variables X_1, X_2 are also normally distributed.

If the correlation coefficient $\rho_{x_1x_2}$ vanishes, then $K_{x_1x_2}$ vanishes, too, and from (25.70) and (25.71) it follows that

$$f(x_1, x_2) = f_1(x_1) f_2(x_2),$$

hence X_1 and X_2 are independent. As already mentioned above, this result shows that the vanishing of the correlation coefficient is a necessary and sufficient condition for the independence of the components of a normally distributed random vector.

We close this section by two remarks concerning some of the concepts discussed above.

1. Random variables considered so far were real-valued functions defined on the sample space. This restriction is by no means necessary. One may easily introduce a complex-valued random variable $Z : \Omega \to C$, by setting $Z = X + iY$, where $X : \Omega \to R$, $Y : \Omega \to R$ are two real-valued random variables. Such a representa-

tion proves sometimes convenient. We mention here only that the mean value and the dispersion of a complex-valued random variable $Z = X + iY$ are

$$m_z = E[Z] = \int_{-\infty}^{\infty}\int_{-\infty}^{\infty} (x + iy) f(x, y)\, dx\, dy = m_x + im_y, \tag{25.72}$$

$$\sigma_z^2 = E[|Z - m_z|^2] = \sigma_x^2 + \sigma_y^2, \tag{25.73}$$

and that the covariance of two complex-valued random variables, Z_1 and Z_2, is

$$K_{z_1 z_2} = E[(Z_1 - m_{z_1})(\overline{Z_2 - m_{z_2}})]. \tag{25.74}$$

These definitions obviously reduce to (25.35), (25.41), and (25.52), respectively, when Z, Z_1, and Z_2 are real random variables.

2. We have defined above several types of convergence for sequences of random variables, namely: convergence almost certainly (Definition 5), convergence in probability (Definition 6), and convergence in r-mean (Definition 22). It may be proved [76] that both convergence almost certainly and convergence in r-mean imply convergence in probability (the converse is not true). Neither convergence almost certainly nor convergence in r-mean implies the other.

§ 26. RANDOM FUNCTIONS

a) Statistical characteristics of random functions

Before introducing the notion of random function, let us examine some examples of physical random phenomena which may give this concept a more concrete support.[1)]

Suppose that one records the height of ocean waves over a fixed reference horizontal plane during a day, at the same place, and that this experiment is repeated n days in rather similar weather conditions. Denoting by $X_r(t)$ the height function recorded on the rth day, one observes that the functions $X_1(t), \ldots, X_n(t)$ are generally different. There are several sources of these differences, such as the variable frequency and intensity of the wind gusts, small changes in the weather situation, and so on.

A similar situation is encountered when recording the pressure variation with time at a certain place on an aircraft structure during the flight on a given route, and successively repeating this experiment.

1) Other significant examples of random phenomena of mechanical origin are discussed, e.g., by PORITSKY [150].

Another example of the same type is provided by the measurement of the surface irregularities of n different segments of the same road. The dependence of the height of the road surface over some fixed horizontal plane on the distance from the origin of the road segment will be generally different from one segment to another.

The main common feature of the above examples is an indeterminancy in the expected behavior of any single record, coupled with statistical properties of large collections of records. To each record one associates a deterministic function called sample function, depending on a parameter t (e.g., time, distance, etc.), which varies continuously within a certain interval $T = (a, b)$.

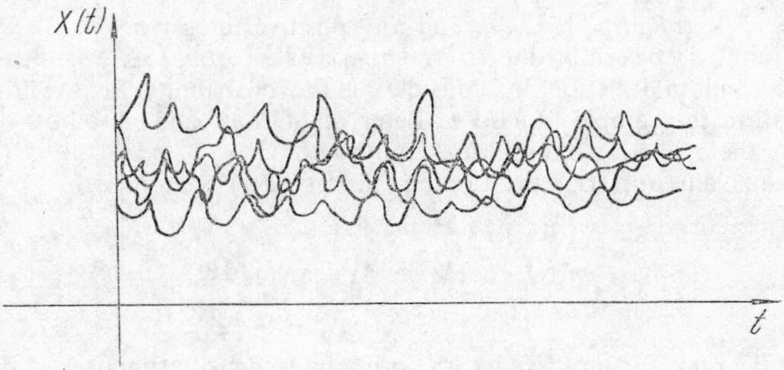

Fig. 75

The physical random phenomenon is fully characterized by the collection of all possible sample functions, which is denoted by $X(t)$ and commonly termed *random function*.[1] Such a function may be graphically illustrated by a family of curves, as suggested by Fig. 75. It is readily seen that the set of the values taken on by all sample functions for a fixed value of the parameter t defines a random variable.

Having intuitively introduced the concept of random function, we are now prepared to rigorously define it. Let (Ω, K, P) be a probability field, $\mathfrak{S}$ the set of random variables defined on this field, and T an interval of R, which may eventually be the whole real axis.

Definition 1. *A function* $X : T \to \mathfrak{S}$ *is said to be a* random function *and is denoted by* $X(t)$, $t \in T$. *When t is physical time, the random function is said to be a* random process.

According to the definition above, a random function associates to any $t \in T$ a random variable $X(t) \in \mathfrak{S}$, and hence is a one-parameter family of random variables. We may also interpret a random function as a function $X(\omega, t)$ defined on the set $\Omega \times T$, which for any fixed $\omega \in \Omega$ is a sample function, and for any fixed $t \in T$ is a random variable.

[1] At the end of this section we shall return to the problem of statistically characterizing a random function by processing of the collection of records.

It may be proved (see, e.g., IOSIFESCU [76]) that a random function $X(t)$ defined on a probability field (Ω, K, P) is completely determined by knowledge of the distribution functions

$$F_{t_1 \ldots t_n}(x_1, \ldots, x_n) = P(X(t_1) < x_1, \ldots, X(t_n) < x_n), \tag{26.1}$$

or of the probability density functions

$$f_{t_1 \ldots t_n}(x_1, \ldots x_n) = \frac{\partial^n F_{t_1 \ldots t_n}(x_1, \ldots, x_n)}{\partial x_1 \ldots \partial x_n} \tag{26.2}$$

for any $t_i \in T$, $x_i \in R$, $i = 1, \ldots, n$, and any positive integer n.

Definition 1 is basically due to WIENER [192]. DOOB [52, 53] introduced the concept of random function by considering the probability field whose elementary events are the sample functions. Later on, DOOB and AMBROSE [54] have shown that these two definitions are equivalent.

The mean value $m_x(t)$ of the random process $X(t)$ is defined by

$$m_x(t) = E[X(t)] = \int_{-\infty}^{\infty} x f_t(x) \mathrm{d}x, \tag{26.3}$$

where $f_t(x)$ denotes, for any fixed t, the probability density function of the random variable $X(t)$.

By making use of the n-dimensional probability density functions (26.2), one may calculate various moments of the random vector $(X(t_1), \ldots X(t_n))$ corresponding to any values $t_1, \ldots, t_n \in T$. Denoting $X(t_i) = X_i$, we have for instance

$$E[X_1^{r_1} \ldots X_n^{r_n}] = \int_{-\infty}^{\infty} \ldots \int_{-\infty}^{\infty} x_1^{r_1} \ldots x_n^{r_n} f_{t_1 \ldots t_n}(x_1, \ldots, x_n) \, \mathrm{d}x_1 \ldots \mathrm{d}x_n, \tag{26.4}$$

$$E[(X_1 - m_{x_1})^{r_1} \ldots (X_n - m_{x_n})^{r_n}] = \int_{-\infty}^{\infty} \ldots \int_{-\infty}^{\infty} (x_1 - m_{x_1})^{r_1} \ldots (x_n - m_{x_n})^{r_n} \times$$
$$\times f_{t_1 \ldots t_n}(x_1, \ldots, x_n) \, \mathrm{d}x_1 \ldots \mathrm{d}x_n, \tag{26.5}$$

where $m_{x_i} = m_x(t_i)$, $i = 1, \ldots, n$.

Definition 2. *The* $(1 + 1)^{\text{th}}$ *central moment which results by setting* $n = 2$, $r_1 = r_2 = 1$ *in* (26.5) *is called the* correlation function *of the random function* $X(t)$ *and is denoted by* $K_x(t_1, t_2)$.

Hence

$$K_x(t_1, t_2) = E[(X_1 - m_{x_1})(X_2 - m_{x_2})]$$
$$= \int_{-\infty}^{\infty} \int_{-\infty}^{\infty} (x_1 - m_{x_1})(x_2 - m_{x_2}) f_{t_1 t_2}(x_1, x_2) \, \mathrm{d}x_1 \, \mathrm{d}x_2. \tag{26.6}$$

In the case of a complex-valued random function, say $Z(t)$, definition (26.6) is to be replaced by

$$K_z(t_1, t_2) = E[(Z_1 - m_{z_1}) \overline{(Z_2 - m_{z_2})}], \tag{26.7}$$

where $Z_i = Z(t_i)$.

The theory based solely on consideration of the second moments is sometimes called the *correlation theory*. Since the random functions admitting first and second moments are said to be *random functions of second order*, the properties which result from the correlation theory are also called *properties to the second order*.

The correlation function of a complex-valued random function has the following properties:

$$\sigma_z^2(t) = K_z(t, t) \geqslant 0, \tag{26.8}$$

$$K_z(t_2, t_1) = \overline{K_z(t_1, t_2)}. \tag{26.9}$$

Equation (26.8) immediately follows from (26.7) and (25.72). From (26.7) we also see that

$$\overline{K_z(t_1, t_2)} = \overline{E[(Z_1 - m_{z_1}) \overline{(Z_2 - m_{z_2})}]} =$$

$$= E[\overline{(Z_1 - m_{z_1})} (Z_2 - m_{z_2})] = K_z(t_2, t_1).$$

For real random variables we directly derive from (26.9):

$$K_x(t_2, t_1) = K_x(t_1, t_2). \tag{26.10}$$

The fundamental algebraic operations with random functions are defined as in the classical theory of real and complex-valued functions.

The notions of continuity, differentiability, and integrability are introduced according to the type of convergence adopted, as follows.

Definition 3. *A random function $X(t)$: $T \to \mathscr{E}$ is said to be* continuous in probability (or almost certainly, *or* in r-mean) *at the point $t_0 \in T$ if the sequence of random variables $X_n(t_n)$ converges in probability (or almost certainly, or in r-mean) to the random variable $X(t_0)$ for any sequence $\{t_n\}$, $t_n \in T$, $n = 1, 2, \ldots$ that converges to t_0.*

Definition 4. *A random function $X(t) : T \to \mathscr{E}$ is said to be* differentiable in probability (*or* almost certainly, *or* in r-mean) *at the point $t = t_0 \in T$, and to have the derivative $\dot{X}(t_0)$ at this point if the sequence of random variables*

$$\left\{ \frac{X(t_n) - X(t_0)}{t_n - t_0} \right\} \tag{26.11}$$

converges in probability (or almost certainly, or in r-mean) to the random variable $X(t_0)$ *for any sequence* $\{t_n\}$, $t_n \in T$, $n = 1, 2, \ldots$ *that converges to* t_0.

Derivatives of random functions are also denoted by

$$\dot{X}(t_0) = \left.\frac{\mathrm{d}X(t)}{\mathrm{d}t}\right|_{t=t_0}. \tag{26.12}$$

indicating within parentheses the type of convergence used.

Definition 5. *A random function* $X(t)$ *is said to be* continuous (*or* differentiable) in an interval *if it is continuous (or differentiable) at each point of the interval.* This definition is meant for each type of convergence.

The Riemann integral of a random function is defined as in the case of deterministic functions. Let $X(t) : T \to \mathscr{E}$ be a random function and let

$$a = t_0 < \ldots < t_i \ldots < t_n = b \tag{26.13}$$

be a partition of the interval $[a, b] \subset T$. The sum

$$I_n = \sum_{i=1}^{n-1} X(\tau_i)(t_{i+1} - t_i) \tag{26.14}$$

is a random variable for any choice of the partition (26.13) and for any $\tau_i \in [t_i, t_{i+1}]$.

Definition 6. *A random function* $X(t)$ *is said to be* Riemann integrable in probability (*or* almost certainly, *or* in r-mean) *in the interval* $[a, b]$ *if there exists a random variable I such that the sequence of random variable* $\{I_n\}$ *converges in probability (or almost certainly, or in r-mean) to I as* $n \to \infty$ *and* $\max_{i=1,\ldots,n-1} |t_{i+1} - t_i| \to 0$.

The integral of a random variable will be denoted by the usual symbol

$$\mathrm{I} = \int_a^b X(t)\,\mathrm{d}t, \tag{26.15}$$

indicating within parentheses the type of convergence employed.

If one of the limits of integration is variable, the integral (26.15) becomes a random function, e.g.,

$$I(t) = \int_a^t X(s)\,\mathrm{d}s. \tag{26.16}$$

Remarks

1. According to the second remark at the end of § 25, the continuity, differentiability, and integrability in r-mean or almost certain imply the same properties in probability. The converse statement is not true. In general there is no relation between almost certain continuity, differentiability, and integrability and the same properties in r-mean.

2. As already mentioned, a random function $X(\omega, t)$ becomes for any fixed $\omega \in \Omega$ a deterministic function called a sample function. Therefore, it is possible to reduce the notions of continuity, differentiability, and integrability of a random function to the corresponding notions for deterministic functions, by making use of the probability field whose elementary events are the sample functions. More precisely, according to Definition 3, § 25, we may say that a random function $X : \Omega \times T \rightarrow R$ is continuous at $t_0 \in T$ (continuous in T, differentiable at $t_0 \in T$, differentiable in T, Riemann integrable in $[a, b] \subset T$) if all sample functions are continuous at $t_0 \in T$ (continuous in T, differentiable at $t_0 \in T$, differentiable in T, Riemann integrable in $[a, b] \subset T$) except, possibly, on a set of sample functions of probability zero.

The following four theorems may be proved to hold for random functions of second order (see, e.g., [57]).

Theorem 1. *The random function $X(t)$ is continuous in mean square at $t_0 \in T$ if and only if its mean value $m_x(t)$ is continuous at t_0 and its correlation function $K_x(t_1, t_2)$ is continuous at $(t_0, t_0) \in T \times T$.*

Theorem 2. *The random function $X(t)$ is differentiable in mean square at $t_0 \in T$ if and only if its mean value $m_x(t)$ is differentiable at t_0 and if the mixed second order partial derivative of its correlation function $K_x(t_1, t_2)$ exists and is finite at $(t_0, t_0) \in T \times T$.*

Theorem 3. *If $X(t)$ is differentiable in T, and if $\dot{X}(t)$ denotes its first derivative, then*

$$m_{\dot{x}}(t) = \frac{\mathrm{d}}{\mathrm{d}t} m_x(t), \tag{26.17}$$

$$K_{\dot{x}}(t_1, t_2) = \frac{\partial^2 K_x(t_1, t_2)}{\partial t_1 \partial t_2} \tag{26.18}$$

for any t, t_1, $t_2 \in T$, irrespective of the type of differentiability adopted.

Theorem 4. *The random function $X(t)$ is Riemann integrable in mean square in $[a, b] \subset T$ if and only if $m_x(t)$ is Riemann integrable in $[a, b]$ and $K_x(t_1, t_2)$ is Riemann integrable in $[a, b] \times [a, b]$.*

Theorem 5. *If the random function $X(t)$ is Riemann integrable in mean square in $[a, b] \subset T$, then*

$$m_y(t) = \int_{t_0}^{t} m_x(s)\, \mathrm{d}s, \tag{26.19}$$

$$K_y(t_1, t_2) = \int_{t_0}^{t_1} \int_{t_0}^{t_2} K_x(s_1, s_2)\, \mathrm{d}s_1\, \mathrm{d}s_2, \tag{26.20}$$

where $Y(t)$ is the random function defined as

$$Y(t) = \int_{t_0}^{t} X(s)\, ds. \tag{26.21}$$

Equations (26.19) and (26.20) hold regardless of the type of integrability adopted.

b) Stationary random processes

When the conditions under which an experiment is performed are essentially unchanging, it is to be expected that the statistical characteristics of the random process generated by the experiment will be little altered by a translation in time. This leads to the concept of stationary random process.

Definition 7. *A random process $X(t)$ is said to be* strongly stationary *or* stationary in the strict sense *if*

$$F_{t_1+\tau,\ldots,t_n+\tau}(x_1,\ldots,x_n) = F_{t_1\ldots t_n}(x_1,\ldots,x_n) \tag{26.22}$$

for any $t_1,\ldots,t_n \in T$, any $\tau \in R$ such that $t_1+\tau,\ldots,t_n+\tau \in T$, and for any positive integer n.

It is easily seen from this definition that each distribution function of a strongly stationary random process depends upon the time instants involved only through their differences. Indeed, by choosing $\tau = -t_1$ in (26.22), we obtain

$$F_{t_1\ldots t_n}(x_1,\ldots,x_n) = F_{0,t_2-t_1,\ldots,t_n-t_1}(x_1,\ldots,x_n). \tag{26.23}$$

By simply differentiating (26.23) and considering (26.2), we deduce that

$$f_{t_1\ldots t_n}(x_1,\ldots,x_n) = f_{0,t_2-t_1,\ldots,t_n-t_1}(x_1,\ldots,x_n). \tag{26.24}$$

In particular, setting $n = 1$ and then $n = 2$ in Eq. (26.23) yields

$$F_{t_1}(x_1) = F_0(x_1), \tag{26.24'}$$

$$F_{t_1t_2}(x_1, x_2) = F_{0,\,t_2-t_1}(x_1, x_2). \tag{26.25}$$

These relations show that the one-dimensional distribution function of a strongly stationary random process is time-independent and its two-dimensional distribution function depends on the times t_1 and t_2 only through their difference. In what follows we shall constantly assume that the random process considered is at least of second order, i.e., at least the first and second moments are finite. An immediate consequence of Eqs. (26.24′) and (26.25) is that the mean value of a second order strongly stationary random process is constant, while its correlation function depends only on the difference of its arguments, i.e.,

$$m_x(t) = \text{const}, \tag{26.26}$$

$$K_x(t_1, t_2) = k_x(t_2 - t_1) = k_x(\tau), \tag{26.27}$$

where $\tau = t_2 - t_1$.

Definition 8. *A random process $X(t)$ is said to be* weakly stationary *or* stationary in the wide sense *if the conditions (26.26) and (26.27) are fulfilled.*

A strongly stationary random process is obviously also weakly stationary; the converse is generally not true. However, for normally distributed random processes [1] weak stationarity implies strong stationarity, too, since for such processes all probability distribution functions depend only on the first and second moments. Moreover, since usually only the mean values and the correlation functions are used in the analysis of random processes, verification of weak stationarity will justify an assumption of strong stationarity for many practical applications.

From (26.8), (26.9), and (26.27) it follows for a complex-valued weakly stationary random process that

$$\sigma_z^2(t) = K_z(t, t) = k_z(0) = \text{const} \tag{26.28}$$

$$k_z(-\tau) = \overline{k_z(\tau)}. \tag{26.29}$$

Moreover, since by (25.33) and (25.54) we have

$$|K_z(t_1, t_2)| \leqslant \sigma_z(t_1)\, \sigma_z(t_2),$$

it results that

$$|k_z(\tau)| = |K_z(t_1, t_2)| \leqslant \sigma_z^2 = k_z(0). \tag{26.30}$$

Hence, the modulus of the correlation function has an absolute maximum for $\tau = 0$.

Theorem 6. *A real-valued weakly stationary random process is continuous in mean square if and only if its correlation function $k_x(\tau)$ is continuous at $\tau = 0$.*

This theorem is an immediate consequence of theorem 1, since for a weakly stationary random process $K_x(t_0, t_0) = k_x(0)$ for any $t_0 \in T$.

It is apparent from Eqs. (26.17) and (26.18) that the derivative in mean square of a weakly stationary random process is weakly stationary, too. Moreover, by taking also into account Eqs. (26.26) and (26.27), we deduce that

$$m_{\dot{x}}(t) = 0, \tag{26.31}$$

$$k_{\dot{x}}(\tau) = \frac{\partial^2 k_x(t_1 - t_2)}{\partial t_1 \partial t_2} = -\frac{\mathrm{d}^2 k_x(\tau)}{\mathrm{d}\tau^2}. \tag{26.32}$$

[1] A random function is said to be normally distributed if all probability density functions (26.2) are normal, according to the n-dimensional generalization of (25.70).

The last relation may be easily generalized for a p times differentiable, weakly stationary random process, to obtain

$$k_{x^{(p)}}(\tau) = (-1)^p \frac{d^{2p} k_x(\tau)}{d\tau^{2p}}. \tag{26.33}$$

It is easily seen from (26.19) and (26.20) that in general the integral of a weakly stationary random process is no longer weakly stationary. For instance, if $m_x(t)$ is a nonzero constant, $m_y(t)$ depends linearly on t. However, if $X(t)$ is weakly stationary, the correlation function (26.20) of its integral may be given a simpler form. Indeed, by (26.27), Eq. (26.20) becomes

$$K_y(t_1, t_2) = \int_0^{t_1} \int_0^{t_2} k_x(t'' - t')\, dt' dt''.$$

By applying the transformation

$$t' = (\xi + \tau)/2, \qquad t'' = (\xi - \tau)/2,$$

we obtain

$$K_y(t_1, t_2) = \int_0^{t_1} (t_1 - \tau)\, k_x(\tau)\, d\tau + \int_0^{t_2} (t_2 - \tau)\, k_x(\tau)\, d\tau -$$

$$- \int_0^{t_2 - t_1} (t_2 - t_1 - \tau)\, k_x(\tau)\, d\tau. \tag{26.34}$$

Setting in the last equation $t_1 = t_2 = t$ yields the variance of $Y(t)$:

$$\sigma_y^2(t) = K_y(t, t) = 2 \int_0^t (t - \tau)\, k_x(\tau)\, d\tau. \tag{26.35}$$

In analyzing the effect of a periodic excitation it is common to decompose it into its harmonic components, i.e., into the terms of its Fourier series expansion. If the excitation is not periodic it cannot be decomposed into discrete harmonic components; however, if it has a Fourier transform, this transform provides a continuous frequency spectrum. This Fourier analysis can be usefully extended to stationary random processes, as will be shown in what follows.

Let $k_x(\tau)$ be the correlation function of a weakly stationary random process. Assume that $k_x(\tau)$ and its derivative are piecewise continuous in every finite interval, and that $\int_{-\infty}^{\infty} |k_x(\tau)|\, d\tau$ exists. Then, as shown in standard mathematical books, $k_x(\tau)$ may be represented by

$$k_x(\tau) = \frac{1}{2} \int_{-\infty}^{\infty} S_x(\nu)\, e^{i\nu\tau} d\nu. \tag{26.36}$$

The integral in the right-hand side converges to $k_x(\tau)$ at the points of continuity and to its arithmetic mean at the points of discontinuity. The function $S_x(\nu)$, which is the Fourier transform of $k_x(\tau)$ to within a scale factor, and which is uniquely defined by Eq. (26.36), is called the *power spectral density* or simply the *spectral density* of the random process $X(t)$. Inverting Eq. (26.36) gives

$$S_x(\nu) = \frac{1}{\pi} \int_{-\infty}^{\infty} k_x(\tau) e^{-i\nu\tau} \, d\tau. \tag{26.37}$$

It must be pointed out that $S_x(\nu)$ may be given a definition which is independent of the correlation function, by first introducing the power spectral densities for deterministic functions, and then defining $S_x(\nu)$ as the mean value of the power spectral densities of the sample functions (see, e.g., LANING and BATTIN [95], Sect. 3.6, and BENDAT [10], Sect. 2.5). When using this approach, Eqs. (26.36) and (26.37) are to be proved, as has been done for the first time by KHINCHIN [86] in 1934. Since their connection with harmonic analysis of $X(t)$ was already implicit in earlier work by WIENER [192], these equations are called the *Wiener-Khinchin relations.*

By replacing ν with $-\nu$ and τ with $-\tau$ in Eq. (26.37), and considering (26.29), we see that

$$S_x(-\nu) = S_x(\nu). \tag{26.38}$$

Since both $k_x(\tau)$ and $S_x(\nu)$ are even functions of their arguments, the Wiener-Khinchin relations may be rewritten as

$$k_x(\tau) = \int_0^{\infty} S_x(\nu) \cos \nu\tau \, d\nu, \tag{26.39}$$

$$S_x(\nu) = \frac{2}{\pi} \int_0^{\infty} k_x(\tau) \cos \nu\tau \, d\tau. \tag{26.40}$$

Since the power spectral densities of the sample functions are by definition nonnegative, it follows that their mean value is also nonnegative, i.e.,

$$S_x(\nu) \geqslant 0. \tag{26.41}$$

For a direct proof of this property, starting from Eq. (26.36) taken as definition of $S_x(\nu)$, see DAVENPORT and ROOT [41].

Setting $\tau = 0$ into (26.39) yields

$$\sigma_x^2 = k_x(0) = \int_0^{\infty} S_x(\nu) \, d\nu. \tag{26.42}$$

This formula suggests the following physical interpretation of the spectral density. Assume that the sample records are filtered by a band-pass filter having sharp cutoff characteristics, such that the output practically consists only in the sample components with frequencies in the band $[\nu, \nu + \Delta\nu]$. Then, as shown by Eq. (26.42), the variance of the output will give the approximate value of $S_x(\nu)\,\Delta\nu$.

For differentiable, weakly stationary random processes, the spectral densities of the successive derivatives may be easily calculated from the spectral density of the process. Indeed, if $X(t)$ is a p times differentiable, weakly stationary random process, then, by Eqs. (26.33) and (26.36), we have

$$k_{x^{(p)}}(\tau) = (-1)^p \frac{\mathrm{d}^{2p} k_x(\tau)}{\mathrm{d}\tau^{2p}} = \frac{1}{2} \int_{-\infty}^{\infty} \nu^{2p} S_x(\nu)\, e^{i\nu\tau}\, \mathrm{d}\nu.$$

On the other hand, applying the definition (26.36) to the random process $X^{(p)}$ gives

$$k_{x^{(p)}}(\tau) = \frac{1}{2} \int_{-\infty}^{\infty} S_{x^{(p)}}(\nu)\, e^{i\nu\tau}\, \mathrm{d}\nu.$$

Comparing the last two relations and considering the uniqueness of the spectral density function, it follows that

$$S_{x^{(p)}}(\nu) = \nu^{2p} S_x(\nu). \tag{26.43}$$

In particular,

$$S_{\dot{x}}(\nu) = \nu^2 S_x(\nu), \qquad S_{\ddot{x}}(\nu) = \nu^4 S_x(\nu). \tag{26.44}$$

Note that the above relations hold for any type of differentiability (almost certainly, in probability, or in r-mean).

The considerations above show that knowledge of the spectral density of a second order weakly stationary random process allows the calculation of several important statistical characteristics of the process itself, of its derivatives, and of its indefinite integral. In some particular cases, the quantity of statistical information contained in the spectral density is even larger, as will become apparent from the applications examined in the following sections.

c) Sampling theory. Ergodicity

We have assumed so far that the distribution functions or the probability density functions of the random processes dealt with were known. Our main concern was then the calculation of various statistical characteristics of the process. In practice, however, we often face a quite different problem, namely the statistical description and analysis of the data obtained by making a set of measurements of a random process. In this case the probability density functions are not known *a priori* and we may ask whether it is possible or not to determine some of the statistical properties of the random process from the results of the measurements.

For a comprehensive treatment of this problem, including an extensive discussion of the statistical methods for evaluating data and of the analogue and digital computer techniques for data processing, the reader is referred to the books by BENDAT and PIERSOL [11], KORN [90], ROTKOP [175], and SREIDER [183].

In this section we confine ourselves to expounding some of the basic statistical techniques used for the processing of random data.

Consider a random variable X that may take on values in the range $-\infty$ to ∞, and let $x_1, \ldots, x_N$ be a set of N observed values of X. This limited information about the random variable does not allow, of course, the exact determination of its statistical characteristics, such as mean value, variance, etc. Hence, one must be content with an approximate evaluation of these characteristics. The set of N values of the random variable upon which the evaluation is based is said to be a *sample*, and the number N is called the *sample size*.

Let now λ be any statistical parameter of the random variable. An evaluation of λ based upon a sample of size N is said to be an *estimate* of λ. If a series of different samples of size N were selected from the same random variable X, the estimates computed from each sample would generally be different, thus generating a random variable λ_N, which is said to be a *sample parameter* or an *estimator* of λ. There is a certain degree of arbitrariness in choosing estimators for a given statistical characteristic. Therefore, it is important to have some criteria enabling one to judge the quality of an estimator and to compare different estimators with each other. It is, of course, desirable that the mean value of the estimator be equal to the true value of the statistical parameter being estimated or at least approach this value as the sample size tends to infinity. Second, it is desirable that the estimator converge in probability to the parameter being estimated as $N \to \infty$. These requirements will be made more precise by means of the following definitions.

Let m_{λ_N} and $\sigma^2_{\lambda_N}$ denote, respectively, the mean value and the variance of an estimator λ_N, i.e.,

$$m_{\lambda_N} = E[\lambda_N], \qquad \sigma^2_{\lambda_N} = E[(\lambda_N - m_{\lambda_N})^2]. \tag{26.45}$$

Definition 9. *An estimator λ_N is said to be* unbiased *if*

$$m_{\lambda_N} = \lambda. \tag{26.46}$$

Definition 10. *An estimator λ_N is said to be* consistent *if, for any $\varepsilon > 0$,*

$$\lim_{N\to\infty} P(|\lambda_N - \lambda| \geqslant \varepsilon) = 0, \tag{26.47}$$

i.e., if λ_N converges in probability to λ as $N \to \infty$.

Since convergence in mean square implies convergence in probability, a sufficient condition for an estimator λ_N to be consistent is

$$\lim_{N\to\infty} E[(\lambda_N - \lambda)^2] = 0. \tag{26.48}$$

Definition 11. *The estimator* λ'_N *is said to be* more efficient *than the estimator* λ''_N *if*

$$E[(\lambda'_N - \lambda)^2] \leqslant E[(\lambda''_N - \lambda)^2]. \tag{26.49}$$

Clearly, an estimator which is more efficient than a consistent estimator is consistent, too.

Definition 12. *An estimator* λ'_N *is said to be* correct *if*

$$\lim_{N\to\infty} m_{\lambda_N} = \lambda, \qquad \lim_{N\to\infty} \sigma^2_{\lambda_N} = 0. \tag{26.50}$$

Definition 13. *An estimator* λ_N *is said to be* absolutely correct *if*

$$m_{\lambda_N} = \lambda, \qquad \lim_{N\to\infty} \sigma^2_{\lambda_N} = 0. \tag{26.51}$$

It is easily seen that *a correct estimator is also consistent.* Indeed, considering (26.45), we have

$$E[(\lambda_N - \lambda)^2] = E[(\lambda_N - m_{\lambda_N} + m_{\lambda_N} - \lambda)^2$$

$$= \sigma^2_{\lambda_N} + 2(m_{\lambda_N} - \lambda)\, E[\lambda_N - m_{\lambda_N}] - (m_{\lambda_N} - \lambda)^2,$$

whence

$$E[(\lambda_N - \lambda)^2] = \sigma^2_{\lambda_N} + (m_{\lambda_N} - \lambda)^2. \tag{26.52}$$

If now λ_N is correct, then from Eqs. (26.50) and (26.52) it results that λ_N satisfies the condition (26.48), and hence is consistent.

Since, obviously, an absolutely correct estimator is also correct, we infer from the reasoning above and by comparing (26.51) with (26.46) that *an absolutely correct estimator is unbiased and consistent.*

As an illustration for the definitions above let us consider the estimator of the mean value of a random variable X defined by

$$m_{xN} = \frac{1}{N} \sum_{i=1}^{N} x_i, \tag{26.53}$$

where the values x_i of X are obtained by N statistically independent repetitions of the basic experiment. If the sample of size N were chosen in all possible ways, then for each fixed $i = 1, \ldots, N$ the values of x_i would generate a random variable with the same distribution function as X. Denoting for the sake of simplicity this random variable also by x_i, we have

$$E[x_i] = E[X] = m_x, \qquad E[x_i x_j] = E[x_i]\,E[x_j] = m_x^2,$$
$$E[(x_i - m_x)(x_j - m_x)] = 0, \qquad E[(x_i - m_x)^2] = \sigma_x^2, \tag{26.54}$$

for any $i, j = 1, \ldots, n$, $i \neq j$.

Theorem 7. *The sample mean value m_{xN} is an absolutely correct estimator of m_x.*
Proof. From (26.53) and $(26.54)_1$ it follows that

$$m_{m_{xN}} = E[m_{xN}] = \frac{1}{N}\sum_{i=1}^{N} E[x_i] = m_x, \tag{26.55}$$

which proves $(26.51)_1$. To prove $(26.51)_2$ we have to show that

$$\lim_{N\to\infty} \sigma_{m_{xN}}^2 = 0. \tag{26.56}$$

In view of (26.53), we may write

$$(m_{xN} - m_x)^2 = \frac{1}{N^2}\sum_{i=1}^{N}(x_i - m_x)^2 - \frac{1}{N^2}\sum_{\substack{i,j=1\\ i\neq j}}^{N}(x_i - m_x)(x_j - m_x),$$

whence, by considering Eqs. (26.54), it results that

$$\sigma_{m_{xN}}^2 = E[(m_{xN} - m_x)^2] = \frac{\sigma_x^2}{N}, \tag{26.57}$$

which implies (26.56).

The following theorem concerns the important problem of estimating the higher-order central moments μ_k of a random variable.

Theorem 8. *The sample central moment of kth order defined by*

$$\mu_{kN} = \frac{1}{N}\sum_{i=1}^{N}(x_i - m_{xN})^k \tag{26.58}$$

is a correct estimator of μ_k, $k = 2, 3, \ldots$

For a proof of this theorem see Mihoc and Firescu [120], § 5.9. The idea of the proof is to determine the asymptotic behavior of the mean value and of the variance of the estimator with respect to the sample size N. It is shown that

$$m_{\mu_{kN}} = \mu_k + 0\left(\frac{1}{N}\right), \qquad \sigma^2_{\mu_{kN}} = 0\left(\frac{1}{N}\right), \tag{26.59}$$

from which it follows that

$$\lim_{N\to\infty} m_{\mu_{kN}} = \mu_k, \qquad \lim_{N\to\infty} \sigma^2_{\mu_{kN}} = 0. \tag{26.60}$$

By slightly modifying the definition (26.58) of the sample central moments, it is possible to obtain absolutely correct estimators of μ_k. Since the modification required depends on k, we indicate here only the expressions of the absolutely correct estimators for μ_2 and μ_3:

$$\sigma^2_{xN} = \mu_{2N} = \frac{1}{N-1} \sum_{i=1}^{N} (x_i - m_{xN})^2, \tag{26.61}$$

$$\mu_{3N} = \frac{N}{(N-1)(N-2)} \sum_{i=1}^{N} (x_i - m_{xN})^3. \tag{26.62}$$

Let us consider now a random process $X(t)$ defined for $t \in T$, and let $t_1, \ldots, t_n$ be an arbitrary partition of the interval T. Suppose that the records of N time histories of the process, i.e., the graphs of N sample functions, are known (Fig. 76).

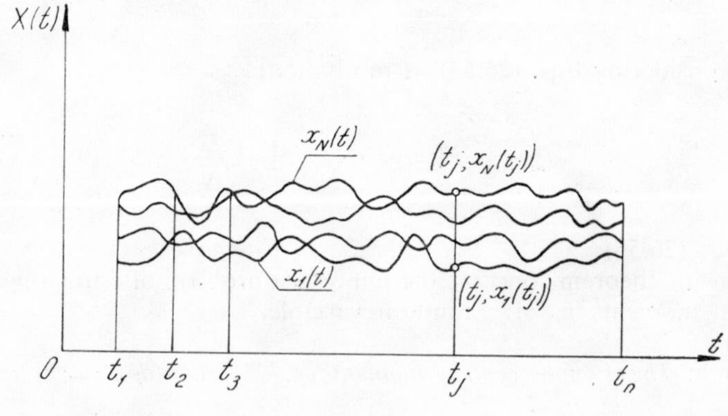

Fig. 76

For any t_j, $j = 1, \ldots, n$, these records provide a sample of size N for the random variable $X(t_j)$, say $x_1(t_j), \ldots, x_N(t_j)$. By Theorem 7, we see that the sample mean value

$$m_{x_j N} = \frac{1}{N} \sum_{i=1}^{N} x_i(t_j) \tag{26.63}$$

is an absolutely correct estimator for the mean value $m_x(t)$ of the random process $X(t)$ at $t = t_j$, as defined by Eq. (26.3).

It can be shown that the sample correlation function defined by

$$K_{xN}(t_j, t_l) = \frac{1}{N-1} \sum_{i=1}^{N} (x_i(t_j) - m_{x_jN})(x_i(t_l) - m_{x_lN}) \tag{26.64}$$

is an absolutely correct estimator of the correlation function $K_x(t, t')$ of the random process $X(t)$ at $t = t_j$, $t' = t_l$, as defined by Eq. (26.6).

The use of the estimators defined by Eqs. (26.63) and (26.64) is conditioned by knowledge of a sufficiently large number of history records of the process. In practice, however, this condition may be difficult to fulfill. Moreover, we sometimes face the problem of evaluating some statistical parameters of a random process from a single available record. Therefore, we will investigate this possibility in some detail.

Let $X(t)$ be a weakly stationary random process of second order. Suppose that λ is the true value of an unknown statistical parameter of the process, and denote by λ_T an estimator of λ which provides an estimate of λ for each sample time history record $x(t)$ extending over a finite time interval $[0, T]$. Since the estimates computed for different sample records will vary randomly, the estimator λ_T is, of course, a random variable. For instance, the quantity

$$m_{xT} = \frac{1}{T} \int_0^T x(t)\,dt \tag{26.65}$$

provides an estimator for the (constant) true mean value m_x of the stationary process $X(t)$, whenever the sample function $x(t)$ considered is Riemann integrable in $[0, T]$.

There exists a formal similarity between estimators using individual sample records on a finite time interval $[0, T]$ and estimators using a finite number N of sample records. The concepts introduced at the beginning of this subsection by Definitions 9—13 regarding the quality of the estimators λ_N and their asymptotic behavior for $N \to \infty$, can be applied to the estimators λ_T by simply replacing N with T.

Let us prove that m_{xT} is an unbiased estimator in the sense of Definition 9. Indeed, since mean values commute with linear operations, we have from (26.65)

$$m_{m_{xT}} = E[m_{xT}] = \frac{1}{T} \int_0^T E[x(t)]\,dt = \frac{1}{T} \int_0^T m_x\,dt = m_x. \tag{26.66}$$

From Eqs. (26.35) and (26.65) we successively obtain

$$\sigma_{m_{xT}}^2 = 2 \int_0^T (T - \tau)\, k_{x/T}(\tau)\,d\tau = \frac{2}{T^2} \int_0^T (T - \tau)\, k_x(\tau)\,d\tau,$$

or

$$\sigma^2_{m_{xT}} = \frac{2}{T}\int_0^T (1 - \tau/T)\, k_x(\tau)\, d\tau. \tag{26.67}$$

In order that m_{xT} be an absolutely correct estimator of m_x it is sufficient that the integral in the right-hand side of this relation approaches zero as $T \to \infty$. An important particular case, frequently occurring in practice, in which this condition is satisfied, is that when $|k_x(\tau)| < \tau^{-\alpha}$ for $\tau \geqslant \tau_0$, with $\alpha > 0$, $\tau_0 > 0$. Indeed, we then have for $T > \tau_0$

$$\sigma^2_{m_{xT}} = \frac{2}{T}\int_0^T k_x(\tau)\, d\tau - \frac{2}{T^2}\int_0^T \tau k_x(\tau)\, d\tau =$$

$$\leqslant \frac{2}{T}\int_0^{\tau_0} |k_x(\tau)|\, d\tau + \frac{2}{T}\int_{\tau_0}^T |k_x(\tau)|\, d\tau - \frac{2}{T^2}\int_0^{\tau_0} \tau|k_x(\tau)|\, d\tau - \frac{2}{T^2}\int_{\tau_0}^T \tau|k_x(\tau)|\, d\tau.$$

The first and third terms in the right-hand side of this inequality approach zero as $T \to \infty$, since the integrals involved are finite. Next, it is easily proved that the second and fourth terms are bounded by $AT^{-\alpha}$, where A is some positive constant, and hence they also approach zero as $T \to \infty$. Consequently, in this case,

$$\lim_{T\to\infty} \sigma^2_{m_{xT}} = 0. \tag{26.68}$$

The random processes for which m_{xT} is an absolutely correct estimator of m_x are said to be *ergodic*. The reasoning above shows that a weakly stationary random process is ergodic if its correlation function tends to zero more rapidly than some negative power of τ as $\tau \to \infty$.

It may be also shown that when a second order stationary random process $X(t)$ is ergodic and normally distributed, its correlation function is absolutely correct estimated by

$$k_{xT}(\tau) = \frac{1}{T - \tau}\int_0^{T-\tau} [x(t) - m^*_{xT}]\,[x(t + \tau) - m^*_{xT}]\, dt, \tag{26.69}$$

where

$$m^*_{xT} = \frac{1}{T - \tau}\int_0^{T-\tau} x(t)\, dt. \tag{26.70}$$

Chapter VIII General methods of study

§ 27. EXISTENCE, UNIQUENESS, AND STABILITY

Let us consider the equation of motion

$$\ddot{x} + f(\dot{x}) + g(x) = p(t), \tag{27.1}$$

where $p(t)$ is a random process defined on a probability field (Ω, K, P), and assume that $f(x)$ and $g(x)$ satisfy hypotheses $H_1 - H_5$.

Since $p(t)$ is a random process, the unknown function $x(t)$ as well as its time derivatives $\dot{x}(t)$ and $\ddot{x}(t)$ will be also random processes defined on the same probability field as $p(t)$.

As already mentioned in Part I, the equation (27.1) is equivalent with the system of differential equations

$$\dot{x} = v, \tag{27.2}$$

$$\dot{v} = p(t) - f(v) - g(x).$$

For convenience, we shall use this system to define various types of solutions and to examine their qualitative properties.

Definition 1. *A pair of random processes* $(x(t), v(t))$ *is said to be a* solution in mean square (*or* almost certain) *in* $[t_0, t]$ *of system (27.2) if* $v(t)$ *and* $p(t) - f(v(t)) - g(x(t))$ *are, respectively, the derivatives in mean square (or almost certain) of* $x(t)$ *and* $v(t)$ *in* $[t_0, t]$ *in the sense of Definition 5, § 26.*[1)]

Initial value problems of Cauchy type may be considered for system (27.7) by introducing initial conditions of the form

$$x(t_0) = x_0, \qquad v(t_0) = v_0, \tag{27.3}$$

where x_0 and v_0 are given random variables.

[1)] It is also possible to define solutions with respect to other types of convergence, but we confine ourselves to those introduced above.

Definition 2. *A solution* $(x(t), v(t))$ *in mean square (or almost certain) of system (27.2) is said to be a* solution of the Cauchy problem (27.2), (27.3) in mean square (*or* almost certain) *if it satisfies almost certainly the initial conditions (27.3).*

Similarly to the case of deterministic systems, we are interested in the existence and uniqueness of the solution of the Cauchy problem (27.2), (27.3). The answer to these questions is given by the following theorem, which will be stated without proof. [1)]

Theorem 1. *Suppose that*

(1) $|v| + |p(t) - f(v) - g(x)| \leqslant L_1(t, r)$ *for* $|x| + |v| \leqslant r$ *and any* $t \in [t_0, \infty)$

(2) $|v_1 - v_2| + |f(v_1) - f(v_2) + g(x_1) - g(x_2)| \leqslant L_2(t, r) \{|v_1 - v_2| + |x_1 - x_2|\}$

for $|x_1| + |v_1| \leqslant r$, $|x_2| + |v_2| \leqslant r$, *and any* $t \in [t_0, \infty)$, *where* L_1 *and* L_2 *are integrable with respect to* t *in any finite closed subinterval of* $[t_0, \infty)$, *increasing with respect to* r, *and* $L_1(t, r) > 0$ *for* $r > 0$. *Then:*

(1) *If the random vector* (x_0, v_0) *is essentially bounded, i.e., if* $|x_0| + |v_0| < M$ *for some positive constant* M, *then there exists an almost certain solution of the Cauchy problem (27.2), (27.3) defined in a certain interval* $[t_0, t_1]$; *this solution is also a solution in mean square.*

(2) *If* $(x_1(t), v_1(t))$, $(x_2(t), v_2(t))$ *are two almost certain solutions of the Cauchy problem (27.2), (27.3) defined in the intervals* $[t_0, t_1]$ *and* $[t_0, t_2]$, *respectively, then, almost certainly,*

$$x_1(t) = x_2(t), \qquad v_1(t) = v_2(t)$$

for any $t \in [t_0, t_1] \cap [t_0, t_2]$.

(3) *If, in addition,* $L_1(t, r) = L(t)\, H(r)$, *where* $\int_{r_0}^{\infty} \mathrm{d}r/H(r) = \infty$ *for some finite* $r_0 > 0$, *then every almost certain solution and also every solution in mean square of the Cauchy problem (27.2), (27.3) exists in the whole interval* $[t_0, \infty)$.

For a proof of this theorem, as well as for a detailed discussion of the relations between various types of solutions and of their qualitative properties, we refer to the book by MOROZAN [127]. We content ourselves here to noting that our hypotheses H_1 — H_5 assure the fulfillment of the conditions of Theorem 1 above.

To close this subsection, we will introduce some basic types of stochastic stability in connexion with the solutions of the Cauchy problem (27.2), (27.3). Let $\bar{x}(t)$, $\bar{v}(t)$ be a solution in mean square of system (27.2) that exists in a maximal interval (t^-, ∞).

1) In the remaining part of this subsection we assume, for the sake of simplicity, that x and v are nondimensional quantities (cf. also § 1c).

Definition 3. *The solution* $\bar{x}(t)$, $\bar{v}(t)$ *is said to be* stable in mean square *if, given any* t_0 *and* ε, $t_0 > t^-$, $\varepsilon > 0$, *there exists a* $\delta(\varepsilon, t_0) > 0$ *such that for any solution* $x(t)$, $v(t)$ *of (27.2) satisfying*

$$\sqrt{E[(x(t_0) - \bar{x}(t_0))^2] - E[(v(t_0) - \bar{v}(t_0))^2]} < \delta \tag{27.4}$$

the inequality

$$\sqrt{E[(x(t) - \bar{x}(t))^2] - E[(v(t) - \bar{v}(t))^2]} < \varepsilon \tag{27.5}$$

holds for any $t \geqslant t_0$. If $\delta(\varepsilon, t_0)$ depends only on ε, the solution $\bar{x}(t)$, $\bar{v}(t)$ is called *uniformly stable in mean square.*

Definition 4. *The solution* $\bar{x}(t)$, $\bar{v}(t)$ *is said to be* asymptotically stable in mean square *if it is stable in mean square and if, given any* $t_0 > t^-$, *there exists a* $\delta(t_0) > 0$ *such that*

$$\lim_{t\to\infty} \sqrt{E[(x(t) - \bar{x}(t))^2] + E[(v(t) - \bar{v}(t))^2]} = 0 \tag{27.6}$$

for any solution $x(t)$, $v(t)$ satisfying

$$\sqrt{E[(x(t_0) - \bar{x}(t_0))^2] - E[(v(t_0) - \bar{v}(t_0))^2]} < \delta \tag{27.7}$$

If $\delta(t_0)$ does not depend on t_0, the solution $\bar{x}(t)$, $\bar{v}(t)$ is called *uniformly asymptotically stable in mean square.* If the condition (27.6) is satisfied without any restriction of the type (27.7), the solution $\bar{x}(t)$, $\bar{v}(t)$ is said to be *asymptotically stable in the large in mean square.*

One can still define several other types of stochastic stability and establish the connection between them, but this would be beyond the scope of this book. For a comprehensive treatment of various problems concerning stochastic stability, including the use of LYAPUNOV's method, the reader is referred to MOROZAN [127—130] and KUSHNER [93].

§ 28. LINEAR OSCILLATORS

In this section we will apply the correlation method to study random vibrations of linear oscillators subjected to second order random excitations. The governing equation of motion has the form

$$\ddot{x} + 2\zeta\nu_n\dot{x} + \nu_n^2 x = p(t), \tag{28.1}$$

where $p(t)$ is a random process which is almost certainly continuous and bounded in $[t_0, \infty)$, and $0 < \zeta < 1$. This obviously assures that the hypotheses of Theorem 1, § 27, are satisfied. Furthermore, any almost certain solution of Eq. (28.1) may be obtained in the same way as in the theory of linear differential equations with constant coefficients. Indeed, as already remarked in § 26 a, the random process

$p(t)$ may be viewed as a function $p(\omega, t)$ defined on the set $\Omega \times (t_0, \infty)$, where Ω is the sample space associated with the process. Let Ω' be the subset of Ω in which $p(\omega, t)$ is continuous and bounded as a function of $t \in [t_0, \infty)$. Then, for any $\omega \in \Omega'$, the general solution of Eq. (28.1) may be written in the form

$$x(\omega, t) = x_0(t) + x_p(\omega, t),$$

where

$$x_0(t) = e^{-\zeta\nu_n t}(A \cos \nu_n\sqrt{1 - \zeta^2}\, t + B \sin \nu_n\sqrt{1 - \zeta^2}\, t), \tag{28.2}$$

with A, B arbitrary constants, is a general solution of the homogeneous equation corresponding to (28.1), and

$$x_p(\omega, t) = -\frac{1}{\nu_n\sqrt{1 - \zeta^2}} \int_0^\infty p(\omega, t - \theta)\, e^{-\zeta\nu_n\theta} \sin \nu_n \sqrt{1 - \zeta^2}\, \theta\, d\theta$$

is a particular solution of Eq. (28.1). From the reasoning above it is apparent that $x_p(\omega, t)$ with $\omega \in \Omega'$ are the sample functions of the random process

$$x_p(t) = -\frac{1}{\nu_n\sqrt{1 - \zeta^2}} \int_0^\infty p(t - \theta)\, e^{-\zeta\nu_n\theta} \sin \nu_n \sqrt{1 - \zeta^2}\, \theta\, d\theta, \tag{28.3}$$

which satisfies almost certainly Eq. (28.1), the integral in (28.3) being also meant in the almost certain sense. Hence, the random process

$$x(t) = x_0(t) + x_p(t) \tag{28.4}$$

where $x_0(t)$ and $x_p(t)$ are given by Eqs. (28.2) and (28.3), respectively, is the general almost certain solution of Eq. (28.1).[1] According to Theorem 1, § 27, $x(t)$ is also the general solution in mean square of Eq. (28.1).

Theorem 1. *The particular solution (28.3) of Eq. (28.1) is asymptotically stable in the large in mean square.*

Proof. Let $x(t)$ be a solution in mean square of Eq. (28.1) satisfying arbitrary initial conditions. From (28.4) it follows that

$$E[(x - x_p)^2] = E[x_0^2] = e^{-2\zeta\nu_n t} E[(A \cos \nu_n \sqrt{1 - \zeta^2}\, t - B \sin \nu_n \sqrt{1 - \zeta^2}\, t)^2],$$

$$E[(\dot{x} - \dot{x}_p)^2] = E[\dot{x}_0^2] = \nu_n^2 e^{-2\zeta\nu_n t} E[\{(B \sqrt{1 - \zeta^2} - A\zeta) \cos \nu_n \sqrt{1 - \zeta^2}\, t -$$

$$- (B\zeta + A \sqrt{1 - \zeta^2}) \sin \nu_n \sqrt{1 - \zeta^2}\}^2],$$

[1] In this statement "general" means as usual that any almost certain solution of the Cauchy problem (28.1), (27.3) may be obtained from (28.4) under a suitable choice of the random variables A and B.

and hence

$$\lim_{t\to\infty} \sqrt{E[(x - x_p)^2] + E[(\dot{x} - \dot{x}_p)^2]} = 0.$$

Thus, condition (27.6) is fulfilled regardless of the initial conditions (27.3), provided the random variables A and B determined by them have finite first and second moments.

In the remaining part of this section, we shall disregard the transitory part of the solution (28.4), by focusing our attention on the particular solution (28.3), which is asymptotically stable in the large.

By introducing the function

$$h(t) = \begin{cases} 0 & \text{for } t \in (-\infty, 0) \\ -\dfrac{e^{-\zeta\nu_n t} \sin \nu_n \sqrt{1-\zeta^2}\, t}{\nu_n \sqrt{1-\zeta^2}} & \text{for } t \in [0, \infty), \end{cases} \tag{28.5}$$

the solution (28.3) may be rewritten as

$$x_p(t) = \int_{-\infty}^{\infty} p(t - \theta)\, h(\theta)\, d\theta. \tag{28.6}$$

THEOREM 2. *The function $h(t)$ defined by Eq. (28.5) satisfies the differential equation*

$$\ddot{x} + 2\zeta\nu_n \dot{x} + \nu_n^2 x = \delta(t) \tag{28.7}$$

and the initial conditions

$$x(-\infty) = \dot{x}(-\infty) = 0, \tag{28.8}$$

where $\delta(t)$ is the Dirac delta function.

Proof. Let

$$H(\nu) = \int_{-\infty}^{\infty} x(t)\, e^{-i\nu t}\, dt \tag{28.9}$$

be the Fourier transform of the solution $x(t)$ of the Cauchy problem (28.7), (28.8). Multiplying both sides of Eq. (28.7) by $e^{-i\nu t}\, dt$, integrating from $-\infty$ to ∞, and considering (28.8), we obtain

$$H(\nu) = \frac{1}{\nu_n^2 - \nu^2 + 2\, i\zeta\nu_n\nu}. \tag{28.10}$$

Inverting (28.9) yields

$$x(t) = \frac{1}{2\pi} \int_{-\infty}^{\infty} \frac{e^{i\nu t} \, d\nu}{\nu_n^2 - \nu^2 + 2i\zeta\nu_n}. \tag{28.11}$$

The integral in the right-hand side of this equation can be calculated in a complex plane, by observing that the integrand has two poles of order one, namely $\nu_{1,2} = \nu_n(i\zeta \pm \sqrt{1-\zeta^2})$. The result reads

$$x(t) = \begin{cases} 0 \text{ for } t \in (-\infty, 0), \\ -\dfrac{e^{-\zeta\nu_n t} \sin \nu_n \sqrt{1-\zeta^2}\, t}{\nu_n \sqrt{1-\zeta^2}} \quad \text{for } t \in [0, \infty), \end{cases}$$

hence $x(t) \equiv h(t)$, Q.E.D.

Replacing now $x(t)$ by $h(t)$ in Eqs. (28.9) and (28.11), and taking into account (28.10), we obtain

$$H(\nu) = \int_{-\infty}^{\infty} h(t)\, e^{-i\nu t}\, dt, \tag{28.12}$$

$$h(t) = \frac{1}{2\pi} \int_{-\infty}^{\infty} H(\nu)\, e^{i\nu t}\, d\nu. \tag{28.13}$$

The function $H(\nu)$, i.e., the Fourier transform of the response of the oscillator to a unit impulse, is called the *complex frequency response function* or the *transfer function* of the oscillator. It is worth noting that the expression (28.10) of the transfer function could be obtained at once, by setting $p(t) = e^{i\nu t}$, $x(t) = H(\nu)\, e^{i\nu t}$ into Eq. (28.1), dividing out a factor of $e^{i\nu t}$, and solving for $H(\nu)$ the algebraic equation resulted.

We will calculate now some statistical characteristics of the response, $x_p(t)$, in terms of the corresponding statistical characteristics of the excitation $p(t)$, assuming that the random process $p(t)$ is weakly stationary.

The mean value of the response

Since mean values commute with integrals, it follows from (28.6) that

$$m_{x_p}(t) = E[x_p(t)] = \int_{-\infty}^{\infty} E[p(t-\theta)\, h(\theta)]\, d\theta = \int_{-\infty}^{\infty} E[p(t-\theta)]\, h(\theta)\, d\theta. \tag{28.14}$$

If now $p(t)$ is stationary, we have

$$E[p(t)] = m_p = \text{const}, \tag{28.15}$$

and Eq. (28.14) gives

$$m_{x_p} = m_p \int_{-\infty}^{\infty} h(\theta)\, d\theta. \tag{28.16}$$

On the other hand, from Eqs. (28.12) and (28.10) it results that

$$\int_{-\infty}^{\infty} h(\theta)\, d\theta = H(0) = 1/\nu_n^2,$$

and hence, finally,

$$m_{x_p} = m_p/\nu_n^2. \tag{28.17}$$

In particular, if the excitation has zero mean value, then so does the response of the oscillator.

The correlation function of the response

Assume, for the sake of simplicity, that $m_p = 0$. Then, from Eqs. (26.6) and (28.17), we have

$$K_{x_p}(t, t + \tau) = E[x_p(t)\, x_p(t + \tau)]. \tag{28.18}$$

On the other hand, in view of (28.6),

$$x_p(t)\, x_p(t + \tau) = \int_{-\infty}^{\infty} p(t - \theta_1)\, h(\theta_1)\, d\theta_1 \int_{-\infty}^{\infty} p(t + \tau - \theta_2)\, h(\theta_2)\, d\theta_2$$

$$= \int_{-\infty}^{\infty}\int_{-\infty}^{\infty} p(t - \theta_1)\, p(t + \tau - \theta_2)\, h(\theta_1)\, h(\theta_2)\, d\theta_1\, d\theta_2,$$

and hence Eq. (28.18) becomes

$$K_{x_p}(t, t + \tau) = \int_{-\infty}^{\infty}\int_{-\infty}^{\infty} E[p(t - \theta_1)\, p(t + \tau - \theta_2)]\, h(\theta_1)\, h(\theta_2)\, d\theta_1\, d\theta_2$$

$$= \int_{-\infty}^{\infty} h(\theta_1)\, h(\theta_2)\, K_p(t - \theta_1,\ t + \tau - \theta_2)\, d\theta_1\, d\theta_2. \tag{28.19}$$

Since $p(t)$ is weakly stationary,

$$K_p(t - \theta_1,\ t + \tau - \theta_2) = k_p(\tau + \theta_1 - \theta_2),$$

hence

$$K_{x_p}(t, t+\tau) = \int_{-\infty}^{\infty}\int_{-\infty}^{\infty} h(\theta_1)\, h(\theta_2)\, k_p(\tau + \theta_1 - \theta_2)\, d\theta_1 d\theta_2.$$

This relation shows that the correlation function of the response depends only on the difference of its arguments. Therefore, we may write

$$k_{x_p}(\tau) = \int_{-\infty}^{\infty}\int_{-\infty}^{\infty} h(\theta_1)\, h(\theta_2)\, k_p(\tau + \theta_1 - \theta_2)\, d\theta_1\, d\theta_2. \qquad (28.20)$$

From Eqs. (28.17) and (28.20) we see that $x_p(t)$ is also weakly stationary. It may be proved that when the input is a strongly stationary random process, then so does the output.

The spectral density of the response

Since $x_p(t)$ is weakly stationary, we may combine Eqs. (26.37) and (28.20) to obtain the spectral density of the response:

$$S_{x_p}(\nu) = \frac{1}{\pi}\int_{-\infty}^{\infty}\left\{\int_{-\infty}^{\infty}\int_{-\infty}^{\infty} h(\theta_1)\, h(\theta_2)\, k_p(\tau + \theta_1 - \theta_2)\, d\theta_1\, d\theta_2\right\} e^{-i\nu\tau}\, d\tau.$$

By conveniently changing the order of integration, we may rewrite this equation as

$$S_{x_p}(\nu) = \frac{1}{\pi}\int_{-\infty}^{\infty} h(\theta_1)\, e^{i\nu\theta_1} d\theta_1 \int_{-\infty}^{\infty} h(\theta_2)\, e^{i\nu\theta_2} d\theta_2\, k_p(\tau + \theta_1 - \theta_2)\, e^{-i\nu(\tau+\theta_1-\theta_2)}\, d\tau.$$

Furthermore, since

$$\frac{1}{\pi}\int_{-\infty}^{\infty} k_p(\tau + \theta_1 - \theta_2)\, e^{-i\nu(\tau+\theta_1-\theta_2)}\, d\tau = S_p(\nu),$$

$$\int_{-\infty}^{\infty} h(\theta_1)\, e^{-i\nu\theta_1} d\theta_1 = H(-\nu), \quad \int_{-\infty}^{\infty} h(\theta_2)\, e^{i\nu\theta_2} d\theta_2 = H(\nu),$$

the last relation takes the form

$$S_{x_p}(\nu) = H(-\nu)\, H(\nu)\, S_p(\nu).$$

Next, by (28.10), $H(-\nu) = \bar{H}(\nu)$, and hence we finally obtain

$$S_{x_p}(\nu) = |H(\nu)|^2 S_p(\nu). \tag{28.21}$$

This relation shows us that the spectral density of the output is completely determined by the spectral density of the input and by the modulus of the transfer function of the system. It should be noted that, unlike (28.20), Eq. (28.21) is an algebraic equation. For this reason it is more frequently used for obtaining some quantitative information about the response.

The variances of the response and of its derivatives

According to Eqs. (26.42) and (28.12), the variance of the response is given by

$$\sigma^2_{x_p} = \int_0^\infty |H(\nu)|^2 S_p(\nu)\, d\nu. \tag{28.22}$$

Furthermore, if $x_p(t)$ is n-times differentiable, the variances of its derivatives may be expressed in view of Eqs. (26.43) and (28.21) by

$$\sigma^2_{x_p^{(k)}} = \int_0^\infty \nu^{2k} |H(\nu)|^2 S_p(\nu)\, d\nu \tag{28.23}$$

for any $k = 1, \ldots, n$, irrespective of the type of differentiability adopted.

§ 29. NONLINEAR OSCILLATORS

We will discuss in this section two approximate techniques, namely the method of statistical linearization and the perturbation method, which have been developed to study the random vibrations of nonlinear oscillators. These methods will be illustrated on the oscillator whose governing equation is

$$\ddot{x} + G(x, \dot{x}) = p(t), \tag{29.1}$$

where $p(t)$ is a weakly stationary random process. We assume that $G(x, \dot{x})$ and $p(t)$ satisfy certain conditions assuring the existence of a weakly stationary solution $x(t)$ and of a corresponding steady-state probability density function $f(x, \dot{x})$. General conditions of this type have been recently given by KUSHNER [94] (see also CAUGHEY [31]).

For a comparative discussion of the accuracy of a number of approximate methods we refer to PAYNE [142, 143].

a) The method of statistical linearization

This method was introduced by KAZAKOV [83—85] and BOOTON [18, 19], and has been independently developed by CAUGHEY [26]. It consists in replacing Eq. (29.1) by the linear equation

$$\ddot{x} + 2\zeta_e \nu_e \dot{x} - \nu_e^2 x = p(t), \tag{29.2}$$

in such a way that the mean square of the equation deficiency is minimized, i.e.,

$$E[G(x, \dot{x}) - 2\zeta_e \nu_e \dot{x} - \nu_e^2 x^2] = \min. \tag{29.3}$$

The vanishing of the derivatives of the left-hand side with respect to ζ_e and ν_e yields the system

$$\left.\begin{aligned} &E[\dot{x}G(x, \dot{x})] - 2\zeta_e \nu_e E[\dot{x}^2] - \nu_e^2 E[x\dot{x}] = 0, \\ &E[xG(x, \dot{x})] - 2\zeta_e \nu_e E[x\dot{x}] - \nu_e^2 E[x^2] = 0, \end{aligned}\right\} \tag{29.4}$$

where all mean values are calculated with respect to the probability density function $f(x, \dot{x})$. In particular, according to (25.44),

$$\left.\begin{aligned} E[\dot{x}G(x, \dot{x})] &= \int_{-\infty}^{\infty}\int_{-\infty}^{\infty} \dot{x}G(x, \dot{x}) f(x, \dot{x})\, \mathrm{d}x\, \mathrm{d}\dot{x}, \\ E[xG(x, \dot{x})] &= \int_{-\infty}^{\infty}\int_{-\infty}^{\infty} xG(x, \dot{x}) f(x, \dot{x})\, \mathrm{d}x\, \mathrm{d}\dot{x}. \end{aligned}\right\} \tag{29.5}$$

For a differentiable, stationary random process, the mean value of $x\dot{x}$ is zero. Indeed, by differentiating the relation

$$E[(x(t) - m_x)(x(t+\tau) - m_x)] = k_x(\tau)$$

with respect to τ and setting $\tau = 0$, we obtain

$$E[(x(t) - m_x)\,\dot{x}(t)] = k_x'(0) = 0.$$

since, as shown in § 26, $k_x(\tau)$ is differentiable and has an absolute maximum for $\tau = 0$. Moreover, we have $E[\dot{x}(t)] = m_{\dot{x}} = 0$, and hence

$$E[x(t)\,\dot{x}(t)] = 0. \tag{29.6}$$

Taking into account (29.5), system (29.4) gives

$$\nu_e^2 = \frac{E[xG(x, \dot{x})]}{E[x^2]}, \qquad 2\zeta_e\nu_e = \frac{E[\dot{x}G(x, \dot{x})]}{E[\dot{x}^2]}. \tag{29.7}$$

A simple calculation shows that the values of ζ_e and ν_e determined by these relations do indeed minimize the mean square of the equation deficiency.

In fact, as the probability density function $f(x, \dot{x})$ is unknown, the parameters ζ_e and ν_e cannot be calculated by Eqs. (29.7) unless some approximation of $f(x, \dot{x})$ is used. The easiest way to break this deadlock is to replace $f(x, \dot{x})$ by the probability density function corresponding to the solution of the linearized equation (29.2). Since this last one depends on ζ_e and ν_e, Eqs. (29.7) become a system of nonlinear algebraic equations in ζ_e and ν_e.

As an illustration of the method let us consider the equation

$$\ddot{x} + 2\zeta\nu_n\dot{x} + \varepsilon\varphi_1(\dot{x}) + \nu_n^2 x + \varepsilon\psi_1(x) = p(t), \tag{29.8}$$

where ε is a small parameter and $p(t)$ is a normally distributed, weakly stationary random process with zero mean value. We now have

$$G(x, \dot{x}) = 2\zeta\nu_n\dot{x} + \varepsilon\varphi_1(\dot{x}) + \nu_n^2 x + \varepsilon\psi_1(x),$$

and hence Eqs. (29.7) become, by considering (29.6),

$$\left.\begin{aligned} \nu_e^2 &= \nu_n^2 + \varepsilon\,\frac{E[x\varphi_1(\dot{x})] + E[x\psi_1(x)]}{E[x^2]}, \\ 2\zeta_e\nu_e &= 2\zeta\nu_n + \varepsilon\,\frac{E[\dot{x}\varphi_1(\dot{x})] + E[\dot{x}\psi_1(x)]}{E[\dot{x}^2]}. \end{aligned}\right\} \tag{29.9}$$

To calculate the mean values involved in (29.9), we replace the unknown probability density function $f(x, \dot{x})$ by the probability density function of the steady-state solution x_p of the linearized equation (29.2). Since $p(t)$ is normally distributed, so is $x_p(\tau)$ [129]. Moreover, in view of (29.6), the covariance $K_{x_p\dot{x}_p}$ of the random variables $x_p(t)$ and $\dot{x}_p(t)$ vanishes for any fixed t. Hence, by (25.70), we may take

$$f(x, \dot{x}) = f_1(x)\, f_2(\dot{x}), \tag{29.10}$$

where

$$f_1(x) = \frac{1}{\sigma_{x_p}\sqrt{2\pi}}\, e^{-x^2/2\sigma_{x_p}^2}, \qquad f_2(\dot{x}) = \frac{1}{\sigma_{\dot{x}_p}\sqrt{2\pi}}\, e^{-\dot{x}^2/2\sigma_{\dot{x}_p}^2}. \tag{29.11}$$

By (28.10), (28.22), and (28.23), the variances of x_p and $\dot{x}_p$ are

$$\sigma_{x_p}^2 = \int_0^\infty \frac{S_p(\nu)\,\mathrm{d}\nu}{(\nu_e^2 - \nu^2)^2 + 4\zeta_e\nu_e^2\nu^2}, \qquad \sigma_{\dot{x}_p}^2 = \int_0^\infty \frac{\nu^2 S_p(\nu)\,\mathrm{d}\nu}{(\nu_e^2 - \nu^2)^2 + 4\zeta_e\nu_e^2\nu^2}. \tag{29.12}$$

In view of (29.10), we may write

$$\left.\begin{aligned} E[x\varphi_1(\dot{x})] &= \int_{-\infty}^{\infty}\int_{-\infty}^{\infty} x\varphi_1(\dot{x})\,f_1(x)\,f_2(\dot{x}) = m_{x_p} m_{\varphi_1} = 0,\\ E[\dot{x}\psi_1(x)] &= \int_{-\infty}^{\infty}\int_{-\infty}^{\infty} \dot{x}\psi_1(x)\,f_1(x)\,f_2(\dot{x}) = m_{\dot{x}_p} m_{\psi_1} = 0, \end{aligned}\right\} \tag{29.13}$$

since, by (28.17) and (26.17), $m_{x_p} = m_{\dot{x}_p} = 0$.

Introducing now (29.10) and (29.11) into (29.9), and considering (29.13), we finally obtain

$$\left.\begin{aligned} \nu_e^2 &= \nu_n^2 + \frac{\varepsilon}{\sigma_{x_p}^3\sqrt{2\pi}} \int_{-\infty}^{\infty} x\psi_1(x)\, e^{-x^2/2\sigma_{x_p}^2}\,\mathrm{d}x,\\ 2\zeta_e\nu_e &= 2\zeta\nu_n + \frac{\varepsilon}{\sigma_{\dot{x}_p}^3\sqrt{2\pi}} \int_{-\infty}^{\infty} \dot{x}\varphi_1(\dot{x})\, e^{-\dot{x}^2/2\sigma_{\dot{x}}^2}\,\mathrm{d}\dot{x}. \end{aligned}\right\} \tag{29.14}$$

When the expressions (29.12) of $\sigma_{x_p}^2$ and $\sigma_{\dot{x}_p}^2$ are substituted into (29.14), a system of two coupled nonlinear algebraic equations results for the determination of ζ_e and ν_e.

b) The perturbation method

This method, which is based on classical perturbation theory, has been developed by CRANDALL [39] for weakly nonlinear oscillators. Let us consider the equation

$$\ddot{x} + 2\zeta\nu_n\dot{x} + \nu_n^2 x + \varepsilon g(x, \dot{x}) = p(t), \tag{29.15}$$

where ε is a small parameter and $p(t)$ is a weakly stationary random process with zero mean value.

Assume that there exists a solution of (29.15) having a steady-state probability density function $f(x, \dot{x})$, and which can be expanded in a power series of ε, for ε

sufficiently small, that is

$$x(t) = x_0(t) + \varepsilon x_1(t) + \varepsilon^2 x_2(t) + \ldots \tag{29.16}$$

Substituting (29.16) into (29.15) and equating coefficients of like powers of ε yields the recursive system of differential equations:

$$\begin{aligned} &\ddot{x}_0 + 2\zeta\nu_n\dot{x}_0 + \nu_n^2 x_0 = p(t), \\ &\ddot{x}_1 + 2\zeta\nu_n\dot{x}_1 + \nu_n^2 x_1 = -\, g(x_0, \dot{x}_0), \\ &\ddot{x}_2 + 2\zeta\nu_n\dot{x}_2 + \nu_n^2 x_2 = -\, x_1 \frac{\partial g}{\partial x}(x_0, \dot{x}_0) - \dot{x}_1 \frac{\partial g}{\partial \dot{x}}(x_0, \dot{x}_0), \text{ etc.} \end{aligned} \tag{29.17}$$

By using (28.6), the solutions of the linear differential equations (29.17) may be readily obtained. Thus

$$\begin{aligned} x_0(t) &= \int_{-\infty}^{\infty} p(t-\theta)\, h(\theta)\, \mathrm{d}\theta, \\ x_1(t) &= -\int_{-\infty}^{\infty} g(x_0(t-\theta),\ \dot{x}_0(t-\theta))\, h(\theta)\, \mathrm{d}\theta, \end{aligned} \tag{29.18}$$

where $h(\theta)$ is given by (28.5).

Equations (29.18) enable us to compute the various statistics of the response. However, the computational difficulties increase rapidly with the number of iterations considered. On the other hand, unlike the deterministic case, the convergence of the series of type (29.16) has not been rigorously investigated so far. Therefore, we will consider here only the nonlinear correction of first order in ε, i.e.,

$$x(t) = x_0(t) + \varepsilon x_1(t). \tag{29.19}$$

In this case, neglecting all terms of second and higher order in ε, the statistics of $x(t)$ to the second order are given by

$$m_x = \varepsilon m_{x_1} = -\frac{\varepsilon}{\nu_n^2} \int_{-\infty}^{\infty} \int_{-\infty}^{\infty} g(x_0, \dot{x}_0)\, f(x_0, \dot{x}_0)\, \mathrm{d}x_0 \mathrm{d}\dot{x}_0, \tag{29.20}$$

$$\begin{aligned} k_x(\tau) &= E[x(t)\, x(t+\tau)] \\ &= k_{x_0}(\tau) + \varepsilon\{E[x_0(t)\, x_1(t+\tau)] + E[x_1(t)\, x_0(t+\tau)], \end{aligned} \tag{29.21}$$

where, by (29.18),

$$\left.\begin{aligned} k_{x_0}(\tau) &= E[x_0(t)\, x_0(t+\tau)] = \\ &= \int_{-\infty}^{\infty}\int_{-\infty}^{\infty} E[p(t-\theta_1)\, p(t+\tau-\theta_2)]\, h(\theta_1)\, h(\theta_2)\, \mathrm{d}\theta_1 \mathrm{d}\theta_2 \\ &= \int_{-\infty}^{\infty}\int_{-\infty}^{\infty} k_p(\theta_1-\theta_2+\tau)\, h(\theta_1)\, h(\theta_2)\, \mathrm{d}\theta_1 \mathrm{d}\theta_2, \\ E[x_0(t)\, x_1(t+\tau)] &= E[x_0(t-\tau)\, x_1(t)] \\ &= -\int_{-\infty}^{\infty} E[x_0(t-\tau)\, g(x_0(t-\theta),\, \dot{x}_0(t-\theta))]\, h(\theta)\, \mathrm{d}\theta. \end{aligned}\right\} \tag{29.22}$$

In particular, from (29.21) and (29.22) it follows that

$$\sigma_x^2 = k_x(0) = \sigma_{x_0}^2 + 2\varepsilon E[x_0(t)\, x_1(t)], \tag{29.23}$$

where

$$\left.\begin{aligned} \sigma_{x_0}^2 &= \int_{-\infty}^{\infty}\int_{-\infty}^{\infty} k_p(\theta_1-\theta_2)\, h(\theta_1)\, h(\theta_2)\, \mathrm{d}\theta_1\, \mathrm{d}\theta_2, \\ E[x_0(t) x_1(t)] &= -\int_{-\infty}^{\infty} E[x_0(t)\, g(x_0(t-\theta),\ \dot{x}_0(t-\theta))]\, h(\theta)\, \mathrm{d}\theta. \end{aligned}\right\} \tag{29.24}$$

The perturbation technique has been extensively used by Crandall [39] and Crandall et al. [40] in the study of nonlinear random vibrations of one- and two-degree-of-freedom systems.

To illustrate this method we consider the Duffing oscillator excited by a weakly stationary, normally distributed random process. In this case we have

$$g(x, \dot{x}) = x^3. \tag{29.25}$$

We will calculate the mean value and the variance of the response to the first order in ε, by using Eqs. (29.20) and (29.23).

Introducing (29.25) into (29.20) and $(29.24)_2$ and considering $(29.18)_1$, we obtain

$$m_x = -\frac{\varepsilon}{\nu_n^2} E[x_3^0]$$

$$= -\frac{\varepsilon}{\nu_n^2}\int_{-\infty}^{\infty} h(\theta_1)\, \mathrm{d}\theta_1 \int_{-\infty}^{\infty} h(\theta_2)\, \mathrm{d}\theta_2 \int_{-\infty}^{\infty} E[p(t-\theta_1)\, p(t-\theta_2)\, p(t-\theta_3)]\, h(\theta_3)\, \mathrm{d}\theta_3,$$

$$E[x_0(t)\,x_1(t)] = -\int_{-\infty}^{\infty} E[x_0(t)\,x_0^3(t-\theta)]\,h(\theta)\,\mathrm{d}\theta$$

$$= -\int_{-\infty}^{\infty} h(\theta_1)\,\mathrm{d}\theta_1 \int_{-\infty}^{\infty} h(\theta_2)\,\mathrm{d}\theta_2 \int_{-\infty}^{\infty} h(\theta_3)\,\mathrm{d}\theta_3 \int_{-\infty}^{\infty} h(\theta_4)\,\mathrm{d}\theta_4 \int_{-\infty}^{\infty} h(\theta)\,\mathrm{d}\theta \times$$

$$\times E[p(t-\theta_1)\,p(t-\theta-\theta_2)\,p(t-\theta-\theta_3)\,p(t-\theta-\theta_4)]. \quad (29.26)$$

On the other hand, it may be shown [39] that a weakly stationary, normally distributed random process, $z(t)$, with zero mean value, satisfies the following identities:

$$E[z(t_1)\,z(t_2)\,z(t_3)] = 0,$$

$$E[z(t_1)\,z(t_2)\,z(t_3)\,z(t_4)] = k_z(t_1-t_2)\,k_z(t_3-t_4) + k_z(t_1-t_3)\,k_z(t_2-t_4) +$$

$$+ k_z(t_1-t_4)\,k_z(t_2-t_3).$$

Consequently,

$$m_x = 0, \quad (29.27)$$

$$E[p(t-\theta_1)\,p(t-\theta-\theta_2)\,p(t-\theta-\theta_3)\,p(t-\theta-\theta_4)] =$$

$$= k_p(\theta-\theta_1+\theta_2)\,k_p(\theta_3-\theta_4) + k_p(\theta-\theta_1+\theta_3)\,k_p(\theta_2-\theta_4) +$$

$$+ k_p(\theta-\theta_1+\theta_4)\,k_p(\theta_2-\theta_3), \quad (29.28)$$

and hence (29.26) is reduced to quadratures of the correlation function.

To write Eq. (29.26) in terms of the input correlation function, we make repeatedly use of Eq. $(29.22)_1$, thus obtaining

$$E[x_0(t)\,x_1(t)] = -3\sigma_{x_0}^2 \int_{-\infty}^{\infty} h(\theta)\,k_{x_0}(\theta)\,\mathrm{d}\theta. \quad (29.29)$$

Finally, by introducing this relation into (29.93), we find

$$\sigma_x^2 = \sigma_{x_0}^2 \left\{1 - 6\varepsilon \int_{-\infty}^{\infty} h(\theta)\,k_{x_0}(\theta)\,\mathrm{d}\theta\right\}. \quad (29.30)$$

Chapter IX

Applications of the correlation theory to the linear oscillator

§ 30. STATISTICAL CHARACTERISTICS OF THE RESPONSE

In this chapter we will consider the linear oscillator with the governing equation

$$\ddot{x} + 2\zeta\nu_n\dot{x} + \nu_n^2 x = p(t), \tag{30.1}$$

where $0 < \zeta < 1$, and $p(t)$ is a weakly stationary random process with zero mean value. The random process $p(t)$ will be characterized by its correlation function or, equivalently, by its spectral density. Various standard forms of these functions that proved efficient in modelling random excitations will be considered in the following.

The main application we have in mind is to random vibrations of road vehicles. In this case, as already shown in the introduction, Eq. (30.1) assumes the form [1)]

$$\ddot{x} + 2\zeta\nu_n\dot{x} + \nu_n^2 x = -\ddot{x}_0(t), \tag{30.2}$$

where $x_0(t)$ is the height above some fixed level of the road irregularity surmounted by the wheel at time t, and

$$x = x_1 - x_0, \tag{30.3}$$

where x_1 is the displacement of the sprung mass. Substituting (30.3) into (30.2) yields

$$\ddot{x}_1 + 2\zeta\nu_n x_1 + \nu_n^2 x_1 = q(t), \tag{30.4}$$

[1)] Denoting by $h(d)$ the height of the irregularity about some reference level as a function of the covered distance d, and assuming that the vehicle rides at a constant speed V, we have $x_0(t) \equiv h(Vt)$.

where

$$q(t) = 2\zeta \nu_n \dot{x}_0 + \nu_n^2 x_0. \tag{30.5}$$

Since only steady-state solutions will be considered, we shall denote them simply by $x(t)$, instead of $x_p(t)$ as in § 28. By (28.21) and (26.43), we obtain the spectral density of $x(t)$ and of its time derivatives as

$$S_{x^{(k)}}(\nu) = \nu^{2k} |H(\nu)|^2 S_{\ddot{x}_0}(\nu), \qquad k = 0, 1, 2, \ldots \tag{30.6}$$

with the convention that $x^{(0)}(t) \equiv x(t)$. If the spectral density of $x_0(t)$ rather than that of $\ddot{x}_0(t)$ is known, we write by Eq. $(26.44)_2$:

$$S_{x^{(k)}}(\nu) = \nu^{2k+4} |H(\nu)|^2 S_{x_0}(\nu), \qquad k = 0, 1, 2, \ldots \tag{30.7}$$

Similarly, starting with (30.4), we find

$$S_{x_1^{(k)}}(\nu) = \nu^{2k} |H(\nu)|^2 S_q(\nu), \qquad k = 0, 1, 2, \ldots \tag{30.8}$$

Denoting by $H_1(\nu)$ the transfer function corresponding to Eq. (30.5), we may write

$$S_{x_0}(\nu) = |H_1(\nu)|^2 S_q(\nu),$$

and hence

$$S_{x_1^{(k)}}(\nu) = \frac{\nu^{2k} |H(\nu)|^2}{|H_1(\nu)|^2} S_{x_0}(\nu), \qquad k = 0, 1, 2, \ldots \tag{30.9}$$

By $(26.44)_2$ we also have

$$S_{x_1^{(k)}}(\nu) = \frac{\nu^{2k-4} |H(\nu)|^2}{|H_1(\nu)|^2} S_{\ddot{x}_0}(\nu), \qquad k = 0, 1, 2, \ldots \tag{30.10}$$

Putting $q(t) = e^{i\nu t}$, $x_0(t) = H_1(\nu) e^{i\nu t}$ into Eq. (30.5), dividing out a factor of $e^{i\nu t}$, and solving the algebraic equation obtained, it results that

$$H_1(\nu) = \frac{1}{2i\zeta \nu_n \nu + \nu_n^2}. \tag{30.11}$$

Making use of (28.10) and (30.11), Eqs. (30.6), (30.7), (30.9) and (30.10) may be rewritten as

$$\left.\begin{aligned} S_{x^{(k)}}(\nu) &= \frac{\nu^{2k} S_{\ddot{x}_0}(\nu)}{(\nu_n^2 - \nu^2)^2 + 4\zeta^2\nu_n^2\nu^2}, \\ S_{x^{(k)}}(\nu) &= \frac{\nu^{2k+4} S_{x_0}(\nu)}{(\nu_n^2 - \nu^2)^2 + 4\zeta^2\nu_n^2\nu^2}, \end{aligned}\right\} \tag{30.12}$$

$$\left.\begin{aligned} S_{x_1^{(k)}}(\nu) &= \frac{\nu^{2k-4}\nu_n^2(\nu_n^2 + 4\zeta^2\nu^2)\, S_{\ddot{x}_0}(\nu)}{(\nu_n^2 - \nu^2)^2 + 4\zeta^2\nu_n^2\nu^2}, \\ S_{x_1^{(k)}}(\nu) &= \frac{\nu^{2k}\nu_n^2(\nu_n^2 + 4\zeta^2\nu^2)\, S_{x_0}(\nu)}{(\nu_n^2 - \nu^2)^2 + 4\zeta^2\nu_n^2\nu^2} \end{aligned}\right\} \tag{30.13}$$

for $k = 0, 1, 2, \ldots$

Of particular interest in road vehicle dynamics are the variances of $x(t)$, $\dot{x}(t)$, and $\ddot{x}_1(t)$. Therefore, we give the explicit expressions of these quantities, as obtained from Eqs. (26.42), (30.12) and (30.13):

$$\left.\begin{aligned} \sigma_x^2 &= \int_0^\infty \frac{S_{\ddot{x}_0}(\nu)\, d\nu}{(\nu_n^2 - \nu^2)^2 + 4\zeta^2\nu_n^2\nu^2}, \\ \sigma_{\dot{x}}^2 &= \int_0^\infty \frac{\nu^2 S_{\ddot{x}_0}(\nu)\, d\nu}{(\nu_n^2 - \nu^2)^2 + 4\zeta^2\nu_n^2\nu^2}, \\ \sigma_{\ddot{x}_1}^2 &= \int_0^\infty \frac{\nu_n^2(\nu_n^2 + 4\zeta^2\nu^2)\, S_{\ddot{x}_0}(\nu)\, d\nu}{(\nu_n^2 - \nu^2) + 4\zeta^2\nu_n^2\nu^2}, \end{aligned}\right\} \tag{30.14}$$

$$\left.\begin{aligned} \sigma_x^2 &= \int_0^\infty \frac{\nu^4 S_{x_0}(\nu)\, d\nu}{(\nu_n^2 - \nu^2)^2 + 4\zeta^2\nu_n^2\nu^2}, \\ \sigma_{\dot{x}}^2 &= \int_0^\infty \frac{\nu^6 S_{x_0}(\nu)\, d\nu}{(\nu_n^2 - \nu^2)^2 + 4\zeta^2\nu_n^2\nu^2}, \\ \sigma_{\ddot{x}_1}^2 &= \int_0^\infty \frac{\nu^4\nu_n^2(\nu_n^2 + 4\zeta^2\nu^2)\, S_{x_0}(\nu)\, d\nu}{(\nu_n^2 - \nu^2)^2 + 4\zeta^2\nu_n^2\nu^2}. \end{aligned}\right\} \tag{30.15}$$

§ 31. THE LINEAR OSCILLATOR EXCITED BY WHITE NOISE

A stationary random process whose spectral density has significant values over a band of frequencies which is of the same order of magnitude as the center frequency of the band is said to be a *wide-band* process (Fig. 77). A useful idealization of a very broad band-width process is obtained by considering that its spectral density is constant over all frequencies. Such a stationary random process is called a *white noise* (Fig. 78).

In a strict sense, white noise is a physically unrealizable phenomenon, since, by (26.42), it would possess an infinite variance. In many practical problems, however, the use of a white noise model simplifies the mathematical calculation without introducing any significant inaccuracy in the end result. It is also worth noting that any given random process can be generated by passing a white noise through a suitable filter.

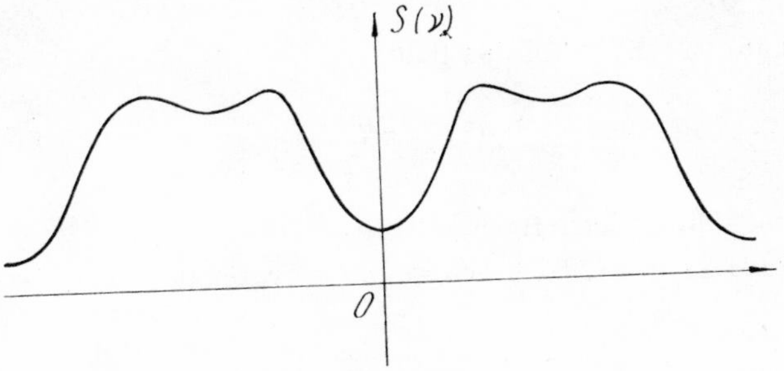

Fig. 77

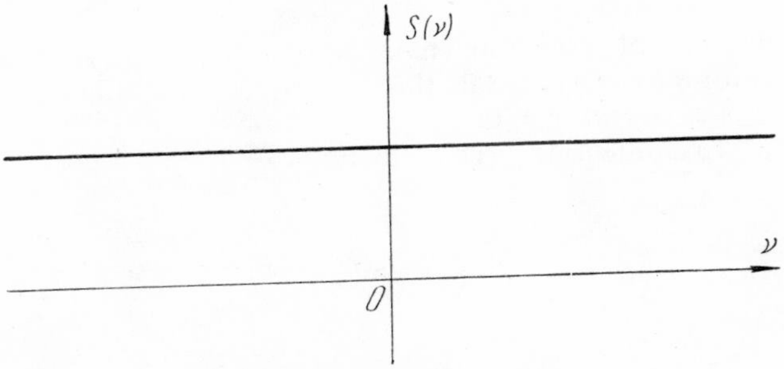

Fig. 78

Sometimes, a more realistic approximation of a wide-band process is provided by the so-called *band-limited white noise*, whose spectral density is constant over a certain band of frequencies and zero elsewhere (Fig. 79). Obviously, the variance of a band-limited white noise is finite.

Let us consider first the case when $\ddot{x}_0(t)$ in Eq. (30.2) is a white noise with zero mean value, i.e.,

$$m_{\ddot{x}_0} = 0, \qquad S_{\ddot{x}_0}(\nu) = S_{\ddot{x}_0} = \text{const.} \tag{31.1}$$

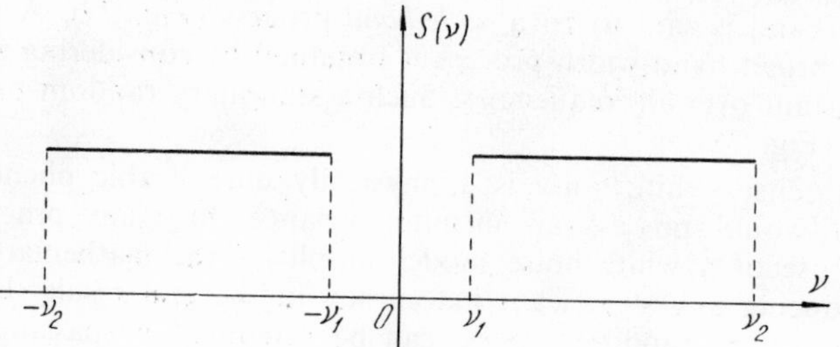

Fig 79

From Eqs. (26.36), (26.37) it follows that

$$k_{\ddot{x}_0}(\tau) = \pi S_{\ddot{x}_0}\, \delta(\tau), \tag{31.2}$$

where $\delta(\tau)$ is the Dirac delta function.

Introducing (31.2) into (28.20) we obtain the correlation function of the steady-state response

$$k_x(\tau) = \pi S_{\ddot{x}_0} \int_0^\infty \int_0^\infty h(\theta_1)\, h(\theta_2)\, \delta(\tau + \theta_1 - \theta_2)\, \mathrm{d}\theta_1 \mathrm{d}\theta_2, \tag{31.3}$$

where the lower limits of integration have been set equal to zero in view of the expression (28.5) for $h(\theta)$. To calculate this integral, it is necessary to consider separately the cases $\tau > 0$ and $\tau < 0$.

For $\tau > 0$ we first integrate over θ_2 to obtain

$$k_x(\tau) = \pi S_{\ddot{x}_0} \int_0^\infty h(\theta_1)\, h(\tau + \theta_1)\, \mathrm{d}\theta_1$$

$$= \frac{\pi S_{\ddot{x}_0}}{\nu_n^2(1 - \zeta^2)} \int_0^\infty \mathrm{e}^{-\zeta \nu_n(\tau + 2\theta_1)} \sin \nu_n \theta_1 \sqrt{1 - \zeta^2} \sin \nu_n(\tau + \theta_1) \sqrt{1 - \zeta^2}\, \mathrm{d}\theta_1,$$

whence

$$k_x(\tau) = \frac{\pi S_{\ddot{x}_0}}{4\zeta \nu_n^3} \mathrm{e}^{-\zeta \nu_n \tau} \left(\cos \nu_n \tau \sqrt{1 - \zeta^2} + \frac{\zeta}{\sqrt{1 - \zeta^2}} \sin \nu_n \tau \sqrt{1 - \zeta^2} \right), \qquad \tau > 0. \tag{31.4}$$

For $\tau < 0$ we can avoid repeating the integration, by using the fact that $k_x(\tau)$ is an even function of τ, thus obtaining

$$k_x(\tau) = \frac{\pi S_{\ddot{x}_0}}{4\zeta\nu_n^3} e^{\zeta\nu_n\tau}\left(\cos \nu_n\tau\sqrt{1-\zeta^2} - \frac{\zeta}{\sqrt{1-\xi^2}}\sin \nu_n\tau\sqrt{1-\zeta^2}\right), \quad \tau < 0. \quad (31.5)$$

By using Eqs. (31.4) and (31.5), it may be shown that $k_x(\tau)$ is continuous together with its first and second derivatives at $\tau = 0$, but the third derivative is discontinuous there. Hence, by Theorem 6, § 26, and Eq. (26.33), we deduce that the response $x(t)$ is continuous in mean square together with its first derivative. From Eqs. (30.2) and (30.3) we also derive that $\ddot{x}_1(t)$ is continuous in mean square.

When the damping ratio ζ is sufficiently small, Eqs. (31.4) and (31.5) may be written in the simpler, approximate form:

$$k_x(\tau) = \frac{\pi S_{\ddot{x}_0}}{4\zeta\nu_n^3} e^{-\zeta\nu_n|\tau|}\cos \nu_n\tau\sqrt{1-\zeta^2}. \quad (31.6)$$

This relation shows that, in the case of a light damping, the correlation function of $x(t)$ is approximately given by a damped harmonic with practically the same circular frequency as the natural frequency of the oscillator (Fig. 80).

The spectral density of the response may be directly obtained by setting $k = 0$ in Eq. $(30.12)_1$:

$$S_x(\nu) = \frac{S_{\ddot{x}_0}}{(\nu_n^2 - \nu^2)^2 + 4\zeta^2\nu_n^2\nu^2}. \quad (31.7)$$

This function assumes the minimum $S_{\ddot{x}_0}/\nu_n^4$ for $\nu = 0$ and the equal maxima $S_{\ddot{x}_0}/\{4\zeta^2(1-\zeta^2)\nu_n^4\}$ for $\nu = \pm\nu_n\sqrt{1-2\zeta^2}$ (see Fig. 81). Thus, for light damping, the maximum point of the spectral density for $\nu \in (0, \infty)$ corresponds approximately to the natural frequency of the oscillator.

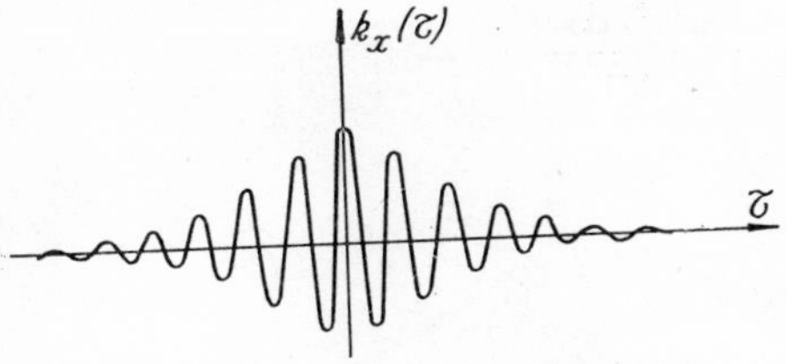

Fig. 80

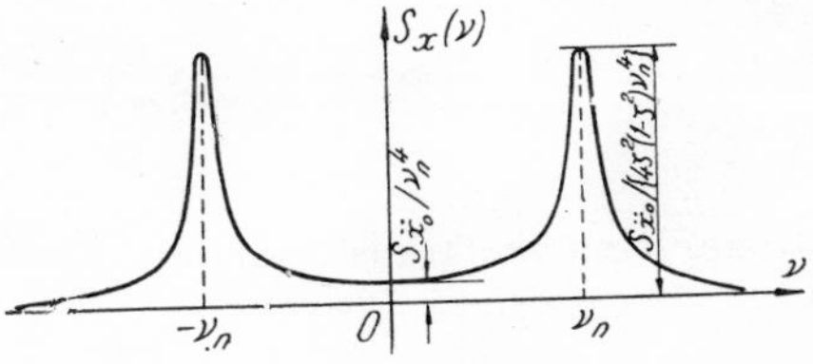

Fig. 81

According to (26.28), the variance of the response may be obtained by letting τ approach zero in (31.4) or (31.5). The result is

$$\sigma_x^2 = k_x(0) = \frac{\pi S_{\ddot{x}_0}}{4\zeta\nu_n^3}. \quad (31.8)$$

To obtain the variances of $\dot{x}(t)$ and $\ddot{x}_1(t)$, we use Eqs. (30.14) with $S_{\ddot{x}_0}(\nu) = S_{\ddot{x}_0}$. It follows that

$$\sigma_{\dot{x}}^2 = \frac{\pi S_{\ddot{x}_0}}{4\zeta\nu_n}, \tag{31.9}$$

$$\sigma_{\ddot{x}_1}^2 = \frac{\pi\nu_n S_{\ddot{x}_0}(1 + 4\zeta^2)}{4\zeta}. \tag{31.10}$$

We remark that $\sigma_{\dot{x}}^2$ and $\sigma_{\dot{x}}^2$ are monotonically decreasing functions of ζ. Considering $\sigma_{\ddot{x}_1}^2$ as a function of ζ for $\zeta \in (0, 1)$, we find that it has the absolute minimum $\pi\nu_n S_{\ddot{x}_0}$ at $\zeta = 0.5$.

Consider next the case when $x_0(t)$ is a white noise. Then, it is readily proved that none of the integrals involved in Eqs. (30.15) is convergent, and hence all variances σ_x^2, $\sigma_{\dot{x}}^2$, and $\sigma_{\ddot{x}_1}^2$, are infinite. To obtain a more realistic model we assume, therefore, that $x_0(t)$ is a band-limited white noise with the spectral density

$$S_{x_0}(\nu) = \begin{cases} S_{x_0} & \text{for } |\nu| \leqslant \nu_1, \\ 0 & \text{for } |\nu| > \nu_1, \end{cases} \tag{31.11}$$

where ν_1 is some cut-off frequency.

In view of (31.11), the variances (30.15) may be written as

$$\begin{aligned} \sigma_x^2 &= S_{x_0} \int_0^{\nu_1} \frac{\nu^4 \, d\nu}{(\nu_n^2 - \nu^2)^2 + 4\zeta^2\nu_n^2\nu^2}, \\ \sigma_{\dot{x}}^2 &= S_{x_0} \int_0^{\nu_1} \frac{\nu^6 \, d\nu}{(\nu_n^2 - \nu^2)^2 + 4\zeta^2\nu_n^2\nu^2}, \\ \sigma_{\ddot{x}}^2 &= \nu_n^2 S_{x_0} \int_0^{\nu_1} \frac{\nu^4(\nu_n^2 + 4\zeta^2\nu^2) \, d\nu}{(\nu_n^2 - \nu^2)^2 + 4\zeta^2\nu_n^2\nu^2}. \end{aligned} \tag{31.12}$$

By applying the transformation

$$\nu = \xi\nu_n, \tag{31.13}$$

Eqs. (31.12) become

$$\begin{aligned} \sigma_x^2 &= \nu_n S_{x_0} \eta_x(\zeta, \xi_1), \\ \sigma_{\dot{x}}^2 &= \nu_n^3 S_{x_0} \eta_{\dot{x}}(\zeta, \xi_1), \\ \sigma_{\ddot{x}_1}^2 &= \nu_n^5 S_{x_0} \eta_{\ddot{x}_1}(\zeta, \xi_1), \end{aligned} \tag{31.14}$$

where $\xi_1 = \nu_1/\nu_n$, and

$$\eta_x(\zeta, \xi_1) = \int_0^{\xi_1} \frac{\xi^4 d\xi}{(1 - \xi^2)^2 + 4\zeta^2\xi^2},$$

$$\eta_{\dot{x}}(\zeta, \xi_1) = \int_0^{\xi_1} \frac{\xi^6 d\xi}{(1 - \xi^2)^2 + 4\zeta^2\xi^2}, \tag{31.15}$$

$$\eta_{\ddot{x}_1}(\zeta, \xi_1) = \int_0^{\xi_1} \frac{\xi^4 (1 + 4\zeta^2\xi^2)\, d\xi}{(1 - \xi^2)^2 + 4\zeta^2\xi^2}.$$

Calculating the integrals (31.15) yields

$$\eta_x(\zeta, \xi_1) = \xi_1 - \frac{3 - 4\zeta^2}{8\sqrt{1 - \zeta^2}} \ln \frac{\xi_1^2 + 2\xi_1\sqrt{1 - \zeta^2} + 1}{\xi_1^2 - 2\xi_1\sqrt{1 - \zeta^2} + 1} + \frac{1 - 4\zeta^2}{4\zeta} \tan^{-1} \frac{2\zeta\xi_1}{1 - \xi_1^2},$$

$$\eta_{\dot{x}}(\zeta, \xi_1) = \frac{\xi_1^3}{3} + 2(1 - 2\zeta^2)\xi_1 - \frac{3 - 4\zeta^2}{8\sqrt{1 - \zeta^2}} \ln \frac{\xi_1^2 + 2\xi_1\sqrt{1 - \zeta^2} + 1}{\xi_1^2 - 2\xi_1\sqrt{1 - \zeta^2} + 1} +$$

$$+ \frac{1 - 12\zeta^2 + 16\zeta^4}{4\zeta} \tan^{-1} \frac{2\zeta\xi_1}{1 - \xi^2}, \tag{31.16}$$

$$\eta_{\ddot{x}_1}(\zeta, \xi_1) = \frac{4\zeta^2\xi_1^3}{3} + (1 + 8\zeta^2 - 16\zeta^4)\xi_1 -$$

$$- \frac{3 + 16\zeta^2 - 80\zeta^4 + 64\zeta^6}{8\sqrt{1 - \zeta^2}} \ln \frac{\xi_1^2 + 2\xi_1\sqrt{1 - \zeta^2} + 1}{\xi_1^2 - 2\xi_1\sqrt{1 - \zeta^2} + 1} +$$

$$+ \frac{1 - 48\zeta^4 + 64\zeta^6}{4\zeta} \tan^{-1} \frac{2\zeta\xi_1}{1 - \xi_1^2},$$

where

$$\tan^{-1} \frac{2\zeta\xi_1}{1 - \xi_1^2} = \begin{cases} \arctan \dfrac{2\zeta\xi_1}{1 - \xi_1^2} & \text{for } 0 \leqslant \xi_1 < 1, \\[2ex] \dfrac{\pi}{2} & \text{for } \xi_1 = 1, \\[2ex] \pi + \arctan \dfrac{2\zeta\xi_1}{1 - \xi_1^2} & \text{for } \xi_1 > 1. \end{cases} \tag{31.17}$$

Remarks

1. For any fixed $\xi_1 > 0$, the variances σ_x^2 and $\sigma_{\dot{x}}^2$ are monotonically decreasing functions of ζ. Indeed, if $\zeta_1 < \zeta_2$, then, for any $\xi \in (0, \xi_1)$ and any positive integer n, we have

$$\frac{\xi^n}{(1 - \xi^2)^2 + 4\zeta_1^2\xi^2} > \frac{\xi^n}{(1 - \xi^2)^2 + 4\zeta_2^2\xi^2},$$

and from (31.15) it follows that

$$\eta_x(\zeta_1, \xi_1) > \eta_x(\zeta_2, \xi_1), \quad \eta_{\dot{x}}(\zeta_1, \xi_1) > \eta_{\dot{x}}(\zeta_2, \xi_1), \qquad \text{Q.E.D.}$$

2. As will be shown below in a more general case, corresponding to any $\xi_1 > 0$, there exists a minimum point ζ_1 of $\sigma_{\ddot{x}_1}(\zeta)$ belonging to the interval (0.1). This property may be also verified by making use of Table 9, where the values of $\eta_{\ddot{x}_1}(\zeta, \xi_1)$ have been listed for $\zeta \in [0.1, 0.9]$ and $\xi_1 \in [2, 10]$. Inspection of this table reveals that $\zeta_1 \approx 0.4$ for $\xi_1 = 2$, and $\zeta_1 \approx 0.2$ for $\xi_1 = 3$ and $\xi_1 = 4$. For larger values of ξ_1, the variance $\sigma_{\ddot{x}_1}^2$ monotonically increases with $\zeta \in [0.1, 0.9]$. Finding in this case the minimum of $\sigma_{\ddot{x}_1}^2$ would require to calculate the values of $\sigma_{\ddot{x}_1}^2$ as a function of $\zeta \in [0, 0.1]$.

Table 9

ξ_1 \ ζ	0.1	0.2	0.3	0.4	0.5	0.6	0.7	0.8	0.9
2	8.891	5.408	4.646	4.540	4.650	4.831	5.016	5.191	5.457
3	10.680	8.181	8.958	10.796	13.111	15.623	18.166	20.638	22.983
4	12.448	11.663	15.191	20.721	27.520	35.140	43.247	51.600	59.380
5	14.453	16.326	24.280	35.710	49.897	66.200	84.090	103.100	122.867
6	16.815	22.551	36.810	57.047	82.270	111.667	144.550	180.250	218.142
7	19.638	30.680	53.692	86.060	126.170	169.930	228.545	288.200	352.300
8	23.006	41.030	75.610	124.012	184.967	257.400	340.000	431.840	531.780
9	27.005	53.915	103.270	172.210	259.400	363.400	482.900	616.600	763.200
10	31.970	69.640	137.400	231.932	351.680	495.170	660.870	847.120	1052.750

§ 32. THE LINEAR OSCILLATOR EXCITED BY A STATIONARY RANDOM PROCESS WITH CORRELATION FUNCTION OF EXPONENTIAL TYPE

We will assume in this section that the spectral density of the process $x_0(t)$ occurring in Eq. (30.2) may be obtained by limiting the band of the spectral density corresponding to the correlation function

$$k^*_{x_0}(\tau) = \sigma^2_{x_0}\, e^{-\alpha|\tau|}, \tag{32.1}$$

where α is a real positive constant.

Obviously, $k^*_{x_0}(\tau)$ satisfies the conditions for the existence of its Fourier transform. Consequently, we may apply Eq. (26.37) to obtain the spectral density associated with $k^*_{x_0}(\tau)$. It results that

$$S^*_{x_0}(\tau) = \frac{2\sigma^2_{x_0}}{\pi} \frac{\alpha}{\nu^2 + \alpha^2}. \tag{32.2}$$

By limiting the frequency band of this spectral density, we obtain

$$S_{x_0}(\tau) = \begin{cases} \dfrac{2\sigma^2_{x_0}}{\pi} \dfrac{\alpha}{\nu^2 + \alpha^2} & \text{for} \quad |\nu| \leqslant \nu_1, \\ 0 & \text{for} \quad |\nu| > \nu_1, \end{cases} \tag{32.3}$$

where ν_1 is some cut-off frequency.

Conversely, it may be shown that the correlation function of the random process $x_0(t)$, i.e.,

$$k_{x_0}(\tau) = \frac{\alpha\sigma^2_{x_0}}{\pi} \int_{-\nu_1}^{\nu_1} \frac{e^{i\nu\tau}\, d\nu}{\nu^2 + \alpha^2}, \tag{32.4}$$

actually approaches $k^*_{x_0}(\tau)$ as $\nu_1 \to \infty$.

Figure 82 illustrates the family of band-limited spectral densities given by Eq. (32.1), corresponding to a few values of the parameter α.

It is apparent from Eq. (32.3) that for $\alpha \gg \nu_1$ the random process $x_0(t)$ may be approximated by a band-limited white noise with spectral density

$$S_{x_0}(\nu) = \begin{cases} \sigma^2_{x_0}/\pi\alpha & \text{for} \quad |\nu| \leqslant \nu_1, \\ 0 & \text{for} \quad |\nu| > \nu_1. \end{cases}$$

That is why we shall restrict our attention mainly to the case of small values of the parameter α, thus obtaining a possible model for random processes which are not representable by a white noise.

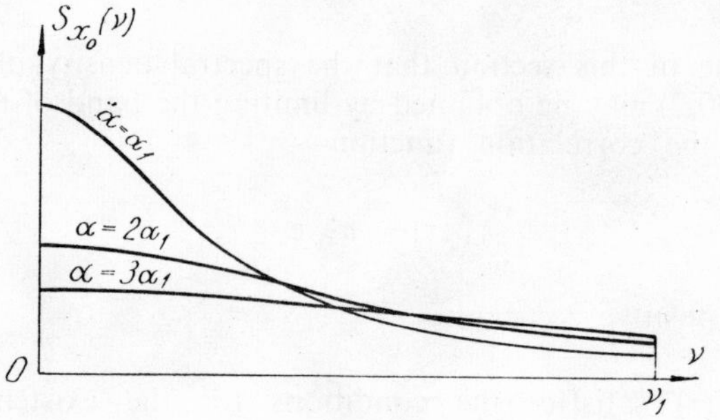

Fig. 82

Substituting (32.3) into (30.15), we obtain

$$\sigma_x^2 = \frac{2\alpha\sigma_{x_0}^2}{\pi}\int_0^{\nu_1}\frac{\nu^4 \mathrm{d}\nu}{(\alpha^2+\nu^2)\,[(\nu_n^2-\nu^2)^2+4\zeta\nu_n^2\nu^2]},$$

$$\sigma_{\dot{x}}^2 = \frac{2\alpha\sigma_{x_0}^2}{\pi}\int_0^{\nu_1}\frac{\nu^6 \mathrm{d}\nu}{(\alpha^2+\nu^2)\,[(\nu_n^2-\nu^2)^2+4\zeta^2\nu_n^2\nu^2]}, \tag{32.5}$$

$$\sigma_{\ddot{x}_1}^2 = \frac{2\alpha\nu_n^2\sigma_{x_0}^2}{\pi}\int_0^{\nu_1}\frac{\nu^4(\nu_n^2+4\zeta^2\nu^2)\,\mathrm{d}\nu}{(\alpha^2+\nu^2)\,[(\nu_n^2-\nu^2)^2+4\zeta^2\nu_n^2\nu^2]}.$$

By introducing the dimensionless parameter

$$\mu = \alpha/\nu_n \tag{32.6}$$

and applying the transformation (31.13), these formulas may be rewritten as

$$\sigma_x^2 = \frac{2\mu\sigma_{x_0}^2}{\pi}\,\eta_x(\zeta,\,\xi_1,\,\mu),$$

$$\sigma_{\dot{x}}^2 = \frac{2\mu\nu_n^2\,\sigma_{x0}^2}{\pi}\,\eta_{\dot{x}}\,(\zeta,\,\xi_1,\,\mu), \tag{32.7}$$

$$\sigma_{\ddot{x}_1}^2 = \frac{2\mu\nu_n^4\,\sigma_{x_0}^2}{\pi}\,\eta_{\ddot{x}_1}(\zeta,\,\xi_1,\,\mu),$$

where

$$\left.\begin{aligned}
\eta_x(\zeta,\xi_1,\mu) &= \int_0^{\xi_1} \frac{\xi^4 \mathrm{d}\xi}{(\mu^2+\xi^2)\,[(1-\xi^2)^2+4\zeta^2\xi^2]}\,,\\
\eta_{\dot{x}}(\zeta,\xi_1,\mu) &= \int_0^{\xi_1} \frac{\xi^6 \mathrm{d}\xi}{(\mu^2+\xi^2)\,[(1-\xi^2)^2+4\zeta^2\xi^2]}\,,\\
\eta_{\ddot{x}_1}(\zeta,\xi_1,\mu) &= \int_0^{\xi_1} \frac{\xi^4(1+4\zeta^2\xi^2)\,\mathrm{d}\xi}{(\mu^2+\xi^2)\,[(1-\xi^2)^2+4\zeta^2\xi^2]}\,.
\end{aligned}\right\} \tag{32.8}$$

Calculating these integrals gives

$$\left.\begin{aligned}
\eta_x(\zeta,\xi_1,\mu) &= \frac{1}{1+2\mu^2(1-2\zeta^2)+\mu^4}\times\\
&\times\left\{-\frac{1+\mu^2(3-4\zeta^2)}{8\sqrt{1-\zeta^2}}\ln\frac{\xi_1^2+2\xi_1\sqrt{1-\zeta^2}+1}{\xi_1^2-2\xi_1\sqrt{1-\zeta^2}+1}+\right.\\
&\left.+\frac{1+\mu^2(1-4\zeta^2)}{4\zeta}\tan^{-1}\frac{2\zeta\xi_1}{1-\xi_1^2}+\mu^3\arctan\frac{\xi_1}{\mu}\right\},\\
\eta_{\dot{x}}(\zeta,\xi_1,\mu) &= \xi_1+\frac{1}{1+2\mu^2(1-2\zeta^2)+\mu^4}\times\\
&\times\left\{-\frac{3-4\zeta^2+\mu^2(5-20\zeta^2+16\zeta^4)}{8\sqrt{1-\zeta^2}}\ln\frac{\xi_1^2+2\xi_1\sqrt{1-\zeta^2}+1}{\xi_1^2-2\xi_1\sqrt{1-\zeta^2}+1}+\right.\\
&\left.+\frac{1-4\zeta^2+\mu^2(1-12\zeta^2+16\zeta^4)}{4\zeta}\tan^{-1}\frac{2\zeta\xi_1}{1-\xi_1^2}-\mu^5\arctan\frac{\xi_1}{\mu}\right\},\\
\eta_{\ddot{x}_1}(\zeta,\xi_1,\mu) &= 4\zeta^2\xi_1+\frac{1}{1+2\mu^2(1-2\zeta^2)+\mu^4}\times\\
&\times\left\{-\frac{1+12\zeta^2-16\zeta^4+\mu^2(3+16\zeta^2-80\zeta^4+64\zeta^6)}{8\sqrt{1-\zeta^2}}\ln\frac{\xi_1^2+2\xi_1\sqrt{1-\zeta^2}+1}{\xi_1^2-2\xi_1\sqrt{1-\zeta^2}+1}+\right.\\
&+\frac{1+4\zeta^2-16\zeta^4+\mu^2(1-48\zeta^4+64\zeta^6)}{4\zeta}\tan^{-1}\frac{2\zeta\xi_1}{1-\xi_1^2}+\\
&+\mu^3(1-4\zeta^2\mu^2)\arctan\frac{\xi_1}{\mu},
\end{aligned}\right\} \tag{32.9}$$

where the function $\tan^{-1}\dfrac{2\zeta\xi_1}{1-\xi_1^2}$ is given by (31.17).

Assuming that $\alpha \ll \nu_n$, and hence $\mu \ll 1$, the terms containing second and higher powers of μ can be dropped out from Eqs. (32.9), and $\eta_x(\zeta, \xi_1, \mu)$, $\eta_{\dot{x}}(\zeta, \xi_1, \mu)$, $\eta_{\ddot{x}_1}(\zeta, \xi_1, \mu)$ can be replaced in (32.7) by the approximate expressions

$$\eta_x(\zeta, \xi_1, 0) = -\frac{1}{8\sqrt{1-\zeta^2}} \ln \frac{\xi_1^2 + 2\xi_1\sqrt{1-\zeta^2}+1}{\xi_1^2 - 2\xi_1\sqrt{1-\zeta^2}+1} + \frac{1}{4\zeta}\tan^{-1}\frac{2\zeta\xi_1}{1-\xi_1^2},$$

$$\eta_{\dot{x}}(\zeta, \xi_1, 0) = \xi_1 - \frac{3-4\zeta^2}{8\sqrt{1-\zeta^2}} \ln \frac{\xi_1^2 + 2\xi_1\sqrt{1-\zeta^2}+1}{\xi_1^2 - 2\xi_1\sqrt{1-\zeta^2}+1} + \frac{1-4\zeta^2}{4\zeta}\tan^{-1}\frac{2\zeta\xi_1}{1-\xi_1^2}, \tag{32.10}$$

$$\eta_{\ddot{x}_1}(\zeta, \xi_1, 0) = 4\zeta^2\xi_1 - \frac{1+12\zeta^2-16\zeta^4}{8\sqrt{1-\zeta^2}} \ln \frac{\xi_1^2 + 2\xi_1\sqrt{1-\zeta^2}+1}{\xi_1^2 - 2\xi_1\sqrt{1-\zeta^2}+1} + \frac{1+4\zeta^2-16\zeta^4}{4\zeta}\tan^{-1}\frac{2\zeta\xi_1}{1-\xi_1^2}.$$

Table 10

ξ_1 \ ζ	0.1	0.2	0.3	0.4	0.5	0.6	0.7	0.8	0.9
2	7.586	4.016	3.038	2.678	2.538	2.487	2.477	2.482	2.495
3	7.888	4.473	3.743	3.688	3.892	4.203	4.556	4.913	5.256
4	8.036	4.760	4.253	4.498	5.066	5.791	6.593	7.420	8.251
5	8.136	4.988	4.698	5.236	6.168	7.321	8.600	9.953	11.341
6	8.213	5.190	5.113	5.941	7.235	8.921	10.596	12.498	14.481
7	8.283	5.388	5.514	6.628	8.285	10.303	12.581	15.046	17.651
8	8.341	5.543	5.901	7.301	9.321	11.776	15.558	17.598	20.836
9	8.396	5.751	6.283	7.968	10.353	13.238	16.533	20.151	24.033
10	8.445	5.926	6.663	8.628	11.371	14.701	18.506	22.706	27.250

By reasoning similar to that given at the end of the preceding section, it may be shown that, for any fixed $\xi_1 > 0$, the variances σ_x^2 and $\sigma_{\dot{x}}^2$ are monotonically decreasing functions of ζ. By using table 10, where the values of $\eta_{\ddot{x}_1}(\zeta, \xi_1, 0)$ are listed for $\zeta \in [0.1, 0.9]$, $\xi_1 \in [2, 10]$, one can see that, corresponding to any fixed value of ξ_1, there exists a minimum point $\zeta_1 \in (0,1)$ of $\sigma_{\ddot{x}_1}(\zeta)$. Thus, approximately,

$\zeta_1 \approx 0.7$ for $\xi_1 = 2$, $\zeta_1 \approx 0.4$ for $\xi_1 = 3$, $\zeta_1 \approx 0.3$ for $\xi_1 = 4$, 5, and 6, $\zeta_1 \approx 0.2$ for $\xi_1 = 7$, 8, 9, 10. This also shows that ζ_1 is a decreasing function of ξ_1 in the interval considered.

The variation of $\eta_{\ddot{x}_1}(\zeta, \xi_1, 0)$ as a function of ζ is graphically illustrated in Fig. 83 for various values of ξ_1.

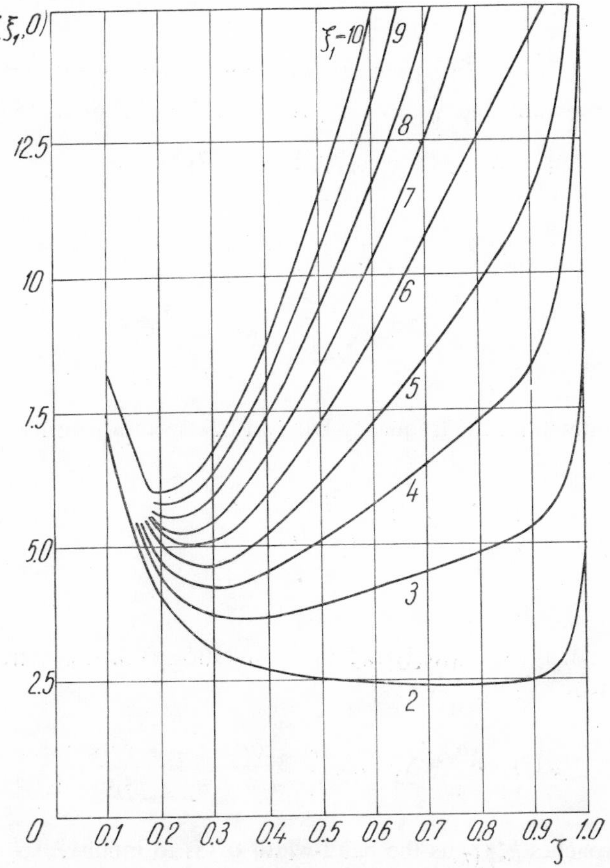

Fig. 83

§ 33. THE LINEAR OSCILLATOR EXCITED BY A STATIONARY RANDOM PROCESS WITH CORRELATION FUNCTION OF DAMPED HARMONIC TYPE

We consider in this section the more general case of a stationary random process whose spectral density may be obtained by limiting the frequency band of the spectral density corresponding to the correlation function

$$k^*_{x_0}(\tau) = \sigma^2_{x_0} \sum_{j=1}^{p} A_j e^{-\alpha_j|\tau|} \cos \beta_j \tau, \tag{33.1}$$

where α_j, β_j are real nonnegative constants, and A_j are real constants satisfying the condition

$$\sum_{j=1}^{p} A_j = 1. \tag{33.2}$$

It may be easily shown that the function (33.1) and its derivative are piecewise continuous and that $\int_{-\infty}^{\infty} |k^*_{x_0}(\tau)|\, d\tau$ exists. Consequently, there exists a spectral density associated with $k^*_{x_0}(\tau)$ given by Eq. (26.37), namely

$$S^*_{x_0}(\tau) = \frac{\sigma^2_{x_0}}{\pi} \sum_{j=1}^{p} A_j \int_{-\infty}^{\infty} e^{-\alpha_j|\tau|}\, e^{-i\nu\tau} \cos \beta_j \tau\, d\tau$$

$$= \frac{2\sigma^2_{x_0}}{\pi} \sum_{j=1}^{p} A_j \frac{\alpha_j (\nu^2 + \alpha_j^2 + \beta_j^2)}{(\nu^2 - \alpha_j^2 - \beta_j^2)^2 + 4\alpha_j^2\nu^2}. \tag{33.3}$$

We shall now limit the frequency band of this spectral density, by taking

$$S_{x_0}(\nu) = \begin{cases} \dfrac{2\sigma^2_{x_0}}{\pi} \displaystyle\sum_{j=1}^{p} A_j \frac{\alpha_j (\nu^2 + \alpha_j^2 + \beta_j^2)}{(\nu^2 - \alpha_j^2 - \beta_j^2)^2 + 4\alpha_j^2\nu^2} & \text{for} \quad |\nu| \leqslant \nu_1, \\ 0 & \text{for} \quad |\nu| > \nu_1. \end{cases} \tag{33.4}$$

Conversely, it may be proved that the correlation function corresponding to the spectral density (33.4), i.e.,

$$k_{x_0}(\tau) = \frac{\sigma^2_{x_0}}{\pi} \sum_{j=1}^{p} \alpha_j A_j \int_{-\nu_1}^{\nu_1} \frac{(\nu^2 + \alpha_j^2 + \beta_j^2)\, e^{i\nu\tau}\, d\nu}{(\nu^2 - \alpha_j^2 - \beta_j^2)^2 + 4\alpha_j^2\nu^2}, \tag{33.5}$$

actually approaches $k^*_{x_0}(\tau)$ as the band-width tends to infinity.

Introducing (33.4) into (30.15) we find

$$\sigma_x^2 = \frac{2\sigma^2_{x_0}}{\pi} \sum_{j=1}^{p} \alpha_j A_j \int_0^{\nu_1} \frac{\nu^4 (\nu^2 + \alpha_j^2 + \beta_j^2)\, d\nu}{[(\nu^2 - \alpha_j^2 - \beta_j^2)^2 + 4\alpha_j^2\nu^2]\, [(\nu_n^2 - \nu^2)^2 + 4\zeta^2\nu_n^2\nu^2]},$$

$$\sigma_{\dot{x}}^2 = \frac{2\sigma^2_{x_0}}{\pi} \sum_{j=1}^{p} \alpha_j A_j \int_0^{\nu_1} \frac{\nu^6 (\nu^2 + \alpha_j^2 + \beta_j^2)\, d\nu}{[(\nu^2 - \alpha_j^2 - \beta_j^2)^2 + 4\alpha_j^2\nu^2]\, [(\nu_n^2 - \nu^2)^2 + 4\zeta^2\nu_n^2\nu^2]}, \tag{33.6}$$

$$\sigma_{\ddot{x}_1}^2 = \frac{2\nu_n^2\sigma^2_{x_0}}{\pi} \sum_{j=1}^{p} \alpha_j A_j \int_0^{\nu_1} \frac{\nu^4 (\nu^2 + \alpha_j^2 + \beta_j^2)(\nu_n^2 + 4\zeta^2\nu^2)\, d\nu}{[(\nu^2 - \alpha_j^2 - \beta_j^2)^2 + 4\alpha_j^2\nu^2]\, [(\nu_n^2 - \nu^2)^2 + 4\zeta^2\nu_n^2\nu^2]}.$$

By using the same reasoning as in § 31, it may be proved that σ_x^2 and $\sigma_{\dot{x}}^2$ are monotonically decreasing functions of ζ when all other parameters are held constant. An interesting extremal property of $\sigma_{\ddot{x}_1}^2$ as a function of ζ is given by the following theorem. Denote as in § 31:

$$\xi = \nu/\nu_n, \qquad \xi_1 = \nu_1/\nu_n. \tag{33.7}$$

Theorem 1. *There exists a $\bar{\xi}_1 > \sqrt{2}$ such that, corresponding to any fixed $\xi_1 > \bar{\xi}_1$, there exists at least one minimum point $\zeta_0 \in (0,1)$ of $\sigma_{\ddot{x}_1}(\zeta)$.*

Proof. Let us first transform Eq. (33.6)$_3$, by using (33.7) and introducing the nondimensional parameters

$$\mu_j = \alpha_j/\nu_n, \qquad \lambda_j = \beta_j/\alpha_j = \beta_j/(\mu_j\nu_n), \qquad j = 1,..,p. \tag{33.8}$$

Then we obtain

$$\sigma_{\ddot{x}_1}^2(\zeta) = \frac{2\nu_n^4\sigma_{x_0}}{\pi} \sum_{j=1}^{p} \mu_j A_j \frac{\xi^4(1 + 4\zeta^2\xi^2) [\xi^2 + \mu_j^2 (1 + \lambda_j^2)] \, d\xi}{[(1 - \xi^2)^2 + 4\zeta^2\xi^2] \{[\xi^2 - \mu_j^2 (1 + \lambda_j^2)]^2 + 4\mu_j^2\xi^2]} \cdot \tag{33.9}$$

Differentiating with respect to ζ yields

$$\frac{d\sigma_{\ddot{x}_1}^2(\zeta)}{d\zeta} = \frac{16\nu_n^2\sigma_{x_0}^2\zeta}{\pi} \sum_{j=1}^{p} \mu_j A_j \frac{\xi^8(\xi^2 - 2) [\xi^2 + \mu_j^2(1 + \lambda_j^2)] \, d\xi}{[(1 - \xi^2)^2 + 4\zeta^2\xi^2] \{[\xi^2 - \mu_j^2(1 + \lambda_j^2)]^2 + 4\mu_j^2\xi^2\}}, \tag{33.10}$$

Next, defining the functions

$$\left. \begin{aligned} f(\xi) &= \sum_{j=1}^{p} \mu_j A_j \frac{\xi^8 [\xi^2 + \mu_j^2(1 + \lambda_j^2)]}{[\xi^2 - \mu_j^2(1 + \lambda_j^2)]^2 + 4\mu_j^2\xi^2}, \\[2mm] g(\xi, \zeta) &= \frac{\xi^2 - 2}{[(1 - \xi^2)^2 + 4\zeta^2\xi^2]^2}, \end{aligned} \right\} \tag{33.11}$$

Eq. (33.10) may be rewritten as

$$\frac{d\sigma_{\ddot{x}_1}^2}{d\zeta} = \frac{16 \, \nu_n^4\sigma_{x_0}^2\zeta}{\pi} \int_0^{\xi_1} f(\xi) \, g(\xi, \zeta) \, d\xi. \tag{33.12}$$

From (33.11) we see that $f(\xi) > 0$ for any $\xi \in R$, and that $g(\xi, \zeta)$ vanishes for $\xi = \sqrt{2}$, being negative for any $\xi < \sqrt{2}$ and positive for any $\xi > \sqrt{2}$. Taking now $\xi_1 > \sqrt{2}$, we can write Eq. (33.12) in the equivalent form

$$\frac{d^2\sigma_{\ddot{x}_1}(\zeta)}{d\zeta} = \frac{16 \, \nu_n^4\sigma_{x_0}^2\zeta}{\pi} \left\{ \int_0^{\sqrt{2}} f(\xi) \, g(\xi, \zeta) \, d\xi + \int_{\sqrt{2}}^{\xi_1} f(\xi) \, g(\xi, \zeta) \, d\xi \right\} \cdot \tag{33.13}$$

By applying the mean-value theorem to either of the integrals in the right-hand side, we obtain

$$\frac{\mathrm{d}\sigma^2_{\ddot{x}_1}(\zeta)}{\mathrm{d}\zeta} = \frac{16\nu_n^4\sigma^2_{x_0}\zeta}{\pi}\left\{ f(\theta_1\sqrt{2})\int_0^{\sqrt{2}} g(\xi,\zeta)\,\mathrm{d}\xi + f(\sqrt{2}+\theta_2(\xi-\sqrt{2}))\int_{\sqrt{2}}^{\xi_1} g(\xi,\zeta)\,\mathrm{d}\xi \right\}, \qquad (33.14)$$

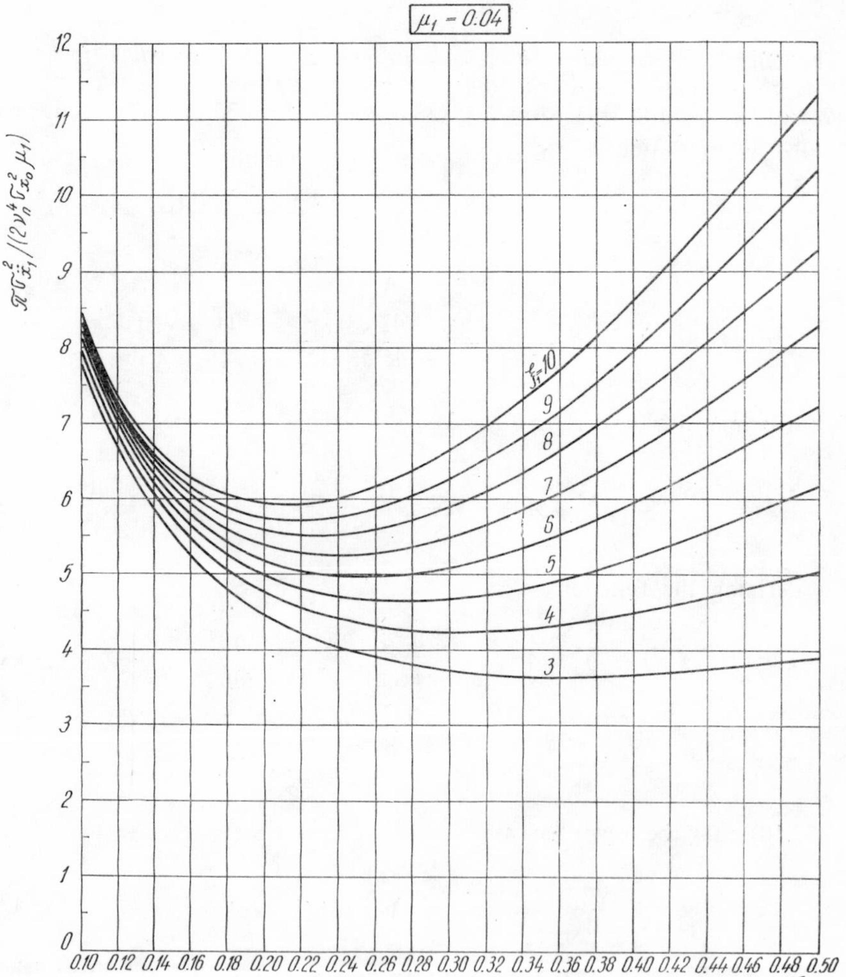

Fig. 84

where $\theta_1, \theta_2 \in (0,1)$. In view of $(33.11)_2$, the first integral in the right-hand side tends to $-\infty$ as $\zeta \to 0$, and the second one is positive and finite for any ζ. Therefore, there exists some $\bar{\zeta} \in (0,1)$ such that $\mathrm{d}\sigma^2_{\ddot{x}_1}(\zeta)/\mathrm{d}\zeta$ is negative for $0 < \zeta < \bar{\zeta}$. On the other hand, for $\zeta = 1$, the first integral is finite and independent of ξ_1, whereas the second integral is positive and tends to ∞ together with ξ_1. Hence, there exists some $\sqrt{2} < \bar{\xi}_1 < \infty$ such that $d\sigma^2_{\ddot{x}_1}(\zeta)/d\zeta$ is positive for $\zeta = 1$ and $\xi_1 > \bar{\xi}_1$. Since $\sigma^2_{\ddot{x}_1}(\zeta)$ is a continuous function of ζ, we conclude that it has at

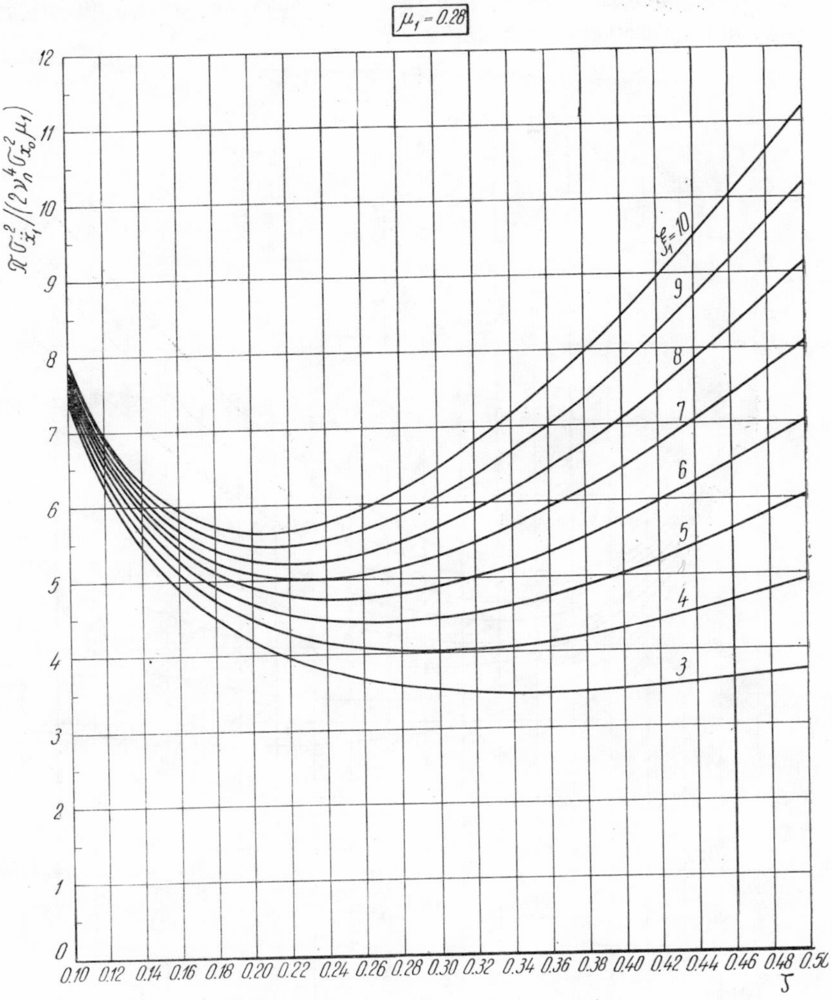

Fig. 85

least one minimum point ζ_0 in the interval (0,1) for any fixed $\xi_1 > \bar{\xi}_1$. This completes the proof.

To illustrate the variation of $\sigma^2_{\ddot{x}_1}$ with ζ, two particular cases are considered, namely:

(1) $A_1 = 1$, $A_j = 0$ for $j \geqslant 2$. Figures 84 and 85 show the variation with ζ of the quantity $\pi\sigma^2_{\ddot{x}_1}/(2\nu_n^4\sigma^2_{x_0}\mu_1)$ for $\mu_1 = 0.04$ and 0.28, respectively.

(2) $A_1 = 0$, $A_2 = 1$, $A_j = 0$ for $j \geqslant 3$, $\lambda_2 = 2$. Figures 86 and 87 show the variation with ζ of the quantity $\pi\sigma^2_{\ddot{x}_1}/(2\nu_n^4\sigma^2_{x_0}\mu_2)$ for $\mu_2 = 0.04$ and 0.28, respectively.

Theorem 2. *If the random process $x_0(t)$ is normally distributed then any value $\zeta \in (0,1)$ which minimizes the variance $\sigma^2_{\ddot{x}_1}$ also minimizes the function $P(|\ddot{x}_1(t, \zeta)| > a)$ for any fixed acceleration level a, and any t.*

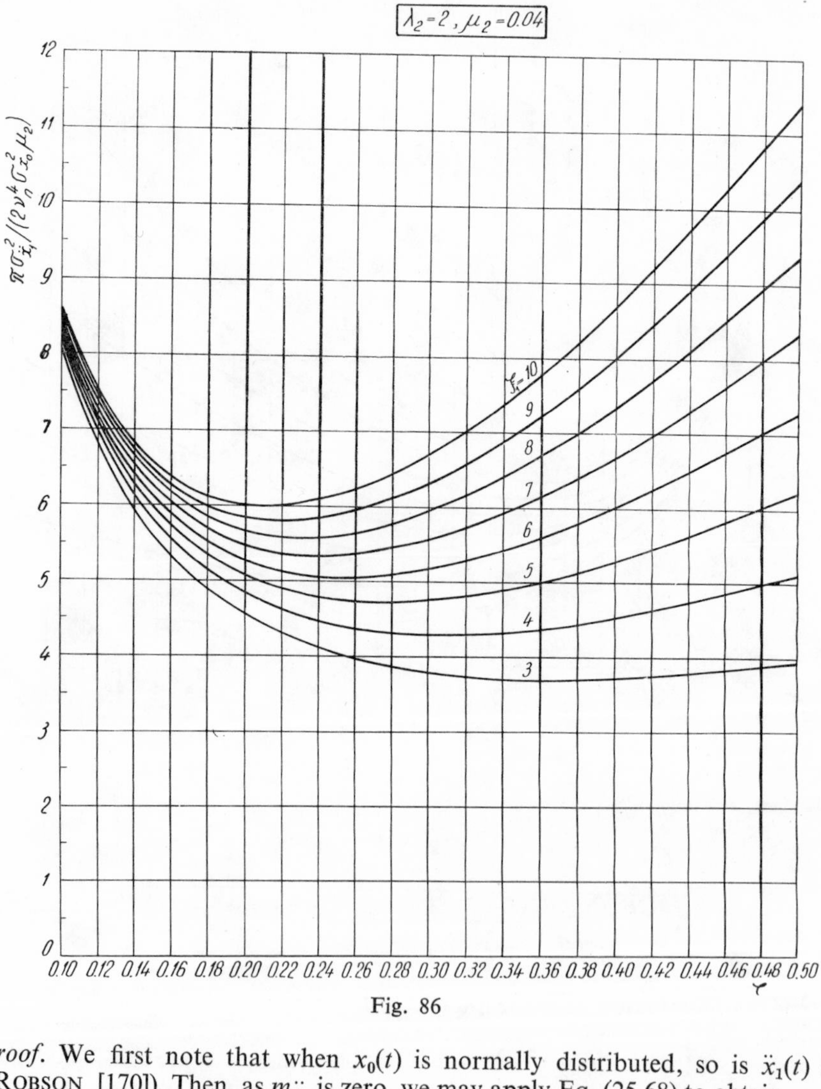

Fig. 86

Proof. We first note that when $x_0(t)$ is normally distributed, so is $\ddot{x}_1(t)$ (see, e.g., Robson [170]). Then, as $m_{\ddot{x}_1}$ is zero, we may apply Eq. (25.68) to obtain

$$P(|\ddot{x}_1(t, \zeta)| > a) = 1 - 2\Phi(a/\sigma_{\ddot{x}_1}(\zeta)). \tag{33.15}$$

Since the Laplace integral function Φ is a monotonically increasing function of its argument, it follows from (33.15) that any minimum point of $\sigma^2_{\ddot{x}_1}(\zeta)$ is also a minimum point of $P(|\ddot{x}_1(t, \zeta)| > a)$.

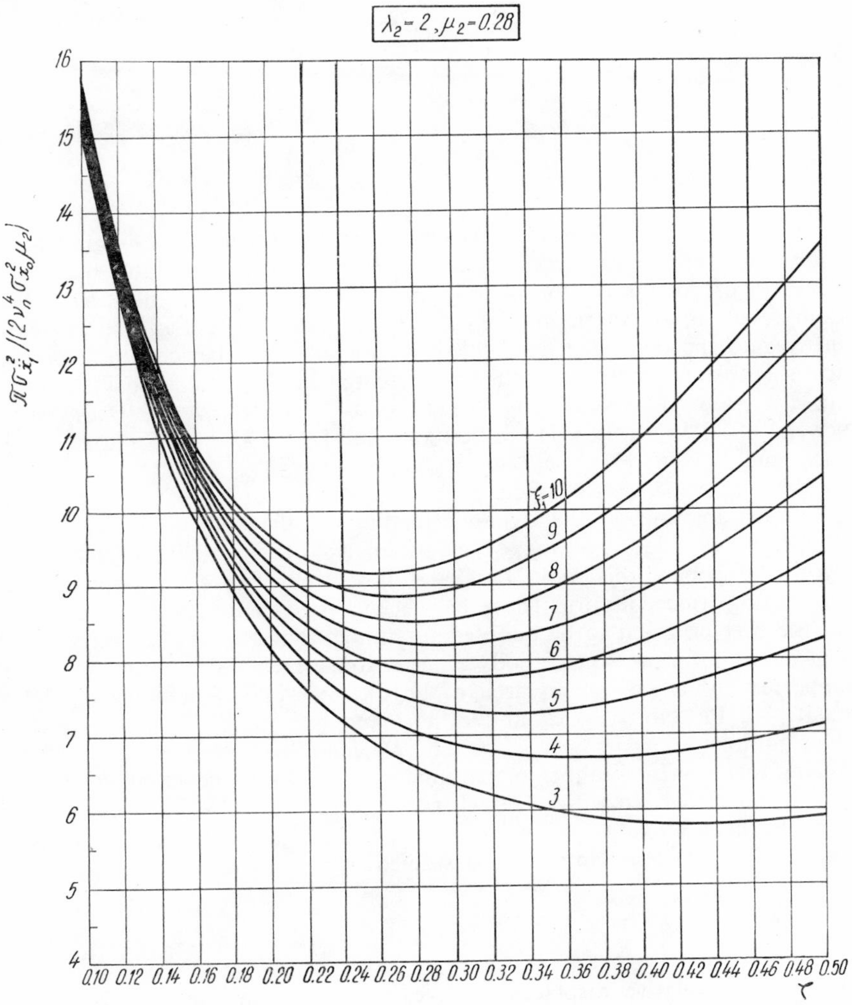

Fig. 87

Chapter X

Some applications of the theory of random vibrations in road vehicle dynamics

One of the most important problems that the theory of random vibrations can solve in road vehicle dynamics is optimizing ride comfort and road holding.

The way of solving this problem is to model the road uneveness by a suitable random function, to adopt an appropriate simplified model of the vehicle, and to investigate the dependence of the response on some basic parameters of the vehicle suspension. For recent extended research in this area, also including non-linear cases, see ARIARATNAM [6], SILAEV [181], MITSCHKE [125], BENDER [12], DINCA et al. [51], ROBSON [171], DINCA and SIRETEANU [50], THOMPSON [190], BRUNS and RÖNITZ [20].

Optimizing suspension characteristics is mainly a problem of finding a compromise between minimizing the dynamic loading on the body and passengers and maximizing the road holding. The solving of this rather complex problem is beyond the scope of this book. In what follows we will approach only the problem of optimizing ride comfort. There has been some discussion in the literature about the best criterion to be adopted for this optimization. The use of the criterion of minimum mean square body acceleration, which will be used in this book, is supported by extensive research of DIECKMANN [45], ROTENBERG [174], VAN DEUSEN [44], PRUCHKOV [152] and others, and is nowadays widely adopted by various authors.

Owing to the great number of parameters that affect ride comfort, it is necessary to use a rather simplified model, taking into account only some basic parameters. We have already seen in the introduction that, under certain simplifying hypotheses, vehicle bouncing may be studied by the equation

$$\ddot{x} + f(\dot{x}) + g(x) = -\ddot{x}_0(t),$$

where x is the relative displacement between the sprung and unsprung masses, $x_0(t)$ is the height above some fixed level of the road uneveness surmounted by the wheel at time t, $f(\dot{x})$ and $g(x)$ are, respectively, the damping force and the elastic restoring force per unit sprung mass.

In our approach to the optimization problem, we shall first consider the linear case, i.e.,

$$\ddot{x} + 2\zeta\nu_n\dot{x} + \nu_n^2 x = -\ddot{x}_0(t),$$

and we shall determine the optimum damping ratio, i.e., the value of ζ that minimizes mean square acceleration. Then, the results obtained will be applied to find the optimum value of the damping coefficient k in the nonlinear equation

$$\ddot{x} + k\,|\dot{x}|^r \operatorname{sgn} \dot{x} - \nu_n^2 x = -\ddot{x}_0(t),$$

where r is a positive number.

§ 34. MODELLING THE EXCITATION INDUCED BY ROAD IRREGULARITIES BY MEANS OF STATIONARY RANDOM PROCESSES

Assume that the height h of the surface of a road segment over some reference plane is recorded as a function of the distance d from the segment origin and that this experiment is repeated for N segments of length L belonging to roads of the same pavement type. These records provide a sample of size N which may be used to approximately determine some basic statistical characteristics of the random function $h(d)$.

Measurements performed on various types of roads show that the road roughness $h(d)$ may be satisfactorily approximated by some normally distributed, stationary random functions. Since any stationary function has a constant mean value, we can always choose a suitable reference plane such that $m_h = 0$.

As has been shown in the preceding chapter, the most useful statistical characteristic of the excitation is its spectral density, or, equivalently, its correlation function. That is why much work has been done to find analytical expressions which are appropriate for modelling the experimentally determined correlation functions of the road profiles. One of these representations, which proved to be adequate for approximating road correlation functions for various types and qualities of the pavement, is

$$k_h^*(s) = \sigma_h^2(A_1 e^{-a_1|s|} + A_2 e^{-a_2|s|} \cos b_2 s), \tag{34.1}$$

where $A_1 + A_2 = 1$. This correlation function has been extensively used in the Russian literature (see, e.g., PARHILOVSKY [106], [205], SILAEV [139], and PEVZNER and TIHONOV [108]). Table 11 gives a compilation of experimental values of the parameters involved in Eq. (34.1).

Obviously, (34.1) is a particular case of the correlation function (33.1). As in § 33, to avoid the occurrence of infinite variances, we shall limit the frequency band of the spectral density corresponding to (34.1), by taking

$$S_h(\Omega) = \begin{cases} \dfrac{2\sigma_h^2}{\pi}\left[\dfrac{A_1 a_1}{\Omega^2 + a_1^2} + \dfrac{A^2 a_2(\Omega^2 + a_2^2 + b_2^2)}{(\Omega^2 - a_2^2 - b_2^2)^2 + 4a_2^2\Omega^2}\right] & \text{for } |\Omega| \leqslant \Omega_1, \\ 0 & \text{for } |\Omega| > \Omega_1, \end{cases} \tag{34.2}$$

where Ω_1 is some cut-off distance frequency to be also experimentally determined.

The limit correlation function for $\Omega_1 \to \infty$ and the spectral density of the excitation $x_0(t) \equiv h(Vt)$ induced by the road irregularities when the vehicle travels at the speed V can be obtained by setting

$$s = V\tau, \qquad \Omega = \nu/V \tag{34.3}$$

in Eqs. (34.1) and (34.2). It results that

$$k^*_{x_0}(\tau) = \sigma^2_{x_0}(A_1 e^{-\alpha_1|\tau|} + A_2 e^{-\alpha_2|\tau|}\cos\beta_2\tau), \tag{34.4}$$

$$S_{x_0}(\nu) = \begin{cases} \dfrac{2\sigma^2_{x_0}}{\pi}\left[\dfrac{A_1\alpha_1}{\nu^2+\alpha_1^2} + \dfrac{A^2\alpha_2(\nu^2+\alpha_2^2+\beta_2^2)}{(\nu^2-\alpha_2^2-\beta_2^2)^2+4\alpha_2^2\nu^2}\right] & \text{for } |\nu| \leqslant \nu_1, \\ 0 & \text{for } |\nu| > \nu_1, \end{cases} \tag{34.5}$$

where

$$A_1 + A_2 = 1, \qquad \alpha_1 = a_1 V, \qquad \alpha_2 = a_2 V, \qquad \beta_2 = b_2 V, \qquad \nu_1 = \Omega_1 V. \tag{34.6}$$

If ν_1 is sufficiently large, we may still use for the parameters involved in (34.5) the values obtained by approximating the real correlation function by (34.4), which actually corresponds to $\nu_1 = \infty$. We shall make use of this remark in our numerical applications, where the spectral density (34.5) will be employed in conjunction with data in table 11.

§ 35. OPTMIZING THE VISCOUS LINEAR DAMPING IN ROAD VEHICLE SUSPENSIONS

Let us consider the linear equation

$$\ddot{x} + 2\zeta\nu_n\dot{x} + \nu_n^2 x = -\ddot{x}_0(t). \tag{35.1}$$

We will try to determine the optimum value of the damping ratio ζ that minimizes the mean square of the body acceleration, $\ddot{x}_1(t)$*). In parallel we shall also calculate the mean square of $x(t)$ and $\dot{x}(t)$, which are also of some practical interest. The calculation will be performed for two particular cases of the spectral density (34.5) of $x_0(t)$.

Case 1. Let us first set

$$A_1 = 0, \quad A_2 = 1, \quad a_2 = a, \quad b_2 = b, \quad \alpha_2 = \alpha = aV, \quad \beta_2 = \beta = bV. \tag{35.2}$$

*) Since the mean values of $x(t)$, $\dot{x}(t)$, and $\ddot{x}_1(t)$ are zero, their mean squares obviously coincide with their variances.

According to Table 11, the resulting correlation function may be used to statistically describe the roughness of asphalt roads, boulder paved roads, and country roads.

Table 11

Road type	A_1	A_2	a_1 [m^{-1}]	a_2 [m^{-1}]	b_2 [m^{-1}]	σ_{x_0} [m]	Ref
Concrete road	1	0	0.15	—	—	0.0050—0.0124	[108]
Asphalt road	0.85	0.15	0.20	0.05	0.60	0.0080—0.0126	[108]
	0	1	—	0.22	0.44	0.012	[106]
Even, stone block paved road	1	0	0.45	—	—	0.0135—0.0225	[108]
Waved, stone block paved road	0.85	0.15	0.50	0.20	2.00	0.0250—0.0380	[108]
Boulder paved road	0	1	—	0.32	0.64	0.017	[106]
Country road	0	1	—	0.47	0.94	0.019	[106]
	0	1	—	0.11	0.146	0.067—0.220	[139]

Inserting (35.2) into (33.6) and setting $A_j = 0$ for $j \geqslant 3$ yields

$$\sigma_x^2 = \frac{2\sigma_{x_0}^2 \alpha}{\pi} \int_0^{\nu_1} \frac{\nu^4(\nu^2 + \alpha^2 + \beta^2)\, d\nu}{[(\nu^2 - \alpha^2 - \beta^2)^2 + 4\alpha^2\nu^2]\,[(\nu_n^2 - \nu^2)^2 + 4\zeta^2\nu_n^2\nu^2]},$$

$$\sigma_{\dot{x}}^2 = \frac{2\sigma_{x_0}^2 \alpha}{\pi} \int_0^{\nu_1} \frac{\nu^6\,(\nu^2 + \alpha^2 + \beta^2)\, d\nu}{[(\nu^2 - \alpha^2 - \beta^2)^2 + 4\alpha^2\nu^2]\,[(\nu_n^2 - \nu^2)^2 + 4\zeta^2\nu_n^2\nu^2]}, \tag{35.3}$$

$$\sigma_{\ddot{x}_1}^2 = \frac{2\nu_n^2\sigma_{x_0}^2 \alpha}{\pi} \int_0^{\nu_1} \frac{\nu^4(\nu^2 + \alpha^2 + \beta^2)(\nu_n^2 + 4\zeta^2\nu^2)\, d\nu}{[(\nu^2 - \alpha^2 - \beta^2)^2 + 4\alpha^2\nu^2]\,[(\nu_n^2 - \nu^2)^2 + 4\zeta^2\nu_n^2\nu^2]}.$$

By putting

$$\xi = \nu/\nu_n, \qquad \xi_1 = \nu_1/\nu_n = \Omega_1 V/\nu_n \tag{35.4}$$

and introducing the nondimensional parameters

$$\mu = \alpha/\nu_n = aV/\nu_n, \qquad \lambda = \beta/\alpha = b/a, \tag{35.5}$$

Eqs. (35.3) may be rewritten as

$$\sigma_x^2 = \frac{2\mu\sigma_{x_0}^2}{\pi}\eta_x(\zeta, \xi_1, \mu, \lambda),$$

$$\sigma_{\dot{x}}^2 = \frac{2\mu\nu_n^2\sigma_{x_0}^2}{\pi}\eta_{\dot{x}}(\zeta, \xi_1, \mu, \lambda),$$

$$\sigma_{\ddot{x}_1}^2 = \frac{2\mu\nu_n^4\sigma_{x_c}^2}{\pi}\eta_{\ddot{x}_1}(\zeta, \xi_1, \mu, \lambda),$$

where

$$\begin{aligned} \eta_x(\zeta, \xi_1, \mu, \lambda) &= \int_0^{\xi_1} \frac{\xi^4[\xi^2 + \mu^2(1+\lambda^2)]\,\mathrm{d}\xi}{\{[\xi^2 - \mu^2(1+\lambda^2)]^2 + 4\mu^2\xi^2\}[(1-\xi^2)^2 + 4\zeta^2\xi^2]}, \\ \eta_{\dot{x}}(\zeta, \xi_1, \mu, \lambda) &= \int_0^{\xi_1} \frac{\xi^6[\xi^2 + \mu^2(1+\lambda^2)]\,\mathrm{d}\xi}{\{[\xi^2 - \mu^2(1+\lambda^2)]^2 + 4\mu^2\xi^2\}[(1-\xi^2)^2 + 4\zeta^2\xi^2]}, \\ \eta_{\ddot{x}_1}(\zeta, \xi_1, \mu, \lambda) &= \int_0^{\xi_1} \frac{\xi^4(1+4\zeta^2\xi^2)[\xi^2 + \mu^2(1+\lambda^2)]\,\mathrm{d}\xi}{\{[\xi^2 - \mu^2(1+\lambda^2)]^2 + 4\mu^2\xi^2\}[(1-\xi^2)^2 + 4\zeta^2\xi^2]}. \end{aligned} \tag{35.6}$$

As shown in § 33, σ_x^2 and $\sigma_{\dot{x}}^2$ are monotonically decreasing functions of ζ. Also, by Theorem 2, there exists a $\bar{\xi}_1 > \sqrt{2}$, such that for any $\xi_1 > \bar{\xi}_1$ there exists an optimum damping ratio $\zeta_0 \in (0,1)$ which minimizes $\sigma_{\ddot{x}_1}^2(\zeta)$. This optimum damping depends on the vehicle speed through the parameters μ and ξ_1 [cf. Eqs. (35.4), (35.5)].

The variation of σ_x^2, $\sigma_{\dot{x}}^2$, and $\sigma_{\ddot{x}_1}^2$ with the parameters ζ, μ, ξ_1, and λ has been numerically analyzed with the help of a digital computer using a set of values that covers the data in table 11, namely

$$\begin{aligned} \xi_1 &= 2, 3, \ldots, 10; \\ \mu &= 0.04, 0.08, 0.16, 0.24, 0.32, 0.40, \ldots, 0.96; \\ \lambda &= 0, 1.4, 1.6, 1.8, 2.\text{[1)]} \end{aligned} \tag{35.7}$$

The functions (35.6) have been plotted versus ζ for each combination of the parameter values (35.7), and the optimum damping ratio $\zeta_0(\zeta_1, \mu, \lambda)$, which minimizes $\eta_{\ddot{x}_1}(\zeta, \xi_1, \mu, \lambda)$, was determined on each plot.

Tables 12—16 and Figs. 88—107 show the variation of the function $\zeta_0(\xi_1, \mu, \lambda)$, $\bar{\eta}_x(\zeta_0, \xi_1, \mu, \lambda) = 2\mu\eta_x(\zeta_0, \xi_1, \mu, \lambda)$, $\bar{\eta}_{\dot{x}}(\zeta_0, \xi_1, \mu, \lambda) = 2\mu\eta_{\dot{x}}(\zeta_0, \xi_1, \mu, \lambda)$, $\bar{\eta}_{\ddot{x}_1}(\zeta_0, \xi_1, \mu, \lambda) = 2\mu\eta_{\ddot{x}_1}(\zeta_0, \xi_1, \mu, \lambda)$ for all combinations of the parameter values (35.7).

From Figs. 95, 99, 103, and 107 one can see that when the limit correlation function $k_{x_0}^*(\tau)$ contains a harmonic factor, i.e., when

$$k_{x_0}^*(\tau) = \sigma_{x_0}^2\, e^{-\alpha|\tau|} \cos\beta\tau, \quad \beta \neq 0, \tag{35.8}$$

1) The case $\lambda = 0$ has been already considered in § 32 for small values of μ.

Table 12

ξ_1	μ	0.04	0.08	0.12	0.16	0.20	0.24	0.28	0.32	0.36	0.48	0.60	0.72	0.84	0.96
2	ζ_0	0.70	0.70	0.70	0.70	0.70	0.70	0.70	0.70	0.70	0.70	0.60	0.60	0.60	0.60
	$\bar{\eta}_x$	0.0500	0.0992	0.1472	0.1934	0.2376	0.2795	0.3190	0.3561	0.3907	0.4797	0.6954	0.7595	0.8044	0.8339
	$\bar{\eta}_{\dot{x}}$	0.0754	0.1504	0.2244	0.2959	0.3680	0.4369	0.5035	0.5676	0.6290	0.7956	1.1634	1.3028	1.4117	1.4936
	$\bar{\eta}_{\ddot{x}_1}$	0.1978	0.3940	0.5871	0.7757	0.9590	1.1360	1.3060	1.4684	1.6234	2.0390	2.3707	2.6355	2.8372	2.9841
3	ζ_0	0.36	0.36	0.36	0.36	0.36	0.35	0.35	0.35	0.35	0.34	0.33	0.32	0.31	0.30
	$\bar{\eta}_x$	0.1461	0.2906	0.4321	0.5697	0.7024	0.8574	0.9822	1.1002	1.2112	1.5530	1.8516	2.1083	2.3271	2.5131
	$\bar{\eta}_{\dot{x}}$	0.2831	0.5648	0.8438	1.1188	1.3887	1.6906	1.9530	2.2074	2.4530	3.2057	3.8975	4.5222	5.0786	5.5695
	$\bar{\eta}_{\ddot{x}_1}$	0.2928	0.5834	0.8696	1.1497	1.4222	1.6858	1.9392	2.1818	2.4132	3.0354	3.5493	3.9606	4.2795	4.5181
4	ζ_0	0.30	0.30	0.30	0.30	0.30	0.30	0.30	0.29	0.29	0.28	0.28	0.27	0.26	0.25
	$\bar{\eta}_x$	0.1884	0.3749	0.5579	0.7358	0.9076	1.0723	1.2291	1.4301	1.5751	2.0314	2.3429	2.6435	2.9741	3.2285
	$\bar{\eta}_{\dot{x}}$	0.4202	0.8386	1.2535	1.6632	2.0662	2.4613	2.8473	3.2915	3.6636	4.8060	5.7492	6.6145	7.5796	8.3674
	$\bar{\eta}_{\ddot{x}_1}$	0.3968	0.6768	1.0092	1.3346	1.6514	1.9584	2.2542	2.5373	2.8076	3.5385	4.1459	4.5725	5.0237	5.3201
5	ζ_0	0.27	0.27	0.27	0.27	0.27	0.27	0.27	0.27	0.26	0.26	0.25	0.24	0.24	0.23
	$\bar{\eta}_x$	0.2160	0.4298	0.6397	0.8439	1.0412	1.2303	1.4106	1.5814	1.8138	2.2525	2.7078	3.1088	3.3173	3.6194
	$\bar{\eta}_{\dot{x}}$	0.5368	1.0714	0.6021	2.1268	2.6438	3.1517	3.6491	4.1348	4.6988	6.0595	7.4258	8.6970	9.4554	10.6086
	$\bar{\eta}_{\ddot{x}_1}$	0.3725	0.7423	1.1069	1.4641	1.8121	2.1494	2.4747	2.7871	3.0843	3.8911	4.5642	5.1126	5.5490	5.8840
6	ζ_0	0.25	0.25	0.25	0.25	0.25	0.25	0.25	0.25	0.24	0.24	0.24	0.23	0.22	0.21
	$\bar{\eta}_x$	0.2374	0.4725	0.7032	0.9278	1.1448	1.3529	1.5613	1.7392	1.9999	2.4838	2.9933	3.4458	3.6753	4.0115
	$\bar{\eta}_{\dot{x}}$	0.6445	1.2868	1.9246	2.5559	3.1787	3.7915	4.3927	4.9811	5.6594	7.3191	8.9912	10.5607	11.8205	13.1466
	$\bar{\eta}_{\ddot{x}_1}$	0.3985	0.7942	1.1844	1.5668	1.9395	2.3008	2.4695	2.9845	3.3038	4.1701	4.8959	5.4903	5.9622	6.3304
7	ζ_0	0.24	0.24	0.24	0.24	0.23	0.24	0.24	0.23	0.23	0.22	0.22	0.21	0.20	0.20
	$\bar{\eta}_x$	0.2498	0.4973	0.7401	0.9766	1.2598	1.4889	1.7072	1.9139	2.1086	2.7408	3.1612	3.4810	3.8875	4.2499
	$\bar{\eta}_{\dot{x}}$	0.7407	1.4790	2.2126	2.9394	3.7234	4.4428	5.1495	5.8421	6.5196	8.6024	0.4042	12.0462	13.7503	15.3475
	$\bar{\eta}_{\ddot{x}_1}$	0.4205	0.8380	1.2499	1.6539	2.0477	2.4290	2.7967	3.1501	3.4881	4.4062	5.1755	5.8133	6.3131	6.7056
8	ζ_0	0.23	0.23	0.23	0.23	0.22	0.23	0.23	0.22	0.22	0.21	0.21	0.20	0.19	0.19
	$\bar{\eta}_x$	0.3727	0.5228	0.7782	1.0269	1.3269	1.5682	1.7981	2.0159	2.2210	2.8917	3.3346	3.8507	4.3132	4.4963
	$\bar{\eta}_{\dot{x}}$	0.8370	1.6714	2.5009	3.3232	4.2076	5.0224	5.8239	6.6106	7.3813	9.7705	11.8205	13.9404	15.9459	17.5584
	$\bar{\eta}_{\ddot{x}_1}$	0.4398	0.8765	1.3074	1.7301	2.1415	2.5405	2.9256	3.2956	3.6500	4.6117	5.4197	6.0813	6.6158	7.0318
9	ζ_0	0.22	0.22	0.22	0.22	0.21	0.21	0.21	0.21	0.21	0.20	0.20	0.19	0.19	0.18
	$\bar{\eta}_x$	0.2762	0.5498	0.8184	1.0799	1.3980	1.6522	1.8944	2.1238	2.3398	3.0518	3.5179	4.0696	4.3369	4.7580
	$\bar{\eta}_x$	0.9337	1.8648	2.7907	3.7090	4.6947	5.6056	6.5023	7.3836	8.2479	10.9063	13.2452	15.6447	17.6533	19.7856
	$\bar{\eta}_{\ddot{x}_1}$	0.4570	0.9108	1.3587	1.7980	2.2261	2.6411	3.0411	3.4262	3.7997	4.7969	5.6371	6.3287	6.8862	7.3221
10	ζ_0	0.21	0.21	0.21	0.21	0.21	0.21	0.20	0.20	0.20	0.20	0.19	0.19	0.18	0.17
	$\bar{\eta}_x$	0.2907	0.5786	0.8613	1.1366	1.4025	1.6576	1.9980	2.2077	2.4673	3.0628	3.7153	4.0859	4.5859	5.0412
	$\bar{\eta}_{\dot{x}}$	1.0313	2.0589	3.0829	4.0980	5.1028	6.0951	7.1871	8.1635	9.1223	11.8836	14.6289	17.1065	19.6290	22.0355
	$\bar{\eta}_{\ddot{x}_1}$	0.4721	0.9430	1.4051	1.8595	2.3026	2.7328	3.1479	3.5459	3.9269	4.9642	5.8355	6.5540	7.1298	7.5884

Table 13

ξ_1	μ	0.04	0.08	0.12	0.16	0.20	0.24	0.28	0.32	0.36	0.48	0.60	0.72	0.84	0.96
2	ζ_0	0.70	0.70	0.70	0.80	0.80	0.80	0.80	0.80	0.80	0.80	0.60	0.46	0.46	0.41
	$\bar{\eta}_x$	0.0507	0.1048	0.1641	0.1866	0.2020	0.3048	0.3672	0.4297	0.4908	0.6523	1.2167	1.6916	18.672	2.0711
	$\bar{\eta}_{\dot{x}}$	0.0759	0.1541	0.2370	0.2664	0.2867	0.4320	0.5252	0.6249	0.7296	1.0573	2.0843	3.0563	3.5582	4.0588
	$\bar{\eta}_{\dot{x}_1}$	0.1995	0.4069	0.6285	0.8686	1.1312	1.4107	1.7119	2.0294	2.3587	3.3591	4.2181	4.7477	4.8794	4.8002
3	ζ_0	0.36	0.36	0.37	0.37	0.38	0.39	0.39	0.40	0.40	0.39	0.36	0.32	0.29	0.26
	$\bar{\eta}_x$	0.1476	0.3025	0.4557	0.6353	0.8069	0.9833	0.1971	1.3729	1.5857	2.1576	2.9023	3.5414	3.9085	4.1468
	$\bar{\eta}_{\dot{x}}$	0.2845	0.5758	0.8607	0.1777	1.4843	1.8022	2.1801	2.5238	2.9370	4.2321	5.9868	7.7410	9.1160	10.2007
	$\bar{\eta}_{\dot{x}_1}$	0.2931	0.6011	0.9270	1.2802	1.6642	2.0797	2.5234	2.9882	3.4654	4.8661	6.0059	6.7121	6.9752	6.9051
4	ζ_0	0.30	0.30	0.31	0.31	0.32	0.32	0.33	0.33	0.34	0.33	0.31	0.28	0.25	0.22
	$\bar{\eta}_x$	0.1903	0.3896	0.5838	0.8132	1.0289	1.2954	1.5246	1.8091	2.0206	2.8608	3.6164	4.2818	4.7529	15.0903
	$\bar{\eta}_{\dot{x}}$	0.4220	0.8528	1.2735	1.7369	2.1843	2.7025	3.1868	3.7610	4.2679	6.2368	8.3714	10.5600	12.5232	4.2195
	$\bar{\eta}_{\dot{x}_1}$	0.3422	0.6966	1.0734	1.4809	1.9236	2.4024	2.9128	3.4473	3.9941	5.5776	6.8334	7.5936	7.8837	7.8432
5	ζ_0	0.27	0.27	0.28	0.28	0.29	0.29	0.30	0.30	0.30	0.30	0.28	0.26	0.23	0.20
	$\bar{\eta}_x$	0.2181	0.4469	0.6653	9.9282	1.1714	1.4759	1.7342	2.0606	2.3880	3.2664	4.1404	4.7216	5.2555	5.6563
	$\bar{\eta}_{\dot{x}}$	0.5387	1.0877	1.6236	2.2094	2.7745	3.4228	4.0293	4.7414	5.4847	7.7913	10.3969	12.8139	15.2238	17.3892
	$\bar{\eta}_{\dot{x}_1}$	0.3752	0.7635	1.1757	1.6211	2.1049	2.6273	3.1848	3.7670	4.3624	6.0713	7.4011	8.1864	8.4769	8.4386
6	ζ_0	0.25	0.25	0.26	0.26	0.27	0.27	0.28	0.28	0.28	0.28	0.26	0.24	0.21	0.19
	$\bar{\eta}_x$	0.2397	0.4903	0.7304	1.0171	1.2812	1.6148	1.8960	2.2529	2.6137	3.5791	4.5467	5.1849	5.7864	5.9813
	$\bar{\eta}_{\dot{x}}$	0.6468	1.3047	1.9469	2.6448	3.3180	4.0842	3.8017	5.6373	6.5070	9.1959	12.2138	15.0188	17.8374	20.0690
	$\bar{\eta}_{\dot{x}_1}$	0.4014	0.8165	1.2569	1.7323	2.2487	2.8057	3.4007	4.0208	4.6543	6.4629	7.8493	8.6452	8.9331	8.8821
7	ζ_0	0.24	0.24	0.24	0.25	0.25	0.26	0.26	4 27	0.27	0.26	0.25	0.23	0.20	0.18
	$\bar{\eta}_x$	0.2522	0.5159	0.8010	1.0685	1.4020	1.6948	2.0716	2.3636	2.7433	3.9256	4.7804	5.4528	6.0964	6.3120
	$\bar{\eta}_{\dot{x}}$	0.7430	1.4978	2.3718	3.0319	3.8074	4.6695	5.5861	6.4260	7.4028	10.6215	13.7771	16.9145	20.0829	22.6489
	$\bar{\eta}_{\dot{x}_1}$	0.434	0.8610	1.2534	1.8265	2.3696	2.9574	3.5821	4.2374	4.9019	6.7978	8.2248	9.0320	9.3097	9.2473
8	ζ_0	0.23	0.23	0.23	0.23	0.24	0.24	0.25	0.25	0.25	0.25	0.24	0.22	0.20	0.18
	$\bar{\eta}_x$	0.2652	0.5423	0.8420	1.1726	1.4717	1.8568	2.1732	2.5879	3.0069	4.1248	5.0232	5.7309	6.1295	6.3504
	$\bar{\eta}_{\dot{x}}$	0.8394	1.6912	2.5674	3.4797	4.3586	5.3486	6.2758	7.3444	8.4516	11.8524	15.3326	18.7933	21.9263	24.7941
	$\bar{\eta}_{\dot{x}_1}$	0.4428	0.9002	1.3853	1.9090	2.4759	3.0891	3.7422	4.4240	5.1198	7.0880	8.5559	9.3692	9.6377	9.5637
9	ζ_0	0.22	0.22	0.22	0.22	0.23	0.23	0.24	0.24	0.24	0.24	0.23	0.21	0.19	0.17
	$\bar{\eta}_x$	0.2788	0.5702	0.8853	0.2330	1.4732	1.9503	2.2802	2.7167	3.1583	4.3355	5.2802	6.0257	6.4453	6.6841
	$\bar{\eta}_{\dot{x}}$	0.9363	1.8856	2.8607	3.8736	4.8392	5.9436	6.9688	8.1450	9.3614	13.0885	16.8916	20.6708	24.1065	27.2765
	$\bar{\eta}_{\dot{x}_1}$	0.4601	0.9353	1.4391	1.9830	2.5713	3.2079	3.8858	4.5933	5.3152	7.3511	8.8544	9.6720	9.9264	9.8371
10	ζ_0	0.21	0.21	0.21	0.22	0.22	0.22	0.23	0.23	0.24	0.24	0.22	0.20	0.18	0.16
	$\bar{\eta}_x$	0.2934	0.6001	0.9317	1.2366	1.6237	2.0502	2.3942	2.8543	3.1665	4.3465	5.5555	6.3428	6.7867	7.0464
	$\bar{\eta}_{\dot{x}}$	1.0340	2.0817	3.1565	4.2006	5.3441	6.5438	7.6673	8.9543	10.1009	14.0798	18,4615	22.1816	26.2923	29.7581
	$\bar{\eta}_{\dot{x}_1}$	0.4758	0.9673	1.4885	2.0499	2.6583	3.3171	4.0165	4.7486	5.4937	7.5905	9.1237	9.9553	10.1941	10.0938

Tabel 14

ξ_1	μ	0.04	0.08	0.12	0.16	0.20	0.24	0.28	0.32	0.36	0.48	0.60	0.72	0.84	0.96
2	ζ_0	0.70	0.70	0.80	0.80	0.80	$\geqslant$ 0.90	$\geqslant$ 0.90	$\geqslant$ 0.90	$\geqslant$ 0.90	0.80	0.60	0.48	0.41	0.37
	$\bar{\eta}_x$	0.0510	0.1066	0.1376	0.1970	0.2620	0.2739	0.3326	0.3914	0.4484	0.7288	1.3383	1.9028	2.2038	2.2268
	$\bar{\eta}_{\dot{x}}$	0.0760	0.1553	0.1968	0.2744	0.3613	0.3799	0.4687	0.5652	0.6643	1.2009	2.3735	3.5739	4.3156	4.4792
	$\bar{\eta}_{\dot{x}_1}$	0.2000	0.4110	0.6423	0.8994	1.1870	1.5049	1.8512	2.2265	2.6128	3.8029	4.7562	5.1967	5.1055	4.6796
3	ζ_0	0.36	0.37	0.37	0.38	0.39	0.40	0.41	0.41	0.42	0.40	0.36	0.30	0.26	0.28
	$\bar{\eta}_x$	0.1481	0.2967	0.4684	0.6436	0.8330	1.0332	1.2366	1.4797	1.6669	2.4175	3.1792	4.0175	4.4288	4.5498
	$\bar{\eta}_{\dot{x}}$	0.2849	0.5662	0.8723	1.1784	1.5034	1.8501	2.2174	2.6607	3.0607	4.7259	6.7124	8.9889	10.5983	11.5738
	$\bar{\eta}_{\dot{x}_1}$	0.2959	0.6068	0.9459	1.3242	1.7477	2.2172	2.7275	3.2687	3.8265	5.4420	6.6588	7.2534	7.2947	6.9988
4	ζ_0	0.30	0.31	0.31	0.32	0.32	0.33	0.34	0.35	0.35	0.34	0.31	0.27	0.23	0.20
	η_x	0.1909	0.3803	0.5992	0.8194	1.0946	1.3544	1.6192	1.8763	2.1873	3.0906	3.9479	4.6767	5.1988	5.4038
	$\bar{\eta}_{\dot{x}}$	0.4225	0.8392	1.2884	1.7354	2.2534	2.7620	3.2971	3.8524	4.5144	6.7327	9.2670	11.8404	14.1228	15.7968
	$\bar{\eta}_{\dot{x}_1}$	0.3430	0.7029	1.0944	1.5302	2.0177	2.5575	3.1438	3.7640	4.3994	6.2038	7.5102	8.1294	8.1871	7.9315
5	ζ_0	0.27	0.28	0.28	0.29	0.29	0.30	0.31	0.31	0.32	0.31	0.28	0.24	0.21	0.19
	$\bar{\eta}_x$	0.2188	0.4343	0.6838	0.9323	1.2456	1.5387	1.8378	2.2096	2.4841	3.5211	4.5097	5.3603	5.7489	5.7669
	$\bar{\eta}_{\dot{x}}$	0.5394	1.0712	1.6406	2.2058	2.8537	3.4880	4.1523	4.9397	5.6716	8.3470	11.4040	14.5181	17.0166	18.8842
	$\bar{\eta}_{\dot{x}_1}$	0.3760	0.7702	1.1982	1.6743	2.2056	2.7944	3.4339	4.1083	4.7984	6.7297	8.0860	8.7052	8.7506	8.4935
6	ζ_0	0.25	0.26	0.26	0.26	0.27	0.28	0.28	0.29	0.29	0.29	0.26	0.23	0.20	0.18
	η_x	0.2404	0.4759	0.7491	1.0611	1.3619	1.6801	2.0864	2.4133	2.8220	3.8519	4.9445	5.6465	6.0651	6.0948
	$\eta_{\dot{x}}$	0.6474	1.2856	1.9656	2.6901	3.4047	4.1530	5.0315	5.8511	6.8157	9.7937	13.2972	16.5375	19.4250	11.6897
	$\eta_{\dot{x}_1}$	0.4023	0.8236	1.2806	1.7884	2.3547	2.9825	3.6642	4.3816	5.1149	7.1465	8.5400	9.1459	9.1731	8.9057
7	ζ_0	0.24	0.24	0.24	0.25	0.26	0.26	0.27	0.27	0.28	0.27	0.25	0.22	0.20	0.17
	$\bar{\eta}_x$	0.2530	0.5219	0.8214	1.1144	1.4288	1.8357	2.1881	2.6377	2.9598	4.2179	5.1938	5.9342	6.1118	6.4204
	$\bar{\eta}_x$	0.7438	1.5037	2.2963	3.0796	3.8925	4.8288	5.7264	6.7771	7.7249	11.2597	14.9059	18.4988	21.3815	24.3548
	$\bar{\eta}_{\dot{x}_1}$	0.4243	0.8684	1.3504	1.8843	2.4814	3.1414	3.8580	4.6193	5.3828	7.5013	8.9203	9.5157	9.5328	9.2360
8	ζ_0	0.23	0.23	0.23	0.24	0.24	0.25	0.26	0.27	0.27	0.26	0.24	0.21	0.18	0.16
	$\bar{\eta}_x$	0.2659	0.5486	0.8214	1.1696	1.5644	1.9251	2.2931	2.6497	3.1019	4.4275	5.4532	6.2337	6.7140	6.7609
	$\bar{\eta}_{\dot{x}}$	0.8402	1.6974	2.5889	3.4692	4.4585	5.4220	6.4209	7.4451	8.6325	12.5139	16.5030	20.4356	24.0126	26.9493
	$\bar{\eta}_{\dot{x}_1}$	0.4437	0.9078	1.4111	1.9689	2.5917	3.2806	4.0293	4.8207	5.6191	7.8112	9.2555	9.8387	9.8239	9.5205
9	ζ_0	0.22	0.22	0.22	0.23	0.23	0.24	0.25	0.25	0.25	0.25	0.23	0.20	0.18	0.16
	$\bar{\eta}_x$	0.2796	0.5768	0.8667	1.2276	1.6426	2.0192	2.4034	2.9014	3.4016	3.8808	5.7270	6.5530	6.7394	6.7891
	$\bar{\eta}_{\dot{x}}$	0.9371	1.8921	2.8353	3.8609	4.9543	6.0181	7.1183	8.3914	9.7196	13.7723	18.1014	22.3675	25.8513	29.0978
	$\bar{\eta}_{\dot{x}_1}$	0.4610	0.9431	1.4667	2.0446	2.6909	3.4058	4.1815	4.9993	5.8316	8.0918	9.5578	10.1311	10.0900	9.7697
10	ζ_0	0.21	0.21	0.22	0.22	0.23	0.23	0.24	0.24	0.25	0.24	0.22	0.20	0.17	0.15
	$\bar{\eta}_x$	0.2943	0.6070	0.9103	1.2896	1.6472	2.1197	2.5207	3.0456	3.4099	4.8852	6.0218	6.5694	7.0894	7.1512
	$\bar{\eta}_{\dot{x}}$	1.0349	2.0886	3.1284	4.2558	5.3636	6.6192	7.8211	9.2072	10.4588	15.0406	19.7102	23.8943	28.0772	31.6126
	$\bar{\eta}_{\dot{x}_1}$	0.4768	0.9754	1.5160	2.1199	2.7821	3.5201	4.3226	5.1669	6.0251	8.3507	9.8377	10.3924	10.3352	9.9936

Table 15

ξ_1 \ μ		0.04	0.08	0.12	0.16	0.20	0.24	0.28	0.32	0.36	0.48	0.60	0.72	0.84	0.96
2	ζ_0	0.70	0.70	0.80	0.80	$\geqslant 0.90$	$\geqslant 0.90$	$\geqslant 0.90$	$\geqslant 0.90$	$\geqslant 0.90$	0.70	0.60	0.43	0.37	0.34
	$\bar{\eta}_x$	0.0512	0.1087	0.1441	0.2093	0.2345	0.3000	0.3677	0.4351	0.5000	1.0124	1.4486	2.2722	2.4328	2.2723
	$\bar{\eta}_{\dot{x}}$	0.0762	0.1567	0.2008	0.2839	0.3150	0.4064	0.5095	0.6234	0.7458	1.6851	2.6853	4.4028	4.9066	4.6819
	$\bar{\eta}_{\ddot{x}_1}$	0.2006	0.4159	0.6581	0.9362	1.2552	1.6166	2.0185	2.4549	2.9165	4.3152	5.3126	5.5284	5.1196	4.4373
3	ζ_0	0.36	0.37	0.38	0.39	0.40	0.41	0.42	0.43	0.43	0.40	0.34	0.28	0.24	0.21
	$\bar{\eta}_x$	0.1487	0.3011	0.4683	0.6567	0.8677	1.0957	1.3296	1.5551	1.8143	2.6864	3.6560	4.4283	4.6926	4.6744
	$\bar{\eta}_{\dot{x}}$	0.2854	0.5700	0.8658	1.1834	1.5310	1.9128	2.3267	2.7653	3.2916	5.2976	7.8517	10.2823	11.8006	12.5589
	$\bar{\eta}_{\ddot{x}_1}$	0.2966	0.6133	0.9683	1.3767	1.8475	2.3818	2.9714	3.6003	4.2487	6.0769	7.2868	7.6535	7.4115	6.8897
4	ζ_0	0.30	0.31	0.31	0.32	0.33	0.34	0.35	0.36	0.36	0.34	0.30	0.25	0.21	0.18
	$\bar{\eta}_x$	0.1916	0.3856	0.6174	0.8611	1.1338	1.4301	1.7356	2.0304	2.3764	3.4311	4.3782	5.1519	5.5168	5.5878
	$\bar{\eta}_{\dot{x}}$	0.4232	0.8441	1.3057	1.7767	2.2866	2.8409	3.4373	4.0653	4.8132	7.4580	10.4467	13.4065	15.6887	17.2929
	$\bar{\eta}_{\ddot{x}_1}$	0.3439	0.7462	1.1193	1.5888	2.1299	2.7437	3.4199	4.1382	4.8716	6.8797	8.1390	8.5034	8.2841	7.8288
5	ζ_0	0.27	0.28	0.28	0.29	0.30	0.31	0.32	0.33	0.33	0.31	0.27	0.23	0.20	0.17
	$\bar{\eta}_x$	0.2195	0.4402	0.7041	0.9790	1.2864	1.6208	1.9662	2.2998	2.6958	3.9075	4.9957	5.6688	5.6688	5.9739
	$\bar{\eta}_{\dot{x}}$	0.5401	1.0768	1.6604	2.2529	2.8890	3.5753	4.3093	4.9249	5.9861	9.1649	12.4265	15.9533	18.2843	21.2913
	$\bar{\eta}_{\ddot{x}_1}$	0.3770	0.7779	1.2248	1.7369	2.3264	2.9904	3.7314	4.5119	5.3034	7.4305	8.7067	9.0445	8.8132	8.6219
6	ζ_0	0.25	0.26	0.26	0.27	0.28	0.29	0.30	0.30	0.31	0.29	0.26	0.22	0.19	0.16
	$\bar{\eta}_x$	0.2412	0.4824	0.7711	1.0697	1.4033	1.7713	2.1423	2.6055	2.9396	4.2741	5.2504	5.9633	6.1747	6.3204
	$\bar{\eta}_{\dot{x}}$	0.6482	1.2918	1.9872	2.6908	3.4413	4.2462	5.1024	6.1174	7.0435	10.6793	14.4190	18.0488	20.9660	23.6183
	$\bar{\eta}_{\ddot{x}_1}$	0.4033	0.8317	1.3085	1.8543	2.4825	3.1950	3.9792	4.8078	5.6472	7.8666	9.1494	9.4575	9.2009	8.7389
7	ζ_0	0.24	0.24	0.25	0.25	0.26	0.27	0.28	0.29	0.29	0.28	0.25	0.21	0.18	0.15
	$\bar{\eta}_x$	0.2538	0.5289	0.8099	1.1697	1.6054	1.9265	2.3352	2.7294	3.2074	4.4830	5.5066	6.2588	6.4933	6.6670
	$\bar{\eta}_{\dot{x}}$	0.7446	1.5105	2.2777	3.1360	4.0024	4.9273	5.9073	6.9263	8.1155	11.9631	16.0681	20.0673	23.3372	26.4032
	$\bar{\eta}_{\ddot{x}_1}$	0.4254	0.8769	1.3793	1.9537	2.6141	3.3633	4.1877	5.0595	5.9375	8.2347	9.5236	9.7988	9.5118	9.0432
8	ζ_0	0.23	0.23	0.23	0.24	0.25	0.26	0.27	0.27	0.28	0.27	0.24	0.20	0.17	0.15
	$\bar{\eta}_x$	0.2668	0.5559	0.8884	0.2274	1.6057	2.0180	2.4452	2.9820	3.3598	4.6996	5.7722	6.5667	6.8139	6.8702
	$\bar{\eta}_{\dot{x}}$	0.8411	1.7045	2.6139	3.5285	4.4956	5.5234	6.6085	7.8802	9.0387	13.2403	17.6998	22.0506	25.6546	28.6863
	$\bar{\eta}_{\ddot{x}_1}$	0.4448	0.9166	1.4415	2.0404	2.7296	3.5115	4.3723	5.2798	6.1944	8.5605	9.8503	10.0947	9.7796	9.2895
9	ζ_0	0.22	0.22	0.23	0.23	0.24	0.25	0.26	0.26	0.27	0.26	0.23	0.19	0.17	0.14
	$\bar{\eta}_x$	0.2806	0.5844	0.8918	1.2882	1.6834	2.1142	2.5607	3.1259	3.5199	4.9280	6.0530	6.8944	6.8400	7.0639
	$\bar{\eta}_{\dot{x}}$	0.9380	1.8996	2.8604	3.9232	4.9911	6.2497	7.3124	8.7005	9.9652	14.5202	19.3288	24.0228	27.5303	31.2931
	$\bar{\eta}_{\ddot{x}_1}$	0.4622	0.9522	1.4971	2.1184	2.8333	3.7206	4.5380	5.4785	6.4258	8.8541	10.1430	10.3632	10.0225	9.5172
10	ζ_0	0.21	0.21	0.22	0.22	0.23	0.24	0.25	0.25	0.26	0.25	0.23	0.19	0.16	0.14
	$\bar{\eta}_x$	0.2953	0.6150	0.9367	1.3532	1.7662	2.2164	2.6834	3.2790	3.6900	5.1714	6.3535	6.9116	7.0138	7.0138
	$\bar{\eta}_{\dot{x}}$	1.0358	2.0964	3.1548	4.3219	5.4907	6.7255	8.0216	9.5276	10.8984	15.8089	20.9659	25.5678	29.0726	33.4341
	$\bar{\eta}_{\ddot{x}_1}$	0.4783	0.9848	1.5474	2.1899	2.9280	3.7660	4.6887	5.6609	6.6370	9.1235	10.4124	10.6035	10.2451	9.7091

Table 16

ξ_1 \ μ		0.04	0.08	0.12	0.16	0.20	0.24	0.28	0.32	0.36	0.48	0.60	0.72	0.84	0.96
2	ζ_0	0.70	0.70	0.80	$\geqslant$0.90	$\geqslant$0.90	$\geqslant$0.90	$\geqslant$0.90	$\geqslant$0.90	$\geqslant$0.90	0.70	0.49	0.38	0.33	0.32
	$\bar{\eta}_x$	0.0516	0.1112	0.1508	0.1859	0.2750	0.3300	0.4076	0.4842	0.5567	1.1144	2.0630	2.6310	2.5833	2.1610
	$\bar{\eta}_{\dot{x}}$	0.0764	0.1583	0.2053	0.2448	0.3335	0.4380	0.5585	0.6643	0.8387	1.9210	3.8870	5.2502	5.4624	4.5394
	$\bar{\eta}_{\ddot{x}_1}$	0.2013	0.4214	0.6764	0.9790	1.3355	1.7494	2.2170	2.7299	3.2739	4.8796	5.7961	5.6634	4.9056	4.0205
3	ζ_0	0.36	0.37	0.38	0.39	0.41	0.43	0.44	0.45	0.45	0.40	0.32	0.26	0.22	0.19
	$\bar{\eta}_x$	0.1493	0.3062	0.4853	0.6959	0.9113	1.1362	1.3943	1.6391	1.9112	2.9459	4.1053	4.7396	4.8371	4.6913
	$\bar{\eta}_{\dot{x}}$	0.2859	0.5744	0.8809	1.2202	1.5679	1.9470	2.4048	2.8957	3.2648	5.9354	9.0856	11.5426	12.8656	13.3244
	$\bar{\eta}_{\ddot{x}_1}$	0.2976	0.6208	0.9941	1.4383	1.9655	2.5761	3.2570	3.9846	4.7307	6.7447	7.8268	7.8607	7.3268	6.6153
4	ζ_0	0.30	0.31	0.32	0.33	0.34	0.36	0.37	0.38	0.38	0.34	0.28	0.23	0.19	0.17
	$\bar{\eta}_x$	0.1924	0.3917	0.6166	1.8797	1.1848	1.6984	1.8094	2.1292	2.4896	3.7526	4.9121	5.5117	5.7083	5.4554
	$\bar{\eta}_{\dot{x}}$	0.4239	0.8497	1.2976	1.7864	2.3317	2.8781	3.5315	4.2262	5.0481	8.2569	11.9320	14.9553	17.1731	18.3978
	$\bar{\eta}_{\ddot{x}_1}$	0.3449	0.7189	1.1482	1.6579	2.2630	2.9641	3.7432	4.5702	5.4054	7.5706	8.6540	8.6761	8.1879	7.5807
5	ζ_0	0.27	0.28	0.28	0.29	0.31	0.32	0.33	0.34	0.34	0.31	0.26	0.21	0.18	0.15
	$\bar{\eta}_x$	0.2204	0.4471	0.7280	1.0348	1.3405	1.7227	2.1190	2.4944	2.9224	4.2670	5.3863	6.0601	6.0700	6.0852
	$\bar{\eta}_{\dot{x}}$	0.5409	1.0832	1.6832	2.3083	2.9381	3.6868	4.5028	5.3654	6.3789	10.0581	14.0819	17.6597	20.1440	22.3688
	$\bar{\eta}_{\ddot{x}_1}$	0.3781	0.7868	1.2559	1.8113	2.4699	3.2328	4.0804	4.9754	5.8720	8.1333	9.1940	9.1752	8.6807	8.0984
6	ζ_0	0.25	0.26	0.26	0.27	0.28	0.30	0.31	0.32	0.32	0.29	0.25	0.20	0.17	0.15
	$\bar{\eta}_x$	0.2422	0.4897	0.7970	1.0348	1.5172	1.8749	2.3065	2.7153	3.1845	4.6625	5.6543	6.3727	6.3979	6.1793
	$\bar{\eta}_{\dot{x}}$	0.6491	1.2988	2.0123	2.3083	3.5617	4.3656	5.3113	6.3064	7.4675	11.6396	15.8376	19.8488	22.7771	25.1852
	$\bar{\eta}_{\ddot{x}_1}$	0.4044	0.8405	1.3411	1.9325	2.6342	3.4465	4.3482	5.2984	6.2432	8.5781	9.6137	9.5485	9.0310	8.4461
7	ζ_0	0.24	0.24	0.25	0.26	0.27	0.28	0.29	0.30	0.30	0.28	0.24	0.19	0.16	0.14
	$\bar{\eta}_x$	0.2548	0.5369	0.8369	1.1850	1.5895	2.1531	2.5120	2.9572	3.4726	4.8873	5.9239	6.6879	6.7264	6.5140
	$\bar{\eta}_{\dot{x}}$	0.7455	1.5182	2.3041	3.1429	4.0561	5.0545	6.1316	7.2604	8.5710	12.9635	17.5383	21.9398	25.2477	28.0717
	$\bar{\eta}_{\ddot{x}_1}$	0.4266	0.8867	1.4129	2.0349	2.7722	3.6266	4.5747	5.5709	6.5581	8.9523	9.9646	9.8561	9.3117	8.7174
8	ζ_0	0.23	0.23	0.24	0.25	0.26	0.27	0.28	0.28	0.29	0.27	0.23	0.19	0.16	0.14
	$\bar{\eta}_x$	0.2678	0.5643	0.8780	1.2415	1.6640	2.1364	2.6288	3.2291	2.6361	5.1207	6.2039	6.7174	6.7618	6.5562
	$\bar{\eta}_{\dot{x}}$	1.5265	1.7126	2.5960	2.5344	4.5502	5.6550	6.8415	8.2391	9.5124	14.2766	19.2154	23.5855	27.2382	30.4408
	$\bar{\eta}_{\ddot{x}_1}$	0.4460	0.9266	1.4761	2.1252	2.8943	3.7854	4.7743	5.8129	6.8360	9.2838	10.2698	10.1232	9.5511	8.9812
9	ζ_0	0.22	0.22	0.23	0.24	0.25	0.26	0.27	0.27	0.28	0.26	0.22	0.18	0.15	0.13
	$\bar{\eta}_x$	0.2816	0.5932	0.9757	1.3010	1.7421	2.2361	2.7512	3.3835	3.8079	5.3671	6.5002	7.0407	7.0992	6.8989
	$\bar{\eta}_{\dot{x}}$	0.9391	1.9081	2.8894	3.9276	5.0464	6.2579	7.5541	9.0735	10.4567	15.5927	20.8867	25.5835	29.5561	33.0925
	$\bar{\eta}_{\ddot{x}_1}$	0.4634	0.9630	2.0127	2.2059	3.0037	3.9282	5.5140	6.0293	7.0870	9.5833	10.5438	10.3563	9.7593	9.1359
10	ζ_0	0.21	0.21	0.22	0.23	0.24	0.25	0.26	0.26	0.27	0.25	0.21	0.18	0.15	0.13
	$\bar{\eta}_x$	0.2964	0.6242	0.9675	1.3642	1.8252	2.3418	2.8812	3.5477	3.9902	5.6297	6.8176	7.0581	7.1200	6.9233
	$\bar{\eta}_{\dot{x}}$	1.0369	2.1054	3.1853	4.3240	5.5464	6.8652	8.2720	9.9148	11.4080	16.9178	22.5644	27.1490	31.4247	35.2838
	$\bar{\eta}_{\ddot{x}_1}$	0.4793	0.9956	1.5845	2.2792	3.1030	4.0581	5.1180	6.2287	7.3169	9.8592	10.7978	10.5767	9.9481	9.3085

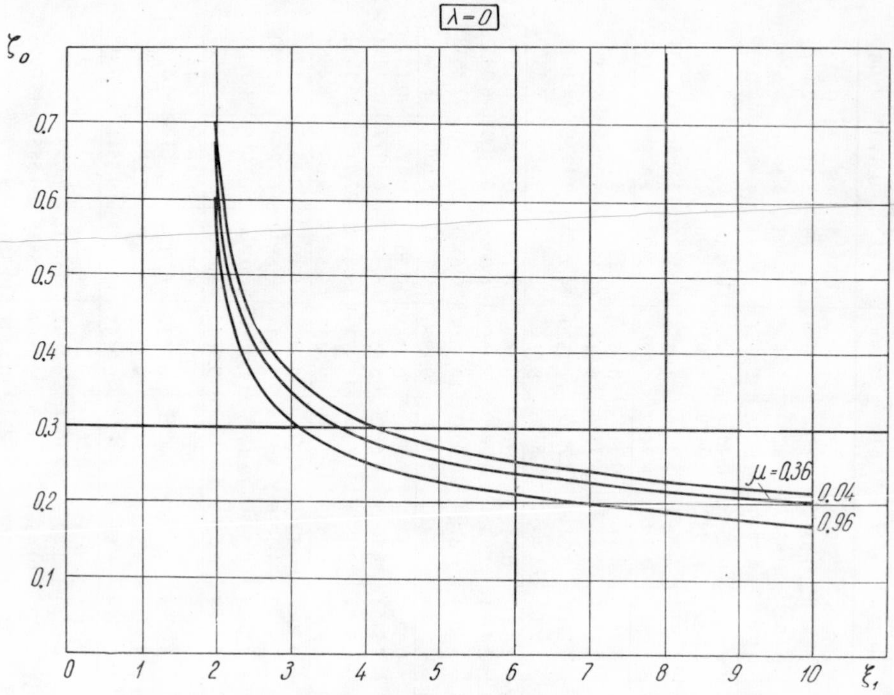
λ=0
ζ0
0.7
0.6
0.5
0.4
0.3
0.2
0.1
0 1 2 3 4 5 6 7 8 9 10
ξ1
μ=0.36
0.04
0.96

Fig. 88

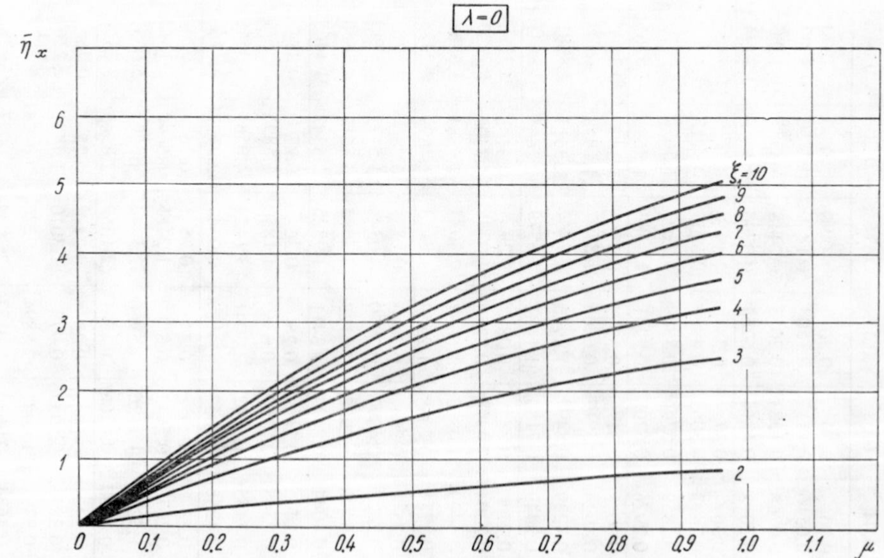
λ=0
η̄x
6
5
4
3
2
1
0 0.1 0.2 0.3 0.4 0.5 0.6 0.7 0.8 0.9 1.0 1.1
μ
ξ1=10
9
8
7
6
5
4
3
2

Fig. 89

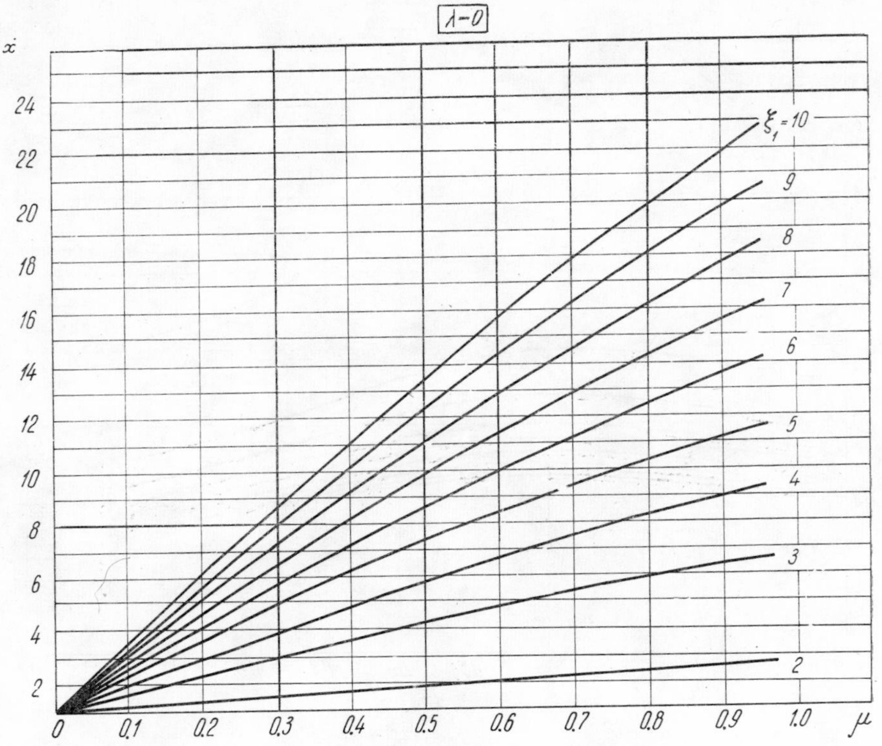
λ=0
η̄ ẋ
24
22
20
18
16
14
12
10
8
6
4
2
0
0.1
0.2
0.3
0.4
0.5
0.6
0.7
0.8
0.9
1.0
μ
ξ₁ = 10
9
8
7
6
5
4
3
2

Fig. 90

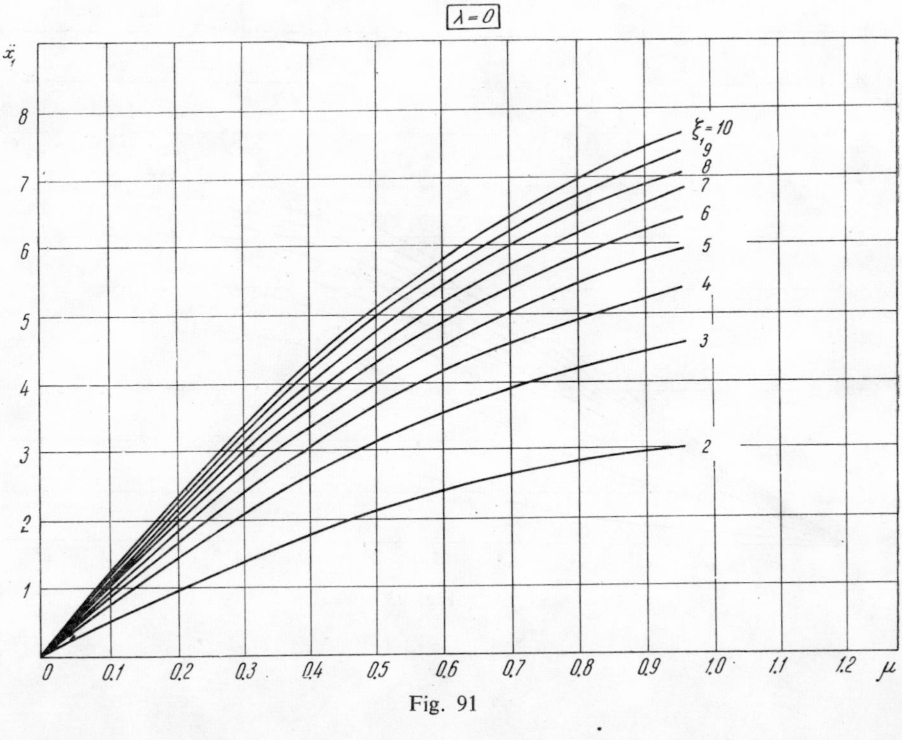
λ=0
η̄ ẍ₁
8
7
6
5
4
3
2
1
0
0.1
0.2
0.3
0.4
0.5
0.6
0.7
0.8
0.9
1.0
1.1
1.2
μ
ξ₁ = 10
9
8
7
6
5
4
3
2

Fig. 91

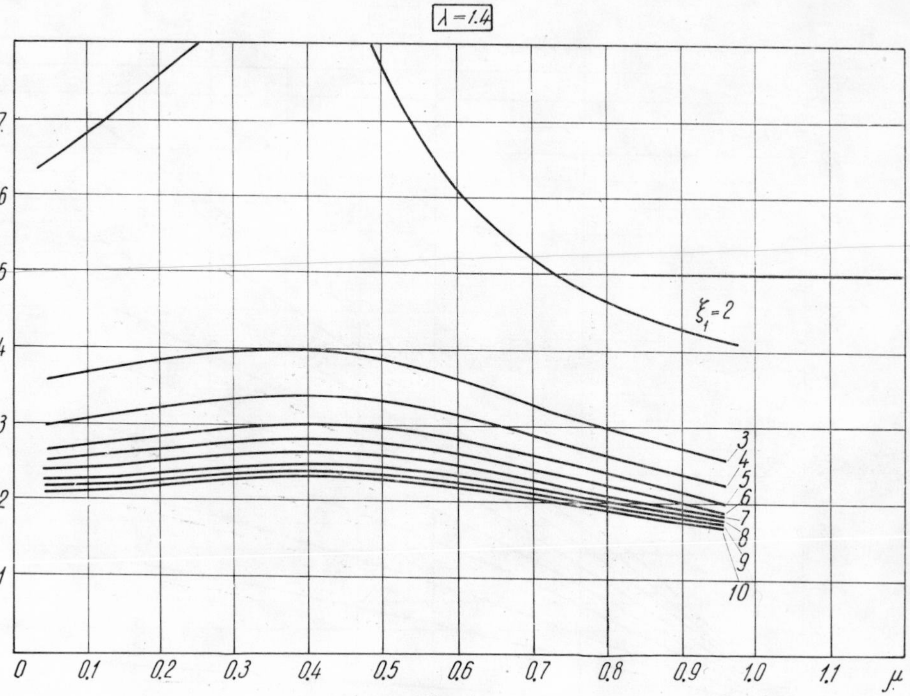
λ=1.4
0.7
0.6
0.5
0.4
0.3
0.2
0.1
0
0.1
0.2
0.3
0.4
0.5
0.6
0.7
0.8
0.9
1.0
1.1
μ
ξ1=2
3
4
5
6
7
8
9
10

Fig. 92

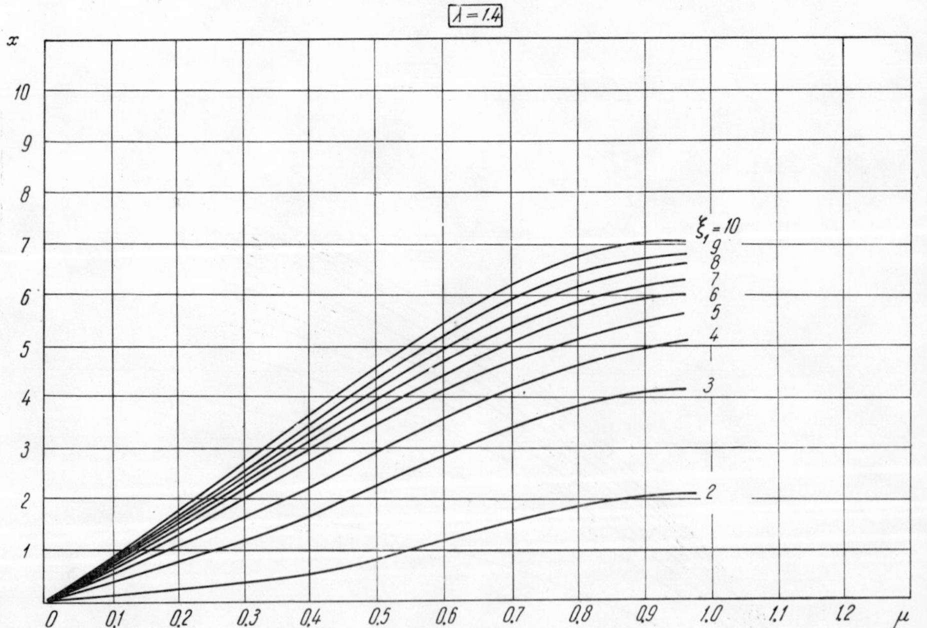
λ=1.4
x
10
9
8
7
6
5
4
3
2
1
0
0.1
0.2
0.3
0.4
0.5
0.6
0.7
0.8
0.9
1.0
1.1
1.2
μ
ξ1=10
9
8
7
6
5
4
3
2

Fig. 93

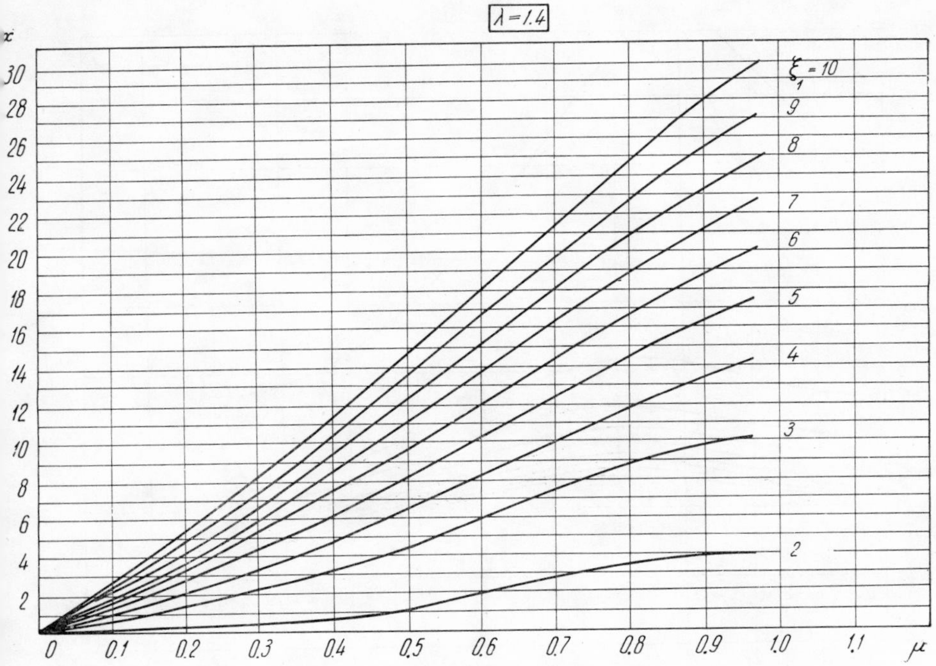
λ=1.4
x
30
28
26
24
22
20
18
16
14
12
10
8
6
4
2
ξ1 = 10
9
8
7
6
5
4
3
2
0
0.1
0.2
0.3
0.4
0.5
0.6
0.7
0.8
0.9
1.0
1.1
μ

Fig. 94

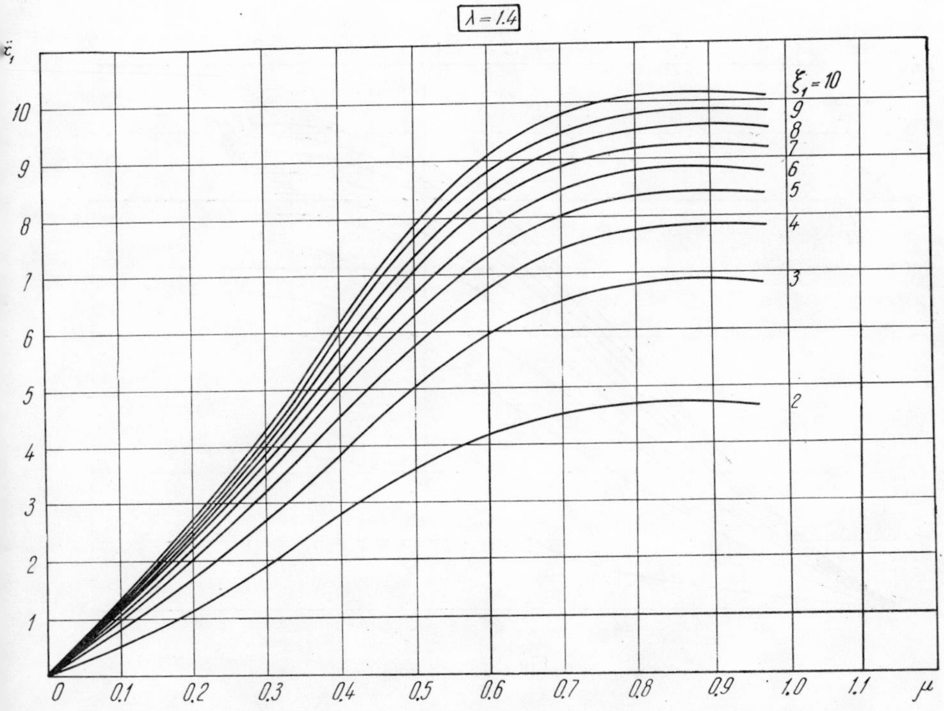
λ=1.4
10
9
8
7
6
5
4
3
2
1
ξ1 = 10
9
8
7
6
5
4
3
2
0
0.1
0.2
0.3
0.4
0.5
0.6
0.7
0.8
0.9
1.0
1.1
μ

Fig. 95

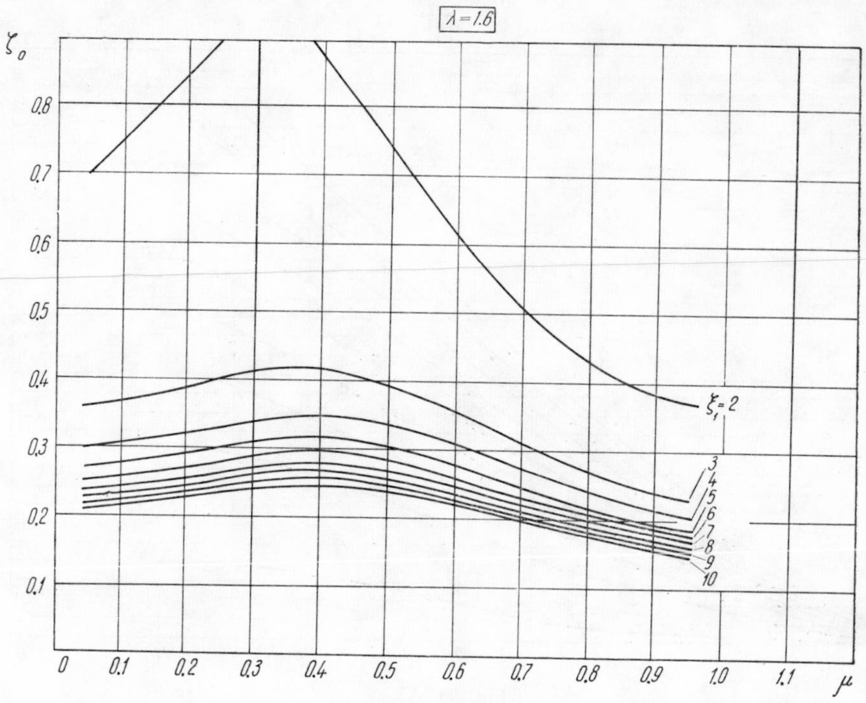
λ=1.6
ζ₀
0.8
0.7
0.6
0.5
0.4
0.3
0.2
0.1
0
0.1
0.2
0.3
0.4
0.5
0.6
0.7
0.8
0.9
1.0
1.1
μ
ξ₁=2
3
4
5
6
7
8
9
10

Fig. 96

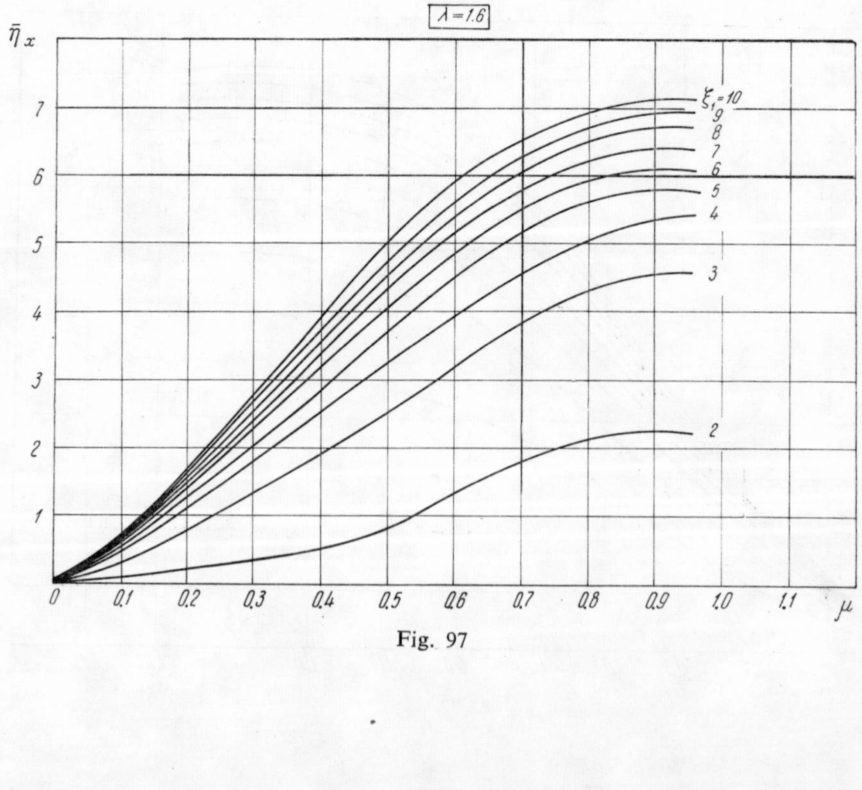
λ=1.6
η̄ₓ
7
6
5
4
3
2
1
0
0.1
0.2
0.3
0.4
0.5
0.6
0.7
0.8
0.9
1.0
1.1
μ
ξ₁=10
9
8
7
6
5
4
3
2

Fig. 97

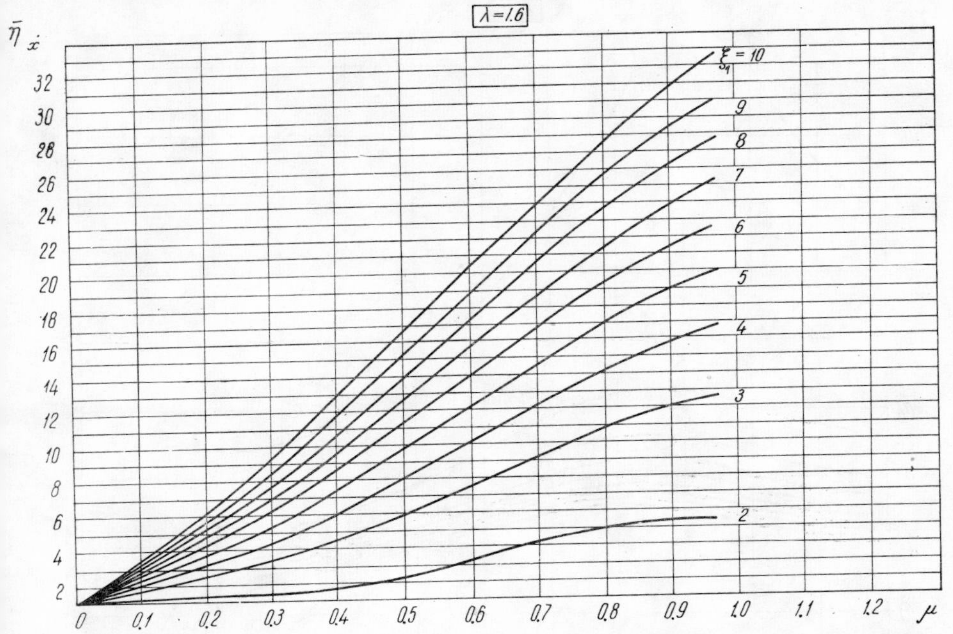
λ = 1.6
$\bar{\eta}_{\dot{x}}$
32
30
28
26
24
22
20
18
16
14
12
10
8
6
4
2
0
0.1
0.2
0.3
0.4
0.5
0.6
0.7
0.8
0.9
1.0
1.1
1.2
μ
ξ₁ = 10
9
8
7
6
5
4
3
2

Fig. 98

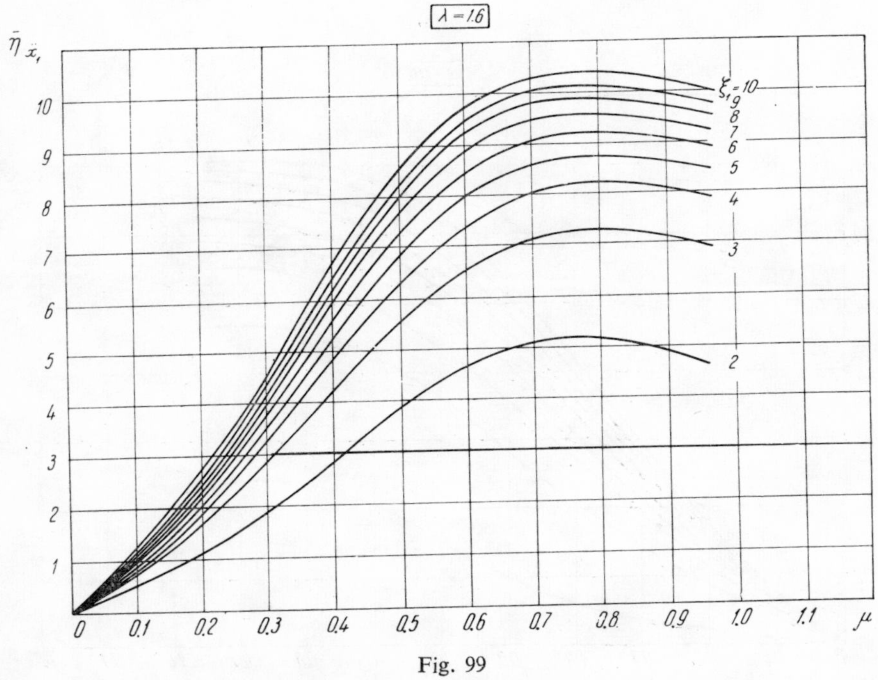
λ = 1.6
$\bar{\eta}_{\ddot{x}_1}$
10
9
8
7
6
5
4
3
2
1
0
0.1
0.2
0.3
0.4
0.5
0.6
0.7
0.8
0.9
1.0
1.1
μ
ξ₁ = 10
9
8
7
6
5
4
3
2

Fig. 99

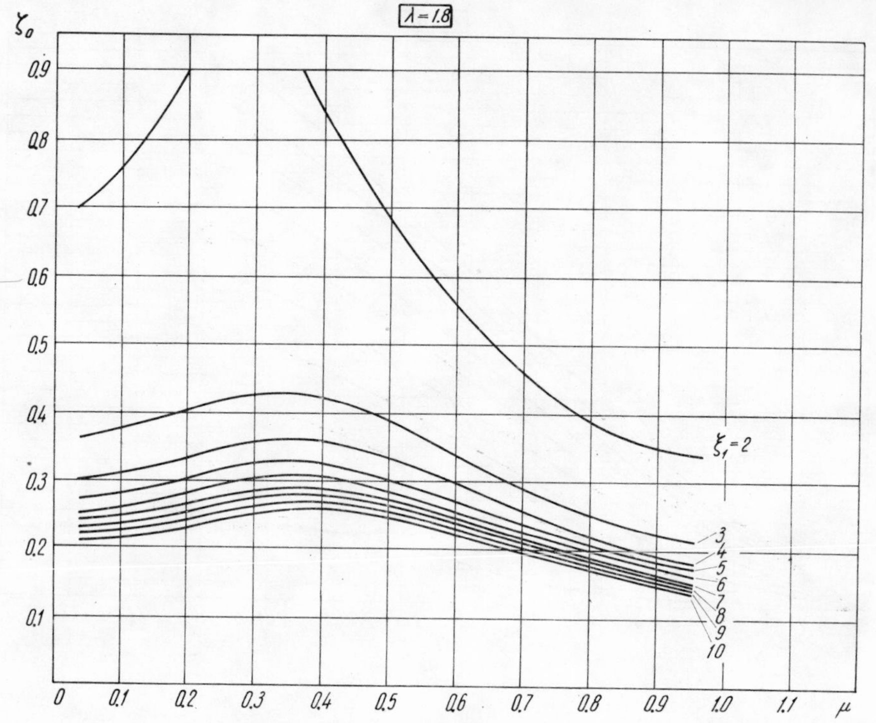

Fig. 100

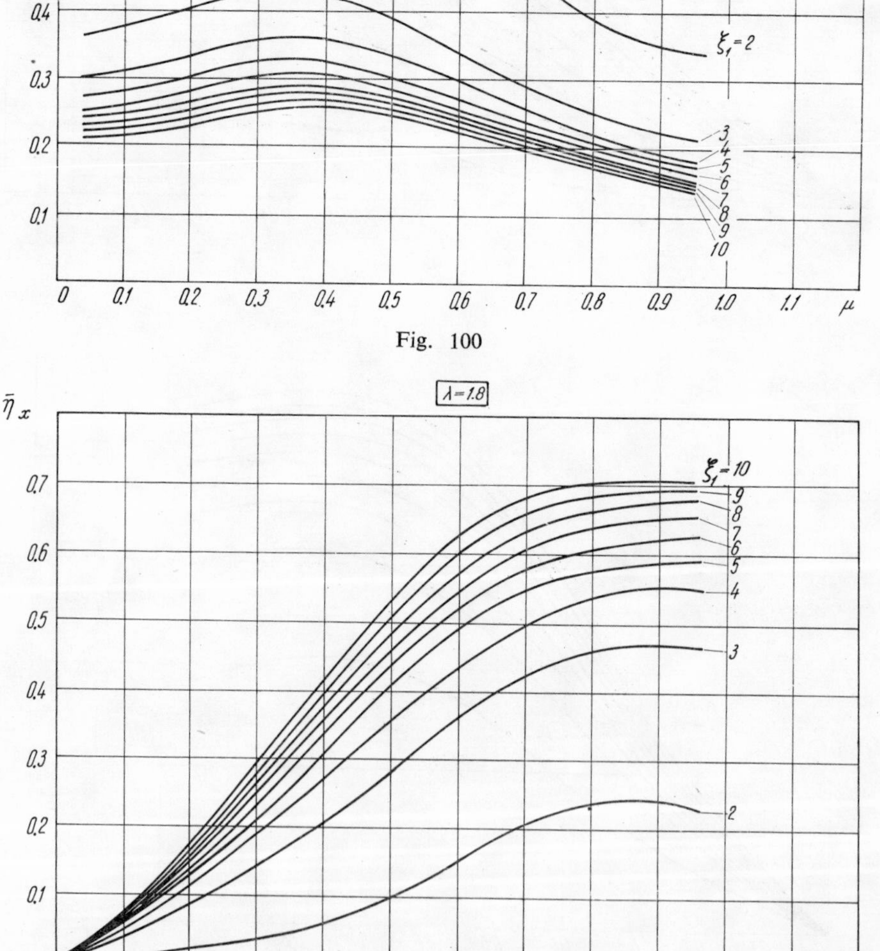

Fig. 101

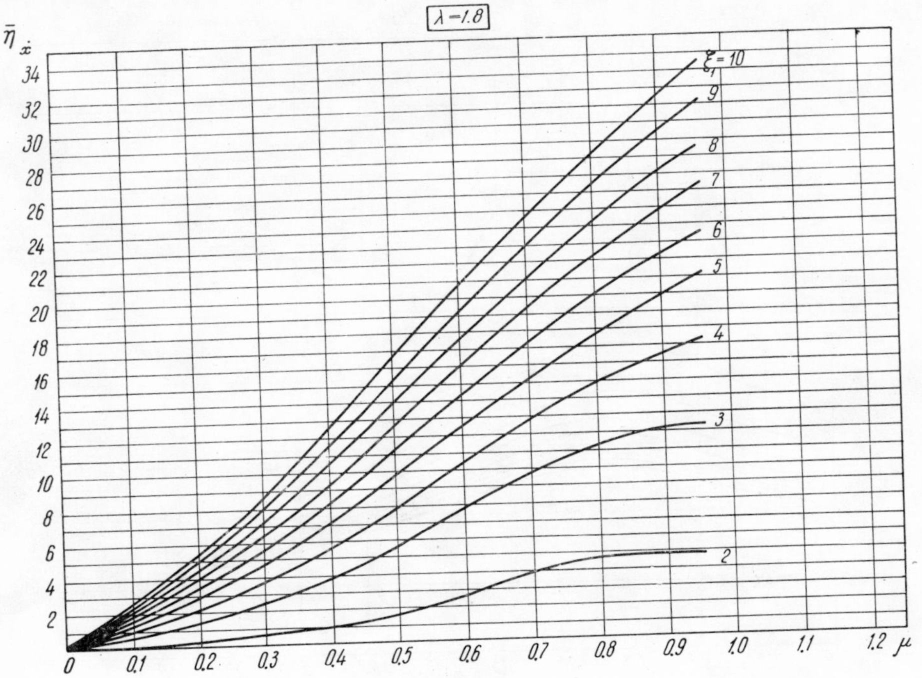

Fig. 102

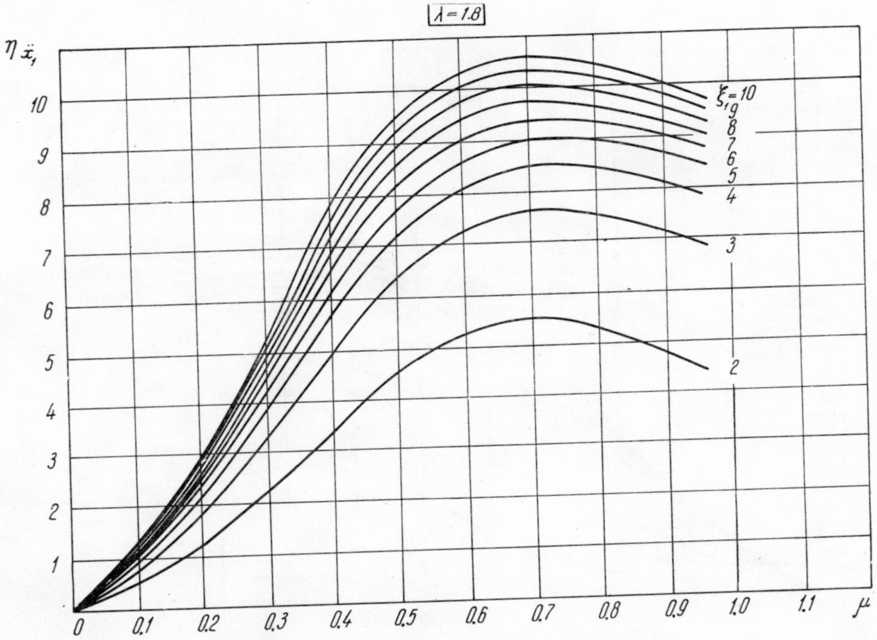

Fig. 103

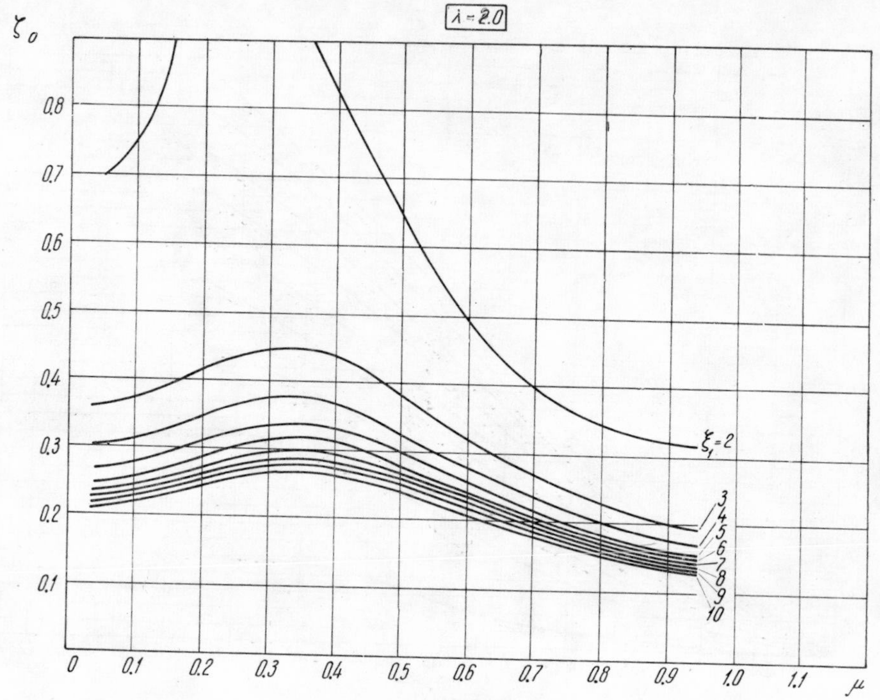

Fig. 104

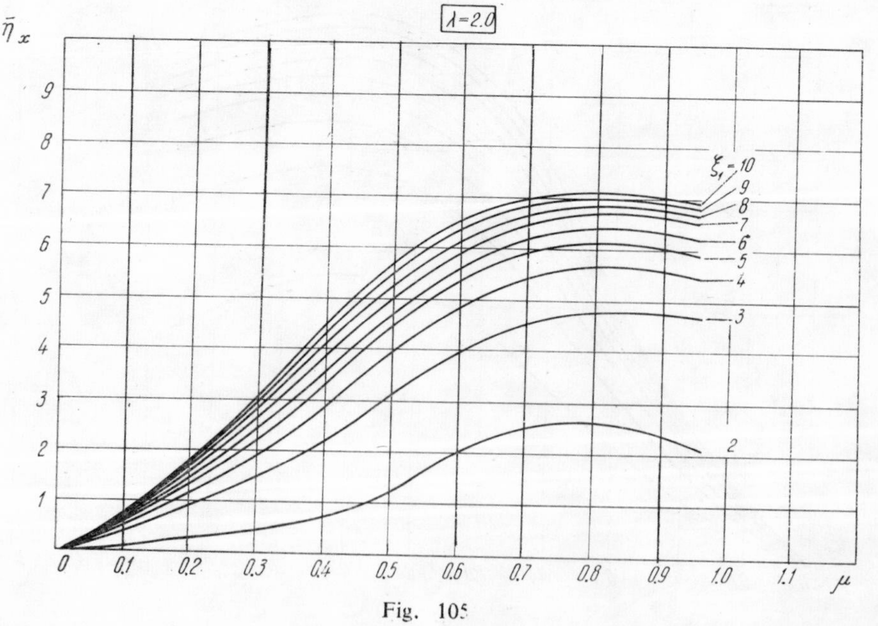

Fig. 105

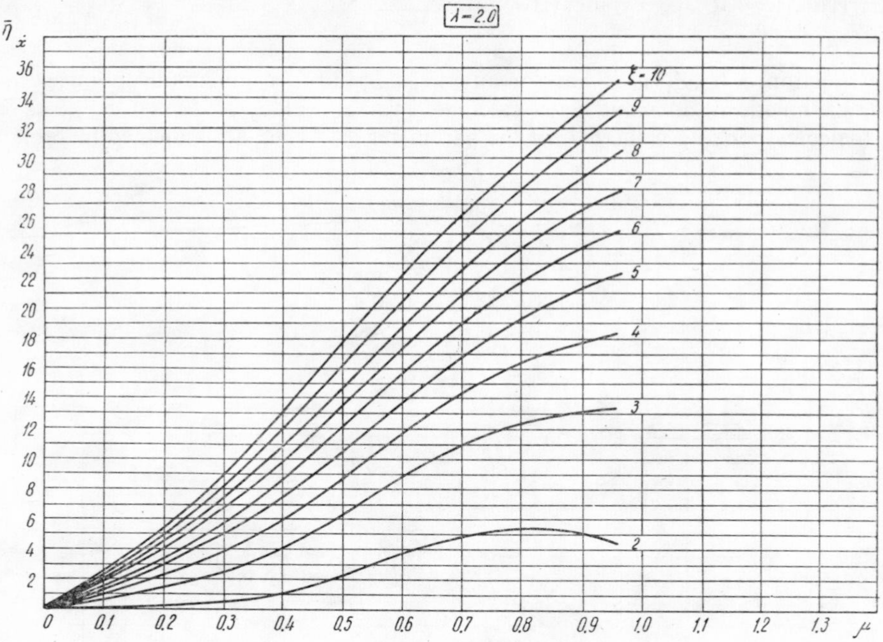

Fig. 106

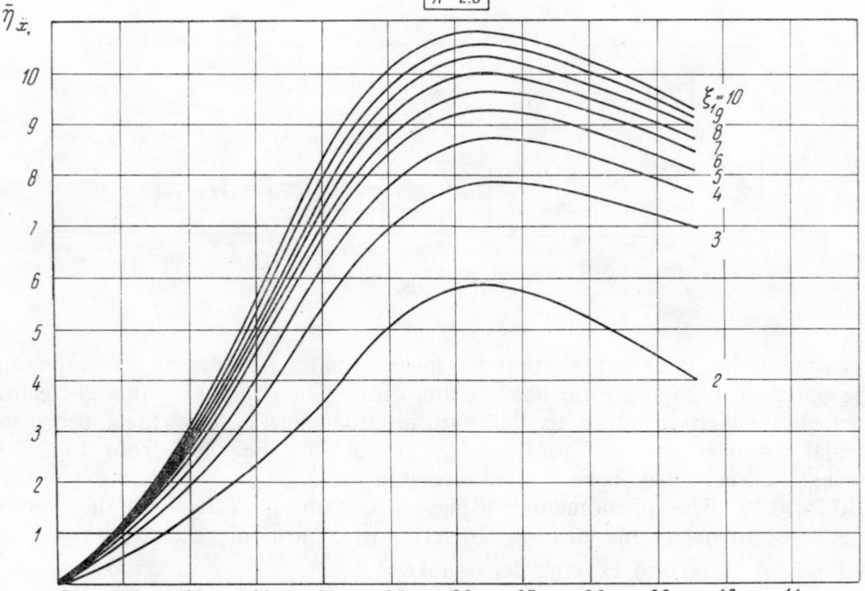

Fig. 107

the function $\bar{\eta}_{\ddot{x}_1}(\zeta_0, \xi_1, \mu, \lambda)$ has for any fixed λ and ξ_1 a maximum point $\mu(\xi_1, \lambda)$. The arithmetic mean value $\bar{\mu}(\lambda)$, calculated for all values of ξ_1 in $(35.7)_1$, as well as the corresponding value $\bar{\beta}(\lambda)$ of the parameter $\beta = \lambda\mu\nu_n$, are indicated in table 17.

Table 17

λ	$\bar{\mu}(\lambda)$	$\bar{\beta}(\lambda)$
1.4	0.84	1.18 ν_n
1.6	0.79	1.26 ν_n
1.8	0.72	1.30 ν_n
2.0	0.62	1.24 ν_n

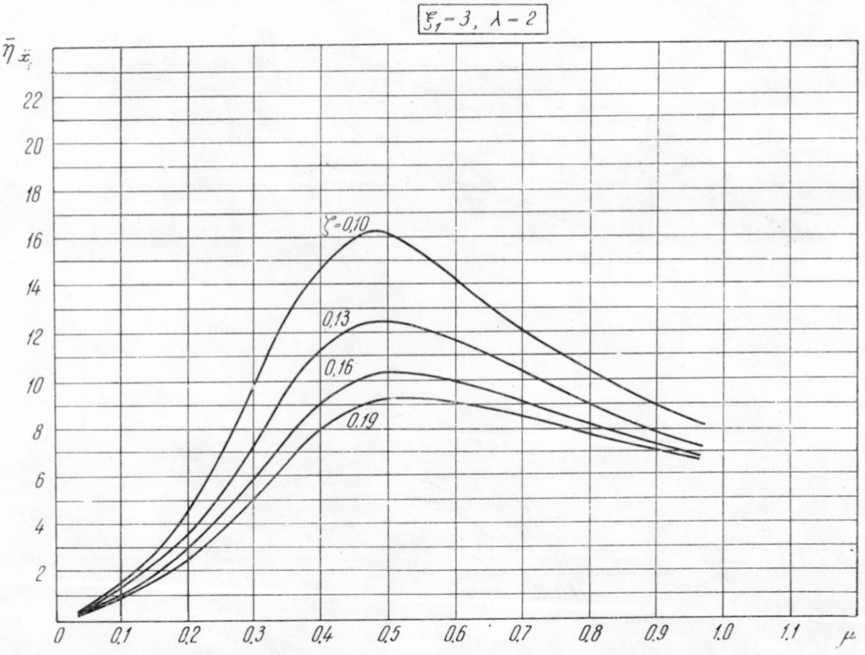

Fig. 108

Inspection of this table reveals that the mean square acceleration corresponding to the optimum damping ratio has maxima for values $\bar{\beta}(\lambda)$ of the correlation frequency β, which are close to the natural frequency ν_n of the suspension. A similar result is obtained for $\zeta \neq \zeta_0$, too, as can be seen from Fig. 108, where $\eta_{\ddot{x}_1}(\zeta, \xi_1, \mu, \lambda)$ has been plotted versus μ for $\xi_1 = 3$, $\lambda = 2$, and $\zeta = 0.10$, 0.13, 0.16, 0.19. This phenomenon displays a certain similarity with the classical resonance occurring in the case of a deterministic harmonic excitation.

Case 2. As a second example let us take

$$A_1 = 0.85, \quad A_2 = 0.15, \quad \beta_1 = 0, \quad A_j = 0 \quad \text{for} \quad j \geqslant 3. \tag{35.9}$$

Inspection of Table 11 reveals that the resulting correlation function may be used to statistically describe the unevenness of asphalt roads and of waved, stone block paved roads.

Introducing (35.9) into (33.6) and setting $A_j = 0$ for $j \geqslant 3$ gives

$$\left.\begin{aligned}
\sigma_x^2 &= \frac{2\sigma_{x_0}^2}{\pi}\int_0^{\nu_1}\left[\frac{A_1\,\alpha_1}{\nu^2+\alpha_1^2}+\frac{A_2\,\alpha_2\,(\nu^2+\alpha_2^2+\beta_2^2)}{(\nu^2-\alpha_2^2-\beta_2^2)^2+4\alpha_2^2\,\nu^2}\right]\frac{\nu^4\,d\nu}{(\nu_n^2-\nu^2)^2+4\zeta^2\nu_n^2\nu^2},\\
\sigma_{\dot x}^2 &= \frac{2\sigma_{x_0}^2}{\pi}\int_0^{\nu_1}\left[\frac{A_1\,\alpha_1}{\nu^2+\alpha_1^2}+\frac{A_2\,\alpha_2\,(\nu^2+\alpha_2^2+\beta_2^2)}{(\nu^2-\alpha_2^2-\beta_2^2)^2+4\alpha_2^2\,\nu^2}\right]\frac{\nu_6\,d\nu}{(\nu_n^2-\nu^2)^2+4\zeta^2\nu_n^2\nu^2},\\
\sigma_{\ddot x}^2 &= \frac{2\nu_n^2\sigma_{x_0}^2}{\pi}\int_0^{\nu_1}\left[\frac{A_1\,\alpha_1}{\nu^2+\alpha_1^2}+\frac{A_2\,\alpha_2\,(\nu^2+\alpha_2^2+\beta_2^2)}{(\nu^2-\alpha_2^2-\beta_2^2)^2+4\alpha_2^2\,\nu^2}\right]\frac{\nu^4(\nu_n^2+4\zeta^2\nu^2)\,d\nu}{(\nu_n^2-\nu^2)^2+4\zeta^2\nu_n^2\nu^2}.
\end{aligned}\right\}\qquad(35.10)$$

By applying the transformation (35.4) and introducing the dimensionless parameters

$$\mu = \alpha_2/\nu_n = a_2 V/\nu_n, \qquad \lambda = \beta_2/\alpha_2 = b_2/a_2, \qquad \gamma = \alpha_1/\alpha_2 = a_1/a_2, \qquad (35.11)$$

Eqs. (35.10) become

$$\left.\begin{aligned}
\sigma_x^2 &= \frac{2\mu\sigma_{x_0}^2}{\pi}\,\eta_x(\zeta, \xi_1, \mu, \lambda, \gamma),\\
\sigma_{\dot x}^2 &= \frac{2\mu\nu_n^2\,\sigma_{x_0}^2}{\pi}\,\eta_{\dot x}(\zeta, \xi_1, \mu, \lambda, \gamma),\\
\sigma_{\ddot x_1}^2 &= \frac{2\mu\,\nu_n^4\,\sigma_{x_0}^2}{\pi}\,\eta_{\ddot x_1}(\zeta, \xi_1, \mu, \lambda, \gamma),
\end{aligned}\right\}\qquad(35.12)$$

where

$$\left.\begin{aligned}
\eta_x(\zeta, \xi_1, \mu, \lambda, \gamma) &= \int_0^{\xi_1}\left\{\frac{\gamma A_1}{\xi^2+\gamma^2\mu^2}+\right.\\
&\quad\left.+\frac{A_2[\xi^2+\mu^2(1+\lambda^2)]}{[\nu^2-\mu^2(1+\lambda^2)]^2+4\mu^2\xi^2}\right\}\frac{\xi^4\,d\xi}{(1-\xi^2)^2+4\zeta^2\xi^2},\\
\eta_{\dot x}(\zeta, \xi_1, \mu, \lambda, \gamma) &= \int_0^{\xi_1}\left\{\frac{\gamma A_1}{\xi^2+\gamma^2\mu^2}+\right.\\
&\quad\left.+\frac{A_2[\xi^2+\mu^2(1+\lambda^2)]}{[\nu^2-\mu^2(1+\lambda^2)]^2+4\mu^2\xi^2}\right\}\frac{\xi^6\,d\xi}{(1-\xi^2)^2+4\zeta^2\xi^2},\\
\eta_{\ddot x_1}(\zeta, \xi_1, \mu, \lambda, \gamma) &= \int_0^{\xi_1}\left\{\frac{\gamma A_1}{\xi^2+\gamma^2\mu^2}+\right.\\
&\quad\left.+\frac{A_2[\xi^2+\mu^2(1+\lambda^2)]}{[\nu^2-\mu^2(1+\lambda^2)]^2+4\mu^2\xi^2}\right\}\frac{\xi^4(1+4\zeta^2\xi^2)\,d\xi}{(1-\xi^2)^2+4\zeta^2\xi^2}.
\end{aligned}\right\}\qquad(35.13)$$

Table 18

ξ_1	μ	0.1	0.2	0.3	0.4	0.5	0.6	0.7	0.8
2	ζ_0	$\geqslant 0.9$	0.4	0.6	0.5	0.5	0.47	0.46	0.44
	$\bar{\eta}_x$	—	1.2734	0.6829	0.9490	0.9618	1.0289	1.0213	1.0337
	$\bar{\eta}_{\dot{x}}$	—	2.4370	1.1997	1.6688	1.7438	1.8906	1.9059	1.9421
	$\bar{\eta}_{\dot{x}_1}$	—	2.8332	2.4105	2.6178	2.7056	2.6994	2.6345	2.5376
3	ζ_0	0.49	0.25	0.26	0.29	0.28	0.27	0.26	0.25
	$\bar{\eta}_x$	0.9413	2.6624	2.6101	2.3060	2.4315	2.4974	2.5208	2.5181
	$\bar{\eta}_{\dot{x}}$	1.6974	6.3096	6.5340	5.2531	5.6423	5.9425	6.1244	6.2129
	$\bar{\eta}_{\dot{x}_1}$	2.5750	4.2400	4.3770	4.0731	4.2009	4.2304	4.1769	4.0713
4	ζ_0	0.40	0.23	0.21	0.21	0.23	0.22	0.21	0.21
	$\bar{\eta}_x$	1.3171	2.9708	3.5730	3.5861	3.1925	3.2831	3.3348	3.2197
	$\bar{\eta}_x$	2.6464	7.7752	11.2482	11.0704	9.0641	9.4238	9.8009	9.8952
	$\bar{\eta}_{\dot{x}_1}$	3.0108	4.6160	5.5572	5.5389	5.1104	5.1076	5.0637	4.9651
5	ζ_0	0.36	0.22	0.19	0.18	0.18	0.19	0.19	0.18
	$\bar{\eta}_x$	1.5564	3.1409	4.0006	4.4251	4.3106	3.9211	3.8039	3.8432
	$\bar{\eta}_{\dot{x}}$	3.4106	8.9756	13.5020	17.2704	16.2010	13.4280	13.4190	13.8184
	$\bar{\eta}_{\dot{x}_1}$	3.3224	4.8784	5.9503	6.6633	6.4103	5.8601	5.7416	5.6351
6	ζ_0	0.34	0.21	0.19	0.17	0.16	0.16	0.17	0.16
	$\bar{\eta}_x$	1.7029	3.3023	4.0534	4.7397	5.0317	4.8094	4.3222	4.3638
	$\bar{\eta}_{\dot{x}}$	3.6477	10.1356	14.0876	20.1790	24.0610	21.9432	18.1580	18.1296
	$\bar{\eta}_{\dot{x}_1}$	3.5766	5.0904	6.2323	7.0717	7.4955	7.0565	6.4213	6.2203
7	ζ_0	0.32	0.20	0.19	0.16	0.15	0.14	0.14	0.15
	$\bar{\eta}_x$	1.8643	3.4657	4.0892	5.0456	5.3991	5.6053	5.3030	4.7227
	$\bar{\eta}_{\dot{x}}$	4.7102	11.2824	16.5876	22.6592	27.9720	31.6884	28.5488	23.6920
	$\bar{\eta}_{\dot{x}_1}$	3.7936	5.2708	6.4848	7.3658	7.9166	8.0982	7.5412	6.8549
8	ζ_0	0.31	0.19	0.18	0.16	0.15	0.13	0.12	0.13
	$\bar{\eta}_x$	1.9564	3.6370	4.3100	5.0821	5.7708	6.0404	6.2154	5.4941
	$\bar{\eta}_{\dot{x}}$	5.2084	12.4284	18.2844	24.6976	31.1970	36.8890	40.5030	35.6736
	$\bar{\eta}_{\dot{x}_1}$	3.9862	5.4316	6.6798	7.6110	8.2167	8.5341	8.5484	7.9058
9	ζ_0	0.29	0.19	0.17	0.15	0.14	0.13	0.12	0.11
	$\bar{\eta}_x$	2.1514	3.6501	4.5471	5.3963	5.8072	6.0906	6.2989	6.4531
	$\bar{\eta}_{\dot{x}}$	5.9710	13.3704	19.9254	27.0010	33.8140	40.4940	46.4590	49.6960
	$\bar{\eta}_{\dot{x}_1}$	4.1600	5.5808	6.8568	7.8264	8.4582	8.8280	8.9748	8.8582
10	ζ_0	0.28	0.18	0.16	0.15	0.14	0.12	0.12	0.11
	$\bar{\eta}_x$	2.2630	3.8311	4.8066	5.4178	5.8350	6.5343	6.3483	6.5355
	$\bar{\eta}_{\dot{x}}$	6.5600	14.5160	21.6850	28.9320	36.3200	44.1970	50.9000	57.0610
	$\bar{\eta}_{\dot{x}_1}$	4.3202	5.7124	7.0272	8.0216	8.6825	9.0799	9.2804	9.2973

A similar numerical analysis of the functions $\eta_x(\zeta, \xi_1, \mu, \lambda, \gamma)$, $\eta_{\dot{x}}(\zeta, \xi_1, \mu, \lambda, \gamma)$, $\eta_{\ddot{x}_1}(\zeta, \xi_1, \mu, \lambda, \gamma)$, $\eta_{\ddot{x}_1}(\zeta, \xi_1, \mu, \lambda, \gamma)$ has been done by adopting for the parameters the combinations of values

$$\begin{aligned} &\lambda = 10, \quad \gamma = 2.5, \quad \mu = 0.1, 0.2, \ldots, 0.8, \\ &\lambda = 12, \quad \gamma = 4.0, \quad \mu = 0.025, 0.050, \ldots, 0.200, \end{aligned} \tag{35.14}$$

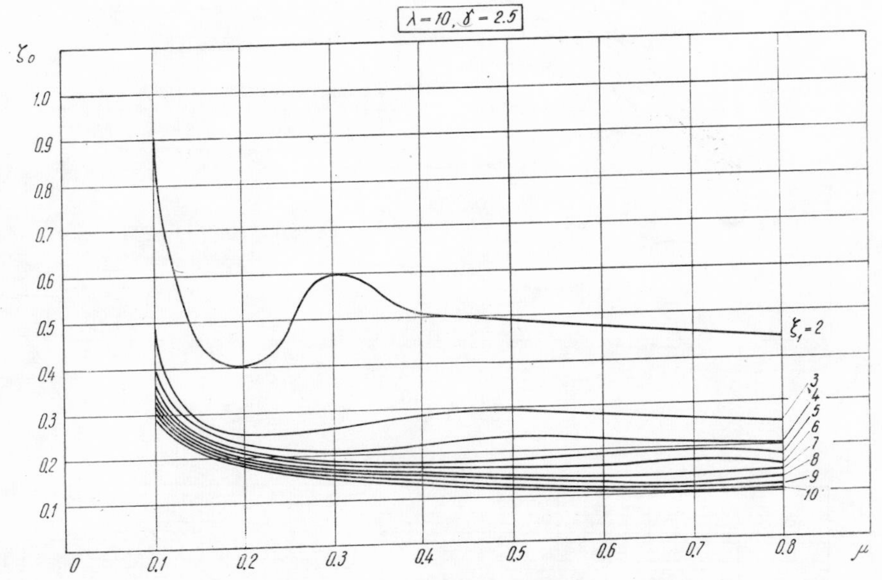

Fig. 109

which correspond, respectively, to the fifth and second lines of Table 11. In both cases the parameter ξ_1 has been given the values 2, 3, . . ., 10.

The variation of the functions $\zeta_0(\xi_1, \mu, \lambda, \gamma)$, $\bar{\eta}_x(\zeta_0, \xi_1, \mu, \lambda, \gamma) = 2\mu\eta_x(\zeta_0, \xi_1, \mu, \lambda, \gamma))$ $\bar{\eta}_{\dot{x}}(\zeta_0, \xi_1, \mu, \lambda, \gamma) = 2\mu\eta_{\dot{x}}(\zeta_0, \xi_1, \mu, \lambda, \gamma)$, and $\bar{\eta}_{\ddot{x}_1}(\zeta_0, \xi_1, \mu, \lambda, \gamma) = 2\mu\eta_{\ddot{x}_1}(\zeta_0, \xi_1, \mu, \lambda, \gamma$ is shown by Table 18 and Figs. 109–112 for the combination $(35.14)_1$, and by Table 19 and Figs. 113–116 for the combination $(35.14)_2$.

To illustrate the variation of ζ_0 with the vehicle speed V, the function $\zeta_0(\xi_1, \mu, \lambda, \gamma)$, which depends on V through ξ_1 and μ, has been graphically represented against V in Fig. 117 for some typical values of the parameters corresponding to various pavement types. One can see from this figure that ζ_0 monotonically decreases with increasing speed and takes on higher values for better pavements. For the range of usual traveling speeds one finds $0.10 < \zeta_0 < 0.50$, a result in good agreement with the values presently adopted for the damping ratio of road vehicle suspensions. It is also worth noting that the optimum value $\zeta_0 = 0.15$ indicated by Rotenberg [174] corresponds in Fig. 117 to the high speed range.

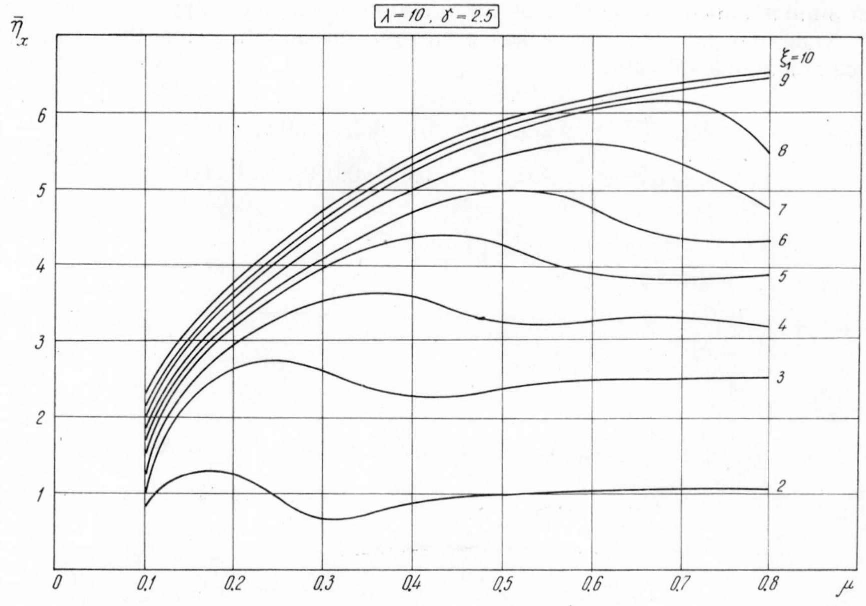

Fig. 110

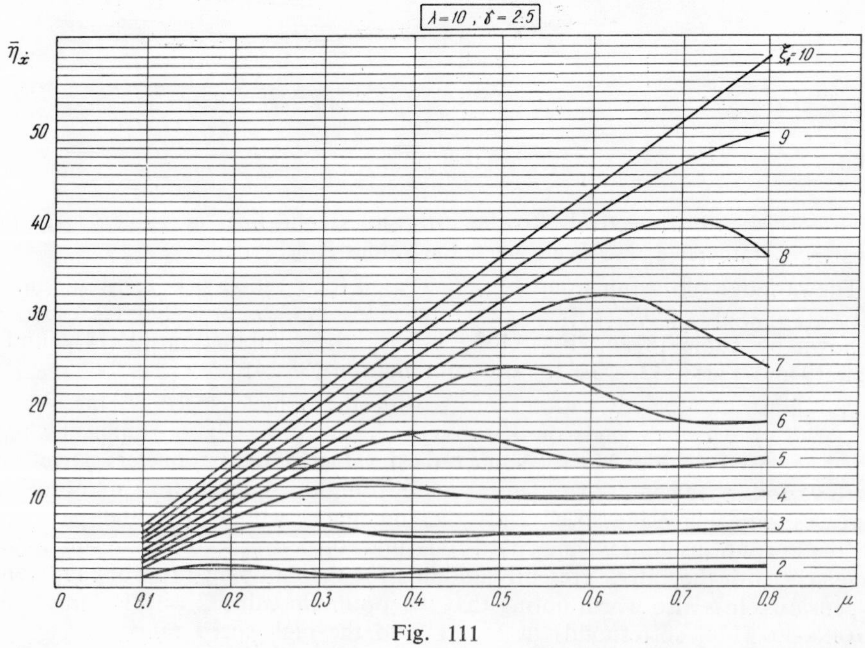

Fig. 111

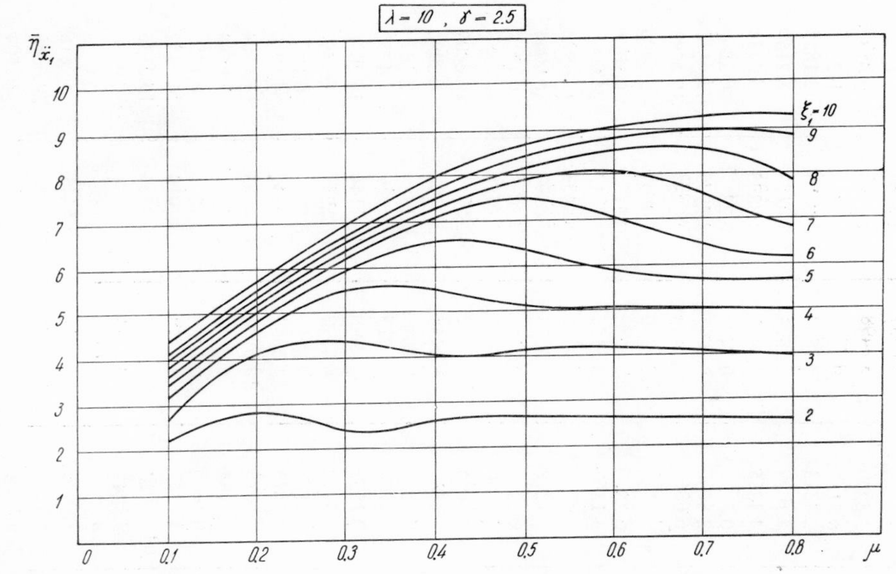

Fig. 112

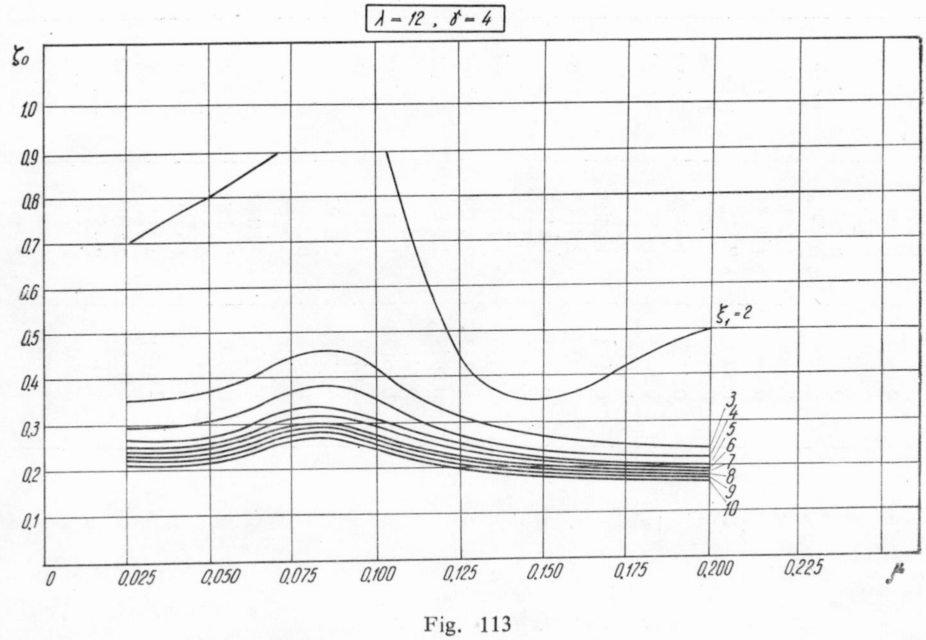

Fig. 113

Table 19

ξ_1 \ μ		0.025	0.050	0.075	0.100	0.125	0.150	0.175	0.200
2	ζ_0	0.70	0.80	$\geqslant$0.90	$\geqslant$0.90	0.45	0.35	0.43	0.50
	$\bar{\eta}_x$	0.1151	0.2185	—	—	1.4116	1.8596	1.2317	0.9463
	$\bar{\eta}_{\dot{x}}$	0.1684	0.2933	—	—	2.6145	3.6537	2.2782	1.6736
	$\bar{\eta}_{\ddot{x}_1}$	0.4451	0.9694	—	—	3.5290	3.6498	2.9167	2.6199
3	ζ_0	0.36	0.38	0.46	0.43	0.31	0.27	0.25	0.24
	$\bar{\eta}_x$	0.3301	0.7092	1.0441	1.5485	2.4213	2.7839	3.0304	3.1855
	$\bar{\eta}_{\dot{x}}$	0.6303	1.2576	1.8511	3.0298	5.1097	6.3873	7.4123	8.2160
	$\bar{\eta}_{\ddot{x}_1}$	0.6569	1.4355	2.6111	3.7892	4.3855	4.6969	4.8835	5.0784
4	ζ_0	0.30	0.31	0.38	0.36	0.27	0.24	0.23	0.22
	$\bar{\eta}_x$	0.4246	0.9295	1.4236	2.0344	2.8700	3.2097	3.3995	3.6112
	$\bar{\eta}_{\dot{x}}$	0.9350	1.8969	2.8560	4.4468	6.6862	8.1135	9.2837	10.4788
	$\bar{\eta}_{\ddot{x}_1}$	0.7613	1.6586	3.0733	4.3378	4.8197	5.0790	5.3637	5.6400
5	ζ_0	0.27	0.28	0.34	0.31	0.25	0.22	0.21	0.20
	$\bar{\eta}_x$	0.4862	1.0563	1.6867	2.4952	3.1367	3.5250	3.7513	4.0068
	$\bar{\eta}_{\dot{x}}$	1.1939	2.4144	3.7102	5.8134	7.9730	9.6399	11.0453	12.4992
	$\bar{\eta}_{\ddot{x}_1}$	0.8343	1.8135	3.4023	4.7298	5.1298	5.3913	5.6997	6.0064
6	ζ_0	0.25	0.26	0.32	0.30	0.23	0.21	0.20	0.19
	$\bar{\eta}_x$	0.5339	1.1539	1.8483	2.6220	3.4180	3.7077	3.9568	4.2388
	$\bar{\eta}_{\dot{x}}$	1.4333	2.8924	4.4382	6.7050	9.2655	10.9692	12.5857	14.2608
	$\bar{\eta}_{\ddot{x}_1}$	0.8923	1.9360	3.6661	5.0358	5.3787	5.6418	5.9703	6.2980
7	ζ_0	0.24	0.25	0.30	0.28	0.22	0.20	0.19	0.18
	$\bar{\eta}_x$	0.5617	1.2101	2.0264	2.8652	3.5798	3.8925	4.1550	4.4752
	$\bar{\eta}_{\dot{x}}$	1.6468	3.3188	5.1806	7.7524	10.3832	12.2901	14.1144	16.0048
	$\bar{\eta}_{\ddot{x}_1}$	0.9411	2.0398	3.8915	5.2964	5.5900	5.8590	6.2031	6.5492
8	ζ_0	0.23	0.23	0.29	0.26	0.21	0.19	0.18	0.18
	$\bar{\eta}_x$	0.5903	1.3235	2.1283	3.1384	3.7472	4.0857	4.3841	4.5016
	$\bar{\eta}_{\dot{x}}$	1.8605	3.8099	5.8333	8.8332	11.5025	13.6137	15.6453	17.4804
	$\bar{\eta}_{\ddot{x}_1}$	0.9840	2.1296	4.0908	5.5270	5.7762	6.0516	6.4116	6.7668
9	ζ_0	0.22	0.22	0.28	0.25	0.20	0.19	0.18	0.17
	$\bar{\eta}_x$	0.6205	1.3890	2.2362	3.2940	3.9242	4.1010	4.4019	4.7452
	$\bar{\eta}_{\dot{x}}$	2.0754	4.2435	6.4902	9.7528	12.6290	14.7080	16.9235	19.2136
	$\bar{\eta}_{\ddot{x}_1}$	1.0223	2.2105	4.2711	5.7322	5.9447	6.2247	6.5951	6.9664
10	ζ_0	0.21	0.22	0.27	0.24	0.20	0.18	0.17	0.17
	$\bar{\eta}_x$	0.6528	1.3930	2.3501	3.4599	3.9342	4.3047	4.6343	4.7616
	$\bar{\eta}_{\dot{x}}$	2.2920	4.6057	7.1529	10.6824	13.5350	16.0350	18.4590	20.6668
	$\bar{\eta}_{\ddot{x}_1}$	1.0571	2.2847	4.4385	5.9206	6.1000	6.3828	6.7680	7.1504

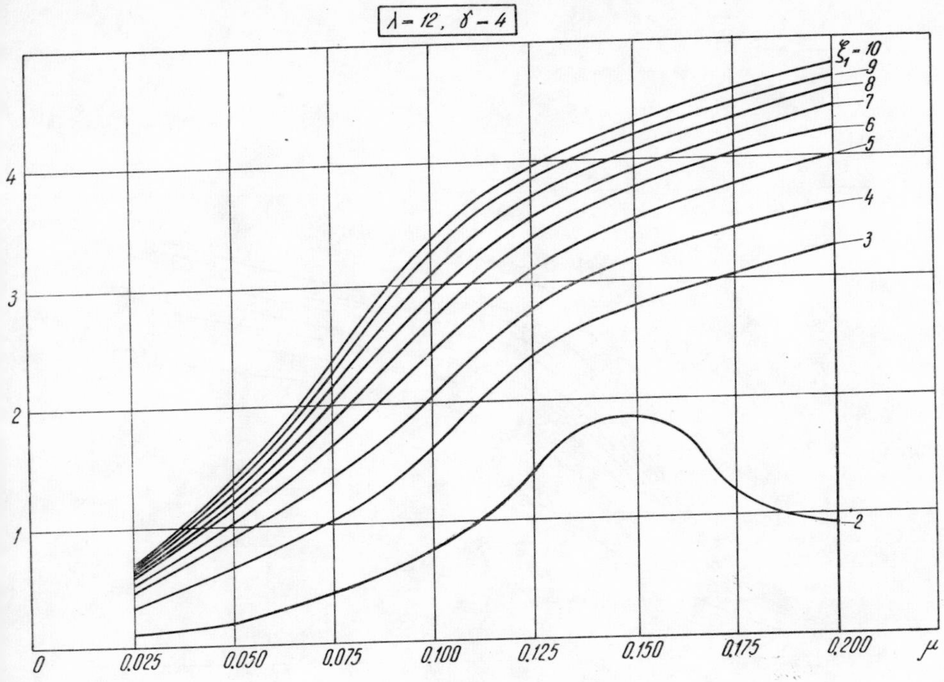

Fig. 114

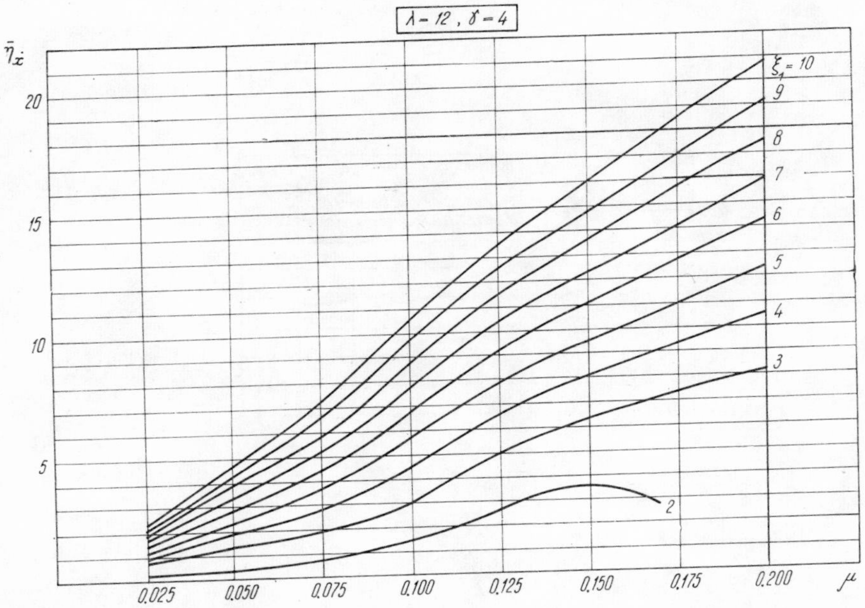

Fig. 115

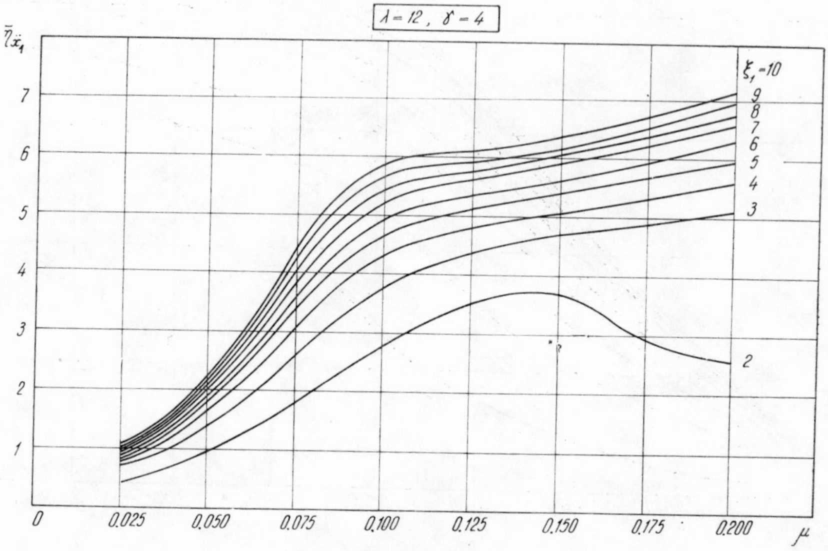

Fig. 116

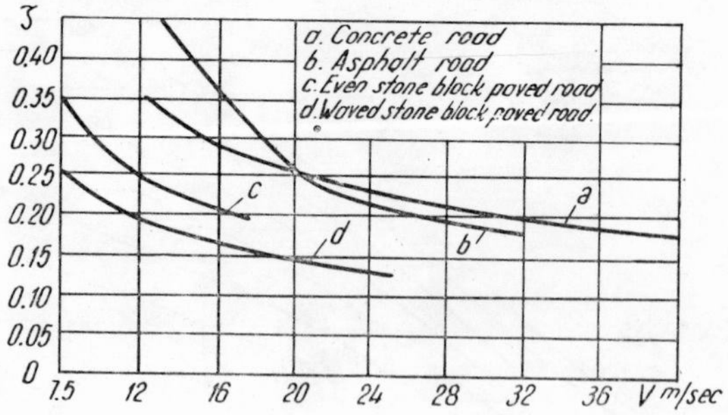

Fig. 117

§ 36. OPTIMIZING VISCOUS NONLINEAR DAMPING IN ROAD VEHICLE SUSPENSIONS

As mentioned in the introduction, when the suspension of a road vehicle is equipped with hydraulic shock absorbers, the bouncing of the sprung mass may be modeled by the equation

$$\ddot{x} + k\,|\dot{x}|^r \operatorname{sgn} \dot{x} + \nu_n^2\, x = -\,\ddot{x}_0(t), \tag{36.1}$$

where, generally, $r \in (0,3)$.[1]

To optimize the viscous nonlinear damping with respect to the criterion of the minimum mean-square body acceleration, we will use the method of the statistical linearization, which allows reducing the problem to the optimization of an equivalent linear damping, enabling us to apply the results obtained in the preceding section.

From Eqs. (29.14) and (36.1) it follows that the parameters of the equivalent linear system are $\nu_e = \nu_n$ and

$$\zeta_e = \frac{k}{\nu_n \sigma_{\dot{x}}^3 \sqrt{2\pi}} \int_0^\infty \dot{x}^{r+1}\, \mathrm{e}^{-\dot{x}^2/2\sigma_{\dot{x}}^2}\, \mathrm{d}\dot{x}, \tag{36.2}$$

whence

$$\zeta_e = \frac{k}{\nu_n \sqrt{\pi}}\, \Gamma(1 + r/2)\, (\sigma_{\dot{x}} \sqrt{2})^{r-1}, \tag{36.3}$$

where

$$\Gamma(s) = \int_0^\infty t^{s-1}\, \mathrm{e}^{-t}\, \mathrm{d}t, \qquad s > 0. \tag{36.4}$$

On the other hand, when the spectral density of $x_0(t)$ is band-limited, the variance of $x(t)$ results from Eq. $(30.15)_2$ as

$$\sigma_{\dot{x}}^2 = \int_0^{\nu_1} \frac{\nu^6 S_{x_0}(\nu)\, \mathrm{d}\nu}{(\nu_n^2 - \nu^2)^2 + 4\zeta_e^2\, \nu_n^2 \nu^2}\,, \tag{36.5}$$

where ν_1 is some cut-off frequency.

Thus, the problem reduces to the solving for ζ_e and $\sigma_{\dot{x}}$ of the coupled equations (36.3) and (36.5). Eliminating $\sigma_{\dot{x}}$ between Eqs. (36.3) and (36.5) yields

$$y_1(\zeta_e) = y_2(\zeta_e), \tag{36.6}$$

[1] The cases $r = 1$ and $r = 2$ correspond, respectively, to the laminar and turbulent flow of the liquid within the hydraulic shock absorber.

where

$$y_1(\zeta_e) \equiv \frac{1}{\sqrt{2}}\left[\frac{\nu_n\sqrt{\pi}}{k\Gamma(1 + r/2)}\right]^{1/(r-1)} \zeta_e^{1/(r-1)}, \tag{36.7}$$

$$y_2(\zeta_e) \equiv \left[\int_0^{\nu_1} \frac{\nu^6 S_{x_0}(\nu)\, d\nu}{(\nu_n^2 - \nu^2)^2 + 4\zeta_e^2 \nu_n^2 \nu^2}\right]^{1/2}. \tag{36.8}$$

Let us investigate now the conditions under which Eq. (36.6) has a unique solution ζ_e in the interval (0,1).

Theorem 1. *If*

$$r > 1 \quad \textit{and} \quad y_2(1) < y_1(1), \tag{36.9}$$

or if

$$0 < r < 1 \quad \textit{and} \quad y_2(1) < y_1(1), \tag{36.10}$$

then Eq. (36.6) has exactly one root in the interval (0,1).

Proof. By the mean-value theorem, we first deduce from (36.8) that

$$y_2(\zeta_e) = (\theta\nu_1)^3 \left[2S_{x_0}(\theta\nu_1)\int_0^{\nu_1} \frac{d\nu}{(\nu_n^2 - \nu^2)^2 + 4\zeta_e^2 \nu_n^2 \nu^2}\right]^{1/2} \tag{36.11}$$

for some $\theta \in (0,1)$. A direct calculation yields

$$\int_0^{\nu_1} \frac{d\nu}{(\nu_n^2 - \nu^2)^2 + 4\zeta_e^2 \nu_n^2 \nu^2} = \frac{1}{4\nu_n^3 \zeta_e}\left(\tan^{-1}\frac{2\zeta_e \xi_1}{1 - \xi_1^2} + \frac{2\zeta_e}{\sqrt{1 - \zeta_e^2}} \ln \frac{\xi_1^2 + 2\xi_1\sqrt{1 - \zeta_e^2} + 1}{\xi_1^2 - 2\xi_1\sqrt{1 - \zeta_e^2} + 1}\right), \tag{36.12}$$

where $\xi_1 = \nu_1/\nu_n$, and $\tan^{-1} 2\zeta_e\xi_1/(1 - \xi_1^2)$ is given by (31.17). By expanding the right-hand side of (36.12) in a power series of ζ_e and introducing the result into (36.11), we find that $y_2(\zeta_e)$ is given for small values of ζ_e by the asymptotic formula

$$y_2(\zeta_e) = C\zeta_e^{-1/2} + f(\zeta_e), \tag{36.13}$$

where

$$C > 0 \quad \text{for} \quad \xi_1 \geqslant 1, \qquad C = 0 \quad \text{for} \quad 0 < \xi_1 < 1, \tag{36.14}$$

$$\lim_{\xi_e \to 0} f(\zeta_e) > 0 \text{ and finite for any } \xi_1 > 0. \tag{36.15}$$

From (36.8) we derive by inspection that y_2 is a monotonically decreasing function of ζ_e. Assume now that (36.9) is satisfied. It then follows from (36.7) that y_1

is a monotonically increasing function of ζ_e. Moreover, from Eqs. (36.7) and (36.13)–(36.15), we deduce that

$$\lim_{\xi_e \to 0} y_2(\zeta_e) > y_1(0) = 0$$

for any ξ_1. From this relation and (36.9) we conclude, in view of the monotonicity of y_1 and y_2, that Eq. (36.6) has exactly one root in (0,1).

Next assume that (36.10) is fulfilled. Then both y_1 and y_2 are monotonically decreasing functions of ζ_e. From (36.6) and (36.13) it follows that for small values values of ζ_e

$$\frac{y_1(\zeta_e)}{y_2(\zeta_e)} = \frac{K\zeta^{(r+1)/2(r-1)}}{C + f(\zeta_e)\,\zeta_e^{1/2}},$$

where $K > 0$. Therefore, $y_1(\zeta_e)/y_2(\zeta_e) \to \infty$ as $\zeta_e \to 0$, and hence $y_1(\zeta_e) > y_2(\zeta_e)$ for ζ_e sufficiently small. Combining this result with (36.10) and taking into account the monotonicity of y_1 and y_2, we again conclude that Eq. (36.6) has exactly one root in (0,1). Thereby the proof is completed.

Since Eq. (36.6) uniquely determines $\zeta_e \in (0,1)$, we may solve the optimization problem simply by setting in Eqs. (36.3) and (36.5) $\zeta_e = \zeta_0$, where ζ_0 is the optimum damping ratio corresponding to a given type of road pavement and traveling speed. It results that

$$\zeta_0 = \frac{k}{\nu_n \sqrt{\pi}} \Gamma(1 + r/2)\,(\sigma_{\dot{x}} \sqrt{2})^{r-1}, \tag{36.16}$$

$$\sigma_{\dot{x}}^2 = \int_0^{\nu_1} \frac{\nu^6 S_{x_0}(\nu)\, d\nu}{(\nu_n^2 - \nu^2)^2 + 4\zeta_0^2\, \nu_n^2 \nu^2}. \tag{36.17}$$

The dependence of $\sigma_{\dot{x}}^2$ on the parameters characterizing the excitation has been analyzed in § 35 for various types of road pavements. Therefore, it will be assumed as known in what follows.

Equations (36.16) and (36.17) lead to a relation between the parameters k and r which characterize the nonlinear damping. Usually, r is determined by the type of shock absorber adopted, and then the optimum value of k results from (36.16) as

$$k_0 = \frac{\nu_n \zeta_0 \sqrt{\pi}}{\Gamma(1 + r/2)\,(\sigma_{\dot{x}} \sqrt{2})^{r-1}}, \tag{36.18}$$

where $\sigma_{\dot{x}}$ is given by (36.17).

However, in practice, it may sometimes happen that the damping forces corresponding to the optimum value (36.18) of the damping coefficient k are larger than certain admissible limits. In such cases it is convenient to approach the optimization problem by first determining from Eq. (36.16) the value of r corresponding to a prescribed limitation of the damping force, and then choosing a suitable type of shock absorber. In the following we will discuss two limitations of this kind.

1. Let us first require that the damping characteristic, i.e., the graph of the function

$$f(\dot{x}) = k\,|\dot{x}|^r \operatorname{sgn} \dot{x}\,, \tag{36.19}$$

pass through a given point (v_0, F_0) with $v_0 F_0 > 0$. It follows that

$$k = F_0/v_0^r, \tag{36.20}$$

and hence Eq. (36.16) may be written in the form

$$z = \frac{a^r}{\Gamma(1 + r/2)} \equiv h_1(r), \tag{36.21}$$

where

$$z = \frac{F_0}{\nu_n \zeta_0 \sigma_{\dot{x}} \sqrt{2\pi}}\,, \qquad a = \frac{v_0}{\sigma_{\dot{x}} \sqrt{2}}\,. \tag{36.22}$$

Equation (36.21) determines the value of r for given values of v_0 and F_0. Differentiating $h_1(r)$ yields

$$h_1'(r) = \frac{a^r\,[2 \ln a - \psi(1 + r/2)]}{2\Gamma(1 + r/2)}\,,$$

where $\psi(u)$ is the logarithmic derivative of the function $\Gamma(u)$. Since $\psi(u)$ is a monotonically increasing function, $h_1'(r)$ does not vanish in (0,3) if and only if

$$0 < a < \mathrm{e}^{-\psi(1)/2} \quad \text{or} \quad a > \mathrm{e}^{\psi(5/2)/2}\,. \tag{36.23}$$

Hence, for the range of a given by (36.23), Eq. (36.21) has at most one real root in (0,3). Furthermore, since $\psi(u)$ is monotonically increasing, $h_1(r)$ has at most one extremum point, and hence, for values of a other than (36.23), Eq. (36.21) has at most two real roots in (0,3). The actual number of real roots depends on the prescribed values of z and a.

Equation (36.21) may be graphically solved by using Fig. 118, where the function $h_1(r)$ is plotted for $a \in [0.1,\ 2]$, $r \in [0,3]$.

2. As a second possible limitation of the damping forces let us take now the case when the statistical mean $m_{|f|}$ of the absolute value of the damping force (36.19) is prescribed, say $m_{|f|} = F_0$. Assuming that $x(t)$ is normally distributed, this condition may be expressed as

$$F_0 = \frac{2k}{\sigma_{\dot{x}} \sqrt{2\pi}} \int_0^\infty \dot{x}^r \mathrm{e}^{-\dot{x}^2/2\sigma_{\dot{x}}^2}\, \mathrm{d}\dot{x}, \tag{36.24}$$

and leads after integrating to

$$k = \frac{F_0 \sqrt{\pi}}{(\sigma_{\dot{x}} \sqrt{2})^r \Gamma\left(\frac{r+1}{2}\right)}. \tag{36.25}$$

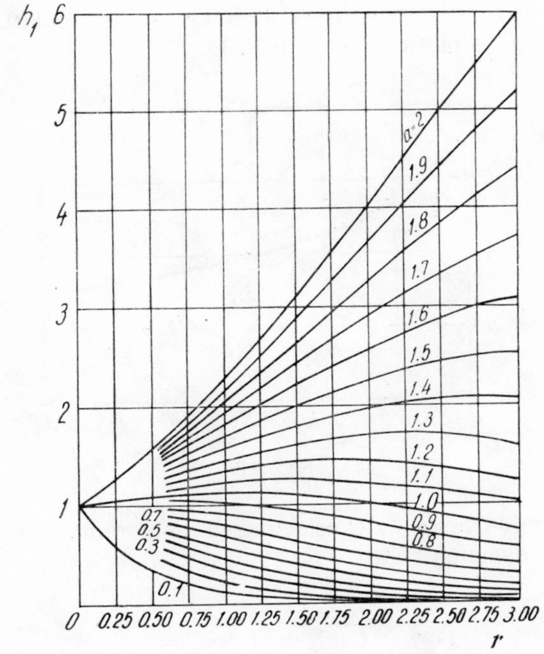

Fig. 118

Introducing (36.25) into (36.16) gives

$$z = \Gamma\left(\frac{1+r}{2}\right) \Big/ \Gamma(1 + r/2) \equiv h_2(r), \tag{36.26}$$

where

$$z = \frac{F_0}{\nu_n \zeta_0 \sigma_{\dot{x}} \sqrt{2}}. \tag{36.27}$$

Since

$$h_2'(r) = \frac{\Gamma\left(\frac{1+r}{2}\right)\left[\psi\left(\frac{1+r}{2}\right) - \psi(1 + r/2)\right]}{2\Gamma(1 + r/2)} < 0$$

for $r \in (0,3)$, it follows that Eq. (36.27) has exactly one real root if $h_2(3) < z < h_2(0)$, and no real root otherwise. Taking into account (36.29), the sufficient condition for the existence of exactly one real root becomes

$$4/(3\sqrt{\pi}) < z < \pi. \tag{36.28}$$

Equation (36.26) may be graphically solved by using Fig. 119, where the graph of the function $h_2(r)$ is plotted for $r \in [0,3]$.

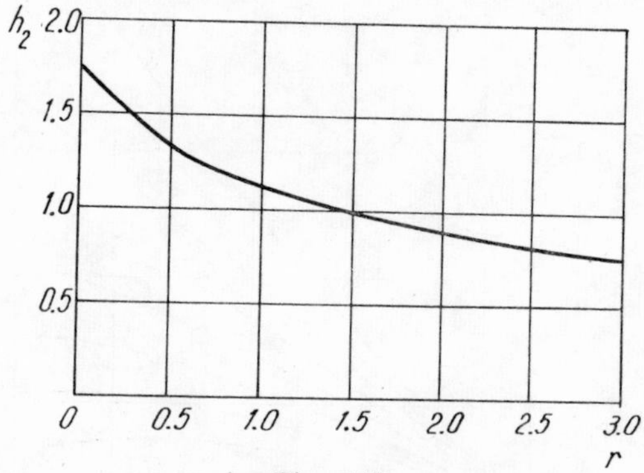

Fig. 119

We conclude this section by considering the important case of an asymmetric viscous quadratic damping. Then

$$f(\dot{x}) = \begin{cases} k\dot{x}^2 & \text{for} \quad \dot{x} \geqslant 0, \\ -(k/\alpha)\,\dot{x}^2 & \text{for} \quad \dot{x} < 0. \end{cases} \tag{36.29}$$

By applying the statistical linearization technique discussed in § 29 and assuming that $\dot{x}(t)$ is normally distributed, we obtain the equivalent damping ratio corresponding to (36.29) as

$$\zeta_e = \frac{k}{2\nu_n\sigma_{\dot{x}}^2\sqrt{2\pi}}\left[\frac{1}{\alpha}\int_{-\infty}^{0}\dot{x}^3 e^{-\dot{x}^2/2\sigma_{\dot{x}}^2}\,d\dot{x} + \int_{0}^{\infty}\dot{x}^3 e^{-\dot{x}^2/2\sigma_{\dot{x}}^2}\,d\dot{x}\right], \tag{36.30}$$

whence

$$\zeta_e = \frac{k(\alpha+1)\,\sigma_{\dot{x}}}{\alpha\nu_n\sqrt{2\pi}}. \tag{36.31}$$

Replacing ζ_e by the optimum damping ratio ζ_0 yields the optimum damping coefficient

$$k_0 = \frac{\alpha\nu_n\sqrt{2\pi}}{\alpha+1}\cdot\frac{\zeta_0}{\sigma_{\dot{x}}}, \tag{36.32}$$

where $\sigma_{\dot{x}}$ is now given by (36.17). Using the results obtained in § 35, we may write Eq. (36.32) in the form

$$k_0 = C\zeta_0/\sqrt{\bar{\eta}_{\dot{x}}(\zeta_0)}, \tag{36.33}$$

where the parameter

$$C = \frac{\alpha\pi\sqrt{2}}{(\alpha+1)\,\sigma_{x_0}} \tag{36.34}$$

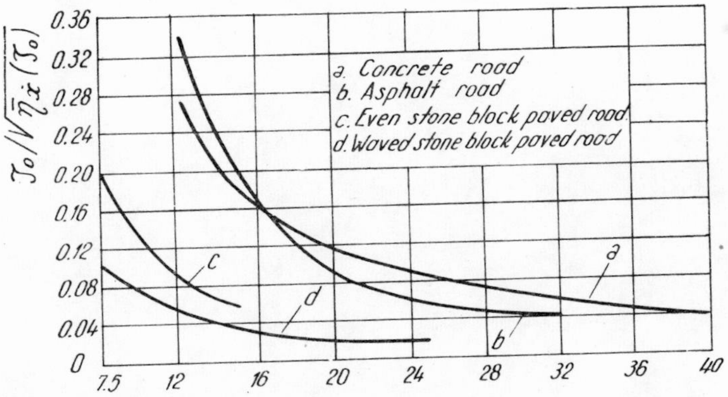

Fig. 120

does not depend on the traveling speed. Figure 120 shows the variation with V of the quantity $\zeta_0/\sqrt{\bar{\eta}_{\dot{x}}(\zeta_0)}$ for various types of road pavement.

References

1. ABRAMSON, H. N., Nonlinear Vibration, in "Shock and Vibration Handbook" (C. M. Harris and C. E. Crede, eds.), Vol I, pp. 4–1 McGraw-Hill, New York, 1961.
2. AMES, W. F., "Nonlinear Partial Differential Equations in Engineering." Academic Press, New York and London, 1965.
3. ANDRONOV, A. A. and CHAIKIN, S. E., "Theory of Oscillations," Princeton Univ. Pres, Princeton, New Jersey, 1949 (Russian, originally published in Moscow, 1937).
4. ANDRONOV, A. A., PONTRYAGIN, L. and WITT, A. A., *Zh. Eksp. Teor. Fiz.* **3**, 165 (1933).
5. APPLETON, E. A., *Philos. Mag.* **47**, 609 (1924).
6. ARIARATNAM, S. T., *J. Mech. Eng. Sci.* **2**, 195 (1960).
7. ASCARI, A., *Rend. Ist. Lomb. Sci. Lett.* **16**, 278 (1952).
8. BABITKY, V. I. and KOLOVSKY, M. Z., *Inz. J. Mech. Tverd. Tela* 147 (1967).
9. BARBĂLAT, I., *Bul. Sti. Acad. RPR, Secţ. Sti. Mat. Fiz.* **5**, 393, 503 (1953).
10. BENDAT, J. S., "Principles and Applications of Random Noise Theory," Wiley, New York, 1958.
11. BENDAT, J. S. and PIERSOL, A. G., "Measurement and Analysis of Random Data," Wiley, New York, 1968.
12. BENDER, E. K., Optimization of the Random Vibration Characteristics of Vehicle Suspensions, D. Sc. Thesis, MIT (1967).
13. BENDIXON, I., *Acta Math.* **24**, 1 (1901).
14. BLAIR, K. W and LOUD, W. S., *J. Soc. Ind. Appl. Math.* **8**, 74 (1960).
15. BLAKE, R. E., Basic Vibration Theory, in "Shock and Vibration Handbook" C. M. Harris and C. E. Crede, eds.), Vol. 2, pp. 2—1 McGraw-Hill, New York, 1961.
16. BOGOLYUBOV, N. N. and MITROPOLSKY, IU. A., "Asymptotic Methods in the Theory of Nonlinear Oscillations," Gordon and Breach, New York, Hindustan Publ., Delhi, 1961 (Russian orig. 2nd. ed. publ. in Moscow, 1955; lst. ed., Moscow, 1955).
17. BOLOTIN, V. V., "Stochastic Methods in Building Mechanics" (in Russian) Iz. Literatury po Stroitel'stvo, Moscow, 1965.
18. BOOTON, R. C., *Proc. Symp. Nonlinear Circuit Analys.* **2**, pp. 369 (1953).
19. BOOTON, R. C., *Trans. I R E* **C.T.-1**, 32 (1954).
20. BRUNS, H. and RÖNITZ, R., *Automobile Eng.* **61**, 26, 45 (1971).
21. BULAND, R. N., *J. Franklin Inst.* **257**, 37 (1954).
22. CACCIOPOLI, R. and GHIZZETTI A., *Mem. Rend. Acc. d'Italia* **3**, 427 (1942).
23. CARTWRIGHT, M. L., Forced Oscillations in Nonlinear Systems, in "Contributions to the Theory of Nonlinear Oscillations" (S. Lefschetz, ed.), Vol. 1, pp. 149 Princeton Univ. Press, Princeton, New Jersey, 1950.

24. Cartwright, M. L. and Littlewood, J. E., *Ann. of Math.* **48**, 472 (1947); **50**, 504 (1949).
25. Caughey, T. K., *J. Appl. Mech.* **21**,.327 (1954).
26. Caughey, T. K., *J. Appl. Mech.* **26**, 341 (1959).
27. Caughey, T. K., *J. Appl. Mech.* **26**, 345 (1959).
28. Caughey, T. K., *J. Appl. Mech.* **27**, 575 (1960).
29. Caughey, T. K., *J. Appl. Mech.* **27**, 649 (1960).
30. Caughey, T. K., *J. Acoust. Soc. Amer.* **35**, 1683 (1963).
31. Caughey, T. K., Nonlinear Theory of Random Vibration, *Advan. Appl. Mech.* 11 (1971).
32. Cesari, L., "Asymptotic Behavior and Stability Problems in Ordinary Differential Equations," Springer, Berlin—Göttingen—Heidelberg, 1963.
33. Chandrasekhar, S., *Rev. Mod. Phys.* **35** (1943).
34. Chung, K. and Kazda, L. F., *Trans. AIEE* **78**, Part II, 100 (1959).
35. Ciucu, G., "Elements of Probability Theory and Mathematical Statistics" (in Romanian), Ed. Didactică și Pedagogică, Bucharest, 1963.
36. Crandall, S. H. and Mark, W. D., "Random Vibration in Mechanical Systems," Academic Press, New York—London, 1963.
37. Crandall, S. H., *Appl. Mech. Rev.* **11**, 739, 1959.
38. Crandall, S. H., *Proc. Internat. Symp. Nonlinear Vibration*, 1963 **1**, 306 Izv. Akad. Nauk SSSR, Kiev, 1963.
39. Crandall, S. H., *J. Acoust. Soc. Amer.* **35**, 1700 (1963).
40. Crandall, S. H., Knabbaz, G. R. and Manning, J. E., *J. Acoust. Soc. Amer.* **36**, 1130 (1964).
41. Davenport, W. B. Jr, and Root, W. L., "An Introduction to Theory of Random Signals and Noise," McGraw-Hill, New York, 1958.
42. Derbaremdiker, A. D., *Automobil'naja Promyslennosti* **24**, 19 (1958).
43. Derbaremdiker, A. D., *Automobil'naja Promyslennosti* **27**, 15 (1961).
44. Deusen, B. D. van, *Automobile Eng.* **53**, 532 (1963).
45. Dieckmann, D., *Internat. Z. Angew. Physiol.* **16**, 519 (1957).
46. Dincă, F. and Teodosiu, C., *St. Cerc. Mec. Apl.* **11**, 879 (1964).
47. Dincă, F. and Teodosiu, C., *St. Cerc. Mec. Apl.* **17**, 1173 (1964).
48. Dincă, F. and Teodosiu, C., Free Oscillations of Systems with Linear Restoring Force and Piecewise Smooth Damping Characteristic, paper submitted to the Conference on Mechanics Bucharest, 1965.
49. Dincă, F. et *al.*, Optimum Shock Absorbers for Road Vehicles. C.M.S. Rep., Bucharest, 1967.
50. Dincă, F. and Sireteanu, T., *Rev. Roumaine Sci. Tech. Sér. Méc.* Appl., **14**, 869 (1969).
51. Dincă, F. *et al.*, Vibration of Mechanical Systems with Applications to Road Vehicle Suspensions. Optimization of Restoring Force and Damping in Road Vehicle Suspensions C.M.S. Rep., Bucharest, 1968, 1969, 1970.
52. Doob, J. L., *Trans. Amer. Math. Soc.* **42** (1937).
53. Doob, J. L., *Trans. Amer. Math. Soc.* **44** (1938).
54. Doob, J. L. and Ambrose W., *Ann. of Math.* **41** (1940).
55. Duffing, G., Erzwungene Schwingungen bei veränderlicher Eigenfrequenz und ihre technische Bedeutung, PhD Thesis, Sammlung Vieweg, Braunschweig, 1918.
56. Einstein, A., "The Theory of Brownian Movement," Dover, New York, 1956.
57. Engel, Z., *Rev. Roumaine Sci. Tech. Sér. Méc. Appl.* **11**, 997 (1966).
58. Förster, H., *Math. Z.* **43**, 271 (1938).

59. FRIEDRICHS, K. O. and STOKER, J. J., *Quart. J. Appl. Math.* **1**, 97 (1943).
60. FROMMER, M., *Math. Ann.* **99**, 222 (1928).
61. FROMMER, M., *Math. Ann.* **109**, 395 (1934).
62. GALERKIN, V. G., *Vestnik Inž. i Techn.* **19**, 897 (1915).
63. HAIMOVICI, A., "Differential Equations and Integral Equations" Ed. Didactică şi Pedagogică, Bucharest, 1965.
64. HALANAY, A., *Bul. Sti. Acad. R.P.R., Secţ. Sti. Mat. Fiz.* **5**, 373 (1953).
65. HALANAY, A., "Qualitative Theory of Differential Equations" Ed. Acad. R.P.R., Bucharest, 1963.
66. HALE, J. K., "Oscillations in Nonlinear Systems," McGraw-Hill, New York, 1963.
67. HAMBURGER, L. and BUZDUGAN, G., "Theory of Vibrations and Applications to Machine Building" (in Romanian) Ed. Techn., Bucharest, 1958.
68. HARTOG, J. P. DEN, *Philos. Mag.* **9**, 801 (1930).
69. HARTOG, J. P. DEN and HEILES, R. M., *J. Appl. Mech.* **3**, 126 (1930).
70. HARTOG, J. P. DEN, "Mechanical Vibrations," 4th ed. McGraw-Hill, Hew York, 1956.
71. HAYASHI, C., "Forced Oscillations in Nonlinear Systems," Nippon, Osaka, 1958.
72. HAYASHI, C., "Nonlinear Oscillations in Physical Systems," McGraw-Hill, New York, 1964.
73. HOBSON, E. W., "The Theory of Spherical and Ellipsoidal Harmonics," Cambridge Univ. Press, London and New York, 1931.
74. HOFMAN, H. J., *ATZ* **60**, 289 (1958).
75. HSU, C. S., *Quart. Appl. Math.* **17**, 102 (1959).
76. IOSIFESCU, M., MIHOC, G. and TEODORESCU R., "Probability Theory and Mathematical Statistics" (in Romanian) Ed. Tehn., Bucharest, 1966.
77. JACOBSEN, L. S., *J. Appl. Mech.* **19**, 543 (1952).
78. JAMES, H. M., NICHOLS N. B. and PHILLIPS, R. S., "Theory of Servomechanisms," MIT Radiat. Lab. Ser., V. 25 McGraw Hill, New York, 1947.
79. JAHNKE, E., EMDE, F. and LÖSCH, F., "Tafeln Höherer Funktionen," B. G. Teubner Verlags - gesellschaft, Stuttgart, 1960.
80. JOHN, F., On Simple Harmonic Vibrations of a System with Nonlinear Characteristic, *in* "Studies on Nonlinear Vibration Theory" (R. Courant, ed.), p. 104 Inst. Math. and Mechn., New York Univ., 1946.
81. JOHN. F., *Comm. Pure Appl. Math.* **1**, 341 (1948).
82. KAUDERER, H., "Nichtlineare Mechanik", Springer, Berlin, Göttingen, Heidelberg, 1958.
83. KAZAKOV, I. E., *Sb. Naucnyh Trudov VVIA im. Zukovskogo* **1**, 394, (1954).
84. KAZAKOV, I. E., *Sb. Naucnyh Trudov VVIA im. Zukovskogo* 1, 95 (1954).
85. KAZAKOV, I. E., *Avtomat. i Telemeh.* **18**, 385 (1956).
86. KHINCHIN, A., *Math. Ann.* **109**, 604 (1934).
87. KLOTTER, K., *Proc. Symp. Nonlinear Circuit Anal.* Polyt. Inst. Brooklyn (1953).
88. KLOTTER, K., *J. Appl. Mech.* **22**, 493 (1955).
89. KOLOVSKI, M. Z., *Izv. Akad. Nauk SSSR, OTN, Meh. i Mašinostroenie* **1**, 3 (1963).
90. KORN, G. A., "Random Process Simulation and Measurements," McGraw-Hill, New York, 1966.
91. KRAMERS, H. A., *Physica* **7**, 284 (1940).
92. KRYLOV, M. N. and BOGOLYUBOV, N.N., "Introduction to Nonlinear Mechanics" (in Russian) Izv. Akad. Nauk USSR, Kiev, 1937.
93. KUSHNER, H. J., "Stochastic Stability and Control," Academic Press, New York, 1967.
94. KUSHNER, H. J., *J. Differential Equations* **6**, 209 (1969).

95. Laning, J. H., Jr. and Battin, R. H., "Random Processes in Automatic Control," McGraw Hill, New York, 1958.
96. Lear, G. A. von and Uhlenbek, G. E., *Phys. Rev.*, **38**, 1583 (1931).
97. Levenson, M. E., *J. Appl. Phys.* **20**, 1045 (1949).
98. Levenson, M. E., *Quart. Appl. Math.* **25**, 11(1967).
99. Levenson, M. E., *Quart. Appl. Math.* **26**, 456 (1968).
100. Levinson, N., *J. Math. Phys.* **22**, 181 (1943).
101. Levinson, N., *J. Math. Phys.* **22**, 41 (1943).
102. Liénard, A., *Rev. Gén. Elec.* **23**, 901, 946 (1928).
103. Lin, C. C., *Quart. Appl. Math.* **1**, 43 (1943).
104. Lindstedt, A., *Mém. Sci. St. Petersbourgh* **31** (1883).
105. Lonn, E. R., *Math. Z.* **44**, 507 (1939).
106. Loud, W. S., *J. Math. Phys.* **34**, 173 (1955).
107. Loud, W. S., *Duke Math. J.* **24**, 63 (1957).
108. Ludeke, C. A., *J. Appl. Phys.* **17**, 603 (1946).
109. Lundquist, S., *Quart. Appl. Math.* **13**, 305 (1955).
110. Lure, A. I. and Chekmarev, A. I., *Prikl. Mat. Meh.* **1** (1938).
111. Lyapunov, M. A., *Ann. Fac. Sci. Univ. Toulouse* **9**, 203 (1907).
112. Lyon, R. H., *J. Acoust. Soc. Amer.* **32**, 716, 953 (1960).
113. Malkin, I. G., "Some Problems of the Theory of Nonlinear Oscillations," (in Russian). Gostehizdat, Moscow, 1956.
114. Mandelstam, G. L. and Papalexi, N., *Z. Phys.* **73**, 225 (1932)
115. Martienssen, O., *Phys. Z.* **11**, 448 (1910).
116. McLachlan, N. W., "Ordinary Nonlinear Differential Equations in Engineering and Physical Science," Oxford Univ. Press (Clarendon), London and New York, 1950.
117. McLachlan, N. W., "Theory of Vibrations," Dover, New York, 1951.
118. Middleton, D., "An Introduction to Statistical Communication Theory," McGraw-Hill, New York, 1960.
119. Mihăilă, N., "Introduction to Probability Theory and Mathematical Statistics," Ed. Didactică şi Pedagogică, Bucharest, 1965.
120. Mihoc, G. and Firescu, D., "Mathematical Statistics" (in Romanian). Ed. Didactică şi Pedagogică, Bucharest, 1966.
121. Minorsky, N., "Introduction to Nonlinear Mechanics," Edwards, Ann Arbor, Michigan, 1947.
122. Minorsky, N., "Nonlinear Oscillation," Van Nostrand Reinhold, Princeton, New Jersey, 1962.
123. Mitropolsky, Iu. A., Nonstationary Processes in Nonlinear Oscillatory Systems, Air Tech. Inteligence Transl. ATIC-270579, F-TS-9085/V (Russian orig. publ. in Izv. Akad. Nauk, Kiev, 1955).
124. Mitropolsky, Iu. A., "Problems of the Asymptotical theory of Nonstationary Oscillations" (in Russian). Izdat. Nauka, Moscow, 1964.
125. Mitschke, M., "Nichtlineare Feder-und Dampferkennungen am Kraftfahrzeug," Vol. 5 Vorträge der Kraftfahrzeug und Motortechnischen Konferenz, 1967.
126. Mizohata, S. and Yamaguti, M., *Mem. Coll. Sci. Kyoto Univ.* Ser. A **27**, 109 (1952).
127. Morozan, T., "Stability of Systems with Stochastic Parameters" (in Romanian). Ed. Acad. RSR, Bucharest, 1969.
128. Morozan, T., *J. Math. Anal. Appl.* **24**, 1 (1968).

129. MOROZAN, T., *J. Math. Anal. Appl.* **24**, 669 (1968).
130. MOROZAN, T., *Rev. Roumaine Math. Pures Appl.* **14**, 829 (1969).
131. NEMYTSKY, V. V. and STEPANOV, V. V., "Qualitative Theory of Differential Equations," (in Russian). Gos. Iz. Tehn.-Teor. Lit. Moscow—Leningrad, 1949.
132. NEWLAND, D. E., *Internat. J. Mech. Sci.* 7, 159 (1965).
133. NICOLESCU, M., DINCULEANU, N. and MARCUS, S., "Manual of Mathematical Analysis" (in Romanian), Vol. 1, 2nd Ed. Ed. Didactică și Pedagogică, Bucharest, 1963.
134. NIZIOL, J., *Proc. Vibration Problems* **2**, 107 (1966).
135. ONICESCU, O., "Probability Calculus" (in Romanian). Ed. Tehn., Bucharest, 1956.
136. ONICESCU, O., MIHOC, G. and IONESCU-TULCEA, C. T., "Probability Calculus and Applications" (in Romanian), Ed. Acad., Bucharest, 1952.
137. OPIAL, Z., *Bull. Acad. Polon. Sci-Sér. Sci. Math. Astronom. Phys*, **7**, 495 (1959).
138. OPIAL, Z., *Bull. Acad. Polon. Sci-Sér. Sci. Math-Astronom. Phys.* **8**, 151 (1960).
139. OPIAL, Z., *Ann. Polon. Math.* **7**, 259 (1960).
140. PARHILOVSKY, J. G., *Avtomobilnaja Promyslennosti* **27**, 25 (1951).
141. PARHILOVSKY, I. G., *Avtomobilnaja Promyslennosti* **35**, 28 (1969).
142. PAYNE, H. J., The Response of Nonlinear Systems to Stochastic Excitation, PhD. Dissertation, California Inst. of Technol. (1967).
143. PAYNE, H. J., *Internat. J. Control* **7**, 451 (1968).
144. PERRON, O., *Math. Z.* **15**, 120 (1922); 16, 273 (1923).
145. PEVZNER, I. M. and TIHONOV, A. A., *Avtomobilnaja Promyslennosti* **30**, 15 (1964).
146. POINCARÉ, H., "Les Méthodes Nouvelles de la Mécanique céleste," Vols. 1—3. Gauthier Villars, Paris, 1892, 1893, 1899.
147. POL, B. VAN DER., *Proc. IRE* **22**, 1051 (1934).
148. PONOMAREV, S. D., BIDERMAN, V. L., LIHAREV, K. K., MAKUSIN, V. M., MALININ, N. N. and FEODOSIEV, V. I., "Strength Calculation in Machine Building" (in Russian). Mashgiz, Moscow, 1958.
149. POPOV, V. M., "Hyperstability of Automatic Systems" (in Romanian). Ed. Acad., Bucharest, 1966.
150. PORITSKY, H., Stochastic Processes of Mechanical Origin, in "Random Vibrations" (S. H. Crandall, ed.), Tech. Press of MIT, Wiley, New York and Chapmann and Hall, London, 1958.
151. PRASIL, F., *Schweiz. Bauz.* **25**, 334 (1908).
152. PRUCHIKOV, O. K., *Avtomobilnaja Promyslennosti* **31**, 30 (1965).
153. PUGACHEV, V. S., "Theory of Random Functions" (in Russian). Fizmatghiz, Moscow, 1962.
154. RAUSCHER, M., *J. Appl. Mech.* **9**, A-169 (1938).
155. REISSIG, R., *Abh. Deutsch. Akad. Wiss. Berlin Math.-Natur. Kl.* **1953**, 1 (1954).
156. REISSIG, R., *Math. Nachr.* **11**, 231 (1954), **12**, 119 (1954).
157. REISSIG, R., *Math. Nachr.* **11**, 345 (1954), **12**, 283 (1954).
158. REISSIG, R., *Math. Nachr.* **13**, 231 (1955), **14**, 17 (1955).
159. REISSIG, R., *Math. Nachr.* **13**, 313 (1955), **14**, 65 (1955), 15, 39, 47 (1956).
160. REISSIG, R., *Math. Nachr.* **15**, 181 (1956).
161. REISSIG, R., Math. Nachr. **15**, 375 (1956).
162. RÉNYI, A., "Probability Theory," Académiai Kiadó, Budapest, 1970.
163. REUTER, G. E. H., *Quart. J. Appl. Math.* **2**, 198 (1949).
164. REUTER, G. E. H., *Proc. Cambridge Philos. Soc.* **47**, 49 (1951).

165. REUTER, G. E. H., *J. London Math. Soc.* **26**, 215 (1952).
166. REUTER, G. E. H., *J. London Math. Soc.* **27**, 48 (1952).
167. RICE, S. O., *Bell System Tech. J.* **23**, 282 (1944), 24, 46 (1945).
168. RICHARDSON, P. D., *Appl. Sci. Res. Sect. A* **11**, 397 (1963).
169. RITZ, W., *J. Reine Angew. Math.* **135**, 1 (1909); Gesammelte Werke, p. 192 Paris, 1911.
170. ROBSON, J. D., "An Introduction to Random Vibration," Elsevier, Amsterdam, 1964.
171. ROBSON J. D., Deduction from Profile-Excited Random Vibration Response, Paper submitted to the *12th Internat. Congr. Appl. Mech., Stanford* (1968).
172. ROSEAU, M., "Vibrations nonlinéaires et théorie de la stabilité," Springer, Berlin—Heidelberg—New York, 1966.
173. ROSENBERG, R. M., *Quart. Appl. Math.* **15**, 341 (1958).
174. ROTENBERG, R. B., "Automobile Suspension and its Oscillations" (in Russian). Masghiz, Moscow, 1960.
175. ROTKOP, L. L., "Statistic Methods of Study on Electronic Models" (in Russian). Energja, Moskow, 1967.
176. SANSONE, G., "Equazioni differenziali nel campo reale," 2nd edit. Vol. 2 Bologna, Zanichelli, 1949.
177. SANSONE, G. and CONTI, R., "Nonlinear Differential Equations," Pergamon, Oxford, 1964.
178. SCHÄFER, M., *Z. Angew. Math. Mech.* **284** (1952).
179. SEIFERT, G., *Ann. Math.* **67**, 83 (1958).
180. SEIFERT, G., *Proc. Amer. Math. Soc.* **10**, 396 (1959).
181. SILAEV, A. A., "Spectral Theory of Road Vehicle Suspensions," Gos. Naucno-Teh-Iz. Masinostr. Lit., Moscow, 1963.
182. SKOWRONSKI, J. and ZIEMBA, S., *Arch. Mech. Stos.* **10**, 699 (1958).
183. SREIDER, I. A., "The Method of Statistical Trials" (in Russian) Fizmatgiz, Moscow, 1962.
184. STOKER, J. J., "Nonlinear Vibrations in Mechanical and Electrical Systems," Wiley (Interscience), New York, 1950.
185. STONE, M. H., *Trans. Amer. Math. Soc.* **4**, 31 (1936).
186. SVESHNIKOV, A. A., "Applied Methods of Random Function Theory" (in Russian) Gos. Soyuz. Iz. Sudostr. Prom., Leningrad, 1961.
187. SZABLEWSKI, W., *Math. Nachr.* **12**, 183 (1954).
188. SZPUNAR, K. and BOGUSZ, W., *Zagad. Drgań. Nielin.* **6**, 21 (1964).
189. TAYLOR, G. I., *Proc. London Math. Soc.* **2**, 196 (1920).
190. THOMPSON, A. G., *Proc. Inst. Mech. Eng.* **184** Pt 2A, 169 (1969).
191. TIMOSHENKO, S., "Vibration Problems in Engineering." 2nd Ed. Van Nostrand Reinhold, Princeton, New Jersey, 1937.
192. WIENER, N., *Acta Math.* **55**, 117 (1930).
193. WINTNER, A., *Amer. J. Math.* **69**, 815 (1947).
194. ZLÁMAL, M., *Czechoslovak Math. J.* **4**, 95 (1954).
195. ZIEMBA, S., *Arch. Mech. Stos.* **10**, 163 (1958).

Author index

Subject index

PRINTED IN ROMANIA